Neurotransmitter Transporters

Contemporary Neuroscience

Neurotransmitter Transporters

Structure, Function, and Regulation

Second Edition

Edited by

Maarten E. A. Reith

College of Medicine, University of Illinois, Peoria, IL

 **Humana Press**
Totowa, New Jersey

© 2002 Humana Press Inc.
999 Riverview Drive, Suite 208
Totowa, New Jersey 07512
humanapress.com

Cover illustration:

Cover design by Patricia F. Cleary.

Production Editor: Mark J. Breaugh.

For additional copies, pricing for bulk purchases, and/or information about other Humana titles, contact Humana at the above address or at any of the following numbers: Tel.: 973-256-1699; Fax: 973-256-8341; E-mail: humana@humanapr.com; or visit our website at www.humanapress.com

Printed in the United States of America. 10 9 8 7 6 5 4 3 2 1

Library of Congress Cataloging in Publication Data

Main entry under title:

Neurotransmitter transporters: structure, function, and regulation / edited by Maarten E. A. Reith.-- 2nd ed.
    p. cm. — (Contemporary neuroscience)
    Includes bibliographical references and index.
    ISBN 0-89603-945-5 (alk. paper)
    1. Neurotransmitters. 2. Carrier proteins. I. Reith, Maarten E. A. II. Series.

    QP364.7 .N4757 2002
    612.8'22—dc21
                                              2001051651

Preface

Neurotransmission is a multicomponent process. Transmitters, released by neuronal activity, act on pre- and postsynaptic receptors, and many books detail advances in the receptor field. In addition, after their release from nerve endings, transmitters are removed from the neuronal vicinity by uptake into neuronal or glial cells by specific transporter proteins that have been studied intensely over the last 30 years; this information is scattered throughout numerous publishing vehicles. Therefore, the primary aim of this second edition of *Neurotransmitter Transporters: Structure, Function, and Regulation* is to offer a comprehensive picture of the characterization of neurotransmitter transporters and their biological roles. The transporter field has moved forward in stages. In the first phase, progress came from the use of substrate or blocker ligands selectively targeting transporters, the application of model systems allowing the study of transmitter transport shielded from storage, and the development of mathematical models for describing transport phenomena. In the second phase, roughly covering the last decade, advances in DNA techniques allowed the cloning of numerous genes coding for different transporter proteins. In the current, third stage, a wealth of information is being accumulated in studies relating transporter structure with function, experiments addressing regulation by posttranslational transformation, investigations into transport modulation by trafficking processes and genomic influences, characterization of channel properties of transporters by electrophysiological approaches, and the creation of transgenic animals under- or overexpressing a given transporter protein.

The first edition of *Neurotransmitter Transporters: Structure, Function, and Regulation* was published at the brink of the above 2nd and 3rd phase; the current, second edition finds the transporter field fully in the 3rd phase. This second edition of *Neurotransmitter Transporters: Structure, Function, and Regulation* has been put together with the goal of retaining most of the original material in describing for each transporter the cloning history as well as including new progress in characterizing its structure, function, regulation, and physiological relevance.

Interest in neurotransmitter transporters has increased tremendously, along with clearer understanding of their roles in the pathophysiology of schizophrenia, Tourette's syndrome, Parkinson's disease, affective disorders, attention deficit/hyperactivity disorder, neurotoxin accumulation and removal, brain ischemia, amyotropic lateral sclerosis, or in the mechanism of action of drugs of abuse, antidepressants, and antiepileptics. Thus, *Neurotransmitter Transporters: Structure, Function, and Regulation, Second Edition* will be of interest to scientists, graduate students, and advanced undergraduates who seek a comprehensive overview of this active field in neuroscience. Selected chapters will also be of interest to physicians who are carrying out imaging and postmortem measurements of neurotransmitter transporters in the human brain, or physicians who study gene linkages and polymorphisms in relation to psychiatric and other complex diseases.

Each chapter of *Neurotransmitter Transporters: Structure, Function, and Regulation* offers a critical summary and synthesis of the progress in characterizing each transporter in terms of structure–function relationships and regulation. Chapters 1–9 focus on various neurotransmitter transporters located in neuronal or glial plasma membranes and in synaptic vesicles. The Na^+, Cl^--dependent plasma membrane transporters are described for monoamines (Chapters 1–5) and for a number of compounds including amino acids (GABA, glycine, proline, betaine, taurine, and creatine (Chapters 6 and 7). The separate family of Na^+-dependent glutamate transporters is discussed (Chapters 7 and 8) as well as the family of vesicular transporters for monoamines, acetylcholine, and GABA/glycine (Chapter 9). Chapters 10–13 cover a variety of issues relevant to transporter structure and function. Chapter 10 describes posttranslational modifications with their important impact on the function of various transporters. Chapter 11 covers the various classes of blockers for the dopamine transporter with detailed discussion of structural determinants, and Chapter 12 describes the use of various in vivo imaging techniques and ligands for biogenic amine transporters, in particular the dopamine transporter in the human brain. In Chapter 13, the final chapter, the focus is on dopamine transporter changes in human brain as a result of cocaine administration, with both in vitro and in vivo imaging approaches.

A number of differences between this second edition and the original book can be highlighted. First, the more basic cloning and pharmacological information on monoamine transporters, previously in Chapter 1, can now be found in Chapters 3 and 4. The previous topic of regulation of serotonin transporters (currently Chapter 1, previously Chapter 2) has been widened to include all three biogenic amine transporters, and it now emphasizes phosphorylation and trafficking phenomena. Three new chapters on biogenic amine transporters have been added: Chapter 3 on chimera and site-directed mutagenesis studies, Chapter 4 on gene organization and the relationship of polymorphisms with psychiatric and other complex human diseases, and Chapter 5 on transgenic animals carrying altered genes for plasma membrane monoamine transporters. Current excitement about channel properties of transporters is covered in Chapters 2 and 8 regarding electrophysiological studies on cloned monoamine and amino acid transporters, respectively. The new chapter on transporter imaging, Chapter 12, entirely focuses on human results including effects of aging and brain injury, and changes in schizophrenia, phobia, drug abuse, and other complex human diseases. Chapter 13 covers changes in dopamine transporters in human brain as a result of cocaine exposure as previously, with added discussion of in vivo imaging approaches for the dopamine transporter included in this update. The previous final chapter detailing the role of biogenic amine transporters in in vivo and in vitro monoamine release studies has been omitted as more recent material on in vivo results is now presented in the new Chapter 5.

The authors of the present chapters have been instrumental in advancing our knowledge of transporters by their experimental and conceptual contributions to the field, and I feel fortunate to have been able to join all their forces together in this second edition. I thank Paul Dolgert, Tom Lanigan, Sr., Elyse O'Grady, Craig Adams, and Mark Breaugh at Humana Press for allowing the opportunity of a second edition, and I hope the book will continue to be used as a popular resource in the field.

Maarten E. A. Reith

Contents

Contributors

URS V. BERGER • *Department of Medicine, Brigham and Women's Hospital, Harvard Institutes of Medicine, Boston, MA*

RANDY D. BLAKELY • *Department of Pharmacology, Center for Molecular Neurosciences, Vanderbilt University Medical Center, Nashville, TN*

MARC G. CARON • *Department of Cell Biology, Howard Hughes Medical Institute Laboratories, Duke University Medical Center, Durham, NC*

F. IVY CARROLL • *Chemistry and Life Sciences, Research Triangle Institute, Research Triangle Park, NC*

NIAN-HANG CHEN • *Department of Pharmacology, School of Basic Medical Sciences, Nanjing Medical University, Nanjing, Jiangsu, China*

SCOTT L. DEKEN • *Department of Neurobiology, University of Alabama at Birmingham, Birmingham, AL*

YU-SHIN DING • *Department of Medicine, Brookhaven National Laboratory, Upton, NY*

CHRISTOPH FELDER • *Department of Chemistry, Brookhaven National Laboratory, Upton, NY*

JOANNA S. FOWLER • *Department of Chemistry, Brookhaven National Laboratory, Upton, NY*

ROBERT T. FREMEAU, JR. • *Department of Neurology & Physiology, University of California at San Francisco School of Medicine, San Francisco, CA*

RAUL R. GAINETDINOV • *Howard Hughes Medical Institute Laboratories, Department of Cell Biology, Duke University Medical Center, Durham, NC*

S. JOHN GATLEY • *Department of Medicine, Brookhaven National Laboratory, Upton, NY*

ANDREW N. GIFFORD • *Department of Medicine, Brookhaven National Laboratory, Upton, NY*

MAUREEN K. HAHN • *Department of Pharmacology, Center for Molecular Neurosciences, Vanderbilt University Medical Center, Nashville, TN*

MATTHIAS A. HEDIGER • *Department of Medicine, Brigham and Women's Hospital, Harvard Institutes of Medicine, Boston, MA*

xi

YOSHIKATSU KANAI • *Department of Pharmacology, Kyorin University, Tokyo, Japan*

BARUCH I. KANNER • *Department of Biochemistry, Hadassah Medical School, The Hebrew University, Jerusalem, Israel*

ANITA H. LEWIN • *Chemistry and Life Sciences, Research Triangle Institute, Research Triangle Park, NC*

JEAN LOGAN • *Department of Chemistry, Brookhaven National Laboratory, Upton, NY*

S. WAYNE MASCARELLA • *Chemistry and Life Sciences, Research Triangle Institute, Research Triangle Park, NC*

DEBORAH C. MASH • *Department of Neurology, University of Miami School of Medicine, Miami, FL*

AMRAT P. PATEL • *Division of Treatment and Research Development, NIDA/NIH, Bethesda, MD*

MICHAEL W. QUICK • *Department of Neurobiology, University of Alabama at Birmingham, Birmingham, AL*

SAMMANDA RAMAMOORTHY • *Department of Physiology/Neuroscience, Medical University of South Carolina, Charleston, SC*

MAARTEN E. A. REITH • *Department of Biomedical and Therapeutic Sciences, University of Illinois College of Medicine, Peoria, IL*

GARY RUDNICK • *Department of Pharmacology, Yale University School of Medicine, New Haven, CT*

SHIMON SCHULDINER • *Alexander Silberman Institute of Life Sciences, The Hebrew University, Jerusalem, Israel*

JULIE K. STALEY • *Department of Psychiatry, Yale University, West Haven, CT*

FRANK W. TELANG • *Department of Chemistry, Brookhaven National Laboratory, Upton, NY*

DAVIDE TROTTI • *Department of Medicine, Renal Division, Brigham and Women's Hospital, Boston, MA*

NORA D. VOLKOW • *Department of Medicine, Brookhaven National Laboratory, Upton, NY*

GENE-JACK WANG • *Department of Chemistry, Brookhaven National Laboratory, Upton, NY*

RODRIGO YELIN • *Novel Genomics, Compugen Ltd., Tel-Aviv, Israel*

Regulation of Monoamine Transporters

Regulated Phosphorylation, Dephosphorylation, and Trafficking

Sammanda Ramamoorthy

1. INTRODUCTION

The monoamines that include dopamine (DA), norepinephrine (NE), and serotonin (5-hydroxytryptamine, 5-HT) act as neurotransmitters in peripheral and central nervous systems (CNS) in a variety of physiological (cognitive, autonomic, and emotional) functions *(1–6)*. Synaptic neu–rotransmission in the central nervous system requires the precise control of the duration and the magnitude of neurotransmitter action at specific molecular targets *(7–9)*. At the molecular level, neurotransmitter signaling is dynamically regulated by a diverse set of macromolecules, including biosynthetic enzymes, secretory proteins, ion channels, pre- and postsynaptic receptors and transporters. The catecholamines DA and NE, which are derived from tyrosine, and the indolamine 5-HT, which is derived from tryptophan, are packaged into synaptic vesicles and released into synapses in response to depolarizing stimuli to activate pre- and postsynaptic receptors and elicit synaptic responses. Monoamine transporters localized near sites of neurotransmitter release remove transmitters from the synaptic cleft and its vicinity, and transport them back into presynaptic neurons. Monoamine transporters belong to the Na^+, Cl^--dependent, gamma-amino butyric acid (GABA)/norepinephrine transporter (GAT1/NET) gene family, which also includes transporters for proline (PROT), taurine (TauT), glycine (GLYT), creatine (CreaT), betaine (BGT), and other "orphan" transporters *(10–12)*. Monoamine transporters are the sites of action for widely used antidepressant drugs and are also high-affinity molecular targets for drugs of abuse including cocaine, amphetamine, and (+)-3,4-methylenedioxymethamphetamine (MDMA or "ecstasy") *(13-15)* (*see* also Chapter 3). Drugs that modulate the activity of biogenic monoamine transporters produce profound behavioral effects, leading to their therapeutic use in the treatment of depres-

From: *Contemporary Neuroscience:*
Neurotransmitter Transporters: Structure, Function, and Regulation, 2nd Edition
Edited by: M. E. A. Reith © Humana Press Inc., Totowa, NJ

sion, obsessive-compulsive disorder (OCD), and other mental diseases, and also to their abuse as stimulants *(16–22)*. Recent molecular and pharmacological analyses of amine-transporter-knockout mice have confirmed the physiological importance and essential expression of presynaptic amine transporters for normal transmitter clearance, presynaptic transmitter homeostasis, and postsynaptic/drug responses *(23–34)* (*see* also Chapter 5). In particular, functional loss of DAT through gene knockout *(24–29,31–33,35,35)* results in profound physical, physiological, and behavioral changes. Altered transporter function or density has been implicated in various types of psychopathology, including depression, suicide, anxiety, aggression, and schizophrenia *(17,21,22,37,38)*. Although there is no evidence yet for an abnormality in the coding regions of DAT or serotonin transporter (SERT), a point mutation in the coding region of human NET has been associated with peripheral Orthostatic Intolerance (OI) *(39)* (*see* also Chapter 4). Two polymorphic regions have been identified in the SERT promoter that have been implicated in anxiety, mood disorders, alcohol abuse, and various neuropsychiatric disorders *(40–44)*.

Reuptake of monoamine neurotransmitters into presynaptic terminals via transporters is the principal mechanism for terminating monoaminergic neurotransmission. Thus, changes in the activity or expression of the transporters should have a significant impact on the duration and concentration of monoamines present in and around the synaptic cleft. These changes, in turn, should influence pre- and postsynaptic responses to released monoamines. This chapter summarizes recent progress in understanding the molecular regulation of neurotransmitter transporters, with a particular focus on intracellular trafficking and post-transcriptional regulation by phosphorylation. Changes in monoamine-transporter function or expression reviewed here suggest that cellular protein kinase(s), protein phosphatase(s), and interacting protein(s) have a potential role in the regulation of transporters for appropriate transporter function and expression. Over the past few years, the scientific research in this area has entered a new age of discovery. Recent studies have revealed that regulation of transporter function and surface expression are rapidly modulated by "intrinsic" transporter activity. Furthermore, drugs that block uptake in previously unidentified ways modulate transporter regulation and cell-surface expression.

2. GENERAL MODELS FOR MONOAMINE-TRANSPORTER REGULATION

Neural activity, hormones, growth factors, environmental factors, and pharmacological agents regulate monoamine uptake, specific radioligand binding, and mRNA levels (discussed for SERT in Chapter 2, 1st edition *[45]*). These studies suggest the presence of endogenous regulatory mechanisms that influence transporter expression and function at the level of gene

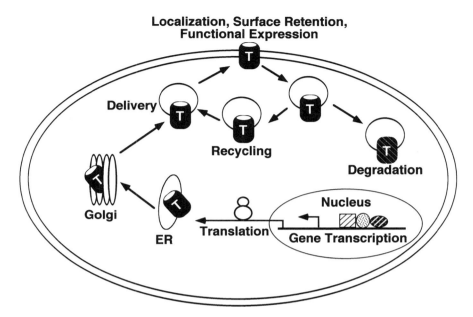

Fig. 1. Life cycle of the monoamine amine transporter. Neurotransmitter clearance is a highly orchestrated process involving regulation of transporter-gene expression and post-translational modifications. Promoter of transporter gene contains several canonical transcription binding sites for constitutive and regulated transcription factors, which drive transcription of transporter in the nucleus. Following transcription, transporters are translated and delivered to their specific membrane sites of expression in cells. At the surface level, the retention and functional expression of transporter proteins are regulated by cellular protein kinases and phosphatases. Depending upon the signals, the transporter may enter into the recycling pool, allowing the transporter to return to the cell surface or enter into lysosomal degradative pathways.

transcription and translation (long-term or chronic regulation requires hours or days) or post-translational protein modifications (short-term or acute regulation requires seconds to minutes) (Fig. 1).

2.1. Long-Term Regulation of Monoamine Transporters at the Gene Level

The structural analysis of transporter-gene-promoter regions reveals a number of canonical transcription binding sites that may be important in controlling the responses of the transporter genes to regulatory factors (46–51). Notably, binding sites for transcription factors including TATA-like motif, an AP1 site, element for CREB binding (CRE), AP-2, NF-IL6, NF-

Kβ, and SP1 sites have been identified in the SERT-promoter region
(46,47,50,51). Relevant investigations performed on JAR cells have shown
that activation of cAMP-dependent (mimicked by cholera toxin, forskolin,
dibutryl cAMP) and independent pathways (mimicked by staurosporine,
interleukin-1β) increases 5-HT uptake. Increases in 5-HT uptake require prolonged stimulation (4-h) and messenger RNA and protein synthesis *(52–54)*.

Since the long-term modulation of SERT expression both in vivo and in
vitro is described in detail in an earlier chapter *(45)*, only recent developments in the long-term regulation of SERT are summarized here. Unlike
SERT, very little is known about NET and DAT gene regulation. Earlier in
vivo and in vitro studies demonstrate modulation of NE uptake by several
factors such as insulin, atrial natriuretic peptide (ANP), angiotensin (ANG
II), dexamethasone, nerve-growth factor (NGF), and pharmacological substances such as desipramine and cocaine *(55–61)*. Altered NET mRNA levels are shown for the NET modulation by insulin, dexamethasone, NGF,
desipramine, and cocaine. Recent reports show upregulation of DAT and NET
genes by cocaine treatment. During pregnancy, cocaine exposure results in
increased DAT mRNA levels in the fetal rhesus monkey brain *(62)*. Another
study revealed increased levels of NET mRNA in the placentas of rats treated
with cocaine at the mRNA level *(63,64)*. Although the actual cellular signaling pathways responsible for such regulation at the genetic level have not yet
been identified, these studies suggest a significant functional role for the
monoamine transporters in fetal development.

2.2. Acute Regulation of Monoamine Transporters

Amino-acid sequence analysis of SERT, DAT, and NET proteins reveals
numerous consensus sites for protein kinases as well as putative interactive
motifs in intracytoplasmic domains, suggesting that second messengers play
a role in the post-translational regulation of monoamine transporters *(65–
70)*. Prior to the cloning of monoamine transporters, many studies suggested
a pivotal role for second messenger-linked pathways in acute modulation of
neurotransmitter uptake *(45,71)*. Potential pathways through which second
messenger-linked signals might regulate transporter function acutely are illustrated in Figs. 1 and 2. Regulation of transporters occurs directly by phosphorylation of transporter proteins. Evidence indicates that phosphorylation
of transporter proteins changes intrinsic transporter activity, alters transporter turnover, regulates fusion of transporter containing vesicles with the
plasma membrane, and regulates sequestration of transporter from the
plasma membrane by modulating endocytic pathways. Alternatively, regulation of transporters could occur through their interaction with other proteins by phosphorylation-dependent/or -independent pathways *(72)*.

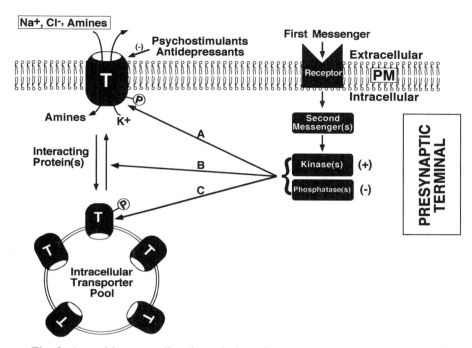

Fig. 2. Acute kinase-mediated regulation of monoamine-transporter expression. Monoamine transporters are regulated by multiple signaling pathways linked to receptor-triggered kinase activation. Transporter phosphorylation may be compartmentalized in a manner, appropriate for its use as a signal for transporter trafficking. Phosphorylation and dephosphorylation cascades regulate transporter expression. Direct phosphorylation may change transporter function by multiple discrete mechanisms. For example, phosphorylation of the transporter may A) change its intrinsic transport activity, B) regulate sequestration of transport from plasma membrane by modulating endocytosis machinery, C) regulate exocytic fusion of transporter-containing vesicles with the plasma membrane. Transporter-associated proteins such as protein phosphatase 2A would govern the stoichiometry of amine-transporter phosphorylation as well as its duration, providing an important trigger for transporter expression and trafficking.

2.3. Regulation of Monoamine-Transporter Surface Expression

A common approach has been employed in studying acute transporter regulation in endogenous and heterologous systems by treatment with various activators and inhibitors of protein kinases and phosphatases. The most consistent finding is that activators of protein kinase C (PKC) or the agents that maintain the phosphorylation state, such as phosphatase inhibitors, rapidly reduce amine transport capacity. The major kinetic alterations typically observed in acute modulation paradigms are changes in transport capacity (V_{max}) with little or no significant change in substrate affinity (K_m). In a variety of preparations, including synaptosomes and cell lines, application of the PKC activator, β-PMA, selectively reduces the V_{max} for transport of DA, NE, and 5-HT *(73–80)*. Several control experiments confirm the specificity of β-PMA on PKC activation and, thus, the modulation of monoamine transporters by PKC is not a generalized effect. First, β-PMA has no, little, or opposite effects on other transporter activities *(78,81)*. Second, the effects of β-PMA are stereospecific; the active stereoisomer of PMA, but not the inactive alpha isomer, modulates monoamine-transporter function *(78)*. Third, the β-PMA effect can be blocked by PKC inhibitors, such as staurosporine and bisindolylmalemide *(78)*. Finally, the β-PMA effect is paralleled by an activation and translocation of PKC from the cytoplasm to the plasma membrane *(82,83)*.

The reduction in transport capacity (V_{max}) suggests silencing of plasma-membrane resident amine transport protein or decreased cell-surface expression. Initial evidence for altered surface expression of transporter protein by PKC has come from homologous GAT1 GABA transporters expressed in *Xenopus* oocytes. In *Xenopus* oocytes, phorbol esters increase GAT1 activity by translocating GAT1 from intracellular compartments to the cell surface *(84,85)*. The first evidence for altered cell-surface expression of a monoamine transporter as the result of PKC activation has come from studies of human SERT stably expressed in Human Embryonic Kidney (HEK-293) cells *(78)*. In this study, immunoprecipitation following cell-surface protein biotinylation was employed to quantify the cell-surface SERT protein pool before and after PKC activation. Exposure to β-PMA leads to 30–40% loss of SERT proteins on the plasma membrane in parallel with reductions in 5-HT uptake capacity. A concomitant increase in the intracellular SERT pool suggests translocation of SERT from the plasma membrane to the intracellular pool. Similar to 5-HT uptake, loss of cell-surface SERT was blocked by the PKC inhibitor staurosporine *(78)*. The protein phosphatase inhibitor, okadaic acid, also regulates SERT activity, suggesting that cellular phosphatases monitor phosphorylation-dependent SERT regulation in this cell model *(86,87)*. Similar observations were made in LLC-PK1 cells stably transfected with human NET *(73,74)*, DAT expressing PC12, African Green Monkey kidney (COS), and Madin-Darby canine kidney (MDCK) cells *(79,88–90)*, and DAT cRNA-injected *Xenopus* oocytes following PKC activation *(80)*. Using PC12 cells stably transfected with human DAT, Melikian and Buckley *(89)* reported that activation of PKC results in decreased uptake capacity in concert with decreased cell-surface DAT density. The study also showed that internalized DAT is colocalized with transferrin receptors, which are known to actively recycle to the plasma membrane, although no supporting data were provided. The authors suggested that the PKC-dependent, internalized DAT may recycle to the plasma membrane. In another study, using green fluorescent protein (GFP)-tagged DAT stably expressed in MDCK cells, Daniels and Amara *(88)* elegantly visualized rapid DAT internalization in live cells following PKC activation. PKC-dependent DAT internalization is not a general increase in vesicular traffic from the cell surface, because another cell-surface protein–E-cadherin–did not internalize under similar conditions. Interestingly, a dominant-negative mutant of dynamin 1, which has been shown to block clathrin-mediated endocytosis, is able to block PKC-dependent DAT internalization. These observations suggest that PKC regulates DAT internalization by clathrin-mediated and dynamin-dependent cellular mechanisms. They also provide evidence that internalized DAT is targeted to endosomal/ lysomal pathways for degradation. The molecular mechanisms responsible for various PKC-dependent DAT internalization pathways (recycling endosomes vs degradative lysosomes) in these studies are unknown. Another signaling pathway mediated by cAMP-dependent protein kinase A (PKA) upregulates striatal DAT by increasing V_{max} *(91)*. Pristupa and colleagues *(77)* showed a bidirectional trafficking of DAT by PKC and PKA pathways in DAT-transfected COS and Sf9 cells *(77, 91–93)*. The studies described above used heterologous expression systems transfected with amine transporter cDNAs or Xenopus oocytes injected with amine transporter cRNA. Studies with isolated preparations such as synaptosomes, brain slices, or platelets also reveal similar changes in kinetic parameters following acute regulation of amine transporter by PKC and other kinases *(90,91,94–99)*. Relatively few studies have demonstrated amine-transporter regulation by presynaptic receptors *(25,100–103)*. These observations establish that rapid shuttling of amine transporters between plasma membrane and intracellular compartments, or distribution of transporters in active and inactive pools, provide a route through which amine uptake may be regulated by second messenger-linked systems, possibly following activation of presynaptic auto/ or hetero receptors (Figs. 1 and 2).

3. TRANSPORTER PHOSPHORYLATION: A SIGNAL FOR AMINE UPTAKE MODULATION

Recent observations have suggested that modulation of amine transporters followed by activation of cellular protein kinases or inhibition of protein phosphatases leads to amine transporter phosphorylation. These observations support the role of amine-transporter phosphorylation in transporter regulation. Recently, Ramamoorthy et al. *(8)* demonstrated that SERT proteins are phosphorylated under basal conditions in HEK-293 cells, and that phosphorylation is increased in a time- and dose-dependent manner by activators of PKC, and PKA, as well as inhibitors of phosphatase 1/2A. β-PMA-triggered SERT phosphorylation is blocked by the PKC inhibitors staurosporine and bisindolylmalemide, but not by PKA inhibitors. However, SERT phosphorylation triggered by cyclic AMP analogs or cholera toxin can be blocked by PKA inhibitors but not by PKC inhibitors. Protein phosphatase (PP1/PP2A) inhibitors such as okadaic acid and calyculin A, but not the calcineurin (PP2B) inhibitor cyclosporin, significantly elevate SERT phosphorylation to a similar extent as observed with β-PMA. The elevation of SERT phosphorylation induced by β-PMA and okadaic acid is positively correlated with losses in 5-HT transport, suggesting an extremely close correlation between SERT phosphorylation and 5-HT uptake capacity. Additivity of PKC-triggered SERT phosphorylation is found after cotreatment with okadaic acid, β-PMA, and cholera toxin at maximally effective concentrations, suggesting that distinct sites are being phosphorylated as a result of the activation of PKC, PKA, and as yet unidentified endogenous kinase(s). Interestingly, the antagonists for PKC, PKA, CaM kinase II, and depletion of intracellular Ca^{2+} fail to reduce SERT phosphorylation observed in the presence of the PP1/PP2A inhibitor. These data suggest that protein kinase(s) other than PKC, PKA, PKG, or Ca^{2+}-dependent kinase—yet to be identified—phosphorylates SERT under basal conditions in this heterologous model *(99)*. Similar evidence for DAT phosphorylation comes from studies using native rat-brain striatal synaptosomes as well as heterologously expressed rat DAT in COS and LLC-PK1 cells *(90,92,104,105)*. Similar to human SERT, DAT exhibits basal phosphorylation that can be augmented by PKC activation and phosphatase inhibition. In both preparations, PKC activation leads to decreased DAT activity, which is manifested primarily as a reduction in V_{max}. PKC-dependent DAT phosphorylation was sensitive to the PKC inhibitors staurosporine and bisindolylmalemide. One important difference between DAT and SERT phosphorylation is that okadaic acid-triggered SERT phosphorylation is not blocked by staurosporine, whereas staurosporine blocks okadaic acid-induced DAT phosphorylation. Another difference between SERT and DAT phosphorylation studies is the evidence that SERT undergoes rapid phosphorylation following PKA activation, whereas DAT is not phosphorylated by PKA activation *(90,92)*. These studies suggest that specific kinase-mediated pathways may influence amine-transporter phosphorylation and regulation in a specific manner, and may not be shared with each other.

3.1. Intrinsic Transporter Activity Governs Amine-Transporter Surface Expression

Another important recent finding is the ability of amine-transporter substrates and antagonists to influence transporter trafficking and plasma-membrane residency *(86)*. A common theme is emerging among transporters such as DAT, GAT1, and GLAST, which are regulated by their substrates *(106–109)*. For example, Ramamoorthy and Blakely *(86)* provided evidence that SERT substrates such as 5-HT, amphetamines, fenfluramine, and antagonists such as antidepressants and cocaine control PKC-dependent SERT phosphorylation and surface redistribution *(86)*. This substrate-transporter feedback loop provides a mechanism by which changes in extracellular neurotransmitter concentrations rapidly modulate neurotransmitter transport capacity. This study showed that extracellular 5-HT significantly blunts the SERT phosphorylation achieved by PKC activation. This inhibitory effect of 5-HT occurs in the presence of 5-HT receptor antagonists, suggesting that the effect of 5-HT on PKC-mediated SERT phosphorylation is not dependent on 5-HT-receptor activation. Moreover, the 5-HT effect displays a concentration-dependence similar to that of 5-HT transport. Like 5-HT transport, the 5-HT effect on PKC-induced SERT phosphorylation requires extracellular Na^+ and Cl^-, and can be blocked by SSRIs but not by DAT- and NET-selective antagonists. Diminished SERT phosphorylation cannot be achieved by preloading the cells with 5-HT prior to PKC activation. Other SERT substrates such as D-amphetamine and fenfluramine, but not dopamine and norepinephrine, have a similar inhibitory effect on PKC-mediated SERT phosphorylation. Cell-surface protein biotinylation experiments reveal that PKC activation results in decreased cell-surface SERT density, and this effect can be abolished by the presence of 5-HT during PKC activation *(86)*. Together, these findings suggest that SERTs in the presence of actively transporting substrates reduce the opportunity for PKC-linked SERT phosphorylation and transporter internalization. Thus, control of SERT cell-surface expression by 5-HT would provide a novel homeostatic mechanism that may serve in the neuron to fine-tune transport capacity to match demands imposed by fluctuating levels of serotonin. Signaling pathways linked to presynaptic auto- and hetero-receptors could provide positive/negative feedback control, and could provide a mechanism by which changes in synaptically released extracellular 5-HT could rapidly modulate SERT capacity. Interestingly, amphetamines can substitute for 5-HT in suppressing PKC-mediated SERT phosphorylation. Such action could override homeostatic transporter sequestration processes and provide for psychostimulant-induced sensitization by increasing the number of psychostimulant targets available to a subsequent stimulus. On the other hand, nonpermeant SERT ligands such as selective serotonin reuptake inhibitors (SSRIs) and cocaine, which prevent 5-HT permeation, block the effect of 5-HT. Thus, SSRIs may have therapeutic utility in disease states, not only by preventing 5-HT uptake, but also by shifting the cellular distribution of SERT.

3.2. Protein Phosphatase 2A: An Intimate Amine Transporter Partner in Regulating the State of Transporter Phosphorylation

As described earlier, agents that maintain the phosphorylated state modulate amine transporter activity. For example, the protein phosphatase 1/2A (PP1/PP2A) inhibitor, known as okadaic acid downregulates SERT, DAT and NET activity *(87,90,92)*. Treatment with okadaic acid or another potent PP1/PP2A inhibitor, calyculin A, results in a rapid increase in SERT phosphorylation in parallel with a decrease in 5-HT uptake *(87)*. Similarly, these inhibitors also promote DAT phosphorylation and functional downregulation in striatal synaptosome preparations *(90)*. Maintenance of the phosphorylation state and regulation of surface amine-transporter expression is a balance between the actions and localization of protein kinases and protein phosphatases. The pronounced effect of PP1/PP2A inhibition on amine uptake and amine-transporter phosphorylation suggests the possible physical association of PP1/PP2A as a regulatory complex with amine-transporter proteins. Desensitization and trafficking of several G-protein-coupled receptors are regulated by phosphorylation-dependent pathways. This involves regulated physical association of receptors with kinases and phosphatases and other adaptor proteins *(110–115)*. For example, the catalytic subunit of PP2A is believed to play an important role in recycling β2-adrenergic receptors by associating with phosphorylated and internalized β2 receptors. The associated PP2Ac catalyzes the dephosphorylation of β2 receptors, facilitating their recycling to the plasma membrane *(113,114)*. Recently, Bauman et al. *(116)* documented the existence of okadaic acid-sensitive phosphatase activity in the immunoisolated SERTs from stably transfected HEK-293 cells. Western blots of SERT immunoprecipitates using monoclonal PP2Ac antibody revealed the presence of PP2A catalytic subunit. Blots of SERT immunoprecipitations from multiple brain regions also showed the presence of PP2Ac with SERT as a complex, suggesting that SERT/PP2Ac exists as an associated complex in vivo. Qualitatively, similar transporter and PP2Ac associations were found for DAT and NET proteins. Evaluation of DAT and NET immunoprecipitates from rat striatal synaptosomes and vas deferens, respectively, shows the presence of PP2Ac as a complex with DAT and NET proteins *(12)*. Together, these findings suggest that PP2Ac associations with biogenic amine transporters are also likely to occur with other transporters of the gene family, and this association plays an important role in their regulation. Another hypothesis is that the associated PP2Ac activity may govern the stoichiometry of amine-transporter phosphorylation, as well as the duration of the phospho-nature of transporter. This will provide an important trigger for transporter-protein expression and trafficking. In addition, the association of PP2Ac with amine transporters may be dynamically regulated by kinases/phosphatases and by the state of transporter phosphorylation. In support of this hypothesis, transporter and PP2Ac associations can be regulated by PKC activation, phosphatase inhibition, and transporter substrates. In 293-hSERT cells, treatment of cells with β-PMA decreases SERT-associated PP2Ac, which is blocked by the PKC inhibitors staurosporine, and bisindolylmalemide. Under similar conditions, β-PMA treatment leads to increased SERT phosphorylation and internalization *(86,87)*. These results suggest that PKC-dependent SERT phosphorylation and internalization occurs by dissociating PP2Ac from a pre-existing PP2Ac/SERT complex. Dissociation of PP2Ac from SERT leads to decreased SERT-dephosphorylation, contributing to enhanced SERT phosphorylation. 5-HT blocks the PKC-triggered dissociation of PP2Ac from SERT, suggesting that transport or the transporter provide a signal for SERT phosphorylation and association of other proteins. However, it is presently not known whether PKC-dependent SERT phosphorylation occurs first—which in turn triggers PP2Ac dissociation—or whether PKC-dependent SERT phosphorylation is dependent on PP2Ac dissociation. In addition to PKC-dependent-regulated association of the SERT/PP2Ac complex, the PP1/PP2A inhibitors, okadaic acid or calyculin A, disassemble the PP2Ac/SERT complex. This regulation is specific and dependent on PP1/PP2A inhibition, but not by inhibiting other phosphatases. Other phosphatase inhibitors—such as cyclosporine A, which is specific for PP2B—have no ability to regulate the PP2Ac/SERT complex. Analogous to SERT studies, PP2Ac association with NET in vas deferens is also decreased by PKC activation and PP1/PP2A inhibition *(116)*. These results suggest several possibilities in the mode of SERT association with PP2Ac. (1) Okadaic acid is known to bind the active site of PP2Ac and, thus, okadaic acid may compete with SERT for the same site on PP2Ac. This also suggests that the PP2Ac active site may be the motif by which PP2Ac assembles with SERT directly or via other adaptor protein(s). (2) Protecting phospho-SERT by inhibiting dephosphorylation of SERT by okadaic acid could trigger release of PP2Ac from the SERT complex. (3) Okadaic acid-linked inhibition of phosphatase leads to activation of kinase, which could trigger the dissociation of PP2Ac from the SERT complex.

3.3. Regulation of Biogenic Amine Transporters by Protein Interactions

Although the direct phosphorylation and dephosphorylation of transporters by cellular protein kinases/phosphatases is involved in the dynamic regulation and expression of transporters, other integral membrane proteins also appear to play an important role in trafficking and the catalytic function of transporters. The first evidence for the physical association of GAT1 GABA transporters with the (t)-SNARE protein, syntaxin-1A, has come from studies by Quick and colleagues using native and heterologous expression model systems *(86,117)*. Interestingly, this association is modulated in a PKC-dependent manner, and the interaction of GAT1 with Syntaxin-1 is required for GAT1 regulation by PKC activation. Syntaxin-1A association not only regulates PKC-dependent trafficking of GAT1, but also appears to influence the catalytic function of the transporter *(85,117)*. Recently, it has been shown that syntaxin-1A interacts with the *N*-terminal cytoplasmic domain of the GAT1 GABA transporter and decreases transport rates *(118)*. Co-immunoprecipitation of syntaxin-1A with the glycine transporter, GLYT2, has been shown in heterologous and native systems *(119)*. Like GAT1 and GLYT2, NET appears to be physically associated with syntaxin-1A. Using noradrenergic tissues such as vas deferens and transfected cell lines, Sung et al. *(120)* reported that syntaxin-1A is found in NET immunoprecipitates. Activation of PKC or phosphatase inhibition decreases the syntaxin-1A level in the NET immunoprecipitates. Altering intracellular Ca^2+ also regulates the syntaxin/NET interaction. Together, these findings suggest an important role for syntaxin-1A in neurotransmission by regulating both transmitter release and reuptake. In addition, signals following presynaptic receptor stimulation may influence transporter function and expression by regulating the stability of transporter heteromeric complexes. Recently, Torrres et al. *(121)* showed the interaction of the PDZ domain-containing protein, PICK1 (protein interacting with C kinase) with C-terminal cytoplasmic domains of DAT and NET *(121)*. Co-expression of DAT and PICK1 in HEK-293 cells enhances DAT activity and PICK1 co-immunoprecipitates with DAT. Furthermore, confocal microscopy analysis revealed colocalization of both DAT and PICK1 in a heterologous system and in dissociated dopaminergic neurons *(104)*. Since PICK1 is a PKC α-binding protein, it is possible that PICK1 acts like an adaptor to bring PKC near DAT proteins, which in turn could regulate PKC-evoked DAT phosphorylation, trafficking, and function. Lee and colleagues *(122)* recently identified the interaction of the C-terminal cytoplasmic domain of DAT with α-synuclein. α-Synuclein is enriched in dopaminergic nerve terminals and has been implicated in Parkinson's and other neurodegenerative disorders. α-Synuclein binds directly to the C-terminal region of DAT and facilitates the plasma-membrane clustering and increase in DAT activity. This study suggests that the α-synuclein/DAT complex regulates normal dopaminergic neurotransmission, and altered interactions may result in abnormal DAT function, causing the dopaminergic neurodegeneration seen in Parkinson's disease *(122)*.

In summary, evidence suggests that amine-transporter levels may be modulated by hormones, growth factors, neuronal activity, and pharmacological agents. The intracellular second messenger systems mimic cellular protein kinases and phosphatases, and subsequently regulate amine-transporter gene and protein expression. The past few years represent an active period in transporter research related to structure, expression, function, and regulation of amine transporters. These investigations undoubtedly provide strong support for the idea that amine transporters are principal players in regulating normal and abnormal amine signaling in the CNS and periphery, and, thus, complex behavioral and physiological functions. However, even more exciting is the discovery that the intrinsic transport capacity of a transporter molecule governs its own plasma-membrane expression and function *(86,106,107)*. There appear to be multiple mechanisms by which transporter substrates and antagonists can influence transporter function and expression. The capacity of transporters to fine-tune their function in response to extracellular neurotransmitter levels would result in maintaining a constant level of neurotransmitter at the synaptic cleft. In other words, increased cell-surface expression of neurotransmitter transporter following an increase in the extracellular level of neurotransmitter would, in turn, promote the rapid clearance of released neurotransmitter from the synaptic cleft and provide a mechanism to maintain the synaptic concentrations of neurotransmitter level within a narrow range (Fig. 3). It is interesting to know how the architecture of trafficking mechanisms is designed in nature at pre- and postsynaptic levels for the termination of neurotransmission (Fig. 3). At the postsynaptic level, receptors for neurotransmitters internalize following agonist-induced phosphorylation. The cascade of events lead to receptor desensitization and termination of neurotransmission. At the presynaptic level, the active termination of neurotransmission is mediated by plasma-membrane transporters. In contrast to what happens at the postsynaptic level (decrease in receptor density caused by internalization following agonist-induced phosphorylation), increased transporter density at the presynaptic membrane in response to synaptic neurotransmitter appears to be another mechanism for rapid termination of neurotransmission. The coordinate regulatory pattern of transporter and receptor trafficking at pre- and postsynaptic levels may provide normal neurotransmission. An altered pattern or a disturbance in the pre- and postsynaptic regulatory mechanisms could lead to abnormal neurotransmission, and thus abnormal behavior or brain disorders. Drugs of abuse such as cocaine and amphetamines bind and inhibit biogenic amine transporters, and may interfere with activity-dependent regulatory mechanisms, and may cause or initiate drug addiction or sensitization. Along with information delineating *cis/trans* acting elements and signals for acute and chronic

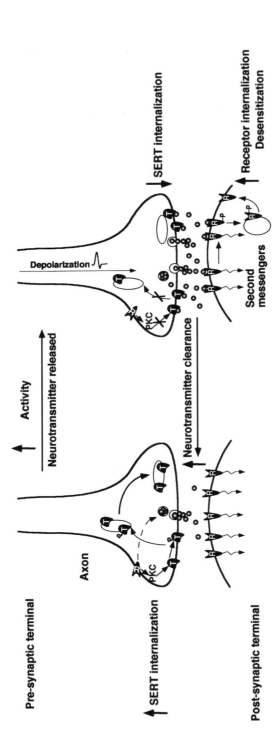

Fig.3. Hypothetical model for the phosphorylation-dependent coordinate trafficking of presynaptic monoamine transporters and postsynaptic receptors in normal neurotransmission. Phosphorylated presynaptic monoamine transporter—for example, SERT—enters into a regulated endocytosis pathway. Internalized SERT could be dephosphorylated by associated protein phosphatase, allowing the return of nonphosphorylated SERT to the cell surface. SERT substrate 5-HT increases SERT plasma-membrane residency time by not allowing the SERT to be phosphorylated and internalized. Increased cell-surface retention of the transporters may regulate synaptic transmission by efficient clearance of released neurotransmitters. This may serve as a built-in "feed-back" mechanism to autoregulate the transporters and suppress the influence of presynaptic receptor-linked stimuli. In contrast to increase in presynaptic transporter following neural activity, postsynaptic receptors are internalized following agonist induced phosphorylation, resulting in receptor desensitization. Thus, the regulation of trafficking and expression of transporters at the presynaptic level and of receptors at the postsynaptic

amine-transporter regulation, over the next few years with hard work and ingenuity, we will expect rapid progress in understanding the role of cellular regulation of amine transporters in amine signaling and behavior. These developments provide opportunities to add further impetus to the drive for psychotherapeutic innovation.

ACKNOWLEDGMENTS

Support from the National Alliance of Research on Schizophrenia and Depression (NARSAD) and the developmental fund from the Medical University of South Carolina are greatly acknowledged. Thanks to Dr. J. F. McGinty for critical review of this chapter. Thanks also go to Dr. Reith for providing this wonderful opportunity to contribute this chapter, and for his encouragement.

REFERENCES

1. Bunney, B. S., Sesack, S. R., and Silva, N. L. (1987) Midbrain dopaminergic systems: neurophysiology and electrophysiological pharmacology, in *Psychopharmacology: The Third Generation of Progress* (Melter, H. Y., ed.), Raven Press, New York, NY, pp. 112–1129.
2. Fozzard, J. (1989) e. *Peripheral actions of 5-hydroxytryptamine*. Oxford University Press, New York, NY.
3. Goldstein, D. S. (1995) *Stress, Catecholamines, and Cardiovascular Disease*. Oxford University Press, New York, NY.
4. Jacobs, B. and E. C. (1992) Azmitia. Structure and function of the brain serotonin system. *Physiol. Rev.* **72,** 165–229.
5. Jacobs, B. L. and Fornal, C. A. (1991) Activity of brain serotonergic neurons in the behaving animal. *Pharmacol. Rev.* **43,** 563–578.
6. Nicoll, R. A., Madison, D. V., and Lancaster, B. (1987) Noradrenergic modulation of neuronal excitability in mammilian hippocampus, in *Psychopharmacology: The Third Generation of Progress*, (Meltzer, H. Y., ed.), Raven Press, New York, NY, pp. 105–112.
7. Barker, E. L. and Blakely, R. D. (1995) Norepinephrine and serotonin transporters: Molecular targets of antidepressant drugs, in *Psychopharmacology: The Fourth Generation of Progress*. (Bloom, F. E. and Kupfer D. J., eds.), Raven Press, New York, NY, pp. 321–333.
8. Iversen, L. L. (1971) Role of transmitter uptake mechanisms in synaptic neurotransmission. *Brit. J. Pharmacol.* **41,** 571–591.
9. Iversen, L. L. (1978) Uptake processes for biogenic amines. in *Handbook of Psychopharmacology* 3rd ed. (Iversen, I., ed.), Penum Press, New York, NY pp. 381–442.
10. Amara, S. G. and Kuhan, M. J. (1993) Neurotransmitter transporters: recent progress. *Annu. Rev. Neurosci.* **16,** 73–93.
11. Miller, J. W., Kleven, D. T., Domin, B. A., and Fremeau, Jr., R. T. (1997)

Cloned sodium- (and chloride-) dependent high-affinity transporters for GABA, glycine, proline, betaine, taurine, and creatine, in *Neurotransmitter Transporters: Structure, Function, and Regulation*, (Reith, M. E. A., ed.), Humana Press, Totowa, NJ, pp. 101–150.

12. Povlock, S. L. and Amara, S. G. (1997) The structure and function of norepinephrine, dopamine, and serotonin transporters, in *Neurotransmitter Transporters: Structure, Function, and Regulation*, (Reith, M. E. A., ed.), Humana Press, Totowa, NJ, pp. 1–28.

13. Gobbi, M., Mennini, T., and Garratini, S. (1997) Mechanism of neurotransmitter release induced by amphetamine derivatives: pharmacological and toxicological aspects. *Curr. Top. Pharmacol.* **3**, 217–227.

14. Rudnick, G. and Wall, S. C. (1992) The molecular mechanism of "ecstasy" [3,4-methylenedioxymethamphetamine (MDMA)]: Serotonin transporters are targets for MDMA-induced serotonin release. *Proc. Natl. Acad. Sci. USA* **89**, 1817–1821.

15. Rudnick, G. and Wall, S. C. (1992) p-Chloroamphetamine induces serotonin release through serotonin transporters. *Biochemistry* **31**, 6710–6718.

16. Arora, R. C., Emery, O. B., and Meltzer, H. Y. (1991) Serotonin uptake in the blood platelets of Alzheimer's disease patients. *Neurology* **41**, 1307–1309.

17. Arora, R. C. and Meltzer, H. Y. (1982) Serotonin uptake by blood platelets of schizophrenic patients. *Psychiatry Res.* **6**, 327–333.

18. Barondes, S. H. (1994) Thinking about Prozac. *Science* **263**, 1102,1103.

19. Klimek, V., Stockmeier, C., Overholser, J., Meltzer, H. Y., Kalka, S., Dilley, G., et al. (1997) Reduced levels of norepinephrine transporters in the locus coeruleus in major depression. *J. Neurosci.* **17**, 8451–8458.

20. Kramer, P. D. (1996) Antidepressants, in *Listening to Prozac*, Penguin Books, pp. 47–67.

21. Meltzer, H. Y. (1990) Role of serotonin in depression. *Ann. NY. Acad. Sci.* **600**, 486–499.

22. Meltzer, H. Y., Arora, R. C., Baber, R., and Tricou, B. J. (1981) Serotonin uptake in blood platelets of psychiatric patients. *Arch. Gen. Psychiatry* **38**, 1322–1326.

23. Bengel, D., Murphy, D. L., Andrews, A. M., Wichems, C. H., Feltner, D., Heils, A., et al. (1998) Altered brain serotonin homeostasis and locomotor insensitivity to 3,4-methylenedioxymetamphetamine ("ecstasy") in serotonin transporter-deficient mice. *Mol. Pharmacol.* **53**, 649–655.

24. Bosse, R., Fumagalli, F., Jaber, M., Giros, B., Gainetdinov, R. R., Wetsel, W. C., et al. (1997) Anterior pituitary hypoplasia and dwarfism in mice lacking the dopamine transporter. *Neuron* **19**, 127–138.

25. Drago, J., Padungchaichot, P., Accili, D., and Fuchs, S. (1998) Dopamine receptors and dopamine transporter in brain function and addictive behaviors: insights from targeted mouse mutants. *Dev. Neurosci.* **20**, 188–203.

26. Fumagalli, F., Gainetdinov, R. R., Valenzano, K. J., and Caron, M. G. (1998) Role of dopamine transporter in methamphetamine-induced neurotoxicity: Evidence from mice lacking the transporter. *J. Neurosci.* **18**, 4861–4869.

27. Gainetdinov, R. R., Fumagalli, F., Jones, S. R., and Caron, M G. (1997) Dopamine transporter is required for in vivo MPTP neurotoxicity: Evidence from mice lacking the transporter. *J. Neurochem.* **69,** 1322–1325.

28. Gainetdinov, R. R., Jones, S. R., Fumagalli, F., Wightman, R. M., and Caron, M. G. (1998) Re-evaluation of the role of the dopamine transporter in dopamine system homeostasis. *Brain. Res. Brain Res. Rev.* **26,** 148–153.

29. Giros, B., Jaber, M., Jones, S. R., Wightman, R. M., and Caron, M. G. (1996) Hyperlocomotion and indifference to cocaine and amphetamine in mice lacking the dopamine transporter. *Nature.* **379,** 606–612.

30. Hirata, H. and Cadet, J. L. (1997) Methamphetamine-induced serotonin neurotoxicity is attenuated in p53-knockout mice. *Brain Res.***768,** 345–348.

31. Jones, S. R., Gainetdinov, R. R., Hu, X. T., Cooper, D. C., Wightman, R. M., White, F. G., et al. (1999) Loss of autoreceptor functions in mice lacking the dopamine transporter. *Nat. Neurosci.* **2,** 649–655.

32. Jones, S. R., Gainetdinov, R. R., Jaber, M., Giros, B., Wightman, R. M., and Caron, M. G. (1998) Profound neuronal plasticity in response to inactivation of the dopamine transporter. *Proc. Natl. Acad. Sci. USA* **95,** 4029.

33. Jones, S. R., Gainetdinov, R. R., Wightman, R. M., and Caron, M. G. (1998) Mechanisms of amphetamine action revealed in mice lacking the dopamine transporter. *J. Neurosci.* **18,** 1979–1986.

34. Rioux, A., Fabre, V., Lesch, K. P., Moessner, R., Murphy, D. L., Lanfumey, L., et al. (1999) Adaptive changes of serotonin 5-HT2A receptors in mice lacking the serotonin transporter. *Neurosci. Lett.* **262,** 112–116.

35. Ralph, R. J., Paulus, M. P., Fumagalli, F., Caron, M. G., and Geyer, M. A. (2001) Prepulse inhibition deficits and perseverative motor patterns in dopamine transporter knock-out mice: differential effects of D1 and D2 receptor antagonists. *J. Neurosci.* **21,** 305–313.

36. Zhuang, X., Oosting, R. S., Jones, S. R., Gainetdinov, R. R., Miller,G. W., Caron, M. G., et al. (2001) Hyperactivity and impaired response habituation in hyperdopaminergic mice. *Proc. Natl. Acad. Sci. USA* **98,** 1982–1987.

37. Meltzer, H. Y. and Arora, R. C. (1987) Genetic control of serotonin uptake in blood platelets: A twin study. *Psychiatry Res.* **24,** 263–269.

38. Owens, M. J., Morgan, W. N., Plott, S. J., and Nemeroff, C. B. (1997) Neurotransmitter receptor and transporter binding profile of antidepressants and their metabolites. *J. Pharmacol. Exp. Ther.* **283,** 1305–1322.

39. Shannon, J. R., Flattem, N. L., Jordan, J., Jacob, G., Black, B. K., Biaggioni, I., et al. (2000) Orthostatic intolerance and tachycardia associated with norepinephrine-transporter deficiency. *N. Engl. J. Med.* **342,** 541–549.

40. Bonnet-Brilhault, F., Laurent, C., Thibaut, F., Campion, D., Chavand, O., Samolyk, D., et al. (1997) Sertonin transporter gene polymorphism and schizophrenia: an association study. *Biol. Psychiatry* **42,** 634–636.

41. Klauck, S. M., Poustka, F., Benner, A., Lesch, K. P., and Poustka, A. (1997) Serotonin transporter (5-HTT) gene variants associated with autism? *Human Mol. Genet.* **6,** 2233–2238.

42. Lesch, K.-P., Bengel, D., Heils, A., Sabol, S. Z., Greenberg, B. D., Petri, S., et al. (1996) Association of anxiety-related traits with a polymorphism in the serotonin transporter gene regulatory region. *Science* **274,** 1527–1531.

43. Lesch, K. P., Gross, J., Franzek, E., Wolozin, B. L., Riederer, P., and Murphy, D. L. (1995) Primary structure of the serotonin transporter in unipolar depression and bipolar disorder. *Biol. Psychiatry* **37**, 215–223.
44. Stöber, G., Heils, A., and Lesch, K. P. (1996) Serotonin transporter gene polymorphism and affective disorder. *Lancet* **347**, 1340–1341.
45. Blakely, R. D., Ramamoorthy, S., Qian, Y., Schroeter, S., and Bradley, C. (1997) Regulation of antidepressant-sensitive serotonin transporters, in *Neurotransmitter Transporters: Structure, Function, and Regulation*, (Reith, M. E. A., ed.), Humana Press, Totowa, NJ, pp. 29–72.
46. Bradley, C. C. and Blakely, R. D. (1997) Alternative splicing of the human serotonin transporter gene. *J. Neurochem.* **69**, 1356–1367.
47. Flattem, N. L. and Blakely, R. D. (2000) Modified structure of the human serotonin transporter promoter. *Mol. Psychiatry* **5**, 110–115.
48. Fritz, J., Lankupalle, J., Thoreson, M., and Blakely, R. D. (1998) Cloning and chromosomal mapping of the murine norepinephrine transporter. *J. Neurochem.* **70**, 2241–2251.
49. Kawarai, T., Kawakami, H., Yamamura, Y., and Nakamura, S. (1997) Structure and organization of the gene encoding human dopamine transporter. *Gene*. **195**, 11–18.
50. Lesch, K. P., Balling, U., Gross, J., Strass, K., Wolozin, B. L., Murphy, D. L., et al. (1994) Organization of the human serotonin transporter gene. *J. Neural. Trans.* **95**, 157–162.
51. Mortensen, O. V., Thomassen, M., Larsen, M. B., Whittemore, S. R., and Wilborg, O. (1999) Functional analysis of a novel human serotonin transporter gene promoter in immmortalized raphe cells. *Mol. Brain Res.* **68**, 141–148.
52. Ramamoorthy, J. D., Ramamoorthy, S., Papapetropoulos, A., Catravas, J. D., Leibach, F. H., and Ganapathy, V. (1995) Cyclic AMP-independent up-regulation of the human serotonin transporter by staurosporine in choriocarcinoma cells. *J. Biol. Chem.* **270**, 17,189–17,195.
53. Ramamoorthy, S., Cool, D. R., Mahesh, V. B., Leibach, F. H., Melikian, H., Blakely, R. D., et al. (1993) Regulation of the human serotonin transporter: Cholera toxin-induced stimulation of serotonin uptake in human placental choriocarcinoma cells is accompanied by increased serotonin transporter mRNA levels and serotonin transporter-specific ligand binding. *J. Biol. Chem.* **268**, 21,626–21,631.
54. Ramamoorthy, S., Ramamoorthy, J. D., Prasad, P., Bhat, G. K., Mahesh, V. B., Leibach, F. H., et al. (1995) Ganapathy. Regulation of the human serotonin transporter by interleukin-1b. *Biochem. Biophys. Res. Commun.* **216**, 560–567.
55. Benmansour, S., Cecchi, M., Morilak, D. A., Gerhardt, G. A., Javors, M. A., Gould, G. G., et al. (1999) Effects of chronic antidepressant treatments on serotonin transporter function, density, and mRNA level. *J. Neurosci.* **19**, 10,494–10,501.
56. Palaic, D. and Khairallah, P. A. (1967) Effect of angiotensin on uptake and release of norepinephrine by brain. *Biochem. Pharmacol.* **16**, 2291–2298.
57. Palaic, D. and Khairallah, P. A. (1967) Inhibition of noradrenaline uptake by angiotensin. *J. Pharmacol.* **19**, 396–397.

58. Palaic, D. and Panisset, J. (1969) Effect of nerve stimulation and angiotensin on the accumulation of ^{3}H-norepinephrine and the endogenous norepinephrine level in guinea pig vas deferens. *Biochem. Pharmacol.* **18**, 2693–2700.
59. Shores, M. M., Szot, P., and Veith, R. C. (1994) Desipramine-induced increase in norepinephrine transporter mRNA is not mediated via a_2 receptors. *Mol. Brain. Res.* **27**, 337–341.
60. Sumners, C. and Raizada, M. K. (1986) Angiotensin II stimulates norepinephrine uptake in hypothalamus-brain stem neuronal cultures. *Am. J. Physiol.* **250**, C236–C244.
61. Szot, P., Ashliegh, E. A., Kohen, R., Petrie, E., Dorsa, D. M., and Veith, R. (1993) Norepinephrine transporter mRNA is elevated in the locus coeruleus following short- and long-term desipramine treatment. *Brain Res.* **618**, 308–312.
62. Fang, Y. and Ronnekleiv, O. K. (1999) Cocaine upregulates the dopamine transporter in fetal rhesus monkey brain. *J. Neurosci.* **19**, 8966–8978.
63. McReynolds, A. M. and Meyer, J. S. (1998) Effects of prenatal cocaine exposure on serotonin and norepinephrine transporter density in the rat brain. *Ann. NY Acad. Sci.* **846**, 412–414.
64. Shearman, L. P. and Meyer, J. S. (1999) Cocaine up-regulates norepinephrine transporter binding in the rat placenta. *Eur. J. Pharmacol.* **386**, 1–6.
65. Blakely, R. D., Berson, H. E., Fremeau, Jr., R. T., Caron, M G., Peek, M. M., Prince, H. K., et al. (1991) Cloning and expression of a functional serotonin transporter from rat brain. *Nature* **354**, 66–70.
66. Hoffman, B. J., Mezey, E. and Brownstein, M. J. (1991) Cloning of a serotonin transporter affected by antidepressants. *Science* **254**, 579,580.
67. Kilty, J. E., Lorang, D., and Amara, S. G. (1991) Cloning and expression of a cocaine-sensitive rat dopamine transporter. *Science* **254**, 578–580.
68. Pacholczyk, T., Blakely, R. D., and Amara, S. G. (1991) Expression cloning of a cocaine- and antidepressant-sensitive human noradrenaline transporter. *Nature* **350**, 350–354.
69. Ramamoorthy, S., Bauman, A L., Moore, K. R., Han, H., Yang-Feng, T., Chang, A. S., et al. (1993). Antidepressant- and cocaine-sensitive human serotonin transporter: molecular cloning, expression, and chromosomal localization. *Proc. Natl. Acad. Sci. USA* **90**, 2542–2546.
70. Shimada, S., Kitayama, S., Lin, C., Patel, A., Nanthakumar, E., Gregor, P., et al. (1991) Cloning and expression of a cocaine-sensitive dopamine transporter complementary DNA. *Science* **254**, 576,577.
71. Blakely, R. D., and Bauman, A. L (2000) Biogenic amine transporters: regulation in flux. *Curr. Opin. Neurobiol.* **10**, 328–836.
72. Blakely, R. D., Ramamoorthy, S., Schroeter, S., Qian, Y., Apparsundaram, S., Galli, A., et al. (1998) Regulated phosphorylation and trafficking of antidepressant-sensitive serotonin transporter proteins. *Biol. Psychiatry* **44**, 169–178.
73. Apparsundaram, S., Galli, A., DeFelice, L. J., Hartzell, H. C., and Blakely, R. D. (1998) Acute regulation of norepinephrine transport: I. PKC-linked muscarinic receptors influence transport capacity and transporter density in SK-N-SH cells. *J. Pharmacol. Exp. Ther.* **287**, 733–743.
74. Apparsundaram, S., Schroeter, S., and Blakely, R. D. (1998) Acute regulation of norepinephrine transport. II. PKC-modulated surface expression of

human norepinephrine transporter proteins. *J. Pharmacol. Exp. Ther.* **287,** 744–751.

75. Bönisch, H., Hammermann, R., and Brüss, M. (1998) Role of protein kinase C and second messengers in regulation of the norepinephrine transporter. *Adv. Pharmacol.* **42,** 183–187.

76. Melikian, H. E. and Buckley, K. M. (1999) Membrane trafficking regulates the activity of the human dopamine transporter. *J. Neurosci.* **19,** 7699–7710.

77. Pristupa, Z. B., McConkey, F., Liu, F., Man, H. Y., Lee, F. J., Wang, Y. T., et al. (1998) Protein kinase-mediated bidirectional trafficking and functional regulation of the human dopamine transporter. *Synapse* **30,** 79–87.

78. Qian, Y., Galli, A., Ramamoorthy, S., Risso, S., DeFelice, L. J., and Blakely, R. D. (1997) Protein kinase C activation regulates human serotonin transporters in HEK-293 cells via altered cell surface expression. *J. Neurosci.* **17,** 45–47.

79. Zhang, L., Coffey, L. L., and Reith, M. E. A. (1997) Regulation of the functional activity of the human dopamine transporter by protein kinase C. *Biochem. Pharmacol.* **53,** 677–688.

80. Zhu, S., Kavanaugh, M. P., Sonders, M. S., Amara, S. G., and Zahniser, N. R. (1997) Activation of protein kinase C inhibits uptake, currents and binding associated with the human dopamine transporter expressed in *Xenopus* oocytes. *J. Pharmacol. Exp. Ther.* **282,** 1358–1365.

81. Miller, K. J. and Hoffman, B. J. (1994) Adenosine A_3 receptors regulate serotonin transoprt via nitric oxide and cGMP. *J. Biol. Chem.* **269,** 27,351–27,356.

82. Myers, C. L., Lazo, J. S., and Pitt, B. R. (1989) Translocation of protein kinase C is associated with inhibition of 5-HT uptake by cultured endothelial cells. *Am. J. Physiol.* **257,** L253–L258.

83. Myers, C. L. and Pitt, B. R. (1988) Selective effect of phorbol ester on serotonin removal and ACE activity in rabbit lungs. *J. App. Physiol.* **65,** 377–384.

84. Corey, J. L., Davidson, N., Lester, H. A., Brecha, N., and Quick, M. W. (1994) Protein kinase C modulates the activity of a cloned g-aminobutyric acid transporter expressed in *Xenopus* oocytes via regulated subcellular redistribution of the transporter. *J. Biol. Chem.* **269,** 14,759–14,767.

85. Quick, M. W., Corey, J. L., Davidson, N., and Lester, H. A. (1997) Second messengers, trafficking-related proteins, and amino acid residues that contribute to the functional regulation of the rat brain GABA transporter GAT1. *J. Neurosci.* **17,** 2967–2979.

86. Ramamoorthy, S. and Blakely, R. D. (1999) Phosphorylation and sequestration of serotonin transporters differentially modulated by psychostimulants. *Science.* **285,** 763–766.

87. Ramamoorthy, S., Giovanetti, E., Qian, Y., and Blakely, R. D. (1998) Phosphorylation and regulation of antidepressant-sensitive serotonin transporters. *J. Biol. Chem.* **273,** 2458–2466.

88. Daniels, G. and Amara S. G. (1999) Regulation trafficking of the human dopamine transporter Clathrin-mediated internalization and lysosomal degradation in response to phorbol esters. *J. Biol. Chem.* **274,** 35,794–35,794.

89. Melikian, H. and Buckley, K. (1999) Membrane trafficking regulates the activity of the human dopamine transporter. *J. Neurosci.* **15,** 7699–7710.

90. Vaughan, R. A., Huff, R. A., Uhl, G. R., and Kuhar, M. J. (1997) Protein kinase C-mediated phosphorylation and functional regulation of dopamine transporters in striatal synaptosomes. *J. Biol. Chem.* **272,** 15,541–15,546.

91. Batchelor, M. and Schenk, J. O. (1998) Protein kinase A activity may kinetically upregulate the striatal transporter for dopamine. *J. Neurosci.* **18,** 10,304–10,309.

92. Huff, R. A., Vaughan, R. A., Kuhar, M. J., and Uhl, G. R. (1997) Phorbol esters increase dopamine transporter phosphorylation and decrease transport V_{max}. *J. Neurochem.* **68,** 225–232.

93. Reith, M. E. A., Xu, C., and Chen, N. H. (1997) Pharmacology and regulation of the neuronal dopamine transporter. *Eur. J. Pharmacol.* **324,** 1–10.

94. Anderson, G. M., Hall, L. M., Horne, W. C., and Yang, J. X. (1996) Adenosine diphosphate inhibits the serotonin transporter. *Biochim. Biophy. Acta.* **1283,** 14–20.

95. Anderson, G. M., and Horne, W. C. (1992) Activators of protein kinase C decrease serotonin transport in human platelets. *Biochim. Biophys. Acta.* **1127,** 331–337.

96. Carli, M. and Reader, T. A. (1997) Regulation of central serotonin transporters by chronic lithium. *Synapse* **27,** 83–89.

97. Gordon, I., Weizman, R., and Rehavi, M. (1996) Modulatory effect of agents active in the presynaptic dopaminergic system on the striatal dopamine transporter. *Eur. J. Pharmacol.* **298,** 27–30.

98. Kiss, J. P., Hennings, E. C., Zsilla, G., and Vizi, E. S. (1999) A possible role of nitric oxide in the regulation of dopamine transporter function in the striatum. *Neurochem. Int.* **34,** 345–350.

99. Zhang, L. and Reith, M. E. A. (1996) Regulation of the functional acitivity of the human dopamine tranporter by the arachidonic acid pathway. *Eur. J. Pharmacol.* **315,** 345–354.

100. Daws, L. C., Gerhardt, G. A., and Frazer, A. (1999) 5-HT1B antagonists modulate clearance of extracellular serotonin in rat hippocampus. *Neurosci Lett.* **266,** 165–168.

101. Daws, L. C., Gould, G. G., Teicher, S. D., Gerhardt, G. A., and Frazer, A. (2000) 5-HT(1B) receptor-mediated regulation of serotonin clearance in rat hippocampus in vivo. *J Neurochem.* **75,** 2112–2122.

102. Mayfield, R. D. and Zahniser, N. R. (2001) Dopamine D2 receptor regulation of the dopamine transporter expressed in Xenopus laevis oocytes is voltage-independent. *Mol. Pharmacol.* **59,** 112–121.

103. Quan, H., Torres, G., Wang, Y.-M., and Caron, M. G. (2000) ATP facilitates dopamine tranporter activity: Potential implication of P2X receptors. *Soc. Neurosci. Abstr.* **26,** 439.13.

104. Vaughan, R. A., Huff, R. A., Uhl, G. R., and Kuhar, M. J. (1998) Phosphorylation of dopamine transporters and rapid adaptation to cocaine. *Adv. Pharmacol.* **42,** 1042–1045.

105. Vrindavanam, N. S., Arnaud, P., Ma, J. X., Altman-Hamamdzic, S., Parratto, N. P., and Sallee, F. R. (1996) The effects of phosphorylation on the func-

tional regulation of an expressed recombinant human dopamine transporter. *Neurosci. Lett.* **216,** 133–136.

106. Bernstein, E. M., and Quick, M. W. (1999) Regulation of gamma-aminobutyric acid (GABA) transporters by extracellular GABA. *J. Biol. Chem.* **274,** 889–895.

107. Duan, S., Anderson, C. M., Stein, B. A., and Swanson, R. A. (1999) Glutamate induces rapid upregulation of astrocyte glutamate transport and cell-surface expression of GLAST. *J. Neurosci.* **19,** 10,193–10,200.

108. Yatin, S. M., Miller, G. M., Goulet, M., Alvarez, X., and Madras, B. K. (2000) Cocaine and dopamine: Differential effects of dopamine transporter distribution. *Soc. Neurosci. Abstr.* **26,** 17.9.

109. Zahniser, N. R., Larson, G. A., and Gerhardt, G. A (1999) In vivo dopamine clearance rate in rat striatum: regulation by extracellular dopamine concentration and dopamine transporter inhibitors. *J. Pharmacol. Exp. Ther.* **289,** 266–277.

110. Barak, L. S., Ferguson, S. S. G., Zhang, J., Martenson, C., Meyer, T., and Caron, M. G. (1997) Internal trafficking and surface mobility of a functionally intact b_2-adrenergic receptor-green fluorescent protein conjugate. *Mol. Pharmacol.* **51,** 177–184.

111. Cao, T. T., Deacon, H. W., Reczek, D., Bretscher, A., and Zastrow, M., (1999) A kinase-regulated PDZ-domain interaction controls endocytic sorting of the beta2-adrenergic receptor. *Nature* **401,** 286–290.

112. Daunt, D. A., Hurt, C., Hein, L., Kallio, J., Feng, F., and Kobilka, B. K (1997) Subtype-specific intracellular trafficking of a_2-adrenergic receptors. *Mol. Pharmacol.* **51,** 711–720.

113. Krueger, K. M., Daaka, Y., Pitcher, J. A., and Lofkowitz, R. J. (1997) The role of sequestration in G protein-coupled receptor resensitization: Regulation of b_2-adrenergic receptor dephosphorylation by vesicular acidification. *J. Biol. Chem.* **272,** 5–8.

114. Lefkowitz, R. J., Pitcher, J., Krueger, K., and Daaka, Y. (1998) Mechanisms of b-adrenergic receptor desensitization and resensitization. *Adv. Pharmacol.* **42,** 416–420.

115. Lin, F. T., Krueger, K. M., Kendall, H. E., Daaka, Y., Fredericks, Z. L., Pitcher, J. A., et al. (1997) Clathrin-mediated endocytosis of the beta-adrenergic receptor is regulated by phosphorylation/dephosphorylation of beta-arrestin1. *J. Biol. Chem.* **272,** 31,051–31,057.

116. Bauman, A. L., Apparsundaram, S., Ramamoorthy, S., Wadzinski, B. E., Vaughan, R. A., and Blakely, R. D. (2000) Cocaine and antidepressant-sensitive biogenic amine transporters exist in regulated complexes with protein phosphatase 2A. *J. Neurosci.* **20,** 7571–7578.

117. Beckman, M. L., Bernstein, E. M., and Quick, M. W. (1998) Protein kinase C regulates the interaction between a GABA transporter and syntaxin 1A. *J. Neurosci.* **18,** 6103–6112.

118. Deken, S. L., Beckman, M. L., Boos, L., and Quick, M. W., (2000) Transport rates of GABA transporters: regulation by the N-terminal domain and syntaxin 1A. *Nat. Neurosci.* **3,** 998–1003.

119. Geerlings, A., Lopez-Corcuera, B., and Aragon, C. (2000) Characterization of the interactions between the glycine transporters GLYT1 and GLYT2 and the SNARE protein syntaxin 1A. *FEBS Lett.* **470,** 51–54.
120. Sung, U., Apparsundaram, S., and Blakely, R. D. (2000) Intracellular calcium regulates association between norepinephrine transporter and syntaxin 1A. *Soc. Neurosci. Abstr.* **26,** 439.10.
121. Torres, G. E., Yao, W. D., Mohn, A. R., Quan, H., and Caron, M. G. (2000) Specific interaction between the dopamine transporter and the PDZ domain-containing protein PICK1. *Soc. Neurosci. Abstr.* **26,** 17.3.
122. Lee, F. J. S., Liu, G., Pristupa, Z. B., and Niznik, H. B. (2000) Direct binding and functional coupling of a-synuclein to the dopamine transporter. *Soc. Neurosci. Abstr.* **26,** 439.16.

2

Mechanisms of Biogenic Amine Neurotransmitter Transporters

Gary Rudnick

1. INTRODUCTION

The biogenic amine transporters, as described in Chapters 1, 3, and 5 of this book, terminate the action of released biogenic amine neurotransmitters. These transporters utilize norepinephrine (NE), dopamine (DA), and serotonin (5-HT), and are referred to as NET, DAT, and SERT, respectively. Interruption of their function by agents such as antidepressants and stimulants causes profound changes in mood and behavior. In addition to their importance in regulating the extracellular concentration of neurotransmitters, these proteins are fascinating molecular machines that utilize the energy from transmembrane ion gradients to accumulate intracellular neurotransmitters. The pharmacology and molecular biology of these proteins is well covered in Chapters 1, 3, 5, and 11. This chapter focuses on the mechanism of neurotransmitter transport. We are still a long way from completely understanding how these proteins work. However, recent advances have given us further insight into this process, and encourage the hope that current and future research will provide a more complete understanding of the transport mechanism.

2. NA, K, AND CL IONS: COFACTORS FOR TRANSPORT

Like many mammalian plasma-membrane transport systems, the biogenic amine transporters require the presence of external Na^+ ions. This phenomenon was observed first for epithelial transporters, such as those for glucose and amino acids (1). The transport of NE into peripheral nerve endings was also found to be Na^+-dependent (2), as was 5-HT transport into platelets (3). When synaptosomes were established as an experimental system for studying presynaptic mechanisms, the Na^+ dependence of neurotransmitter uptake

From: *Contemporary Neuroscience:*
Neurotransmitter Transporters: Structure, Function, and Regulation, 2nd Edition
Edited by: M. E. A. Reith © Humana Press Inc., Totowa, NJ

became firmly established, as each system in turn demonstrated its dependence on extracellular Na^+ *(4–9)*. In all these cases, replacement of Na^+ with any other ion results in a loss of transport activity.

The importance of this Na^+ requirement became more apparent as the energetics of transport were studied. The Na^+ requirement provided a way to understand how the energy of ATP hydrolysis drives transport. ATP is utilized by the Na^+ pump to move Na^+ ions out of—and K^+ into—the cell. By incorporating Na^+ into the transport reaction, the neurotransmitter transporters couple the energy released by ATP hydrolysis to the downhill Na^+ flux that accompanies transmitter accumulation (Fig. 1). Because K^+ is pumped by the ATPase, it is not surprising that some neurotransmitter transporters also utilize internal K^+ in the transport process. The most notable are the 5-HT and glutamate transporters *(10,11)*, although other systems may also take advantage of K^+ concentrated inside cells by the Na^+ pump. The requirement for internal K^+ can be fulfilled by H^+ in the case of the 5-HT transporter *(12)*, but efforts to demonstrate this phenomenon with glutamate transporters have resulted in the opposite conclusion—that H^+ equivalents were moving with the flow of substrate rather than against it *(13)*.

Another consequence of Na^+ pump action is a result of the fact that three Na^+ ions are pumped out for each two K^+ ions pumped into the cell, leading to the generation of a transmembrane electrical potential. In itself, this potential can be used by neurotransmitter transporters, but more specifically, the potential leads to the loss of Cl^- ions from the cell. Many neurotransmitter transporters utilize this asymmetric Cl^- distribution as a driving force. With the exception of glutamate transporters, all of the neurotransmitter transporters and many related transporters require Cl^- as well as Na^+ *(14)*. For this reason, the neurotransmitter transporter gene family is often referred to as the NaCl-coupled transporter family *(15)*. The glutamate transporters represent a separate gene family.

2.1. Transport and Binding Requirements

One advantage to the study of biogenic amine transporters, relative to other members of the family, is that high-affinity ligands are available. Tricyclic antidepressants such as imipramine and desipramine bind to SERT and NET, and have been used to investigate the ion dependence of the binding process. SERT requires both Na^+ and Cl^- ions for maximal [^3H]imipramine binding *(16)*, although repeated attempts to demonstrate any effect of K^+ have been negative. Similar results have been obtained with [^3H]desipramine and NET *(17)*. Antidepressant binding differs from trans-

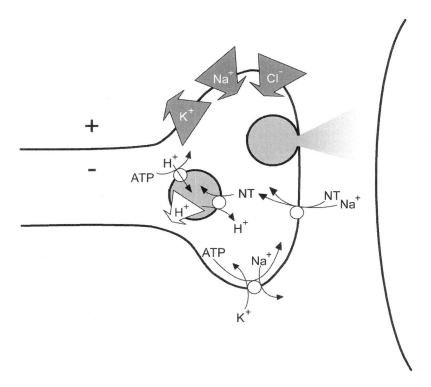

Fig. 1. Neurotransmitter recycling at the nerve terminal. Neurotransmitter (NT) is released from the nerve terminal by fusion of synaptic vesicles with the plasma membrane. After release, the transmitter is transported across the plasma membrane by a Na^+-dependent transporter in the plasma membrane. Transmitter delivered into the cytoplasm is further sequestered in synaptic vesicles by a vesicular transporter using the transmembrane H^+ gradient as a driving force. This driving force, shown by the arrow pointing in the direction of downhill H^+ movement, is generated by an ATP-dependent H^+ pump in the vesicle membrane. The Na^+ and K^+ gradients across the plasma membrane are generated by the Na^+/K^+-ATPase. This enzyme also creates a transmembrane electrical potential (negative inside) that causes Cl^- to redistribute. Neurotransmitter transport across the plasma membrane is coupled to the Na^+, Cl^-, and K^+ gradients and the membrane potential generated by the ATPase.

port, however. The Na^+ dependence of [³H]imipramine binding to SERT and imipramine inhibition of transport were both sigmoidal, suggesting that two or more Na^+ ions participate in the reaction *(18)*. In contrast, the Na^+ dependence of 5-HT transport and 5-HT inhibition of [³H]imipramine binding showed simple saturation behavior consistent with the involvement of only one Na^+ ion in 5-HT binding and transport *(18)*. Similar results were obtained with NET. Transport is a simple saturable function of Na^+ *(19)*, but [^{3}H]desipramine binding shows a sigmoidal Na^+ dependence *(17)*.

When binding of other ligands was examined, the difference between substrates and inhibitors became even more obvious. Paroxetine and the cocaine analogs CFT and β-CIT bind to SERT and inhibit transport. This binding process was stimulated by Na^+ but not by Cl^- *(20–22)*, although Cl^- is required for transport. The DA transporter also binds cocaine and its analogs *(23,24)*, and has demonstrated a different ion dependence for transport and β-CIT binding *(24)*. For both SERT and DAT, β-CIT binding was inhibited when Na^+ was replaced with Li^+ but did not require Cl^-, and for both transporters, β-CIT binding was inhibited at low pH *(24)*. In contrast, the binding of 5-HT or DA, was stimulated by Cl^- and not affected by low pH *(24)*, although neutralization of DA at high pH decreases its binding to DAT *(25)*. The presence of Li^+ ions apparently favors a conformation of SERT with a lower affinity for cocaine and its analogs, but with similar 5-HT affinity *(25a)*. These studies and others clearly indicate that inhibitor binding may be smilar, in some aspects, to substrate binding. However, the two processes are distinct in their ionic dependence. Despite the differences, interactions between inhibitors and substrates at 5-HT, NE, and DA transporters are competitive, at least in equilibrium-binding studies. Thus, a single binding site, or a set of overlapping binding sites, could account for substrate and inhibitor binding.

The sensitivity of β-CIT binding to displacement by substrate has allowed measurements of substrate binding under conditions which preclude direct substrate-binding measurements. Using [^{125}I]β-CIT, 5-HT binding to SERT was measured in the absence of Na^+ and Cl^-, and the individual effects of these ions was determined *(22)*. Although Cl^- stimulated 5-HT binding by itself, Na^+ alone actually decreased 5-HT-binding affinity. Maximal 5-HT affinity was observed only in the presence of both Na^+ and Cl^-, suggesting that 5-HT binds to the transporter together with these two ions *(22)*.

3. ION GRADIENTS DRIVE BIOGENIC AMINES ACROSS THE MEMBRANE

3.1. Influence of Ion Gradients on 5-HT Transport

Studies with synaptosomes and platelets have indicated that the biogenic amine transport systems possess an impressive ability to concentrate DA, NE, and 5-HT. However, these preparations contained intracellular amine storage organelles (synaptic vesicles or dense granules) that sequester most of the intracellular amine. The ability of the plasma-membrane transporters to concentrate their substrates was not appreciated until platelet membrane vesicles were shown to accumulate internal 5-HT to concentrations hundreds of fold higher than the external medium when appropriate transmembrane ion gradients were imposed *(26)*. These vesicle experiments demonstrated conclusively that the plasma-membrane transporters generated gradients of their substrate amines, using the energy of transmembrane Na^+, Cl^-, and K^+ ion gradients.

3.1.1. Na^+

When a Na^+ concentration gradient (out > in) was imposed across the vesicle membrane in the absence of other driving forces, this gradient was sufficient to drive 5-HT accumulation *(26)*. Coupling between Na^+ and 5-HT transport follows from the fact that Na^+ can drive transport only if its own gradient is dissipated. Thus, Na^+ influx must accompany 5-HT influx. Na^+-coupled 5-HT transport into membrane vesicles is insensitive to inhibitors of other Na^+ transport processes such as ouabain and furosemide, supporting the hypothesis that Na^+ and 5-HT fluxes are coupled directly by the transporter *(26,27)*. Many of these results have been reproduced in membrane-vesicle systems from cultured rat basophilic leukemia cells *(28)*, mouse brain synaptosomes *(29)*, and human placenta *(30)*.

3.1.2. Cl^-

The argument that Cl^- is cotransported with 5-HT is somewhat less direct, as it has been difficult to demonstrate 5-HT accumulation with only the Cl^- gradient as a driving force. However, the transmembrane Cl^- gradient influences 5-HT accumulation when a Na^+ gradient provides the driving force. Thus, raising internal Cl^- decreases the Cl^- gradient, and inhibits 5-HT uptake. External Cl^- is required for 5-HT uptake, and Cl^- can be replaced by Br^-; to a lesser extent, by SCN^-, NO_3^-, and NO_2^-; and not at all by SO_4^{2-}, PO_4^{3-}, and isethionate *(27)*. In contrast, 5-HT efflux requires internal but not external Cl^- *(27)*. The possibility that Cl^- stimulated transport by electrically compensating for electrogenic (charge moving) 5-HT transport was ruled out by the observation that a valinomycin-mediated K^+ diffusion potential (interior negative) was unable to eliminate the external Cl^- requirement for 5-HT influx *(27)*.

3.1.3. K^+

The ability of internal K^+ to stimulate 5-HT transport was not immediately obvious, for two reasons. First, there was no absolute requirement for

K^+ in transport, and, second, no Na^+ cotransport system had ever been shown to be coupled also to K^+. Initially, we proposed that a membrane potential generated by K^+ diffusion was responsible for driving electrogenic 5-HT transport *(26)*. However, subsequent studies showed that K^+ stimulated transport even when the membrane potential was close to zero *(11,31)*. In the absence of a K^+ gradient, the addition of 30 mM K^+ simultaneously to both the internal and external medium increased the transport rate 2.5-fold *(31)*. Moreover, hyperpolarization of the membrane by valinomycin in the presence of a K^+ gradient had little or no effect on transport. Two conclusions were drawn from these results. First, the transport process was likely to be electrically silent. Second, since the K^+ gradient did not seem to act indirectly through the membrane potential, it was likely to act directly through exchange with 5-HT.

A study of the pH dependence of 5-HT transport revealed the reason that 5-HT transport still occurred in the absence of K^+. A study of the pH dependence of 5-HT transport. In the absence of K^+, internal H^+ ions apparently fulfill the requirement for a countertransported cation *(12)*. Even when no other driving forces were present (NaCl in=out, no K^+ present), a transmembrane pH difference (ΔpH, interior acid) could serve as the sole driving force for transport. ΔpH-driven 5-HT accumulation required Na^+ and was blocked by imipramine or by high K^+ (in=out), indicating that it was mediated by the 5-HT transporter, and not the result of non-ionic diffusion *(12)*. From all of these data, it was concluded that inwardly directed Na^+ and Cl^- gradients, and outwardly directed K^+ or H^+ gradients served as driving forces for 5-HT transport (Fig. 2).

3.1.4. Electrical Consequences

Although studies using platelet plasma-membrane vesicles provided direct evidence that 5-HT transport was electrically silent *(11,31)*, evidence relating the membrane potential to 5-HT transport has been mixed in other systems. Bendahan and Kanner *(28)* found that 5-HT transport into plasma-membrane vesicles from rat basophilic leukemia cells was stimulated by a K^+ diffusion potential. However, other workers studying plasma-membrane vesicles from mouse brain and human placenta concluded that 5-HT transport in these tissues was not driven by a transmembrane electrical potential ($\Delta\psi$, interior negative) *(32,33)*.

One might expect that electrogenicity could be easily tested if cells expressing the 5-HT transporter could be directly impaled with microelectrodes. This has been done by Mager et al. *(34)*, using Xenopus oocytes injected with 5-HT transporter mRNA, with somewhat surprising results. A simple prediction is that current should not flow across the membrane during

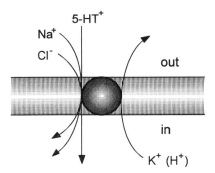

Fig. 2. Driving forces for 5-HT transport. Na^+ and Cl^- on the outside of the cell are transported inward with the cationic form of 5-HT. In the same catalytic cycle, K^+ is transported out from the cytoplasm. In the absence of internal K^+, H^+ ions take the place of K^+. The energy released by downhill movement of Na^+, Cl^-, and K^+ provides the driving force for 5-HT accumulation against its concentration gradient.

5-HT transport if the transporter is electroneutral. In fact, a 5-HT-dependent current has been measured, but closer inspection of its properties suggests that it does not represent electrogenic 5-HT transport, but rather, a conductance that is stimulated by transport. The key finding is that the transport-associated current is voltage-dependent. Thus, the inward current increases as the inside of the cell becomes more negative. In the same cells, however, [^{3}H]5-HT transport is independent of membrane potential. Therefore, it is very unlikely that the voltage-dependent current represents the 5-HT transport process. Instead, as discussed in **Subheading 7**, the current results from a newly discovered ion-channel property of neurotransmitter transporters.

3.2. 5-HT Dopamine and Norepinephrine Transporters use the Protonated Form of the Substrate

Because 5-HT, DA, and NE exist primarily in the protonated form at physiological pH, many researchers have assumed that these substrates are transported as cations. However, a small fraction of these amines exist in the neutral or zwitterionic form at neutral pH, and it is important to assess the possibility that these forms are the true substrates for transport. In the case of the vesicular monoamine transporter (VMAT), the ionic form of the substrate is a matter of some controversy. Different investigators have reached opposite conclusions *(35–37)*. If the neutral form were transported by the plasma membrane biogenic amine transporters, one consequence might be that the K_m for transport would be pH-dependent. As the pH increases, the mole fraction of biogenic amine in the neutral or zwitterionic form will increase sharply, but below pH 8.0–9.0, the majority of the substrate will be in the cationic form and the mole fraction of that form will not change significantly. The K_m for total substrate (cationic and neutral) will therefore appear to decrease if the neutral form is the substrate but will be pH-independent if the cationic form is transported. Results with the 5-HT and NE transporters *(38,39)* show no change in K_m with pH, and suggest that the cationic form is the true substrate. An extensive study of dopamine transporter concluded that the cationic (and possibly also the zwitterionic) form of DA was the likely substrate *(25)*.

There is another consequence if the neutral or zwitterionic form is transported. The substrate would need to dissociate a H^+ ion before transport, and then would bind H^+ after transport. By imposing a transmembrane pH difference (ΔpH), the equilibrium amine distribution across the membrane could be influenced. In seeming agreement with this prediction, 5-HT accumulation by platelet plasma-membrane vesicles is increased (in the absence of K^+) when the vesicle interior is acidified *(12)*. However, this phenomenon represents the ability of H^+ to replace K^+ in 5-HT countertransport, and not an influence on 5-HT protonation. In the presence of K^+, ΔpH does not stimulate 5-HT uptake, although this should occur if the neutral form of 5-HT is the true substrate *(12)*.

3.3. Overall 5-HT Stoichiometry

The number of Na^+, Cl^-, and K^+ ions transported with 5-HT transporter has been estimated by imposing known Na^+, Cl^-, and K^+ concentration gradients across the plasma membrane as a driving force, and measuring the 5-HT concentration gradient accumulated in response to that driving force at equilibrium. This is essentially a thermodynamic measurement, balancing a known driving force against a measured gradient of substrate. Technically, such measurements require that the imposed ion gradients are relatively stable, so that the available driving force is known at a given time after imposition of the ion gradient.

3.4. Transport Kinetics can Suggest, but not Determine, Stoichiometry

Kinetic techniques have also been used to assess the Na^+, Cl^-, and K^+ stoichiometry for 5-HT, NE, and DA transport. One technically simple technique is to measure the dependence of transport rate (or its kinetic determinants K_m or V_{max}) on Na^+, Cl^-, or K^+ concentration, and to calculate a Hill coefficient for that ion. Using this analysis for the 5-HT and NE transporters yields an n of one for both Na^+ and Cl^- in membrane vesicles *(40–42)*, where initial rates of transport showed a simple hyperbolic dependence on Na^+ or Cl^-—consistent with a Na^+:Cl^-:substrate stoichiometry of 1:1:1.

However, steady-state kinetics do not necessarily provide accurate information on cotransport stoichiometry *(43)*. It is possible that more than one Na^+ ion is required for substrate binding or even translocation (as reflected in the Hill coefficient calculated from rate measurements), but that only one of those Na^+ ions is actually cotransported. It is also possible that a substrate is cotransported with two Na^+ ions, but that the affinities or rates of association of the two Na^+ ions are so disparate that the initial rate of transport is dependent on only the weaker binding or slower associating of the two, leading to an apparent Hill coefficient of 1. These difficulties are inherent in any kinetic method, whether transport is measured directly by tracer flux, or indirectly by measurements of electrical currents that may accompany transport. Thus, the dependence of transport rate on the concentration of a given ion may suggest a transport stoichiometry, but cannot provide proof for it.

A still more direct method is to measure the flux of driving ions as well as the flux of substrate. Usually the basal levels of ion fluxes are too fast relative to the rates of substrate transport, but in a purified, reconstituted system, Kanner and colleagues were able to measure Na^+ and Cl^- flux along with GABA flux by the GABA transporter. In this case, in which both thermodynamic and direct kinetic data exist, both methods indicate a Na^+:Cl^-:GABA stoichiometry of 2:1:1 *(63, 39, 40)*.

3.5. Thermodynamic Approach

Because kinetic approaches may be experimentally difficult or misleading, it is essential to confirm the stoichiometry by a thermodynamic measurement. In the thermodynamic method, known Na^+, Cl^-, or K^+ concentration gradients are imposed across the plasma membrane as a driving force, and the substrate concentration gradient in equilibrium with that driving force is measured. By varying the concentration gradient of the driving ion, and measuring the effect on substrate accumulation, the stoichiometry can be calculated. For a simple system in which two solutes, A and B, are cotransported, a plot of $ln(A_{in}/A_{out})$ vs $ln(B_{out}/B_{in})$ gives the B:A stoichiometry as its slope. As a special case, if the stoichiometry is 1:1, then a plot of A_{in}/A_{out} vs B_{in}/B_{out} will be a straight line. Using this method, a 1:1 coupling was determined for 5-HT transport with both Na^+ *(18)* and K^+ *(26)*. The Cl^- stoichiometry was deduced from the fact that 5-HT transport was not affected by imposition of a $\Delta\psi$ (interior negative), and was therefore likely to be electroneutral. Because 5-HT is transported in its cationic form *(12,39)*, only a 5-HT^+:Na^+:Cl^-:K^+ stoichiometry of 1:1:1:1 is consistent with all the known facts. Obviously, this analysis requires an experimental system such as membrane vesicles, where the composition of both internal and external media can be controlled. This method also relies on the ability to measure, or at least to estimate, an equilibrium substrate gradient under conditions in which the ion gradients are known *(43)*.

3.6. Each Transporter has a Characteristic Coupling of Ion Flux to Substrate Flux

3.6.1. Norepinephrine Transport

Although no membrane-vesicle systems containing the DA transporter have been described, two plasma-membrane vesicle systems have emerged for studying NE transport: the placental-brush-border membrane *(42)* and cultured PC-12 cells *(47)*. Harder and Bonisch *(47)* concluded that NE transport into PC12 vesicles was coupled to Na^+ and Cl^- and was electrogenic, but they failed to arrive at a definitive coupling stoichiometry because of uncertainties about the role of K^+. According to their analysis, stimulation of NE influx by internal K^+ resulted either from direct K^+ countertransport as occurrs with SERT *(31)*, or from a K^+ diffusion potential which drives electrogenic NE influx, as with GAT-1 *(48)*. Ganapathy and colleagues *(42)* studied NET-mediated transport of both NE and DA into placental-membrane vesicles (DA is utilized by NET as a substrate) *(49)*. They reached similar conclusions regarding ion coupling, but also were left with some ambiguity regarding K^+. In fact, the effects of ions on NET-mediated DA accumulation were similar to those observed with SERT-mediated 5-HT transport in the same membranes, and the two activities were distinguished only by their inhibitor sensitivities *(19)*. Part of the difficulty in interpreting and comparing these data stems from the fact that they were obtained in various cell types, with unknown and potentially very different conductances to K^+.

Two further problems made it difficult to interpret previous data on NET ion coupling. Both previous studies assumed that the cationic form of the catecholamine substrate was transported *(42,47)*, and did not consider the possibility that the neutral or zwitterionic form was the true substrate. Moreover, previous studies estimated NET stoichiometry using kinetic rather than thermodynamic measurements. The number of Na^+ ions cotransported with each catecholamine substrate was estimated from the dependence of transport rate on Na^+ concentration *(42,47)*.

The author and colleagues established LLC-PK_1 cell lines stably expressing the biogenic amine transporters SERT, NET, and DAT, as well as the GABA transporter GAT-1. Using these cell lines, we characterized and compared the transporters under the same conditions and in the same cellular environment *(49)*. One attractive advantage of LLC-PK_1 cells is that it has been possible to prepare plasma-membrane vesicles that are suitable for transport studies *(50)*. We took advantage of this property to prepare membrane vesicles containing transporters for GABA, 5-HT and NE, all in the same $LLC-PK_1$ background. These vesicles should have identical composition, except for the heterologously expressed transporter. Moreover, these vesicles are suitable for estimating equilibrium substrate accumulation in response to imposed ion gradients. This property allowed us to define the ion-coupling stoichiometry for NET, using the known stoichiometries for GAT-1 and SERT-mediated transport as internal controls.

The results for SERT and GAT-1 are consistent with previously reported determinations of ion-coupling stoichiometry. For NET, accumulation of [³H]dopamine (DA) was stimulated by imposition of Na^+ and Cl^- gradients (out > in) and by a K^+ gradient (in > out). To determine the role that each of these ions and gradients play in NET-mediated transport, we measured the influence of each ion on transport when that ion was absent, present at the same concentration internally and externally, or present asymmetrically across the membrane. The presence of Na^+ or Cl^-, even in the absence of a gradient, stimulated DA accumulation by NET, but K^+ had little or no effect in the absence of a K^+ gradient. Stimulation by a K^+ gradient was markedly enhanced by increasing the K^+ permeability with valinomycin, suggesting that net positive charge is transported together with DA. The cationic form of DA is likely to be the substrate for NET, since varying pH did not affect the K_m of DA for transport. We estimated the Na^+:DA stoichiometry by measuring the effect of internal Na^+ on peak accumulation of DA. Taken together, the results suggest that NET catalyzes cotransport of one cationic substrate molecule with one Na^+ ion, and one Cl^- ion, and that K^+ does not participate directly in the transport process *(38)*.

3.6.2. Dopamine Transport

The DA transporter has a different ion dependence from that of SERT or NET. Although initial rates of DA transport were found to be dependent on a single Cl^-, two Na^+ ions were apparently involved in the transport process *(49,51)*. Thus, the initial rate of DA transport into suspensions of rat striatum was a simple hyperbolic function of $[Cl^-]$, but depended on $[Na^+]$ in a sigmoidal fashion. These data are consistent with a Na^+:Cl^-:DA stoichiometry of 2:1:1. These differences in Na^+ stoichiometry have been reproduced with the cloned transporter cDNAs stably expressed in $LLC-PK_1$ cell lines, indicating that they are intrinsic properties of the transporters and not artifacts caused by the different cell types used *(49)*. Although the precautions discussed here prevent any firm conclusions regarding the ion-coupling stoichiometry of the DA transporter, the ion dependence differs from that of both SERT and NET, suggesting that each of these three biogenic amine transporters has a unique stoichiometry of coupling.

4. AMPHETAMINES ARE SUBSTRATES FOR BIOGENIC AMINE TRANSPORTERS

4.1. Amphetamine Action

Amphetamines represent a class of stimulants that increase extracellular levels of biogenic amines. Their mechanism differs from simple inhibitors such as cocaine, although it also involves biogenic amine transporters. Amphetamine derivatives are apparently substrates for biogenic amine transporters, and lead to transmitter release by a process of transporter-mediated efflux from intracellular stores *(52–54)*. Both catecholamine and 5-HT transporters are affected by amphetamines. In particular, compounds such as p-chloroamphetamine and 3,4-methylenedioxymethamphetamine (MDMA, also known as "ecstasy") preferentially release 5-HT *(55,56)*, and also cause degeneration of serotonergic nerve endings *(57)*. Other amphetamine derivatives, such as methamphetamine, preferentially release catecholamines.

4.2. Actions at the Plasma-Membrane Transporter

The process of exchange stimulated by amphetamines results from two properties of amphetamine and its derivatives. These compounds are substrates for biogenic amine transporters *(53,58–60)* and they are also highly permeant across lipid membranes *(54,58,60)*. As substrates, they are taken up into cells expressing the transporters, and as permeant solutes, they rapidly diffuse out of the cell without requiring participation of the transporter. The result is that an amphetamine derivative will cycle between the cytoplasm and the cell exterior in a process that allows Na^+ and Cl^- to enter the cell and K^+ (in the case of SERT) to leave each time the protonated amphetamine enters. Additionally, a H^+ ion will remain inside the cell if the amphetamine leaves as the more permeant neutral form (Fig. 3). This dissipation of ion gradients and internal acidification may possibly be related to the toxicity of amphetamines in vivo. The one-way utilization of the transporter (only for influx) leads to an increase in the availability of inward-facing transporter-binding sites for efflux of cytoplasmic 5-HT, and, together with reduced ion gradients, results in net 5-HT efflux.

4.3. Actions at the Vesicular Membrane

In addition, the ability of amphetamine derivatives to act as weak base ionophores at the dense granule membrane leads to leakage of vesicular biogenic amines into the cytoplasm *(54,58)*. Weakly basic amines are able to raise the internal pH of acidic organelles by dissociating into the neutral, permeant form, entering the organelle, and binding an internal H^+ ion. Weakly basic amines such as ammonia and methylamine have been used to

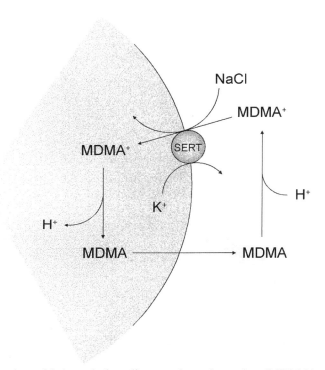

Fig. 3. Interaction of 3,4-methylenedioxymethamphetamine (MDMA) with the 5-HT transporter. MDMA is a substrate for SERT, and, like 5-HT, is transported into cells together with Na^+ and Cl^- and in exchange for K^+. Since it is membrane permeant in its neutral form, MDMA deprotonates intracellularly and leaves the cell, at which time it can reprotonate and serve again as a substrate for SERT. This futile transport cycle may lead to dissipation of cellular Na^+, Cl^-, and K^+ gradients and acidification of the cell interior.

raise the internal pH of chromaffin granules, for example *(61)*. Although amphetamine derivatives are certainly capable of dissipating transmembrane ΔpH by this mechanism, they are much more potent than simple amines when tested in model systems. For example, p-chloroamphetamine (PCA) and 3,4-methylenedioxymethamphetamine (MDMA, ecstasy) are 5–10 times more potent than NH_4Cl in dissipating ΔpH in chromaffin-granule membrane vesicles *(53,58)*. This result suggests that these compounds are crossing the membrane both in their neutral form, and also as protonated species. By cycling into the vesicle as an uncharged molecule and back out in the protonated form, an amphetamine derivative could act as a classical uncoupler to increase the membrane permeability to H^+ ions.

In addition to this uncoupling activity, amphetamine derivatives have affinity to the vesicular monoamine transporter, VMAT. Binding of various amphetamine derivatives to VMAT has been observed with both native and heterologously expressed VMAT *(62)*. Despite the affinity of many amphetamines to VMAT, at least one compound, PCA, has no demonstrable binding to VMAT *(60)*, despite its robust ability to release stored biogenic amines *(58)*. Thus, the ability of amphetamines to dissipate vesicular pH differences is sufficient to explain their effects on vesicular release.

Sulzer et al. *(63)* have extended this hypothesis by measuring the effects of intracellularly injected amphetamine and DA. Using the *Planorbis corneus* giant dopamine neuron, they demonstrated that amphetamine could act intracellularly to release DA from the cell. Moreover, injections of DA directly into the cytoplasm led to DA efflux that was sensitive to nomifensine, suggesting that it was mediated by the plasma-membrane transporter. According to the hypothesis put forth by Sulzer et al. *(63)*, amphetamine action at the vesicular membrane is sufficient to account for amphetamine-induced amine release. These results would also appear to explain the observation that the blockade of plasma-membrane transporters prevents the action of amphetamines *(52,64)*.

4.4. The Amphetamine Permeability Paradox

If amphetamine derivatives act only by uncoupling at the level of biogenic amine storage vesicles, then classical uncouplers such as 2,4-dinitrophenol should act as psychostimulants like amphetamine. However, no such action has been reported for uncouplers or other weakly basic amines. Moreover, all of the amphetamine derivatives that we have tested bind to plasma-membrane amine transporters *(53,58–60,65,66)*. What role could the plasma-membrane transporters play in amphetamine action? One possibility is that they serve merely to allow amphetamine derivatives into the cell.

An alternative possibility is that the ability of amphetamines to serve as substrates for plasma-membrane transporters is important in their action, even if the membrane does not constitute a barrier to amphetamine diffusion. In this view, futile cycling of the plasma-membrane transporter is induced by transporter-mediated influx followed by diffusion back out of the cell. As described previously, this process will lead to dissipation of Na^+, Cl^-, and K^+ gradients, and could possibly acidify the cell interior. As a result of the lower gradients and the appearance of cytoplasmic binding sites following amphetamine dissociation on the cell interior, biogenic amine efflux via the transporter will be stimulated.

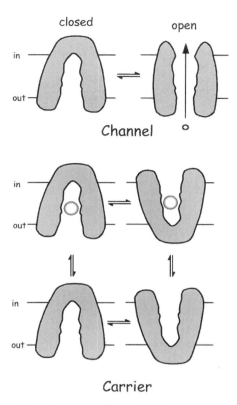

closed   open

Channel

Carrier

Fig. 4. Channels and carriers may have similar structures. A single structural model can account for transport by carriers and channels. In a channel, one or more "gates," or permeability barriers, open to allow free passage for ions from one side of the membrane to the other. A carrier can be thought of as a channel with two gates. Normally, only one is open at a time. By closing one gate and opening another, the carrier allows a solute molecule, bound between the two gates, to cross the membrane.

The availability of knockout mice in which the gene encoding DAT was inactivated has helped to resolve this issue. Jones et al. *(67)* demonstrated that amphetamine caused dopamine efflux from synaptic vesicles, as judged by a decrease in DA released from the cell during exocytosis. However, DA was not released from the cell by amphetamine treatment *(67)*. These results indicate that amphetamine does not require the plasma-membrane transporter for entry into the cell or action on synaptic vesicles. However, the plasma-membrane transporter was required for amphetamine to exchange with cytoplasmic DA.

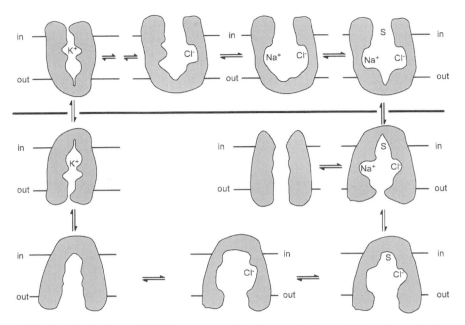

Fig. 5. Mechanism of 5-HT transport. Starting at the lower left and continuing counter-clockwise, the transporter binds Cl⁻, 5-HT (S), and Na⁺. These binding events permit the carrier to undergo a conformational change to the form in the upper right-hand corner. This internal-facing form dissociates Na⁺, Cl⁻, and 5-HT to the cyto-plasm. Upon binding internal K⁺, another conformational change allows the carrier to dissociate K⁺ on the cell exterior, generating the original form of the transporter which can initiate another round of transport by binding external Na⁺, Cl⁻, and 5-HT. Also shown is a putative channel mode of the transporter, which could be responsible for uncoupled currents, in equilibrium with the loaded form of the carrier.

5. CONFORMATIONAL CHANGES AND TRANSLOCATION

5.1. Mechanisms for Ion Coupling

It is interesting to consider how the biogenic amine transporters, with pre-dicted molecular weights of 60–80 Kd, are able to couple the fluxes of sub-strate, Na⁺, Cl⁻, and K⁺ in a stoichiometric manner. The problem faced by a coupled transporter is more complicated than that faced by an ion channel, because a channel can function merely by allowing its substrate ions to flow across the lipid bilayer. Such uncoupled flux will dissipate the ion gradients and will not utilize them to concentrate another substrate. However, the struc-tural similarities between transporters and ion channels may provide a clue to the mechanism of transport (Fig. 4). Just as ion channels are believed to have a central aqueous cavity surrounded by amphipathic membrane-spanning helices, neurotransmitter transporters may have a central binding site which accommodates Na^+, Cl^-, and substrate. The difference in mechanism between a transporter and a channel may be that a channel assumes open (conducting) and closed (nonconducting) states, yet a transporter can also assume two states which differ only in the accessibility of the central binding site. In each of these states, the site is exposed to only one face of the membrane, and the act of substrate translocation represents a conformational change to the state in which the binding site is exposed on the opposite face (Fig. 4). Thus, the transporter may behave like a channel with a gate at each face of the membrane, but only one gate is usually open at any point in time.

For this mechanism to lead to cotransport of ions with substrate molecules, the transporter must obey a set of rules governing the conformational transition between its two states (Fig. 5). For cotransport of Na^+, Cl^-, and 5-HT, the rule would allow a conformational change when the binding site was occupied with Na^+, Cl^-, and substrate. To account for K^+ countertransport with 5-HT, the reverse conformational change to external-facing form would occur only when the binding site contained K^+. This simple model of a binding site exposed alternately to one side of the membrane or the other can explain most carrier-mediated transport. However, it makes specific predictions about the behavior of the transport system.

In particular, this model requires that the substrate is transported in the same step as Na^+ and Cl^-, but in a different step than a countertransported ion such as K^+. As described previously, there is ample evidence from studies measuring binding of 5-HT and inhibitors that Na^+ and Cl^- bind to the transporter together with 5-HT *(22)*. There is also evidence that 5-HT and K^+ are transported in different steps. The exchange of internal and external 5-HT does not require the K^+-dependent step that is rate-limiting for net 5-HT flux. Thus, the steps required for 5-HT binding, translocation, and dissociation do not include the steps (where K^+ is translocated) that become rate-limiting in the absence of K^+ *(31)*.

5.2. Structural Correlates of Conformational Changes

As detailed in Chapters 3, 4, and 6–8, the biogenic amine transporters were proposed to consist of alternating hydrophobic and hydrophilic stretches of amino-acid residues. The hydrophilic external-loop structures in SERT have been identified by site-directed chemical modification *(68)*. The schematic mechanism of transport outlined in Fig. 5 ultimately must be reconciled with a structure that may contain 12 helical transmembrane segments connected by alternating extracellular and cytoplasmic loops. The conformational changes that are believed to convert substrate accessibility from extracellular to cytoplasmic and back again are triggered by binding events that presumably occur within the transmembrane segments of a transport protein. However, the conformational changes themselves may involve more than transmembrane domains. Experiments with chimeric SERT constructs in which external-loop regions have been replaced with corresponding NET sequence emphasize the importance of these loops. Many of these constructs are severely defective in transport activity, although some chimeras that do not transport retain relatively normal ligand-binding activity *(69,70)*. Cell-surface expression of these chimeras was at least as robust as wild-type. The loss of transport but not binding activity suggests that some of the external loops must be critical for steps other than binding, such as conformational changes that change the accessibility of the substrate-binding site. Despite the substitution of NET sequence for SERT, none of the chimeras with residual binding activity gained affinity for NET substrates or ligands. The similar binding selectivity between these mutants and wild-type suggests that the external loops do not play a significant role in forming the binding site for substrates and inhibitors.

In contrast, there is growing evidence that the transmembrane regions contain residues involved with substrate binding. Cysteine-scanning mutagenesis of the third transmembrane domain (TM3) of SERT revealed that at one position, Tyr-176, replacement with cysteine blocked transport activity *(71)*. 5-HT and a cocaine analog were bound by Y176C, but the affinity was reduced. Modification of Cys-176 with the sulfhydryl reagent [2-(trimethylammonium)ethyl]methanethiosulfonate (MTSET) destroyed the remaining binding activity, and that inactivation was decreased in the presence of 5-HT or cocaine. Substitution of Ile-172 (one helical turn away from Tyr-176) with cysteine did not ablate transport activity, but I172C was also sensitive to inactivation of both transport and binding by MTSET, and like Y176C, 5-HT and cocaine protected against inactivation *(71)*. These results suggest that Tyr-176 and Ile-172 may be close to, or within, a binding site shared by 5-HT and cocaine. Recent evidence indicates that after Cys-172 was inactivated by extracellular MTSET, it could be reactivated by intracellular thiols *(72)*. Therefore, this position is accessible from both sides of the plasma membrane, although probably not simultaneously.

On the same predicted helical face of TM3 is another isoleucine, Ile-179, also one turn away from Tyr-176. Substitution at this position with cysteine led to sensitivity to MTSET. However I179C was not protected by 5-HT or cocaine *(71)*. In both SERT and NET, in which the corresponding I155C mutant was occluded from reactivity after translocation of its catecholamine substrate *(72)*, this position is conformationally sensitive. In many ways, SERT Ile-179 and NET Ile-155 behave as if they are part of the gate that closes to prevent substrate from dissociating after it is bound to the transporter.

6. BIOGENIC AMINE TRANSPORTERS ARE RELATED TO PUMPS, RECEPTORS, AND CHANNELS

6.1. What is the Difference Between a Channel and a Carrier?

The proteins responsible for accumulating biogenic amines are commonly called transporters, but they are more precisely referred to as carriers. Carriers and pumps are proteins that move solutes across membranes by a mechanism that requires a conformational change for every molecule or ion transported. Although carriers can couple the transmembrane movement of more than one solute, they are distinguished from pumps by their lack of coupling to metabolic energy. Pumps are quite similar to carriers in mechanism, but in addition to moving solutes, pumps also mediate a chemical reaction such as ATP hydrolysis, decarboxylation, or a redox or photochemical reaction that is coupled to the conformational changes in a way that utilizes the energy source to drive solute transport. Both carriers and pumps have the ability to use an energy source for solute accumulation, and this property sets them apart from channels, which only allow their substrate ion to flow down its electrochemical gradient.

The characteristic that most clearly distinguishes channels from carriers is the phenomenon of counterflow. In carrier-mediated transport it is common for the influx of a substrate into a cell or vesicle to be insensitive to, or even stimulated by, substrate efflux from same cell. The movement of substrate in one direction does not interfere with movement in the opposite direction. This happens because the step that translocates substrate into the cell is distinct from the step that translocates substrate out. In a channel, where the two processes of influx and efflux both require the same aqueous pathway through the open channel, it is inevitable that influx will inhibit efflux and vice versa. An analogous situation in everyday macroscopic life is the difference between a stairway and an elevator in a multi-story building. If many people are rushing down the stairs, it is more difficult to climb upward against the crowd. However, if the same crowd is using the elevator to descend, it is even more likely to find an elevator waiting on the ground floor than when nobody is going down.

In this simplistic discussion, we have assumed that a given protein can be either a carrier or a channel, but recent evidence suggests that some proteins act as both. The 5-HT, NE, and DA transporters all seem to mediate uncoupled ion fluxes in addition to their ability to catalyze substrate accumulation *(34,73–79)*. These transporters may not be unique in possessing more than one activity. There are reports that cystic fibrosis transmembrane regulator (CFTR) and multidrug resistance protein (MDR) can operate as both channels and pumps *(80)*. It is important to distinguish the two activities. The ability of a biogenic amine carrier to conduct ions as a channel may represent a distinct activity of the protein. Although there might be conditions under which carrier and channel activity influence each other, uphill substrate accumulation cannot result from channel activity, and rapid ion conductance is unlikely to be caused by carrier activity.

6.2. How is a Carrier like a Receptor?

The triggering of conformational changes by substrate binding is one of the key events postulated to result in coupling of solute fluxes. For example, SERT does not transport 5-HT in the absence of Na^+ or Cl^-, presumably because all three solutes must be bound together on the transporter to trigger the conformational change that exposes them to the other side of the membrane. This process is not unlike the one in which a surface receptor undergoes a conformational change in response to agonist binding. In a receptor, that conformational change could open an ion channel or stimulate nucleotide exchange or enzymatic activity in an associated intracellular protein. In a carrier, the conformational change acts on the agonist (substrate) itself to change its accessibility from the internal and external faces of the membrane. From the perspective of evolution, carriers were much more important for primitive unicellular organisms (to ingest foodstuffs and excrete wastes) than channels and receptors. It is tempting to speculate that the structure and function of carriers was adapted to carry out the activities of channels and receptors as organisms developed the need for cell-cell communication.

7. TRANSPORTERS MEDIATE UNCOUPLED IONIC CURRENTS IN ADDITION TO TRANSPORT CURRENTS

7.1. What is the Relationship Between Uncoupled and Transport Current?

Recently, it has become apparent that in addition to the coupled flux of Na^+, K^+, and Cl^-, the 5-HT, DA, and NE transporters also catalyze uncoupled ion flux *(34,73,79)*. Recent evidence for each of these three transporters demonstrates the appearance of ion-channel activity either by direct observation of single channels or by analysis of current noise *(74,81,82)*. The uncoupled flux is therefore likely to represent events in which the transporter transiently operates like an ion channel instead of operating like a carrier. From the amount of current that flows with each channel event, and the frequency of channel events, it was possible to estimate that, for the 5-HT transporter, channels very rarely open relative to the number of times that a molecule of substrate is transported. Although there is evidence that substrates can move through the open channels *(76,78)*, these channel events are probably not an integral part of the transport process, but rather are analogous to a side reaction that occurs rarely in an enzymatic reaction. The uncoupled current, however, has important implications for the use of current recording techniques in transporter studies.

Studies that attempt to measure solute transport by recording the electrical current associated with transport can be confounded by the uncoupled ion flux. For example, although 5-HT transport is electroneutral, addition of 5-HT to oocytes expressing SERT leads to an inward current *(34)*. Measurements of substrate accumulation, as opposed to current, depend on imposed ion gradients—and, in some cases, electrical potentials—but they are not likely to be influenced by uncoupled currents carried by the transporter. The total current will be the sum of any current associated with ion-coupled substrate transport, plus the uncoupled current that flows through any channels that open during the time of the measurement. Since the uncoupled currrent is frequently stimulated by substrate binding or transport *(74,81,82)*, it is difficult to estimate the proportion of the total current that is coupled or uncoupled. The danger of assuming that all transporter-mediated current represents substrate flux is illustrated by the case of the glutamate transporter. This transporter has an associated uncoupled anion channel that is stimulated by glutamate *(83,84)*. Previous studies of glutamate-stimulated current unknowingly assumed that the anion current actually represented glutamate influx, and prematurely concluded that anion efflux was coupled stochiometrically to glutamate influx *(85)*.

7.2. What do Electrogenic and Electroneutral Really Mean?

The recent demonstration that neurotransmitter transporters also mediate uncoupled ion flux *(34,73,83,86)*, has cast a measure of confusion on the terms used to describe these proteins and their properties. If, as it seems likely, these transporters transiently form conductive channels through the membrane, a distinction must be made between the types of electrical currents caused by channel activity and substrate transport. The term "electrogenic" has traditionally been used to describe a coupled transport process in which net charge crosses the membrane. In the absence of ion gradients, an electrogenic transporter should generate an electrical potential in response to an imposed transmembrane substrate gradient. In contrast, the channel activity of such a transporter can only mediate energetically downhill ion flux. Thus, although SERT and NET both conduct ions by virtue of their intermittent channel activity, NET is an electrogenic transporter because it transports net charge with substrate *(38)*, and SERT is electroneutral because the 5-HT transport cycle itself does not move net charge across the membrane *(11)*.

7.3. What do Uncoupled Currents Tell Us?

Transporter-mediated currents are divided into three categories. The first and simplest are transport currents themselves, which result from electrogenic movement of ions during the transport reaction. Second, the individual steps in the transport cycle also may be associated with charge movement that appears as a transient current when the transporter redistributes between two states. These transient currents have been very useful in demonstrating and characterizing electrogenic binding of Na^+ to the GABA transporter *(34,82,87–89)*. The third category are the uncoupled currents that seem, at least in some cases, to be the result of a channel activity of a transporter that normally functions only as a carrier. If, as recent results suggest, these channel openings occur very rarely with respect to the normal transport cycle, one might question whether they can tell us anything at all about the transporters. Perhaps they are just an epiphenomenon that has nothing to do with normal transport or physiological function. It is too early to tell for sure, but there are reasons to believe that these uncoupled fluxes may allow insight into the transporters that would be difficult to obtain in any other way.

The most interesting issue is the relationship between the aqueous pore through which ions permeate during channel activity and the pathway that substrates take as they are transported across the membrane. We think of the binding site for substrates as a potential channel through the core of the transporter, which is normally separated from one surface of the membrane or the other by permeability barriers (or gates). It is possible that if the barriers are both open, an aqueous channel through the membrane will be formed. If the ion channel does represent the substrate-transport pathway, it will provide support for the concept that transport substrates bind in a "channel" closed at one end or the other by gates. To learn if the transport pathway and the ion channel are identical, it would be helpful to know if substrates are bound during the time that the transporter is acting like a channel. Substrate stimulates channel activity by 5-HT, NE, and glutamate transporters *(74,82,83)*. On the surface, this result might suggest that the channel opens when substrate is bound, and that the transport pathway, occupied by bound substrate, could not function as the channel. An alternative explanation, however, is that the transport of substrate and its release to the cytoplasm leave the transporter in a conformation with a higher probability of opening as a channel.

To distinguish between these possibilities, we must learn if the requirements for substrates to stimulate channel activity are at all different from those for transport. If conditions dictate that a substrate cannot be transported but does stimulate channel activity, it will suggest that substrate binding opens a separate channel, as in a ligand-gated ion channel. Alternatively, if substrate binds within the ion channel, but dissociates from that site to leave the transporter in a state with a higher probability of opening for uncoupled ion flux, then substrate stimulation of channel activity would be observed only when substrate transport occurs.

Aside from the issue of the ion channel being the substrate site, there are other ways that channel activity may provide useful information about transporters. The properties of the channel—either its conductance, or opening or closing kinetics—may differ between states of the transporter. For example, the conductance of substrate-stimulated channels in a mutant 5-HT transporter is different from that of the channels observed in the absence of substrate *(88)*. It may be possible to identify intermediates in the transport cycle with specific ion-channel properties. These properties could then be used to analyze the presence of those intermediates under specific conditions or in mutant transporters. The true impact of channel activity by transporters will be likely to modify our understanding of transport mechanism as more of its details are explored.

REFERENCES

1. Crane, R. K., Forstner, G., and Eichholz, A. (1965) Studies on the mechanism of the intestinal absorption of sugars. *Biochim Biophys Acta* **109,** 467–477.
2. Iversen, L. L. (1967) *The Uptake and Storage of Noradrenaline in Sympathetic Nerves,* Cambridge University Press, University Printing House Cambridge, U. K.
3. Sneddon, J. M. (1969) Sodium-dependent accumulation of 5-hydroxytryptamine by rat blood platelets. *Br. J. Pharmacol.* **37,** 680–688.
4. Bennett, J. P., Jr, Logan, W. J., and Snyder, S. H. (1972) Amino acid neurotransmitter candidates: sodium-dependent high-affinity uptake by unique synaptosomal fractions. *Science* **178,** 997–999.
5. Curtis, D. and Johnston, G. (1974) Amino acid transmitters in the mammalian central nervous system. *Rev. Physiol. Biochem. Exp. Pharm.* **69,** 97–188.
6. Iversen, L. L. (1970) Neuronal uptake processes for amines and amino acids. *Adv. Biochem. Psychopharmacol.* **2,** 109–132.
7. Krnjevic, K. (1974) Chemical nature of synaptic transmission in vertebrates. *Physiol. Rev.* **54,** 418–540.
8. Kuhar, M. J. (1973) Neurotransmitter uptake: a tool in identifying neurotransmitter-specific pathways. *Life Sci.* **13,** 1623–1634.
9. Levi, G. and Raiteri, M. (1976) Synaptosomal transport processes. *Int. Rev. Neurobiol.* **19,** 51–74.
10. Kanner, B. I. and Sharon, I. (1978) Active transport of L-glutamate by membrane vesicles isolated from rat brain. *Biochemistry* **17,** 3949–3953.
11. Rudnick, G. and Nelson, P. J. (1978) Platelet 5-hydroxytryptamine transport an electroneutral mechanism coupled to potassium. *Biochem.* **17,** 4739–4742.
12. Keyes, S. R. and Rudnick, G. (1982) Coupling of transmembrane proton gradients to platelet serotonin transport. *J. Biol. Chem.* **257,** 1172–1176.

13. Nelson, P. J., Dean, G. E., Aronson, P. S., and Rudnick, G. (1983) Hydrogen ion cotransport by the renal brush border glutamate transporter. *Biochemistry* **22**, 5459–5463.

14. Kuhar, M. J. and Zarbin, M. A. (1978) Synaptosomal transport: a chloride dependence for choline, gaba, glycine and several other compounds. *J. Neurochem.* **31**, 251–256.

15. Jones, E. M. C. (1995) Na+- and Cl−-dependent neurotransmitter transporters in bovine retina - identification and localization by in situ hybridization histochemistry. *Vis. Neurosci.* **12**, 1135–1142.

16. Talvenheimo, J., Nelson, P. J., and Rudnick, G. (1979) Mechanism of imipramine inhibition of platelet 5-hydroxytryptamine transport. *J. Biol. Chem.* **254**, 4631–4635.

17. Lee, C. M., Javitch, J. A., and Snyder, S. H. (1982) Characterization of [3H]desipramine binding associated with neuronal norepinephrine uptake sites in rat brain membranes. *J. Neurosci.* **2**, 1515–1525.

18. Talvenheimo, J., Fishkes, H., Nelson, P. J., and Rudnick, G. (1983) The serotonin transporter-imipramine 'receptor': different sodium requirements for imipramine binding and serotonin translocation. *J. Biol. Chem.* **258**, 6115–6119.

19. Ramamoorthy, S., Prasad, P., Kulanthaivel, P., Leibach, F. H., Blakely, R. D., and Ganapathy, V. (1993) Expression of a cocaine-sensitive norepinephrine transporter in the human placental syncytiotrophoblast. *Biochemistry* **32**, 1346–1353.

20. Cool, D. A., Leibach, F. H., and Ganapathy, V. (1990) High-affinity paroxetine binding to the human placental serotonin transporter. *Am. J. Physiol.* **259**, C196–C204.

21. Rudnick, G. and Wall, S. C. (1991) Binding of the cocaine analog 2-beta-[H-3] carbomethoxy-3-beta-[4-fluorophenyl]tropane to the serotonin transporter. *Mol. Pharmacol.* **40**, 421–426.

22. Humphreys, C. J., Wall, S. C., and Rudnick, G. (1994) Ligand binding to the serotonin transporter: equilibria, kinetics and ion dependence. *Biochemistry* **33**, 9118–9125.

23. Ritz, M. C., Lamb, R. J., Goldberg, S. R., and Kuhar, M. J. (1987) Cocaine receptors on dopamine transporters are related to self-administration of cocaine. *Science* **237**, 1219–1223.

24. Wall, S. C., Innis, R. B., and Rudnick, G. (1993) Binding of the cocaine analog [125I]-2β-carbomethoxy-3β-(4-iodophenyl) tropane (β-CIT) to serotonin and dopamine transporters: different ionic requirements for substrate and β-CIT binding. *Mol. Pharmacol.* **43**, 264-270.

25. Berfield, J. L., Wang, L. C., and Reith, M. E. (1999) Which form of dopamine is the substrate for the human dopamine transporter: the cationic or the uncharged species? *J. Biol. Chem.* **274**, 4876-4882.

25a. Ni,Y.G., Chen, J. G., Androutsellis-Theotokis, A., Huang, C. J., Moczydlowski, E., and Rudnick, G. (2001), A lithium induced condormational change in serotonin transporter alters cocaine binding, ion conductance, and reactivity of cys-109. *J. Biol. Chem.* **276**, 30,942–30,947.

26. Rudnick, G. (1977) Active transport of 5-hydroxytryptamine by plasma membrane vesicles isolated from human blood platelets. *J. Biol. Chem.* **252**, 2170–2174.

27. Nelson, P. J. and Rudnick, G. (1982) The role of chloride ion in platelet serotonin transport. *J. Biol. Chem.* **257,** 6151–6155.

28. Kanner, B. I. and Bendahan, A. (1985) Transport of 5-hydroxytryptamine in membrane vesicles from rat basophillic leukemia cells. *Biochim. Biophys. Acta* **816,** 403–410.

29. O'Reilly, C. A. and Reith, M. E. A. (1988) Uptake of [*3*H]serotonin into plasma membrane vesicles from mouse cortex. *J. Biol. Chem.* **263,** 6115–6121.

30. Balkovetz, D. F., Tirruppathi, C., Leibach, F. H., Mahesh, V. B., and Ganapathy, V. (1989) Evidence for an imipramine-sensitive serotonin transporter in human placental brush-border membranes. *J. Biol. Chem.* **264,** 2195–2198.

31. Nelson, P. J. and Rudnick, G. (1979) Coupling between platelet 5-hydroxytryptamine and potassium transport. *J. Biol. Chem.* **254,** 10,084–10,089.

32. Cool, D. R., Leibach, F. H., and Ganapathy, V. (1990) Modulation of serotonin uptake kinetics by ions and ion gradients in human placental brush-border membrane vesicles. *Biochemistry* **29,** 1818–1822.

33. Reith, M. E. A., Zimanyi, I., and O'Reilly, C. A. (1989) Role of ions and membrane potential in uptake of serotonin into plasma membrane vesicles from mouse brain. *Biochem. Pharmacol.* **38,** 2091–2097.

34. Mager, S., Min, C., Henry, D. J., Chavkin, C., Hoffman, B. J., Davidson, N., and Lester, H. A. (1994) Conducting states of a mammalian serotonin transporter. *Neuron* **12,** 845–859.

35. noth, J., Isaacs, J., and Njus, D. (1981) Amine transport in chromaffin granule ghosts. pH dependence implies cationic form is translocated. *J. Biol. Chem.* **256,** 6541–6543.

36. Kobold, G., Langer, R., and Burger, A. (1985) Does the carrier of chromaffin granules transport the protonated or the uncharged species of catecholamines? *Nauyn-Schmiedegerg's Arch. Pharmacol.* **331,** 209–219.

37. Scherman, D. and Henry, J.-P. (1981) pH Dependence of the ATP-dependent uptake of noradrenaline by bovine chromaffin granule ghosts. *Eur. J. Biochem.* **116,** 535–539.

38. Gu, H. H., Wall, S., and Rudnick, G. (1996) Ion coupling stoichiometry for norepinephrine transporter in membrane vesicles from stably transfected cells. *J. Biol. Chem.* **271,** 6911–6916.

39. Rudnick, G., Kirk, K. L., Fishkes, H., and Schuldiner, S. (1989) Zwitterionic and anionic forms of a serotonin analog as transport substrates. *J. Biol. Chem.* **264,** 14,865–14,868.

40. Friedrich, U. and Bonisch, H. (1986) The neuronal noradrenaline transport system of PC-12 cells: kinetic analysis of the interaction between noradrenaline, Na$^+$ and Cl$^-$ in transport. *Nauynn-Schmiedegerg's Arch Pharmacol.* **333,** 246–252.

41. Humphreys, C. J., Beidler, D., and Rudnick, G. (1991) Substrate and inhibitor binding and translocation by the platelet plasma membrane serotonin transporter. *Biochem. Soc. Trans.* **19,** 95–98.

42. Ramamoorthy, S., Leibach, F. H., Mahesh, V. B., and Ganapathy, V. (1992) Active transport of dopamine in human placental brush-border membrane vesicles. *Am. J. Physiol.* **262,** C1189–C1196.

43. Rudnick, G. (1998) Bioenergetics of neurotransmitter transport. *J. Bioenerg. Biomem.* **30**, 173–185.
44. Keynan, S. and Kanner, B. I. (1988) γ-Aminobutyric acid transport in reconstituted preparations from rat brain: coupled sodium and chloride fluxes. *Biochemistry* **27**, 12–17.
45. Keynan, S., Suh, Y. J., Kanner, B. I., and Rudnick, G. (1992) Expression of a cloned gamma-aminobutyric acid transporter in mammalian cells. *Biochemistry* **31**, 1974–1979.
46. Radian, R. and Kanner, B. I. (1983) Stoichiometry of Na^+ and Cl^- coupled GABA transport by Synaptic plasma membrane vesicles isolated from rat brain. *Biochemistry* **22**, 1236–1241.
47. Harder, R. and Bonisch, H. (1985) Effects of monovalent ions on the transport of noradrenaline across the plasma membrane of neuronal cells (PC-12 cells). *J. Neurochem.* **45**, 1154–1162.
48. Kanner, B. I. (1978) Active transport of γ-aminobutyric acid by membrane vesicles isolated from rat brain. *Biochemistry* **17**, 1207–1211.
49. Gu, H. H., Wall, S. C., and Rudnick, G. (1994) Stable expression of biogenic amine transporters reveals differences in ion dependence and inhibitor sensitivity. *J. Biol. Chem.* **269**, 7124–7130.
50. Brown, C. D., Bodmer, M., Biber, J., and Murer, H. (1984) Sodium-dependent phosphate transport by apical membrane vesicles from a cultured renal epithelial cell line (LLC-PK_1). *Biochim. Biophys. Acta.* **769**, 471–478.
51. McElvain, J. S. and Schenk, J. O. (1992) A multisubstrate mechanism of striatal dopamine uptake and its inhibition by cocaine. *Biochem. Pharmacol.* **43**, 2189–2199.
52. Fischer, J. F. and Cho, A. K. (1979) Chemical release of dopamine from striatal homogenates: evidence for an exchange diffusion model. *J. Pharm. Exp. Therap.* **208**, 203–209.
53. Rudnick, G. and Wall, S. C. (1992) The molecular mechanism of ecstasy [3,4-methylenedioxymethamphetamine (MDMA)]—serotonin transporters are targets for MDMA-induced serotonin release. *Proc. Natl. Acad. Sci. USA* **89**, 1817–1821.
54. Sulzer, D. and Rayport, S. (1990) Amphetamine and other psychostimulants reduce ph gradient in midbrain dopaminergic neurons and chromaffin granules: a mechanism of action. *Neuron* **5**, 797–808.
55. Peyer, M. and Pletscher, A. (1981) Liberation of catecholamines and 5-hydroxytryptamine from human blood-platelets. *Naunyn Schmiedebergs Arch Pharmacol.* **316**, 81–86.
56. Steele, T., Nichols, D., and Yim, G. (1987) Stereochemical effects of 3,4-methylenedioxymethamphetamine (MDMA) and related amphetamine derivatives on inhibition of uptake of [3H]monoamines into synaptosomes from different regions of rat brain. *Biochem. Pharmacol.* **36**, 2297–2303.
57. Mamounas, L. A., Mullen, C., Ohearn, E., and Molliver, M. E. (1991) Dual serotoninergic projections to forebrain in the rat - morphologically distinct 5-HT axon terminals exhibit differential vulnerability to neurotoxic amphetamine derivatives. *J. Comp. Neurol.* **314**, 558–586.

58. Rudnick, G. and Wall, S. C. (1992) *p*-Chloroamphetamine induces serotonin release through serotonin transporters. *Biochemistry* **31,** 6710–6718.
59. Rudnick, G. and Wall, S. C. (1993) Non-neurotoxic amphetamine derivatives release serotonin through serotonin transporters. *Mol. Pharmacol.* **43,** 271–276.
60. Schuldiner, S., Steiner-Mordoch, S., Yelin, R., Wall, S. C. and Rudnick, G. (1993) Amphetamine derivatives interact with both plasma membrane and sectretory vesicle biogenic amine transporters. *Mol. Pharmacol.* **44,** 1227–1231.
61. Johnson, R. G. and Scarpa, A. (1979) Protonmotive force and catecholamine transport in isolated chromaffin granules. *J. Biol. Chem.* **254,** 3750–3760.
62. Peter, D., Jimenez, J., Liu, Y. J., Kim, J., and Edwards, R. H. (1994) The chromaffin granule and synaptic vesicle amine transporters differ in substrate recognition and sensitivity to inhibitors. *J. Biol. Chem.* **269,** 7231–7237.
63. Sulzer, D., Chen, T. K., Lau, Y. Y., Kristensen, H., Rayport, S., and Ewing, A. (1995) Amphetamine redistributes dopamine from synaptic vesicles to the cytosol and promotes reverse transport. *J. Neurosci.* **15,** 4102–4108.
64. Azzaro, A. J., Ziance, R. J., and Rutledge, C. O. (1974) The importance of neuronal uptake of amines for amphetamine-induced release of ^{3}H-Norepinephrine from isolated brain tissue. *J. Pharmacol. Exp. Ther.* **189,** 110–118.
65. Rudnick, G. and Wall, S. C. (1992) The platelet plasma membrane serotonin transporter catalyzes exchange between neurotoxic amphetamines and serotonin. *Ann. NY Acad. Sci.* **648,** 345–347.
66. Wall, S. C., Gu, H. H., and Rudnick, G. (1995) Biogenic amine flux mediated by cloned transporters stably expressed in cultured cell lines: amphetamine specificity for inhibition and efflux. *Mol. Pharmacol.* **47,** 544–550.
67. Jones, S. R., Gainetdinov, R. R., Wightman, R. M., and Caron, M. G. (1998) Mechanisms of amphetamine action revealed in mice lacking the dopamine transporter. *J. Neurosci.* **18,** 1979–1986.
68. Chen, J. G., Liu-Chen, S., and Rudnick, G. (1998) Determination of external loop topology in the serotonin transporter by site-directed chemical labeling. *J. Biol. Chem.* **273,** 12,675–12,681.
69. Smicun, Y., Campbell, S. D., Chen, M. A., Gu, H., and Rudnick, G. (1999) The role of external loop regions in serotonin transport. Loop scanning mutagenesis of the serotonin transporter external domain. *J. Biol. Chem.* **274,** 36,058–36,064.
70. Stephan, M. M., Chen, M. A., Penado, K. M. Y., and Rudnick, G. An extracellular loop region of the serotonin transporter may be involved in the translocation mechanism. *Biochemistry* **36,** 1322–1328.
71. Chen, J. G., Sachpatzidis, A., and Rudnick, G. (1997) The third transmembrane domain of the serotonin transporter contains residues associated with substrate and cocaine binding. *J. Biol. Chem.* **272,** 28,321–28,327.
72. Chen, J.-G. and Rudnick, G. (2000) Permeation and gating residues in serotonin transporter. *Proc. Natl. Acad. Sci. USA* **97,** 1044–1049.
73. Galli, A., DeFelice, L. J., Duke, B. J., Moore, K. R., and Blakely, R. D. (1995) Sodium-dependent norepinephrine-induced currents in norepinephrine-transporter-transfected hek-293 cells blocked by cocaine and antidepressants. *J. Exp. Biol.* **198,** 2197–2212.

74. Galli, A., Blakely, R. D., and DeFelice, L. J. (1996) Norepinephrine transporters have channel modes of conduction. *Proc. Nat. Acad. Sci. USA* **93,** 8671–8676.
75. Galli, A., Petersen, C. I., Deblaquiere, M., Blakely, R. D., and DeFelice, L. J. (1997) Drosophila serotonin transporters have voltage-dependent uptake coupled to a serotonin-gated ion channel. *J. Neurosci.* **17,** 3401–3411.
76. Galli, A., Blakely, R. D., and DeFelice, L. J. (1998) Patch-clamp and amperometric recordings from norepinephrine transporters - channel activity and voltage-dependent uptake. *Proc. Nat. Acad. Sci. USA* **95,** 13,260–13,265.
77. Mager, S., Naeve, J., Quick, M., Labarca, C., Davidson, N., and Lester, H. A. (1993) Steady states, charge movements, and rates for a cloned GABA transporter expressed in Xenopus oocytes. *Neuron* **10,** 177–188.
78. Petersen, C. I. and DeFelice, L. J. (1999) Ionic interactions in the Drosophila serotonin transporter identify it as a serotonin channel. *Nature Neurosci.* **2,** 605–610.
79. Sonders, M., Zhu, S., Zahniser, N., Kavanaugh, M., and Amara, S. (1997) Multiple ionic conductances of the human dopamine transporter—the actions of dopamine and psychostimulants. *J. Neurosci.* **17,** 960–974.
80. Higgins, C. (1995) Volume-activated chloride currents associated with the multidrug resistance P-glycoprotein. *J. Physiol. (Lond)* **482,** 31S–36S.
81. Cammack, J. N. and Schwartz, E. A. (1996) Channel behavior in a gamma-aminobutyrate transporter. *Proc. Nat. Acad. Sci. USA* **93,** 723–727.
82. Lin, F., Lester, H. A., and Mager, S. (1996) Single-channel currents produced by the serotonin transporter and analysis of a mutation affecting ion permeation. *Biophys. J.* **71,** 3126–3135.
83. Fairman, W. A., Vandenberg, R. J., Arriza, J. L., Kavanaugh, M. P., and Amara, S. G. (1995) An excitatory amino-acid transporter with properties of a ligand-gated chloride channel. *Nature* **375,** 599–603.
84. Wadiche, J. I., Amara, S. G., and Kavanaugh, M. P. (1995) Ion fluxes associated with excitatory amino acid transport. *Neuron* **15,** 721–728.
85. Bouvier, M., Szatkowski, M., Amato, A., and Attwell, D. (1992) The glial cell glutamate uptake carrier countertransports pH-changing anions. *Nature* **360,** 471–474.
86. Cammack, J. N., Rakhilin, S. V. and Schwartz, E. A. (1994) A GABA Transporter operates asymmetrically and with variable stoichiometry. *Neuron* **13,** 949–960.
87. Hilgemann, D. W. and Lu, C. C. (1999) GAT1 (GABA : Na^+: Cl^-) cotransport function - Database reconstruction with an alternating access model. *J. Gen. Physiol.* **114,** 459–475.
88. Lu, C. C. and Hilgemann, D. W. (1999) GAT1 (GABA : Na^+: Cl^-) cotransport function - Steady state studies in giant Xenopus oocyte membrane patches. *J. Gen. Physiol.* **114,** 429–444.
89. Lu, C. C. and Hilgemann, D. W. (1999) GAT1 (GABA : Na^+: Cl^-) cotransport function - Kinetic studies in giant Xenopus oocyte membrane patches. *J. Gen. Physiol.* **114,** 445–457.

3

Structure-Function Relationships for Biogenic Amine Neurotransmitter Transporters

Nian-Hang Chen and Maarten E. A. Reith

1. INTRODUCTION

The monoamine transporters or carriers, located on the plasma membrane of nerve terminals, transport monoamine substrates across the membrane. By taking up substrate into neurons, they play a critical role in terminating biogenic amine neurotransmission and in maintaining monoamine homeostasis in the central nervous system (CNS) *(1–6)*. Mounting evidence indicates that monoaminergic transmission, both catecholaminergic and tryptaminergic, occurs, for a large part, by volume transmission involving migration of transmitter through the extracellular space, nonjunctional aminergic innervation of neurons, and location of biogenic amine transporters outside synapses *(7–11)*. Many substances, such as the psychostimulant amphetamine, the dopaminergic neurotoxin 1-methyl-4-phenyl-pyridinium (MPP$^+$), and various sympathomimetic amines, structurally resemble dopamine (DA). They are thus substrates for the DA transporter (DAT) and can be transported *(12,13)*. Likewise, the 5-HT releasers p-chloro-amphetamine and 3,4-methylenedioxymethamphetamine (MDMA, also known as ecstasy), are substrates for the serotonin (5-HT) transporter (SERT), and enter serotonergic nerve terminals partly through translocation by the SERT *(see* Chapter 2), whereas a classical example of an indirectly acting sympathomimetic amine entering norepinephrine (NE) terminals through the NE transporter (NET) is tyramine *(14)*, responsible for the well-known "cheese effect" of MAO-inhibitors *(15)*. The DAT is also a major molecular target for the addictive drug cocaine and, to a lesser extent, antidepressants *(16,17)*. The SERT and NET are targeted by antidepressants—for example, tricyclics of the tertiary and secondary amine type, respectively *(see* refs. *16,21,48,49,93,* also *see* Table 2). These drugs cannot be transported, but can bind to the carriers to block transport of the substrates. There-

From: *Contemporary Neuroscience:*
Neurotransmitter Transporters: Structure, Function, and Regulation, 2nd Edition
Edited by: M. E. A. Reith © Humana Press Inc., Totowa, NJ

fore, interactions with the biogenic amine-transporter proteins can have profound neurobiological, pathophysiological, and pharmacological consequences. In the last decade, since the cloning of the monoamine transporters (*see* **Subheading 2.1.**), major advances have been made in the characterization of biogenic amine carriers. This chapter focuses on the recent progress in elucidating structure-function relationships for the monoamine transporters. The section on DAT was published previously *(18)*, and has been fully updated up to and including the cloning of Drosophila DAT *(19)*.

2. DOPAMINE TRANSPORTER

2.1. Molecular and Pharmacological Characteristics

The DAT has been identified from brains of various species (for references, *see* Table 1). A transporter with neuronal DA-transporter properties and a partial cDNA clone has also been characterized from the African Green Monkey kidney (COS) cell line *(36)*. The mammalian DATs exhibit high sequence identity (Table 1). The reported longer C-terminus of the bovine (b) DAT was caused by an error of the authors in the translation of the cDNA sequence from the corresponding mRNA, introducing a frame shift. The correct translation of the bDAT C-terminal sequence results in a C-terminus of 40 amino acids, not longer than the C-termini of other DATs *(33)*. Among other members of the neurotransmitter transporter gene family, the mammalian DATs display highest amino-acid homology with the NET (~67%), SERT (~49%), and γ-amino butyric acid (GABA) transporter (~45%) (*see* Figs. 1 and 2). DA uptake by all the identified DATs has a similar pharmacological profile, including a highly comparable sensitivity to cocaine analogs (Table 1). It should be noted that high-affinity binding of the phenyltropane analog of cocaine, 2β-carbomethoxy-3β-(4-fluorophenyl) tropane (CFT = WIN 35,428), is not known for the mouse *(33,34)*, *Caenorhabditis elegans* (Ce) *(35)*, and *Drosophila melanogaster* (d) *(19)* DATs; appears to be poor in bDAT *(31)* (however, *see 33*); and is not found in COS DAT *(36)*. Additionally, Ce- and d-DATs have a relatively lower amino-acid identity. The two DATs appears to be highly sensitive to the tricyclic antidepressants desipramine and imipramine, and the NET-selective inhibitor nisoxetine, but relatively insensitive to DAT-selective GBR compounds *(19,35)*. These features distinguish them from mammalian DA carriers (*see* Table 2) (for more details on drug development targeting DAT, *see* Chapter 11 and ref. *60*). There is a strong structural homology and functional similarity between hDAT and the recently cloned monkey DAT *(59)*. Interestingly, two DAT variants have been obtained from monkey substantia nigra (SN), with only a one- or two-amino-acid difference with wild-

Table 1
Comparison of Cloned Dopamine Transporters

Species	Name	Size (amino acid)	Homology to hDAT (%)	DA K_m^a (nM)	Sensitivity to inhibitors (K_i ratio)^b GBR:MAZ:COC:BUP^c	Refs.
Human	hDAT	620	100	1220 (Ltk-, 37°C)	1:0.65:3.4:19	20
				2400 (COS, 22°C)	1:4.3:53:56	21,22
				1500 (COS, 25°C)		23
				2540 (COS, 37°C)		24
Rat	rDAT	619	92	885 (Hela, 37°C)	1:1.4:19:31	25
				300 (COS, 37°C)		26
				1320–2090 (COS, 37°C)		21,27–29
Bovine	bDAT	617	88	31500 (CV-1, 37°C)	1:0.98:4.7:53	30
				990–1910 (COS, 37°C)		31,32
Mouse	mDAT	619	93	930 (HEK293, 35°C)	1:—:56:288	33
				2000 (MDCK, 22°C)		34
C. elegans	CeDAT	615	43	1200 (HeLa, 37°C)	1:0.05:29:—	35
D. melanogaster	dDAT	631	49	4800 (MDCK, 22°C)	1:0.007:3.97:7.2	19
COS cell	COS-DAT		96 (TM^d 1–6)	130 (high, 22°C)	1:0.05:0.41:1.08	36
				8300 (low, 22°C)		

^{a}Except for MDCK, K_m data were taken from references using transiently transfected cells. For comparison, the host cell and the assay temperature are included in parentheses.

^{b}Except for mDAT, the K_i ratios (K_i inhibitor/K_i GBR) for inhibiting DA uptake are calculated according to data reported in the cited reference.

^{c}GBR, GBR 12909; MAZ, mazindol; COC, cocaine; BUP, bupropion.

^{d}TM, transmembrane domain.

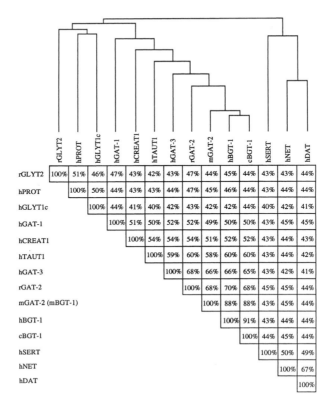

Fig. 1. Amino-acid sequence relationships between various members of the Na⁺- and Cl⁻-dependent transporter family. The percent amino-acid sequence identity between pairs of transporters is presented. Three distinct subfamilies can be distinguished: the subfamily of amino acid (glycine and proline) transporters; the subfamily of GABA, betaine, taurine, and creatine transporters; and the subfamily of biogenic amine (serotonin, norepinephrine, and dopamine) transporters. rGLYT2, rat-brain glycine transporter 2 (37); hPROT, human-brain proline transporter (38); hGLYT1c, human-brain glycine transporter 1c (39); hGAT-1, human-brain GABA transporter 1 (40); hCREAT1, human-brain creatine transporter 1 (41); hTAUT1, human-brain taurine transporter 1 (42); hGAT-3, human-brain GABA transporter 3 (43); rGAT-2, rat-brain GABA transporter 2 (44); mGAT-2, mouse-brain GABA transporter 2 (45); hBGT-1, human-brain betaine/GABA transporter 1 (46); cBGT-1, canine-kidney betaine/GABA transporter 1 (47); hSERT, human-brain serotonin transporter (48); hNET, human-brain norepinephrine transporter (49); and hDAT, human-brain dopamine transporter (21).

type; one of these variants interacts differently with cocaine, and the other has a low surface expression (59).

All of the identified DATs are functionally dependent on the presence of external sodium and chloride (*see* references in Table 1). Recent transport, binding, and electrophysiological investigations reveal that the interactions of ions with the DA carrier are considerably more complex than the simple picture of two Na^+ ions and one Cl^- ion being cotransported with one DA molecule *(61–65)*.

2.2. Structure of DAT

To date, no X-ray crystallographic or high-resolution structural information is available for the topological assignments of the transporters. Hydropathy analysis of the primary sequences of mammalian monoamine transporters predicts a topology with 12 transmembrane segments connected by alternating extracellular and intracellular loops, and the N- and C-termini located in the cytosol (Fig. 3). A study using cysteine/lysine-modifying reagents and biotinylated probe scanning has confirmed the proposed topology on predicting extracellular domains of the SERT *(66)*. For DA carriers, some aspects of the proposed topology have also been experimentally verified. Electron-microscopic immunocytochemical evidence confirms the cytoplasmic location of N-terminal and the extracellular location of the sequence between transmembrane domains (TMs) 3 and 4 *(67,68)*. Proteolytic mapping and epitope-specific immunoprecipitation of photolabeled DATs demonstrate the presence of at least one transmembrane helix or other membrane-anchoring structure in two different regions of the protein predicted to contain TMs, and again identifies the extracellular orientation of the region between putative transmembrane segments 3 and 4 *(69)*. Determination of the accessibility of endogenous cysteines—located on the putative intracellular or extracellular loops of the human (h) DAT, for membrane-permeable and membrane-impermeable sulfhydryl reagents—also supports the originally proposed topology: Cys^{90} and Cys^{306} appear to be extracellular, and Cys^{135} and Cys^{342} appear to be intracellular *(70)*. Additionally, three residues on the extracellular face of the hDAT, His^{193}, His^{375}, and Glu^{396}, have been characterized to form three coordinates in the Zn^{2+}-binding site. The common participation of these residues in binding the small Zn^{2+} ion suggests a close proximity between extracellular loop (EL) 2, TM 7, and TM 8 in the tertiary structure of the hDAT *(71,72)*.

Photoaffinity labeling studies show that DA carriers are glycoproteins with a molecular mass of ~80 kDa on polyacrylamide gels. Prevention of transporter glycosylation by glycosidase reduces the molecular mass to 50 kDa *(73)* (also *see* Chapter 10). All of the DATs cloned to date have at least three consensus sites for N-linked glycosylation in the second EL located between transmembrane segments 3 and 4, with dDAT carrying a remakable number

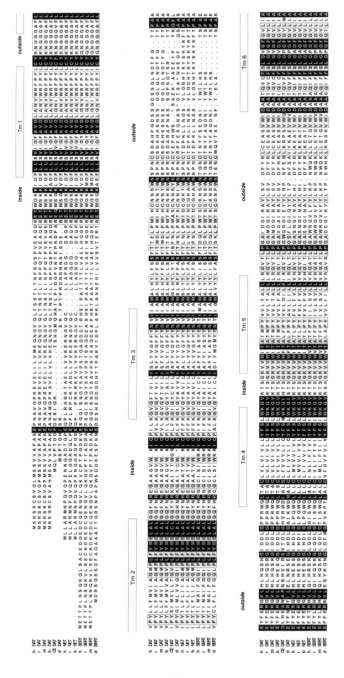

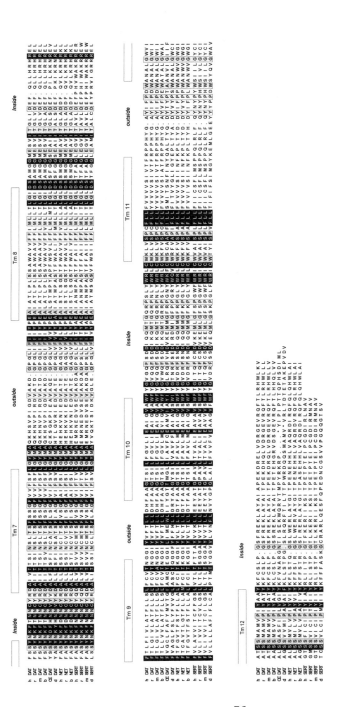

Fig. 2. Alignment of monoamine transporter sequences. The alignment is as per Povlock and Amara (*50*) extended with mDAT (*33*), CeDAT (*35*), dDAT (*19*), mSERT (*19*), rNET (*52*), and bNET (*53*). In this figure, the N-terminal sequence for rSERT (*54*) is partly different from that reported by Blakely et al. (*55*), and the N-terminal sequence for dSERT (*56*) is partly different from that reported by Corey et al. (*57*). The following recently published sequences are not included here: mNET (*57a*), bSERT (*58*), guinea pig SERT (*58a*), frog ET (*168*), and monkey DAT, NET, and SERT (*59*). The amino acids shaded in grey represent residues generally found in all members of the Na+/Cl−-transporters (*see* Chapter 6). Residues indicated by white letters on a black background are conserved only between the monoamine transporters with a few exceptions occurring in the species added compared with the same figure in the first edition of this book. *For additional, inserted sequence at this location, see original publication for respective clone as referenced above.

Table 2

Examples of Inhibitors of Transport in Cells Expressing Human Monoamine Transporters[a]

Inhibitors[b]	K_i, nM		
	DAT	NET	SERT
Cocaine	267	872	392
RTI-55 (=CIT)	3.2	2.5	0.49
WIN 35,428 (=CFT)	26.1	31.9	127
GBR 12935	21.5	225	6514
Bupropion	2,784	1389	45,026
Nisoxetine	477	5.1	383
Desipramine	78,720	4.0	61
Nortryptiline	13,920	3.4	161
Mazindol	27.6	3.2	153
Imipramine	24,576	67	7.7
Amytriptyline	3000[g]	100[e]	14.7[f]
Citalopram	210,000[h]	>1000[e]	5.4[i]
Paroxetine		312[e]	0.25[f]

Substrates[c]		DAT	NET	SERT
DA	K_m	460[c]		
	K_m	2540[c,j]	670[c,j]	
			6.6[d]	30,800[d]
NE	K_m	20,100[c,j]	2600[c,j]	
	K_m		457[c,e]	
		13,700[d]	110[d]	540,000[d]
5-HT	K_m			463[c,f]
		15,200[d]	5420[d]	170[d]
MPP+		>10,000[d,g]		
D-amphetamine		41[d]	23.2[d]	11,000[d]
L-amphetamine		138[d]	30.1[d]	57,000[d]

[a]Except where indicated by superscript, values are from Eshleman et al. *(93)* for HEK-293 cells expressing DAT, NET, or SERT; IC$_{50}$ values were converted to K_i values by considering [substrate]=20 nM and K_m for DAT from Eshleman et al. *(93)*, for NET from Pacholczyk et al. *(49)*, and for SERT from Ramamoorthy et al. *(48)* (*see* K_m values listed under "Substrates").

[b]Values represent K_i for inhibition of uptake of substrate appropriate for the given transporter.

[c]Only where value has superscript c, does it represent K_m for uptake saturation measured with radiolabeled substrate; otherwise, *see* superscript d.

[d]Values represent IC$_{50}$ for uptake inhibition which does not equal the binding affinity K_i (with Cheng-Prusoff correction) in case of a substrate; the IC$_{50}$ value also involves the K_m for the substrate used as inhibitor and the preincubation time applied.

[e]Values from Pacholczyk et al. *(49)*.

[f]Values from Ramamoorthy et al. *(48)*.

[g]Values from Giros et al. *(21)*.

[h]Values from Zhang et al. *(94)*.

[i]Values from Barker et al. *(95)*.

[j]Values from Giros et al. *(22)*.

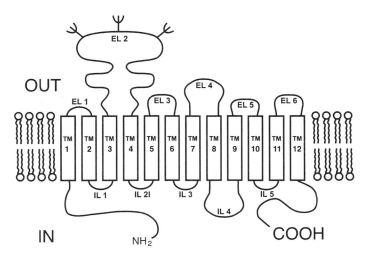

Fig. 3. Proposed topology of plasma-membrane amine transporters. Three N-linked glycosylation sites (y) are depicted on EL2, but the number of such sites on EL2 varies according to transporter and species, with one form (dDAT) having an extra glycosylation site on EL3. Loops (ELs and ILs) and TMs are numbered for location of residues presented in Tables 1–8.

of seven consensus sites for N-linked glycosylation, one of which is on EL3 *(19)*. Preliminary data indicates that removal of more than two of the glycosylation sites dramatically reduces plasma-membrane expression *(74)*. It has been shown that the stability and trafficking/targeting, not ligand recognition, of the hNET to the membrane greatly depend on glycosylation of the consensus sites *(75,76)*. Whether such an effect also applies to the DA carrier remains to be explored. Two cysteines in the second EL are also conserved completely within monoamine transporters, which have been presumed to form a disulfide bridge and to be involved in the cell-surface expression in the case of the rat (r) DA *(77)* and 5-HT *(78,79)* transporter.

DATs themselves undergo in vivo phosphorylation in stably expressing cells and striatal synaptosomes *(80–82)*. Protein kinase C (PKC)-mediated DAT phosphorylation *(81,82)* results in sequestration of the transport protein *(83–85)* and downregulation of transport activity *(81,82,86,87)*. The primary sequence of the DA carrier contains multiple phosphorylation sites in the putative intracellular domains. Although preliminary studies with antibodies reveal that serine residues and N-terminal tail may be involved in the phosphorylation of the protein *(80,88)*, DATs do not appear to be directly phosphorylated at their consensus phosphorylation sites. Removing all the potential PKC sites failed to affect the ability of PKC activators or inhibitors to regulate DA transport *(32)*, whereas mutation of the calcium/calmodulin-dependent protein kinase (CaM kinase) site, Thr[613] in the hDAT, has shown contradictory results on the effect of CaM kinase inactivation *(89)*. Thus, direct DAT phosphorylation may occur at nonconsensus phosphorylation sites. It is also likely that the phosphorylation of the DA carrier is substantially impacted by ligand occupancy *(80)*, as shown for the SERT *(90)*.

2.3. Substrate Structure/Active Form and Substrate Recognition

The structural requirements for the interaction of substrates with the DAT have been examined by comparing the transport of phenethylamine derivatives in striatal-slice preparations *(91)*. These studies indicate that the DAT requires molecules that possess a phenyl ring with a primary ethylamine side chain for optimal activity, and the β rotamer of the extended conformation of catecholamines is transported preferentially *(91)*. It is proposed that the catechol appears to mediate the recognition of the substrate, whereas the amine side chain apparently facilitates the conformational change of the transporter that results in movement of DA across the membrane. Recent transport studies on the cloned hDAT suggest that, although β- or phenolic-ring hydroxylation of a substrate results in change in the K_m over a wide range, the presence of a phenolic hydroxyl group is not essential for optimal function of the transporter. Compounds without a phenolic hydroxyl group such as β-phenethylamine, amphetamines, and MPP[+], can bind to the carrier and be transported with the same V_{max} as dopamine *(92; also see* Table 2). Therefore, phenethylamine appears to be the most important structural element accommodated by the DAT.

DA can exist in the anionic, neutral, cationic, or zwitterionic form, depending on the ambient pH. Data from studies addressing the pH dependence of DA binding/transport *(96)* favor the cationic forms of DA, perhaps including the zwitterion, as the more likely substrate. pH-induced changes in the interaction between DA and its transporter could result from changes at either the ligand level or the transporter level. At the ligand level, the introduction of positive charges on the ligand by protonation may result in conformational changes in the ligand itself. At the transporter level, the protonated amine group of DA could interact with a negatively charged residue, such as Asp[79], as suggested by site-directed mutagenesis experiments *(97)* and molecular modeling techniques *(98)*, or with a binding domain that is susceptible to pH.

In evaluating the effect of a mutation on the binding affinity of a substrate, it should be noted that substrate binding to the DAT cannot be measured directly because of its rapid dissociation within the time-scale of the separation of bound from free ligand in binding assays. Furthermore, the half-saturation substrate concentration (K_m) for transport is not identical to the dissociation constant (K_d) for the binding of the substrate to the transporter. K_m also depends in a complex manner on rate constants for transporter orientation *(99)*. Thus, a change in a step beyond substrate recognition but associated with DA translocation can influence the K_m value, confounding the interpretation for substrate affinity. Indeed, it is not uncommon that mutations showing lower V_{max} values display lower K_m values. In this scenario, the change in K_m cannot be simply interpreted as a change in substrate-binding affinity. For better evaluation of DA affinity for a mutant DAT, the ability of DA to compete for a nontransportable ligand-binding site overlapping with the DA site must be considered.

Almost all DA carrier substrates are phenethylamine derivatives, and are positively charged at physiological pH. These features have been used as a guide to find residues of the substrate-binding site at the carrier. Thus, charged, aromatic, and polar residues become the most attractive targets for mutagenesis. For the DAT, mutations at Asp[79] of the rat variant results in an increase in both K_m for DA uptake and the K_i for DA to inhibit [³H]CFT-binding site *(97)*, suggesting the importance of this residue in substrate recognition. This amino acid is conserved in all monoamine transporters, but not in other members of Na^+/Cl^--dependent neurotransmitter transporters (*see* Fig. 2). As observed in rDAT, replacement of the aligned aspartate in rSERTs (D98) or hNETs (D75) disrupts the transport activity. Mutation at D98 of rSERT also causes a loss of affinity as an inhibitor of the binding of [¹²⁵I] 2β-carbomethoxy-3β-(4-iodophenyl) tropane (RTI-55), a cocaine analog *(100)*. Experiments have been carried out to test the hypothesis that the positively charged amino group of amine substrates form ionic bonds with the negatively charged carboxylic acid of this aspartate. Thus, if this interaction exists, lengthening the carboxylic side chain by one CH_2 unit via site-directed mutagenesis to a glutamate may improve transport of amine analogs whose alkylamine side chain is shortened by one CH_2. The results have been mixed. D79E mutation in rDAT does not effectively enhance the affinity—as an inhibitor of [³H]DA uptake and [³H]CFT binding—of dihydroxy-benzylamine, which is identical to DA, except shorter by one CH_2 at the alkylamine side chain *(101)*. The D98E rSERT mutation selectively enhances the affinity, as an inhibitor of [³H]5HT uptake and at activating substrate-induced currents, of gramine, a serotonin analog with a shorter alkylamine side chain. However, the result from D98E could also be mediated by changes beyond direct contact, such as impacts of the mutation on ion permeation coupled to substrate occupancy (*100*; also *see* **Subheading 4.2.2.**) (Table 7). For hDAT, D79 is not the only acidic residue implicated in substrate recognition. Thus, a neutralizing mutation at D313 or D476 reduces the potency of DA in inhibiting both [^{3}H]DA uptake and [^{3}H]CFT binding (102). Clarification of their mechanisms needs further experimentation.

As cationic molecules, transporter substrates could bind to the π faces of aromatic rings, producing noncovalent cation-π interactions (103). Alternatively, the aromatic ring of substrates could recognize phenyl-ring side chains on the DAT through π-π interactions (104). In search of such interactions, a large number of aromatic residues of the DAT, including most of the phenylalanine and tryptophan residues in TMs, have been mutated (102,105,106). To date, none of those mutations with normal surface expression has been reported to raise both the K_m for DA uptake and the K_i for inhibiting [^{3}H]CFT binding. Mutation of Phe^{155}, a residue located on TM 3 and unique to the DAT family, strikingly raises the K_m for DA uptake and reduces the V_{max}. Intriguingly, it also enhances the potency of DA in inhibiting [^{3}H]CFT binding more than 30-fold (105), indicating that this mutated DAT could still fully recognize DA. A marked dissociation between K_m for DA uptake and K_i for DA to inhibit CFT binding has also been observed in mutations at F76 of rDAT (105) and W311 (W310) of hDAT (rDAT) (102,106). This dissociation could be interpreted as the mutation-induced separation between binding sites for DA and CFT—i.e., DA-binding affinity at the transporter no longer equals the K_i for DA in competition for CFT. Alternatively, there are extreme changes in translocation steps impacting the K_m value.

Simultaneous mutation of the two serines in position 356 and 359 in transmembrane domain 7 of the rDAT reduces DA uptake (97). Conservation of these two serines in catecholamine transporters and DA D$_2$ and β-adrenoceptors raises the hypothesis that the two serines might form hydrogen bonds with the two hydroxyl groups of DA. However, results obtained from single and simultaneous mutation of the aligned serines in the homologous NET argue against a direct hydrogen bond interaction between substrates and the two serines (107; also see **Subheading 3.2.2.**) (Table 5).

2.4. Substrate Transport: Inward and Outward

Among more than 100 single or double/triple DAT mutants reported thus far, 27 display reductions in turnover rates for DA to less than one-third of wild-type values (Table 3), although they show near-normal membrane expression or CFT-binding capacity. Most of the 27 residues are conserved throughout mammal DA and NE carriers and located in regions spanning TMs 1–8 and 10–12 (Table 3). Noticeably, 16 of 17 substitutions of charged residues located on the extracellular surface of the hDAT display little or only modest reduction in turnover rates for DA (71,72). These data generally favor the view that determinants important for substrate translocation appear to be closely associated with transmembrane segments. However, the role of extracellular and intracellular loops remains to be established for the DAT. Experimental evidence indicates that lack of certain conserved loop residues of hDAT (C342, D345, and D436) impairs translocation (*102,110*; Table 3). In the SERT, external loops have been shown to participate in the conformational changes required for translocation, although they are not the primary determinants for ligand recognition (*111*).

Using site-directed chemical modification, the cytoplasmic-loop residue Cys[342] has been identified to be in a conformationally sensitive part associated with the permeation way or cytoplasmic gating of the hDAT (*110*). Cys[342] is located in a region (TMs 5–8) believed to be critical for substrate translocation. m-Tyramine specifically increased the rate of reaction of relatively permeant methane-thiosulfonate-ethylamminum (MTSEA), rather than impermeant methane-thiosulfonate-ethyltrimethylammonium (MTSET), with X-342C, a cysteine-less mutant construct with Cys[342] restored. This effect was observed only under conditions that allowed inward translocation, and it was blocked by cocaine. The observations indicate that substrate translocation involves structural rearrangement of the hDAT protein, which probably exposes either the thiol side chain of Cys[342] or other endogenous cysteine residues, the exposure of which requires the presence of Cys[342] to intracellular MTSEA. Conformationally sensitive and translocation-associated residues have also been reported for homologous GABA and 5-HT transporters (*112,113*).

Studies have also explored the DAT structures that participate in the transport of dopaminergic neurotoxic MPP[+]. Amino-acid substitutions in the rat transporter have suggested that some serine and tyrosine residues located in TMs 7 or 11 appear more influential on interaction with MPP[+] than with DA. Thus, alanine substitution for Ser[350] and Ser[353] preferentially elevate V_{max} for MPP[+] transport, whereas alanine substitution for Ser[527], Tyr[533], and Ser[538] mainly reduce K_m for MPP[+] transport (*28*). Whether this effect is specific for MPP[+] transport—or, as found for the aligned transmembrane serine in TM 11 of the hDAT (*92*), whether this also applies to the transport of other substrates—remains to be clarified. Phenylalanine substitution of Tyr[533] in TM 11 of the rDAT, when the corresponding residue is phenylalanine in the human carrier, increases the velocity of MPP[+] uptake. The difference in MPP[+] uptake kinetics between the wild-type and the Y533F mutant was similar to the difference observed between r- and h-DATs (*114*). Therefore, Tyr[533] may be important for the species difference in the sensitivities to toxicity of MPP[+] between humans and rats. The transport of MPP[+] in 29 rDAT mutants with alanine substitution for phenylalanine has also been ex-

Table 3
Effect of DAT Mutations on Dopamine Transport and Cocaine Analog Binding[a]

Residues[b]	Conservation[c]		Location	DA		Cocaine analogs	Transport	Refs.
	DATs	NET/SERT		$1/K_m^d$	$1/K_i^e$	$1/K_d$ or $1/K_i^f$	Turnover[g]	
W63 (L)	Yes	Yes	hDAT-N	N.D.	N.D.	N.D.	N.D.	102
D68 (N)	Yes	Yes	hDAT-TM1	0	+	–	0	102
F69 (A)	Yes	Yes	rDAT-TM1	+	?	0	–	105
F76 (A)	Yes	No (SERT)	rDAT-TM1	0	+	–	–	105
D79 (AGE)	Yes	Yes	rDAT-TM1	–	–	–	–	97
W84 (A)	Yes	Yes	rDAT-TM1	0	0	+	0	108
W84 (L)	Yes	Yes	hDAT-TM1	0	0	+	0	102
P101 (AG)	Yes	Yes	rDAT-TM2	+	+	0	–	106
F105 (A)	No (d, Ce)	No (SERT)	rDAT-TM2	0	?	0	–	105
F114 (A)	Yes	Yes	rDAT-TM2	0	?	0	–	105
F155 (A)	Yes	No (both)	rDAT-TM3	–	+	0	–	105
E218 (Q)	No (d)	No (SERT)	rDAT-EL2	0	?	–	–	72
Y251 (AS)	Yes	Yes	rDAT-TM4	0(A), +(S)	?	–	0(A), –(S)	29,109
W266 (A)	Yes	Yes	rDAT-TM5	0	–	0	–	108
P272 (AG)	Yes	Yes	rDAT-TM5	–	–	–	0	106
Y273 (AS)	Yes	Yes	rDAT-TM5	0	?	–	–	29,109
P287 (AG)	Yes	Yes	rDAT-EL3	0	0	–	0	106
E307 (Q)	No (d, Ce)	Yes	hDAT-EL3	0	?	–	0	72
W311 (L)	Yes	Yes	hDAT-TM6	0	–	–	–	102
D313 (N)	Yes	Yes	hDAT-TM6	–	–	0 (+)	–	102
T315 (A)	No (b)	No (SERT)	rDAT-TM6	0	?	–	–	109
Q316 (A)	Yes	Yes	rDAT-TM6	+	?	0	–	109
F319 (A)	Yes	Yes	rDAT-TM6	0	?	0	–	105

(continued)

Residues[b]	Conservation[c]		Location	DA		Cocaine analogs	Transport	Refs.
	DATs	NET/SERT		$1/K_m{}^d$	$1/K_i{}^e$	$1/K_d$ or $1/K_i{}^f$	Turnover[g]	
F331 (A)	No (d, Ce)	Yes	rDAT-TM6	+	?	0	–	105
D345 (N)	Yes	Yes	hDAT-IL3	+	?	–	–	102
F361 (A)	Yes	Yes	rDAT-TM7	0	0	– (0)	0	105,109
D385 (N)	No (Ce)	No (NET)	hDAT-EL4	0	?	–	0	72
W406 (A)	Yes	No (SERT)	rDAT-TM8	+	+	–	–	108
F410 (A)	Yes	Yes	rDAT-TM8	0	?	0	–	105
E428 (Q)	Yes	Yes	hDAT-IL4	0	0	–	0	102
D436 (N)	Yes	Yes	hDAT-IL4	+	+	–	–	102
T455 (A)	No (d, Ce)	No (SERT)	rDAT-TM9	–	?	–	0	109
S459 (A)	No (d, Ce)	No (both)	rDAT-TM9	0	?	–	+	109
T464 (A)	No (m)	Yes	rDAT-TM9	0	?	–	+	109
D476 (N)	No (Ce)	No (SERT)	hDAT-EL5	–	–	–	0	72,102
W496 (A)	No (Ce)	Yes	rDAT-TM10	0	?	–	–	108
W520 (L)	Yes	Yes	hDAT-TM10	0	0	0	–	102
W523 (A)	Yes	Yes	rDAT-TM11	+	0	–	–	108
F530 (A)	No (Ce)	Yes	rDAT-TM11	+	?	0	–	105
W555 (A)	No (Ce)	Yes	rDAT-TM12	0	–	–	0	108
W556 (L)	No (Ce)	Yes	hDAT-TM12	0	0	–	–	102

[a]Only mutants are listed that are significantly different from wild-type by approximately threefold or more and that have near-normal membrane expression and/or B_{max} of CFT binding.

[b]Amino-acid substitutes used are indicated in parentheses; sometimes multiple substitutes were used.

[c]For DAT, if a transporter does not contain the aligned residue, it is indicated in parentheses: Ce, *Caenorhabditis elegans* DAT; b, bovine DAT; d, *drosophila melanogaster* DAT; m, mouse DAT. For NET and SERT, only the transporter without the aligned residue is indicated in parentheses that is from the same species as the DAT listed in the "Location" column.

[d]Calculated according to the DA concentration producing 50% maximal rate of [³H]DA uptake.

[e]Calculated according to the DA concentration producing 50% inhibition of [³H]CFT binding.

[f]Calculated according to the K_d for [³H]CFT binding or cocaine K_i in inhibiting [³H]CFT binding.

[g]Defined as the ratio of V_{max} for [³H]DA uptake to B_{max} for [³H]CFT binding. 0, no change or change less than threefold; +, enhance by threefold or more; –, reduce by threefold or more; N.D., not detectable; ?, not reported.

amined *(115)*. Evidence for mutations that strongly inhibit MPP$^+$ transport without affecting DA transport has not yet been found.

A common feature of DA carrier substrates is that they can be transported in both directions. There are essentially two ways to unveil the outward transport of internal substrates: first, by changing transmembrane ion gradients *(116)*, and second, by adding substrates to the external medium *(12,117–120)*. In the latter case, external substrates, by undergoing inward transport, enhance transporter availability at the inner face of the membrane, and thus release of cytosolic substrates by inducing outward transport. This has been demonstrated by recent studies with the NET *(121)*. Additionally, inward Na$^+$ current accompanying substrate uptake may also be the trigger for DAT-mediated release *(61,122)*. Substrate-induced DA efflux and uptake of the inducer appear to depend on common structural features of the inducer. As inferred from the K_m or K_i values for uptake and for inducing DA efflux, both effects are reduced by β-hydroxylation, stereoselective to α-methylation, but relatively insensitive to a switch of a single phenolic hydroxyl group from the m- to the p-position. Thus, individual substrates have similar K_m for uptake and for inducing DA efflux, resulting in an identical rank order of K_m in efflux and uptake assays *(92)*. These findings provide strong support for the concept that outward transport of internal substrates is closely coupled with inward transport of external substrates.

Despite the similarities in the substrate structural-activity relationship between efflux and uptake, the exchange between external substrates and internal DA is extremely unequal. Simultaneous monitoring of tyramine uptake and induced DA efflux reveals that the initial exit rate of internal DA is only 6% of the initial entry rate of external tyramine. Recently, the contribution of the DA carrier protein itself to the slow outward transport has been revealed by studying the human protein mutant S528A *(92)*. Thus, a structural perturbation with alanine substitution for Ser[528] dramatically accelerates DA efflux induced by various substrates. Further analysis reveals that the mutation does not change the structure-activity relationships, V_{max}, or cation dependence for the uptake of external substrates, but selectively enhances the efflux kinetics of internal DA. Additionally, simultaneous monitoring of tyramine uptake and resultant DA efflux demonstrates that S528A facilitates the DA efflux relative to tyramine uptake. These experiments suggest that the structural requirements of the human DA-carrier protein differ for inward and outward transport, which may be partially responsible for the extremely unequal exchange between external substrate and internal DA. Further evidence that the transporter conformation may differ between inward and outward transport comes from the study on effects of covalent sulfhydryl modification on DA uptake by X-342C. The m-tyramine-induced increase in the reactivity of X-342C is not observed under conditions allowing outward transport only *(110)*, implying that Cys-342 is not accessible to sulfhydryl reagents during outward transport. Recent evidence from mutagenesis studies of the hNET indicates that a single mutation, Ser^{357} to alanine, differentially affects DA uptake and tyramine-induced DA efflux *(107;* also *see* **Subheading 3.2.2.**) (Table 5).

In a physiological context, a restricted outward transport would enhance the efficiency of the overall dopamine uptake process. It can be speculated to also have significance in pathological processes. For example, the release of preloaded $[^3H]MPP^+$ is facilitated in COS cells expressing certain mutant rDATs (S350A/S353A, Y273A, and Y533F), which show less sensitivity to MPP^+ toxicity. This suggests that redistribution of MPP^+ resulting from influx/efflux turnover through the DAT may be a subtle factor in determining MPP^+ toxicity *(29)*.

2.5. Inhibitor Structure and Recognition

Among inhibitors, most information is available for cocaine and the related analog CFT, as described in **Subheading 2.6.** General indications as to which DAT domains interact with various inhibitors have been obtained in chimera studies. Thus, chimeras between the transporters for norepinephrine and dopamine constructed in two different laboratories *(22,123)* point to TMs 5–8 for conferring inhibitor sensitivity to a variety of compounds, including the antidepressants desipramine, nisoxetine, nortryptiline, nomifensine; the anorectic drug mazindol; the stimulants amphetamine, cocaine, and the diphenyl-piperazine derivatives that potently inhibit dopamine uptake, such as GBR 12909, GBR 12935, and LR1111 (also *see* **Subheading 3.2.1.**) (Table 4). Proteolysis of the DAT labeled with the photoaffinity label $[^{125}I]1$-[2-(diphenylmethoxy)ethyl]-4-[2-(4-azido-3-iodophenyl)ethyl]-piperazine (DEEP), a derivative of the highly potent DA uptake inhibitor GBR-12935, shows labeling of the amino half of the dopamine transporter, in a region including TMs 1 and 2 and the 6-amino-acid-linker *(69,124)*. Interestingly, the same region interacts with $[^{125}I]GA$ II 34, a benztropine photoaffinity analog *(125)*, contrasting with the region that interacts with the cocaine analog $[^{125}I]4'$-azido-3'-iodophenylethyl ester (RTI-82), which encompasses TMs 4–7 *(69,124)* *(see* **Subheading 2.6.**). However, in contrast to the previously mentioned ligands which show nearly completely specificity in their binding-site incorporation, a recently developed piperidine-based photoaffinity label $[^{125}I]AD$-96-129, 4-[2-(diphenyl-methoxy)ethyl]-1-[(4-azidophenyl)methyl]-piperidine, is incorporated into both sites at comparable levels. These results suggest that the two domains may be in close three-dimensional proximity, and may contribute to the binding of multiple uptake blockers *(126)*.

Protection by compounds against attack of the DAT by sulfhydryl reagents has been monitored in attempts to delineate differences in binding domains for inhibitors and substrates *(127–131)*, but this approach has not revealed specific regions of the transporter involved in binding interactions. A recent extension of the protection strategy has been the use of more selective sulfhydryl reagents targeting cysteine residues in the form of methanethiosulfonates (MTS) such as MTSEA and MTSET (also *see* **Subheading 2.5.**). Exposure to cocaine enhances the accessibility of Cys^{90} to MTS in an hDAT mutant with many cysteines removed, and this paradoxically enhances rather than reduces subsequent $[^3H]CFT$ binding *(70)*. It is likely that Cys^{90} also contributes to binding increases with MTS treatment of wild-type hDAT, in addition to binding decreases mediated by other cysteine residues reacting with MTS, as the C90A mutant displays a much weaker protection by cocaine than the wild-type *(132,133)*.

2.6. Relationship Between Substrate Permeation and Cocaine-Binding Sites

Structural components important for cocaine recognition but not for DA transport, are most critical in the identification of regions essential for function and transport inhibition. This could provide significant insights into a transporter site that may recognize a DA-sparing cocaine antagonist and could even provide information about the structure of the cocaine antagonist, accommodated by the site *(134,135)*. However, there is substantial evidence that domains on the carrier for DA transport and for cocaine-analog binding are closely associated although they are not identical.

Chimeric proteins constructed from domains of the DA and NE transporter *(22,136)* indicate the importance of sequences in TM regions 9–11, and perhaps 1–3 (*see* **Subheading 3.2.1.**), for the apparent affinity of substrates. Sequences in TM regions 5–8 contribute to substrate translocation and inhibitor interaction. In particular, regions encompassing TM 1 as well as regions encompassing TM segments 5–8 appear to be important for the recognition of cocaine and RTI-55, a phenyltropane analog of cocaine (also *see* **Subheadings 3.** and **3.2.1.**) (Table 4). In other studies using chimeras between h- and b-DAT, DA transport and CFT binding are dramatically reduced when the bovine region encompassing TM 3 is substituted for that of the human carrier *(137)*. Later studies have indicated that P152 on TM3 of the hDAT plays a crucial role in the interactions of substrates and CFT with the transporter *(138)*. Likewise, results from cysteine-scanning mutagenesis approaches also suggest that TM 3 of the rSERT contains residues associated with substrate and cocaine binding *(139)* (also *see* **Subheading 4.2.2.**) (Table 7).

These studies indicate that the interactions of cocaine are likely to involve several domains, including those shown to be important for substrate transport. In accordance, photoaffinity-labeling studies have shown that [^{125}I]RTI 82, a cocaine-like compound, is incorporated in a region containing TMs 4–7 *(69,124)*, the same region implicated in DA translocation *(22)*. Consistent with chimeric and photoaffinity-labeling studies, DAT mutations indicate involvement of residues in diverse regions of the primary sequence in both cocaine-analog binding and DA binding/translocation (Table 3). In a recent survey of alanine-substitution mutants altering polar residues in TMs of the rDAT, calculation of changes in Gibbs free energies associated with interactions between combined TM mutants ($\Delta\Delta G°_{int}$) reveals that combined mutations in transmembrane segments 4+5 yield some of the largest $\Delta\Delta G°_{int}$ values for DA uptake, and TM 4+11 mutants provide the largest $\Delta\Delta G°_{int}$ for CFT binding *(109)*. Because replacement of D^{79} of the rDAT, or the aligned aspartate of the rSERT, strongly affects the binding affinity of both amine substrates and cocaine analogs containing an amine nitrogen, a direct ionic interaction of the wild-type aspartate with positive charged amine groups of these compounds remains a possibility *(97,100,140)*.

Evidence for nonidentical domains is also growing. As shown in Table 3, many mutations have displayed selective impact on DA transport or cocaine-analog binding. In general, mutations in TMs 2-3 and 6 preferentially affect DA translocation, and mutations in regions encompass IL4, TM9, and EL5 preferentially affecting cocaine-analog binding (Table 3). These observations suggest that it is possible to dissociate domains that influence DA translocation from those that influence cocaine-analog recognition. However, strong evidence for structures that selectively modulate binding of DA or cocaine analogs remains limited. For example, in 11 mutations that reduce the binding affinity of cocaine analogs but allowed a near-normal turnover rate, only three (D68N, P287A, and E428Q) have been demonstrated to exert no inhibition of DA binding, as inferred from the potency of DA to inhibit [^{3}H]CFT binding. The three residues are putatively located in or close to the hydrophilic loops. Their superficial location and relatively modest impact (threefold changes) on cocaine-analog affinity may favor an indirect rather than a direct participation in cocaine binding. Lin et al. *(105)* reported that mutation of F^{361} in the rDAT substantially reduced CFT binding but preserved DA binding/transport. However, this result was not replicated by the same group *(109)*. Clearly, it remains a challenge to develop small molecules which selectively occlude cocaine recognition without affecting DA transport *(105)*. In this context, it is important to note that Meltzer and colleagues *(141)* highlight the three-dimensional volume of cocaine-like compounds of the bicyclo[3,2,1]octane series rather than specific functionalities for DAT inhibition, based on a wide-ranging set of newly synthesized compounds that lack a protonatable nitrogen in the 8-position or lack an 8-group capable of hydrogen bonding. In this view *(141)*, the possibility of developing a drug that interferes with cocaine binding while allowing the passage of DA depends on the position of the cocaine recognition site in the "pore" which is theorized to be formed by a circular arrangement of the twelve TMs *(98)*.

It has been noted that changes in binding affinity of a cocaine analog (K_d) do not always parallel changes in its potency to inhibit DA uptake (K_i). For example, in mutant hDAT in which the C-terminus is truncated or substituted, both cocaine and CFT inhibit DA uptake with a K_i similar to that in wild-type, while their binding affinities are at least four- to fivefold lower than those in wild-type *(142)*. Likewise, CFT potently inhibits DA uptake by COS-DAT in which the carrier appears prematurely truncated after TM 6, although no specific [³H]CFT binding is observed in those cells *(36)*. Intriguingly, a single mutation (D345N) almost completely abolished cocaine analog binding, as determined in [³H]CFT binding and photoaffinity labeling with [¹²⁵I]RTI-82, but retained high potency of the cocaine analogs in inhibiting DA transport *(102)*. These observations suggest that the apparent lack of cocaine-analog binding does not necessarily mean a poor recognition by the carrier. Changes in DAT structure may alter the association and dissociation rates for cocaine-analog binding so that their binding cannot be detected by classical equilibrium-binding methods. Another possibility is that certain mutants display conformational states during the uptake cycle that have a high affinity for cocaine analogs, whereas such conformational states are not prevalent under the conditions of the binding assay. Binding to the transporter may not be the only way by which cocaine analogs inhibit DA transport. In searching for residues specific for cocaine binding, it is necessary to examine the potency of cocaine analogs to inhibit DA transport in mutant human DA carriers in addition to assessing radiolabeled cocaine-analog binding.

Conformational hypotheses, in addition to models with overlapping and separate domains in the interaction with DA and cocaine analogs, are also raised by information obtained from structure-function studies. Reaction of Cys⁹⁰ or Cys³⁰⁶ with impermeant MTS derivatives enhances DA uptake and CFT binding to a similar extent. Because removal of Cys⁹⁰ or Cys³⁰⁶ does not affect DA uptake and CFT binding, chemical modification of these cysteines seems to induce a conformational change in favor of both DA transport and cocaine binding. A difference between the effects of the covalent sulfhydryl modification of Cys^{90} or Cys^{306} on binding and transport is that the reaction of MTSET with Cys^{90} or Cys^{306} decreases the K_d of [³H]CFT binding *(70)*, but it increases the V_{max} of DA uptake *(110)*. These data suggest that cocaine may bind with higher affinity to a conformational state along the transport cycle, and that a modification that increases the propensity of the transporter to exist in such a state increases CFT binding and also enhances DA transport by facilitating transition along the transport cycle. Interestingly, mutations increasing DA affinity, such as D68N, F76A, P101A, F155A, W406A, and D436N (Table 3), tend to decrease CFT affinity, although some of these reductions are less than threefold. These data may suggest that the conformational states for substrate binding differ from those for cocaine binding. In our preliminary work addressing Na^+-dependence of hDAT mutants, Na^+-induced conformational changes appear to favor CFT binding, but not DA binding (unpublished data).

3. NOREPINEPHRINE TRANSPORTER

3.1. Molecular and Pharmacological Characteristics

hNET cDNA was isolated from the human SK-N-SH cell line by expression cloning *(49)*. As a member of the family of Na^+/Cl^--dependent transporters, the NET contains many residues absolutely conserved with the GABA transporter *(45;* also *see* Figs. 1 and 2). The next NET to be cloned was the bNET, by screening of a bovine adrenal medulla cDNA library under low stringency with a cDNA probe of the hNET *(53)*. bNET is 93.5% identical with hNET, with most of the differences residing in the amino and carboxyl tails. It has three consensus sites for phosphorylation by PKC, as opposed to one site in hNET. The rNET was cloned from PC12 cells by Brüss et al. *(52)* with the use of an antisense rNET primer, based on a known partial rNET sequence, and a series of hNET sense primers in polymerase chain reaction (PCR) experiments. The sequence identity between r- and h-NET is 87%, and between r- and b-NET the sequence identity is 84%. The recently cloned monkey NET is more than 98% homologous with human NET, and a novel exon 5 splice variant of monkey NET has been isolated that does not transport NE and does not bind cocaine with high affinity *(59)*.

rNET, like h- and b-NET, contains a leucine-zipper motif in TM2 (for topography, *see* Fig. 3); it has two PKC consensus sites, and carries a potential casein kinase II phosphorylation site. hNET, as b-NET, has three consensus sites for N-linked glycosylation in EL2, and rNET has two of such sites *(49,52,53)* (also *see* Chapter 10), and hNET carries a number of cytoplasmic consensus phosphorylation sites for protein kinase C (PKC) and Ca^{2+}- CaM kinase on the amino and carboxy tails as well as IL2 (see ref. *53,*

Chapter 1, and Fig. 3). Evidence strongly indicates the glycosylation consensus sites in sugar attachment to the NET *(75,76)* (*see* also Chapter 10), but it is unknown whether the phosphorylation consensus sites are actually used (*see* also Chapter 1). Bönisch et al. *(143)* provide evidence against the involvement of PKC consensus sites in PKC action on the NET (*see* Chapter 1 for more details).

The NET is responsible for high-affinity uptake of NE, but also mediates the transport of DA, actually with higher affinity than NE itself *(22,49,53, 93,144)* (*see* Table 2). The NET is the target for many antidepressant drugs, with secondary tricyclics such as desipramine selectively inhibiting the NET over the DAT and SERT *(21,48,49,93)* (*see* Table 2) (for more details on drug development targeting NET, *see* ref. *60*).

Functional transport by the NET requires the presence of Na^+ and Cl^-, with one Na^+ and one Cl^- being cotransported with NE (*see* Chapter 2). Current transporter models consider the possibility of gating events with transporters acting, in part, as channels (*see* Chapter 2). It has been argued that during substrate translocation, residues forming a gate may be brought into close contact with other residues forming the permeation pathway, and thereby will be sequestered from the medium *(113)*. Thus, gating residues would be conformationally sensitive to MTS reagents that react only with cysteines exposed to the medium. Indeed, I155 in TM3 of rNET could be part of an external gate, with the sensitivity of I155C rNET to MTSET being protected by the substrate DA but potentiated by the blocker cocaine *(113)*.

3.2. Substrate/Inhibitor Structure and Recognition

3.2.1. Chimera Studies

Moore and Blakely *(145)* constructed hNET-rSERT chimeras with the rSERT sequence begining C-terminally from "switch points" located at the N-terminal side of TM1, within IL1, within EL1, and at the N-terminal side of TM5; with the complementary portion coming from hNET. Additional chimeras studied had the hNET sequence C-terminally from switch points at the N-terminal side of TM1 and within TM7 (Table 4). Functionally active constructs were those that had N-terminal NET up to TM1, showing fully intact 5-HT uptake, and N-terminal SERT up to TM1 (chimera 10), displaying normal NE uptake. This indicates that the amino tail does not provide specific information determining substrate selectivity for the NET or SERT. In combination with information from other NET-SERT chimeras with switched amino or carboxyl tails *(148)*, it appears that substrate or blocker selectivity does not depend on either the amino or carboxyl tail. A chimera which included the rSERT-sequence 12 amino acids into TM1 (chimera 11), showed approx 30% of wild-type NE uptake activity *(145)*, suggesting a

Table 4
NET Chimeras

Chimera	"Switch points" determining source of domains for chimera construction	Major findings	Refs.
hNET-rSERT	N-terminal side of TM1 within IL1 within EL1 N-terminal side of TM5 within TM7	Most chimeras not functional, except: N-terminal NET up to TM1 shows 5-HT uptake N-terminal SERT up to TM1 shows NE uptake	145,146
hNET-rDAT	within TM1, 2, 3, 9, 10 within EL1, 6 within IL4 N-terminal side of TM1, 5, 10 C-terminal side of TM7	TMs 1–3: Inhibitor affinity and K_m of substrate uptake TMs 5–8: Inhibitor sensitivity/selectivity and V_{max} of substrate uptake TMs 10–11: K_m of substrate uptake	123,136
hNET-hDAT	within IL1 within TM5 within IL4	TMs 1–5: Functions common to monoamine transporters, such as ion dependence TMs 6–8: Inhibitor interactions TM9 - C-tail: Affinity and stereoselectivity of substrates Na$^+$ and Cl$^-$-dependence	22,147

TM, transmembrane domain; EL, extracellular loop; IL, intracellular loop; DAT, dopamine transporter; NET, norepinephrine transporter; SERT, serotonin transporter; MPP$^+$, 1-methyl-4-phenylpyridinium. For species notation, *see* Table 5.

contribution from within this region of 12 amino acids to substrate recognition or transport. The difference between chimera 10 and 11 is that chimera 11 has I in 93 and Y in position 95 of the rSERT sequence, whereas chimera 10 has V and F, respectively, from the hNET sequence (position 70 and 72 in hNET numbering). Therefore, V70 and/or F72 could contribute significantly to substrate transport by the NET, and F72Y NET has displayed reduced substrate uptake (*see* **Subheading 3.2.2.**).

An extensive collection of hNET-rDAT chimeras was constructed by Buck and Amara (*123,136*; Table 4). Seven chimeras consisted of a C-terminal NET portion with increasing lengths of N-terminal DAT, and seven chimeras were constructed the other way around, C-terminal DAT with increasing portions of N-terminal NET. Substrate uptake was within a factor two compared with wild-type DAT or NET for most chimeras, except those junctioning in a region encompassing TMs 5–8. TMs 1–3 profoundly affected the catecholamine substrate affinity for uptake by the NET and DAT (*136*), and also contributed to the affinity of NET for antidepressants such as desipramine and nisoxetine, and the affinity of DAT for the piperazine derivatives GBR 12909 and LR1111 (*123*). Chimeras junctioning within or near TM 1 displayed a decreased affinity for cocaine, suggesting the importance of TM 1 in cocaine recognition. TMs 5–7 conveyed selectivity for uptake inhibitors, with chimeras including DAT sequences in the TM 5–7 region showing more a DAT-like profile (GBR 12909 potent, nisoxetine weak) and chimeras including NET sequences at that location showing a more NET-like profile (desipramine and nisoxetine potent). Chimeras with junctions within the TMs 5–8 portion, but not those with junctions flanking TMs 5–8, were deficient in substrate uptake, suggesting the importance of TMs 5–8 not only in inhibitor interactions, but also substrate translocation or other events contributing to uptake, such as transporter trafficking. TMs 10 and 11 contributed to the affinity of uptake of substrates—which included the positively charged MPP$^+$, as was the case for TMs 5–8 (*136*).

A different set of hNET-hDAT chimeras was constructed by Giros et al. (*22*). With the switch points as indicated in Table 4, chimeras were designed with varying portions of N-terminal NET and C-terminal DAT and vice versa; in addition, chimeras were constructed with alternating NET/DAT sequences. The results indicate that TMs 6–8 determined tricyclic antidepressant binding and cocaine interactions. The latter conclusion has been weakened by new results from a follow-up study in which single mutations in the TM 6–8 region did not alter cocaine potency (*149*) (*see* **Subheading 3.2.2.**). Based on the results of the study of Giros et al. (*22*), the C-terminal portion starting with TM 9 was believed to determine substrate affinity for uptake and substrate stereoselectivity. A comparison of this report on NET-

DAT chimeras with the studies of Buck and Amara *(123,136)*, reveals that the conclusions regarding inhibitor interactions with TMs 5–8 are similar, as well as the conclusions concerning the contributions to substrate affinity of the C-terminal region that begins with TM 9. The major difference lies in the importance *(123,136)*, or lack of importance *(22)* of TMs 1–3 in determining substrate affinity (also *see* **Suheading 5.1.**). A recent study by Syringas et al. *(147)* examined the Na⁺- and Cl⁻-dependence of DA uptake by the chimeras of Giros et al. *(22)*. Whereas the C-terminal region starting with TM 9 appeared to be important for the Cl⁻-dependence, the N-terminal TM 1–3 region was found to contain major determinants for the Na⁺-dependency, with additional roles for the latter of regions encompassing TM 9 through the C terminus and TM 3–5 in conjunction with a determinant found between EL3 and IL4 *(147)*.

3.2.2. Site-Directed Mutagenesis Studies

Aspartate in the 75 position of hNET in TM1 has generated interest based on the crucial role of the corresponding residue D79 in rDAT (Table 1). Mutation of aspartate to glutamate, alanine, glycine, or asparagine greatly reduced uptake of NE by hNET *(100,146;* Table 5). Reduced surface expression of the mutant transporters was ruled out as a factor *(146)*. Most likely, NE is transported in its positively charged form, abundant at physiological pH *(151;* also *see* Chapter 2), and it is possible that D75 in NET is essential for the interaction of the positively charged amino group of NE with the negatively charged aspartate.

The incentive for examining TM1 residue 72 in hNET arose from studies on another member of the Na⁺,Cl⁻-dependent neurotransmitter transporter family. The corresponding residue in hSERT, Y95, influenced the potency of mazindol and citalopram in an opposite manner, with Y95F displaying enhanced mazindol but reduced citalopram potency *(95;* also *see* Table 7). For the hNET, the opposite was found: F72Y showed a decreased mazindol but increased citalopram potency *(95;* also *see* Table 5). In general, these results implicate TM1 of NET and SERT as part of the binding site for NE and 5-HT uptake inhibitors.

The rationale for studying the serine residues at positions 354 and 357 in the TM7 of hNET *(107;* also *see* Table 5) was based on the known interaction between the two hydroxyl groups of NE with S204 and S207 in TM5 of the β- and α2-adrenergic receptor *(152,153)*. The corresponding serines in rDAT are S356 and S359, which are important for DAT function, as shown by the 65% decrease in DA uptake activity in rDAT mutants with alanine substituted for serine at those locations *(97)*. In hNET, the DA uptake K_m was increased in S354A, S357A, and the double mutant *(107;* also *see* Table

Table 5
NET Site-Directed Mutagenesis Studies[a]

Mutated residues[b]	Conservation in[c]			Location	Changes	Refs.
	NET	DAT	SERT			
D75 (E,A,G)	Yes	Yes	Yes	hNET-TM1	NE uptake ⇩⇩⇩	146
D75 (E,A,G,N)	Yes	Yes	Yes	hNET-TM1	NE uptake ⇩⇩⇩	100
F72 (Y)	Yes	Yes	No (h,r,m)	hNET-TM1	DA uptake ⇩ 2x; mazindol K_i ⇑ 20x; citalopram K_i ⇩7x	95
S354 (A)	Yes	Yes	Yes	hNET-TM7	DA uptake K_m ⇑ 2–4x; nisoxetine K_d ⇑ 70x; substrate K_i against [^{3}H]nisoxetine ⇑ 2–8x	107
S357 (A)	Yes	No (d)	Yes	hNET-TM7	DA uptake K_m ⇑ 2–4x	107
Above double	As above	As above	As above	hNET-TM7	DA uptake K_m ⇑ 4x; nisoxetine K_d ⇑ 16x	107
F316 (C)	Yes	No (h,r,b,m)	Tes	hNET-TM6	Tricyclics[d] K_i ⇑ 8xe	149
V356 (S)	No (b,r)	No (h,r,b,Ce,m,d)	No (d)	hNET-TM7	Tricyclics[d] K_i ⇑ 5x	149
G400 (L)	Yes	No (h,r,b,Ce,m,d)	No (h,r,m)	hNET-TM8	Tricyclics[d] K_i ⇑ 4x[f]	149
Above triple	As above	As above	As above	hNET-TM6,7,8	Desipramine K_i ⇑ 40x	149
G478 (S)	No (Ce)	Yes	Yes	hNET-TM10	NE uptake K_m ⇑ 4x; desipramine K_i unaffected	149a
M566 (F)	Yes	No (Ce,d)	No (h,r,m,d)	hNET-TM12	Cocaine and DA K_i ⇑ 2x; tricyclics unaffected	150

[a]Mutants listed probably had near-normal, or adequate, membrane expression based on measurements of surface expression or uptake or binding activity as compared with wild-type. TM, transmembrane domain; EL, extracellular loop; IL, intracellular loop; DAT, dopamine transporter; NET, norepinephrine transporter; SERT, serotonin transporter.

[b]Amino-acid substitutes used are indicated in parentheses; sometimes multiple substitutes were used.

[c]If a transporter in a given species does not contain the aligned residue, the species is indicated in parentheses. For each transporter, the following species were included in this compilation: human (h), rat (r), and bovine (b) NET; h, r, b, *Caenorhabditis elegans* (Ce), mouse (m), and *Drosophila* (d) DAT; h, r, m, and d SERT.

[d]Desipramine and nortryptiline against DA uptake.

[e]Effect enhanced by D336 → T

5). The affinity for nisoxetine was decreased substantially in S354A and the double mutant, whereas the affinity of DA as measured against [^{3}H]nisoxetine binding was decreased in S354A. Thus, S354 plays a role in nisoxetine interaction with the NET, and probably also in substrate interaction. However, the latter does not appear to be caused by a specific interaction between S354 and a hydroxyl group on the substrate, because DA-binding affinity measured at the double mutant was the same as that at wild-type *(107)*. Moreover, the binding affinity of m- and p-tyramine, mono-hydroxylated DA derivatives, was the same at S357A as wild-type, and transport assays also did not reveal specific interactions between the m- or p-hydroxyl group of tyramine with a particular serine hydroxyl group.

Roubert et al. *(149)* followed up on previous work demonstrating the importance of TMs 5–8 in the interaction between tricyclics and the NET (*22*; also *see* **Subheading 3.2.1.** and Table 4). Thus, tricyclic binding was lost in the NET chimera M that had a TM 6–8 portion of the DAT substituted for the NET sequence. Chimera M has 24 nonconserved residues as compared to hNET, and Roubert et al. *(149)* systematically replaced these residues in hNET, one by one, by their hDAT counterparts. Confirming the idea that ELs and ILs do not confer specific information for compound interactions, mutation of residues in these loops had no effect on the potency of compounds in inhibiting substrate uptake. In contrast, the TM mutants F316C, V356S, and G400L showed decreased potency of the tricyclics desipramine and nortriptyline, but normal potency of cocaine (*149*; also *see* Table 5). The effect of the F-316C mutation was enhanced by an additional D336T mutation. The 316-356-400 triple mutant displayed a greatly reduced desipramine potency. It was speculated that F316 could be involved in stacking or hydrophobic interactions with the rings forming the core of tricyclics, that the hydroxyl group of S356 hampers the positioning of tricyclics in their binding pocket, and that G400L, when combined with S399P, produces a conformationally altered state *(149)*. Furthermore, the lack of effect of the mutations on cocaine potency suggests that the previously reported loss of cocaine activity in chimeras with DAT sequences in the TMs 5–8 region (*22*; also *see* **Subheading 3.2.1.** and Table 4) was caused by interdomain interactions rather than the loss of specific cocaine interaction points.

Because a single phenylalanine in TM12, F586, was found to be responsible for the increased sensitivity of hSERT to tricyclics such as imipramine compared with rSERT (Table 7), the corresponding methionine residue in hNET, M566, was mutated to phenylalanine (*150*; also *see* Table 5). Although no increased potency for tricyclics at M566F NET was found, the mutant displayed a twofold lower potency for cocaine and DA (Table 5).

Runkel et al. *(149a)* examined the functional implication of five exon variants of hNET identified in an earlier study by the same group. One of these variants, G478S, displayed a fourfold elevated K_m for NE uptake, but no change in its affinity for desipramine (Table 5). G478 lies in TM10, part of the region identified by both Buck and Amara *(123,136)* and Giros et al. *(22;* also *see* Table 4) that impacts on substrate uptake. Because the G478S substitution is expected to substantially interfere with the ability of NET to lower extracellular NE, it will be of interest to examine its potential role in diseases with disturbed catecholamine levels, such as heart failure and a number of mental diseases (also *see* Chapter 5).

4. SEROTONIN TRANSPORTER

4.1. Molecular and Pharmacological Characteristics

rSERT was cloned by two groups. Hoffman et al. *(54)* isolated rSERT cDNA from rat basophilic leukemia cells by screening with a degenerate oligonucleotide directed at a region conserved in NE and GABA transporters, whereas Blakely et al. *(55)* used a partial clone encoding for the rSERT to design an antisense oligonucleotide for screening a rat brain stem cDNA library. Subsequently, hSERT was isolated by Ramamoorthy et al. *(48)* by screening a human placental cDNA library with an oligonucleotide probe obtained from PCR fragments resulting from PCR of human placental trophoblastic cell cDNA with degenerate oligonucleotides used for identifying rSERT. Corey et al. *(57)* also used a PCR-based cloning strategy for isolating d-SERT, and m-SERT was cloned by screening a mouse-brain cDNA library with the entire rSERT cDNA as a hybridization probe *(51)*. The rSERT sequence is closely related to that of hNET, with 71% similarity if conservative substitutions are accepted, and 50% identity (Figs. 1 and 2). Compared with rSERT, hSERT is 92% identical, and dSERT is 52%; m-SERT is 88% identical with hSERT. The recently cloned monkey SERT is more than 98% homologous with hSERT *(59)*.

Consensus sites for N-linked glycosylation are present in EL2, with two such sites found in r-, h-, and m-SERT, and one site in d-SERT. Consensus sites for phosphorylation are present, with r-SERT having five sites for protein kinase A (PKA) and PKC, whereas h-SERT carries eight of such sites combined. The two consensus sites for N-linked glycosylation in EL2 of hSERT have been shown to be the sites of attachment of sugars to the transporter protein *(154)*, but the use of the consensus sites for phosphorylation by PKC and cAMP-dependent protein kinase is still under debate. PKC-mediated effects on 5-HT transport do not appear to depend on the presence of consensus sites for PKC *(155,156;* also *see* Chapter 1 for more details).

The SERT is the target for many antidepressant drugs, with tertiary tricyclics such as imipramine and amytriptyline selectively inhibiting the mammalian SERT over the NET and DAT (*21,48,49,93*; also *see* Table 2) (for more information on drug development targeting SERT, *see* ref. *60*). The pharmacological profile of dSERT is a mixture of those observed for mammalian SERT, NET, and DAT (*48,57*). Functional transport by the SERT requires the presence of Na^+ and Cl^-. In this process, one Na^+ and one Cl^- is cotransported with 5-HT and one K^+ is concomittantly countertransported (*see* Chapter 2).

The EL topology as originally proposed (Fig. 3) was confirmed for the rSERT with membrane-impermeant biotin reagents targeting cysteine and lysine residues (*66*). The two cysteine residues in EL2, C200, and C209 in rSERT probably form a disulfide bridge with SERT mutants that showed only the C200 or C209 residue being more sensitive to MTS than mutants with the two cysteines present together (*78*; also *see* Table 7). It has been shown that the external loops contribute to the conformational flexibility of the SERT required during transport, although the external loops do not themselves determine the binding selectivity for substrates and inhibitors (*111,157*; also *see* **Subheading 4.2.1.**).

As described for the NET (*see* **Subheading 3.1.**) residues associated with a gate would be expected to be conformationally sensitive. Isoleucine at position 179 in rSERT, corresponding to I155 in rNET, could be part of an external gate, as suggested by the 5-HT-mediated protection of the sensitivity of I179C rSERT to MTSET (*113*).

Mounting evidence supports the existence of SERT oligomeric complexes. Thus, dimeric and tetrameric SERT complexes have been reported with the use of sulfhydryl oxidizing and crosslinking agents (*158,* and Chang et al., *159*) suggested the existence of SERT tetramers based on deletion mutants and linear concatenates. Recently, Kilic and Rudnick (*160*) provided evidence for dimers and possibly tetramers in immunoprecipitation experiments with two differently tagged SERTs. As visualized by fluorescence resonance energy transfer microscopy, hSERT prefers to exist as an oligomer in the membrane of living cells (*161*).

4.2. Substrate/Inhibitor Structure and Recognition

4.2.1. Chimera Studies

SERT-DAT chimeras have not been reported. SERT-NET chimeras (*145*) are listed in Table 4, and their properties are consonant with a lack of specific information to substrate or blocker selectivity coming from either the amino or carboxyl tail (*see* **Subheading 3.2.1.**). Rudnick et al. (*111*) substituted NET ELs for SERT ELs in rSERT (Table 6). The *selectivity* for substrate and inhibitor binding was not affected in the chimeras, but transport

Table 6
SERT Chimeras

Chimera	"Switch points" determining source of domains for chimera construction	Major findings	Refs.
rSERT-rNET	All transition points between TMs and ELs, with additional points within EL2 and EL4 to correct for loop size differences between SERT and NET, and within EL3 to narrow down effect	The *selectivity* for substrate and inhibitor binding was not affected in the chimeras, but transport and binding *activity* was reduced to varying degrees. Although SERT ELs do not provide the primary information for substrate and inhibitor recognition	*111*
hSERT-rSERT	C-terminal side of TM4 N-terminal side of TM7 within IL5	C-terminal region starting at TM11 determines species difference for interaction with imipramine and d-amphetamine, which can be further narrowed down to TM12	*146,162*
hSERT-rSERT	Between TM12 and carboxyl tail	rSERT with human carboxyl tail shows rat profile for imipramine and d-amphetamine, indicating lack of influence of carboxyl tail for these interactions	*146,150*
hSERT-dSERT	N-terminal side of TM2 C-terminal side of TM2 within IL4	TM1 and TM2 determine species difference for mazindol and citalopram, whereas domain distal to TM2 determines potency of other inhibitors such as RTI-55, fluoxetine, paroxetine, and imipramine	*95*
hSERT-rSERT	C-terminal side of TM2 N-terminal side of TM6 Middle of TM9	C-terminal 155 residues determine effect of low pH in potentiating transport-mediated current	*163*

For rSERT-hNET chimeras, *see* Table 4. TM, transmembrane domain; EL, extracellular loop; IL, intracellular loop; DAT, dopamine transporter; NET, norepinephrine transporter; SERT, serotonin transporter. For species notation, *see* Table 5.

and binding *activity* was reduced to varying degrees in the mutants, despite normal expression at the cell surface. SERT ELs appear to be essential for the functional transport activity of SERT, although the ELs do not provide the primary information for substrate and inhibitor recognition.

Dual-species SERT chimeras have been explored extensively by the group of Blakely *(95,146,150,162)*. hSERT-rSERT chimeras were designed with N- or C-terminal hSERT, with the complementary portion consisting of rSERT, using the C-terminal side of TM4 and the N-terminal side of TM7 as "switch points" (*146,163*; also *see* Table 6). In addition, an alternating hSERT/rSERT/hSERT chimera was constructed with the N-terminal side of TM7 and IL5 as switch points. The results indicated that the C-terminal region begining at TM11 determined the species difference for interaction with imipramine and d-amphetamine. Because all of the sequence differences between h- and r-SERT TM 11 and 12 reside in TM 12, and since the work with SERT-NET chimeras had already ruled out the carboxyl tail as a determining factor (*see* Table 4 and **Subheading 3.2.1.**), the differences were believed to reside somewhere within TM 12. Indeed, rSERT with a human carboxyl tail showed the rat profile for imipramine and d-amphetamine, indicating the lack of influence of the carboxyl tail for these interactions (*146,150*; also *see* Table 6). Additional hSERT-dSERT chimeras were used by Barker et al. *(95)* to track the different potencies of mazindol, citalopram, and other inhibitors for human and *Drosophila* SERTs. Accordingly, dSERT/hSERT and hSERT/dSERT chimeras were constructed with the switch points listed in Table 6. TMs 1 and 2 were concluded to determine the species difference for mazindol and citalopram, whereas the domain distal to TM 2 determines the potency of other inhibitors such as RTI-55, fluoxetine, paroxetine, and imipramine. Differences between human and *Drosophila* TM 1 and 2 sequences were followed up by site-directed mutagenesis studies (*95*; also *see* **Subheading 4.2.2.**). Results from h- and r-SERT chimeras, used by Cao et al. *(163)* in electrophysiological experiments (Table 6), were also the starting point for site-directed approaches (*see* **Subheading 4.2.2.**).

4.2.2. Site-Directed Mutagenesis Studies

Because the results of chimeras of h- and d-SERT indicated that the N-terminal domain encompassing TMs 1 and 2 determined the selectivity toward the 5-HT uptake blockers mazindol and citalopram (*95*; also *see* **Subheading 4.2.1.** and Table 6), the eight amino acid residues that were different in that region in man and *Drosophila* were examined in site-directed mutagenesis studies *(95)*. Tyrosine in position 95 in hSERT was found to contribute both a positive and a negative influence on the inhibitory potency

Table 7
SERT Site-Directed Mutagenesis Studies[a]

Mutated residues[b]	Conservation in[c]			Location	Changes	Refs.
	SERT	DAT	NET			
Y95 (F)	No (d)	No (h,r,b,Ce,m,d)	No (h,r,b)	hSERT-TM1	Mazindol $K_i \Downarrow$ 9x; citalopram $K_i \Uparrow$ 8x	95
F90 (Y)	No (h,r,m)	Yes	Yes	dSERT-TM1	5-HT uptake $\Downarrow\Downarrow\Downarrow$	100
D98 (E,A,G,N,T)	Yes	Yes	Yes	rSERT-TM1	5-HT uptake $\Downarrow\Downarrow\Downarrow$	100
D98 (E)	Yes	Yes	Yes	rSERT-TM1	Imipramine and cocaine $K_i \Uparrow$ 4–5 x; citalopram $K_i \Uparrow$ 123 x; "shorter" substrate $K_i \Downarrow$ 13x; altered ion permeation with 5-HT	100
N177 (G)	Yes	Yes	Yes	rSERT-TM3	Channel conductance $\Uparrow$	164
I172 (C) – C109A	No (d)	No (h,r,Ce,m,d)	No (h,r,b)	rSERT-TM3	Inactivation of transport by MTSET, protected with 5-HT and cocaine	139
Y176 (C) – C109A	Yes	Yes	Yes	rSERT-TM3	Inactivation of binding by MTSET; 5-HT, cocaine $K_i \Uparrow$ 5–8x	139
I179 (C) – C109A	Yes	Yes	Yes	rSERT-TM3	Inactivation of transport by MTSET, <u>not</u> protected with 5-HT and cocaine	139
C209 (S)	Yes	Yes	Yes	rSERT-EL2	5-HT uptake $V_{max} \Downarrow$ 7x; no positive cooperativity of 5-HT transport	79
C200S-C209S	Yes	Yes	Yes	rSERT-EL2	5-HT uptake $\Downarrow$ 7x	78
C109A-C200S-C209S	Yes	Yes	Yes	rSERT-EL2 (C109 in EL1)	5-HT uptake $\Downarrow$ 7x	78
N368(C,I,L,K)	Yes	Yes	Yes	rSERT-TM7	5-HT uptake $\Downarrow$ 3.5 or more and/or altered Na$^+$-sensitivity of 5-HT uptake	165
S372 (A,C,R)	Yes	Yes	Yes	rSERT-TM7	ibid	165

F380 (C,V,L,G)	Yes	Yes	Yes	rSERT-TM7	ibid	165
G384 (A,C,S)	Yes	Yes	Yes	rSERT-TM7	ibid	165
Y385 (T,D)	Yes	Yes	Yes	rSERT-TM7	ibid	165
M386 (T)	Yes	Yes	Yes	rSERT-TM7	ibid	165
T490 (K)[d]	No (h,d)	No (Ce,d)	Yes	rSERT-EL5	Reduction of effect of low pH potentiating transport-mediated current	163
K490 (T)	No (r,d,m)	No (h,r,b,Ce,m,d)	No (h,r,b)	hSERT-EL5	rSERT-like low pH effect	163
E493 (Q)[d]	No (d)	No (h,r,b,Ce,m,d)	No (h,r,b)	rSERT-EL5	Reduction in low pH effect	163
E493 (Q) –C109A[d]	No (d)	No (h,r,b,Ce,m,d)	No (h,r,b)	rSERT-EL5	Reduction in low pH effect	163
E493 (D)[d]	No (d)	No (h,r,b,Ce,m,d)	No (h,r,b)	rSERT-EL5	Reduction in low pH effect and transport-mediated current	163
S545 (A)d	Yes	No (Ce,d)	Yes	rSERT-TM11	5-HT uptake V_{max} and K_m ⇑ 2–3x; 5-HT uptake with Li^+ instead of Na^+; imipramine K_i ⇑ 5x	166
V586 (F)[d]	No (h)	No (h,r,b,Ce,m,d)	No (h,r,b)[e]	rSERT-TM12	Imipramine, desipramine, and nortryptiline K_i ⇓ 2–4 x	150
F586 (V)	No (r,d,m)	No (h,r,b,Ce,m,d)	No (h,r,b)[e]	hSERT-TM12	Imipramine, desipramine, and nortryptiline K_i ⇑ ~4x	150
V586 (D)[d]	No (h)	No (h,r,b,Ce,m,d)	no (h,r,b)[e]	rSERT-TM12	Imipramine, desipramine, nortryptiline, cocaine K_i ⇓ 3–11x	150

[a] Mutants listed probably had near normal, or adequate, membrane expression based on measurements of surface expression or uptake or binding activity as compared with wild-type. TM, transmembrane domain; EL, extracellular loop; IL, intracellular loop; DAT, dopamine transporter; NET, norepinephrine transporter; SERT, serotonin transporter; MTSET, methane thiosulfonate ethyltrimethylammonium.

[b] Amino-acid substitutes used are indicated in parentheses; sometimes multiple substitutes were used.

[c] If a transporter in a given species does not contain the aligned residue, the species is indicated in parentheses. For each transporter, the following species were included in this compilation: human (h), rat (r), and bovine (b) NET; h, r, b, Caenorhabditis elegans (Ce), mouse (m), and Drosophila (d) DAT; h, r, m, and d SERT.

[d] hSERT numbering system is used throughout; according to Corey et al. (57) K490 in hSERT is T489 in rSERT, S545 in hSERT is S543 in rSERT, and F586 in hSERT is V584 in rSERT.

[e] Alignment according to Giros and Caron (166a).

of the two blockers. Thus, substitution of the corresponding *Drosophila* residue phenylalanine in Y95F hSERT showed a ninefold higher potency for mazindol than WT and an eightfold lower potency for citalopram (Table 7). The converse mutation in *Drosophila*, F90Y d-SERT, displayed virtually no 5-HT uptake activity, but the surface expression of this mutant could not be determined because of the lack of dSERT antibodies. The analogous mutation was introduced in hNET, F72Y, and indeed caused a loss of mazindol potency and gain in citalopram potency (*see* **Subheading 3.2.2.** and Table 5).

D98 in TM1 of rSERT, corresponding to D79 in rDAT (Table 3) and D75 in hNET (Table 5) was studied by Barker et al. (*100*; also *see* Table 7). Replacement by glutamate, alanine, glycine, asparagine, or threonine greatly reduced functional 5-HT uptake. The ~30–50% retention of 5-HT uptake activity in D98E rSERT allowed further characterization, showing a four- to fivefold decreased potency for imipramine and cocaine, and a 123-fold decrease for citalopram (*100*; also *see* Table 7). Lengthening the carboxylic acid in the 98 position by one carbon in D98E offset the loss of the potency in inhibiting $[^3H]$5-HT uptake, compared with dimethyltryptamine, of gramine, which has an alkylamine side chain that is one carbon shorter than that of dimethyltryptamine (*100*). These results support an interaction between the negatively charged carboxylic-acid side chain of aspartate and the positively charged amine group of substrate. Similar experiments on D79 in rDAT, involving D79E and dihydroxybenzylamine, which is one carbon shorter than DA, did not show the expected increase in potency upon combining the shorter substrate-carbon chain length with the longer carboxylic acid (*101*; also *see* **Subheading 2.3.**), casting doubt upon the generality of the proposed ion pairing between the TM 1 aspartate and substrate. Other factors may play a role in the reduced transport of 5-HT in D98E rSERT. Indeed, the K_m for Na^+ and Cl^- to stimulate 5-HT uptake was greatly elevated in D98E, suggesting that Asp^{98} confers Na^+ and Cl^- coupling determinants to TM 1. In addition, a component of substrate-activated current in transporter-expressing oocytes was eliminated in D98E, in accord with the possibility that TM 1 determines ion selectivity in SERT channels and participates in coupling 5-HT permeation to ion gradients (*100*).

As mentioned previously in **Subheading 3.1.**, transporters display channel characteristics which can be characterized in voltage-clamped oocytes expressing the transporter protein (also *see* Chapters 2 and 8). Single-channel currents were measured in wild-type rSERT outside-out oocyte patches, and their pharmacological profile was similar to that of 5-HT uptake (164). Replacement of TM 3 Asn^{177} by glycine caused a larger channel conductance (Table 7), suggesting that N177 lies within or close to the permeation pathway unless a propagated structural change has occurred in N177G (*164*).

The group of Rudnick *(139)* examined TM 3 of rSERT with the cysteine-scanning method, replacing amino acids one by one by cysteine to assess their accessibility to MTS reagents. C109A, insensitive to MTS, was used as the background for the cysteine substitutions. C172, C176, and C179 were found to be accessible to MTS, with 5-HT and cocaine affording protection of C172 *(139;* also *see* Table 7). Y176C had a five- to eightfold reduced affinity for 5-HT and cocaine. The results suggested that I172, Y176, and I179 are on one face of an α-helical TM element, and that Ile[172] and Tyr[176] are in proximity to the binding site for 5-HT and cocaine *(139)*.

Endogenously present cysteines in rSERT were mutated by Sur et al. *(79)*. C209S (Table 7) was remarkable in that it showed an ~85% reduction in V_{max} of 5-HT uptake and a lack of positive cooperativity observed in wild-type *(79)*. The mutant was not sensitive to DTT, in contrast to wild-type, in which DTT decreased uptake and abolished the positive cooperativity phenomenon. It was concluded that C209 in EL2 is part of a disulfide bridge, and that this disulfide bond plays a role in the positive cooperativity *(79)*. The former conclusion agrees with results from the Rudnick group *(78)*, indicating a C200-C209 bond (also *see* **Subheading 4.1.**). The latter, concerning cooperativity, remains speculative at this point, as 5-HT cooperativity has not been reported for wild-type SERT by groups other than that of Sur et al. The importance of an intact C200-C209 disulfide bond for functional 5-HT transport is also suggested by the five- to sevenfold reduction in 5-HT uptake displayed by C200S-C209S and C109A-C200S-C209S despite their near-normal surface expression *(78;* also *see* Table 7).

Penado et al. *(165)* subjected TM7 of rSERT to random mutagenesis. Numerous mutants were obtained; and more were added by site-directed mutagenesis to obtain enough nonconservative mutations for each amino acid. Six residues were found to be highly sensitive to mutation, and functionally nonconservative substitutions were shown to be quite effective (reduction in uptake of more than 3.5×). Those are the mutations shown in Table 6, plus another mutation, S372A (C,R), which caused a ~twofold reduction in uptake along with altered Na⁺-sensitivity. The latter was also displayed by three mutants from the total of seven shown in Table 8: N368C, G376S, and F380C. The first five residues shown—positions 368, 372, 376, 380, and 384—all fall on a stripe that runs at an angle down one side of the predicted α-helix of TM 7. This pattern breaks down at the C-terminal end, with positions 384, 385, and 386 all being sensitive to mutation, most likely representing the beginning of the hydrophilic EL4.

Current transport models incorporate an external gate that is open when the substrate enters the pore of the permeation pathway, and an internal gate that opens when the substrate is released into the cytoplasm (also *see* Chapter 2). Normally, only one of the two gates is open at one time, and ions pass through with 5-HT$^+$ in stoichiometric proportions: one Na$^+$ and Cl$^-$ are cotransported, with countertransport of one K$^+$—predicting, stoichiometrically, no movement of net charge. However, in one form of current measured in transporter-expressing oocytes, the external gate is believed to be transiently open when substrate is passing through the open internal gate, allowing transport-mediated current to occur with the two gates open at the same time *(163)*. This transport-mediated current in rSERT, but not hSERT, is increased at low pH, suggesting the involvement of acidic side chains on amino-acid residues in the transporter protein. Cao et al. *(163)* constructed hSERT-rSERT chimeras with the switching points as indicated in Table 6, and the results indicated that the pH effect was localized to the portion of the transporter with the C-terminal 155 residues. From the residues that were different between h- and r-SERT in this region, polar residues in extracellular and TM regions were mutated, one by one. T490K rSERT displayed a seriously reduced effect of low pH in potentiating transport-mediated current, implicating T490 in EL5 in the pH effect *(163;* also *see* Table 7). Conversely, K490T hSERT showed the rSERT-like pH effect. E493Q rSERT had a reduced pH potentiation of current, as opposed to the lack of effect on pH potentiation in E494Q, suggesting that Glu^{493} could become protonated, just downstream from Thre^{490}. The C109A-E493Q mutant behaved just like E493Q in the electrophysiological experiments, but could be used to demonstrate with MTS that E493 is facing extracellularly (Table 7). Aspartate could not take the role of glutamate in the transport-mediated current and in the low pH effect, as E493D showed less transport-mediated current and a lack of effect of low pH *(163;* also *see* Table 7). The latter may be responsible for the lack of low pH potentiation reported for several other neurotransmitter transporters which have aspartate instead of glutamate in the position corresponding to Glu^{493} in rSERT.

Sur et al. *(166)* examined three serine residues in TM11 of rSERT (Table 7). S545 appeared to be important for the cation dependence of 5-HT transport. Thus, substrate transport was driven as efficiently by LiCl as by NaCl without a significant change in relative substrate affinity—i.e., the K_m with LiCl was about 2–3× higher than with NaCl in both wild-type and S545A, but the V_{max} was reduced by 76% in wild-type upon Li$^+$ substitution, whereas it was unchanged in S545A. S545 was proposed to be part of the external gate, constitutively open in the S545A mutant and allowing Li$^+$ to replace Na$^+$ as the cotransported ion *(166)*. S545A also bound imipramine, but not citalopram, with reduced potency (Table 7), indicating a role for S545 in imipramine binding to the SERT.

As discussed previously in **Subheading 4.2.1.**, the h- and r-SERT chimeras studied by Barker and colleagues (*146,150,162*; also *see* Table 6) implicated TM 12 in the species differences observed for the interaction of imipramine and amphetamine with SERT. Substitution of the human counterparts into rSERT positions that differed between man and rat resulted in one mutant, V586F that showed increased affinity for the tricyclics imipramine, desipramine, and nortryptiline, without a change for cocaine (Table 7). Conversely, substitution of the rat residue into F586V hSERT (numbering according to rat position) decreased the affinity for tricyclics, again without altering cocaine affinity. Curiously, substitution of an aspartate instead of phenylalanine at position 586 into rSERT in V586D caused a similar increase in tricyclic potency (Table 7). Therefore, if the phenyl rings of tricyclics interact with Phe via π-π stacking interactions, and if the alkylamine group of tricyclics interact with the negatively charged Asp, these interactions are probably of a secondary, modulatory nature, with critical binding interactions occurring elsewhere. This is believed to be the case because ionic bonds should provide larger increases in binding potency than observed here (*150*). It is of interest that V586D also displayed increased potency for cocaine, implicating TM 12 in cocaine interaction with SERT. In the studies of Vaughan et al. (*69,124,125*) the N_3 group of the cocaine-like photoaffinity ligand [^{125}I]RTI-82 was found to be directed toward TM 4–7 of the DAT, but other portions of the molecule might still interact with other regions of the DAT, and the binding pocket for cocaine may not be identical between DAT and SERT.

5. COMMONALITIES AND DIFFERENCES AMONG MONOAMINE TRANSPORTERS IN SUBSTRATE/INHIBITOR INTERACTIONS

5.1. Substrates

Clearly, it is an early point in time to compare the three biogenic amine transporters in terms of detailed structural determinants of substrate recognition and translocation. As Table 8 shows, a small number of residues that are conserved between the three transporters have been studied by site-directed mutagenesis in more than one biogenic amine transporter. Thus, Phe[76] in rDAT, located one helical turn below Asp[79], as well as its counterpart Phe[72] in hNET and Phe[90] in dSERT, appear to play a role in substrate uptake, as mutation in DAT, NET, and SERT reduced substrate uptake (Table 8). Asp[79] in rDAT, and the corresponding Asp[75] in hNET and Asp[98] in rSERT, also play a role in substrate uptake, with mutation of Asp leading to severely diminished substrate uptake in all three monoamine transporters (Table 8).

Table 8
Mutation of Residues Presented in Tables 3, 5, and 7, and More, that are Common Among Monoamine Transporters[a]

Mutated residues in SERT/NET	Studied mutations of corresponding residues	Effects of mutation	Refs.
F90 dSERT-TM1	F90Y in dSERT F76A in rDAT F72Y in hNET	At SERT, 5-HT uptake was reduced; at DAT, DA turnover was reduced, substrate potency for binding increased, and cocaine affinity reduced; at NET, substrate uptake was reduced, and mazindol and citalopram potency affected in opposite direction	95,105
D98 rSERT-TM1	D98E (A, G, N) in rSERt D79E (A, G, E) in rDAT D75E (A, G, N) in hNET	In all cases substrate uptake was greatly reduced; at DAT (D79A), DA potency, DAT turnover, and cocain potency was reduced; at D79E there was only a small effect of dihydroxybenzylamine potency; at SERT (D98E), DA and cocaine potency was reduced but 5-HT potency unaltered; dihydroxybenzylamine potency was increased	95,97,100, 101,146
C109 rSERT-EL1	C109A in rSERT C90A in hDAT	In SERT, sensitivity of uptake to MTS treatment of cells in Li⁺-media was lost. In both SERT and DAT, sensitivity of binding to MTS treatment of membranes was enhanced, but the effect was much greater in DAT. Li⁺ did not alter sensitivity to MTS of CFT binding to wild-type hDAT (Reith, unpublished observations) as opposed to the enhanced sensitivity of 5-HT uptake by wild-type rSERT. Cocaine accelerated reaction of C90 in hDAT with MTS, as measured by CFT binding, but not C109 in rSERT monitored by 5-HT uptake. Li⁺ did not enhance reaction of C90 in hDAT with MTS judged from CFT-binding assays	70,78,132, 133,139, 167
I179 rSERT-TM3	I179C in rSERT I155C in rNET	In SERT, position 179 appears conformationally sensitive judged from reactivation by free Cys of C179 inactivated by MTS, with Na⁺ and 5-HT cooperatively enhancing reactivation; in NET, inactivation of Cys as the corresponding residue by MTS is protected by substrate but accelerated by cocaine	113
C200	C200S in rSERT	Both in SERT and DAT, transport activity was reduced by diminished	77,78

			References
rSERT-EL2	C180A in rDAT	surface expression of transporter. In SERT, MTS sensitivity was markedly increased, which was counteracted by introducing a second mutation in C200S-C209S but not in C200S-C109A	77,78,139
C209 rSERT-EL2	C209A in rSERT C189A in rDAT	Both in SERT and DAT, transport activity was reduced by diminished surface expression of transporter. In SERT, adding C209S but not C109S to C200S attenuated MTS sensitivity observed for C200S	107,165
S354 hNET-TM7	S354A in hNET S372A(C,R) in rSERT	In NET, DA K_m, DA K_i, and nisoxetine K_i were increased; at SERT, 5-HT uptake was reduced with decreased Na^+-sensitivity	107,165
S357 hNET-TM7	S357A in hNET S375A(T) in rSERT	In NET, DA K_m was increased; at SERT, 5-HT uptake appeared unaffected	107,165
S354-S357 hNET-TM7	S354-S357A in hNET S356-S359A(G) in rDAT S372G-S375Y in rSERT	In NET and DAT, substrate K_m was increased; in SERT, 5-HT uptake was lost; V_{max} was reduced in DAT but not NET. In NET, substrate K_i and nisoxetine K_d were increased, and DAT CFT K_d was increased	97,107, 165
F380 rSERT-TM7	F380C in rSERT F364A in rDAT	In SERT, 5-HT uptake was reduced with lower Na^+-sensitivity; in DAT, DA K_m was increased and V_{max} decreased, and CFT K_d was increased	105,165
S545 rSERT-TM11	S545A in rSERT S528A in hDAT	Cation dependence of substrate uptake was altered in SERT but not DAT	92,166

[a]A number of commonly studied mutations are not shown in Table 3 for the DAT because expression levels were uncertain or because no major changes in function/recognition occurred. Removal of glycosylation sites in EL2 of DAT, SERT, and NET is not included here and is described in Chapter 10.

Of course, substrate uptake consists of several steps, including substrate recognition, ion-coupled translocation, release, and transporter reorientation, and therefore mutations could have effects at different steps in the process. Substrate recognition is generally approximated by measuring the inhibition of radiolabeled inhibitor binding by substrate, with the underlying assumption that there is an overlapping binding domain at the transporter for inhibitors and substrates. However, allosteric interactions underlying mutual inhibition at remote sites have not been rigorously excluded. Substrate affinity is one aspect that determines the K_m for substrate uptake, and from a large collection of DAT mutants, the general conclusion can be made that K_m values for DA uptake do not always correlate well with the K_i values for DA inhibition of radiolabeled CFT binding (see Table 3, D68N, F76A, F155A, W266A, W311L, and W555A). This would suggest that K_m values should be used as measures for recognition with caution, at least in the case of the DAT. The second step in the uptake cycle, ion-coupled translocation, can be addressed in terms of ion-dependence of uptake as measured with varying [Na$^+$] and [Cl$^-$]. The third step, transporter re-orientation, can be approximated with knowledge about the K_m, K_i, and turnover rate as described by Chen et al. *(102)*. Finally, substrate release at the internal side, where [K$^+$] is high and [Na$^+$] low, is generally believed to be extremely rapid, playing no role in the overall process.

Examples for changes at the first three steps by mutations are numerous. Many changes in substrate recognition, as reflected in K_m or K_i values, can be easily seen in Table 3 for the DAT, Table 5 for the SERT, and Table 7 for the NET. Changes in ion coupling have been reported for the rSERT *(see* Table 7) in D98E *(100)*, in S545A *(1626)*, and in N368C, G376S, and F380C *(165,167)*; results with DAT-NET chimeras point to the importance of the N-terminal portion, including TMs 1–3, and the C-terminal region starting with TM9, in the Na$^+$- and Cl$^-$-dependence of substrate uptake *(147)*. Alterations in reorientation rate constants have been reported for hDAT in W267L, D313N, D436N, W520L, and W556L *(102)*. Which of these steps is involved in the uptake changes observed upon mutating Phe[76] and Asp[79] in rDAT, and their counterparts in hNET and rSERT? The reported experiments *(see* Table 8) focused on recognition events: F76A rDAT showed enhanced DA affinity, contrasted by a somewhat increased DA K_m, whereas D79A rDAT displayed reduced DA potency, estimated by both K_i and K_m. As discussed previously, the support for direct contact points between Asp and substrate found in SERT was lacking for DAT in experiments based on the same approach of combining a longer amino-acid residue (Glu for Asp) with a shorter substrate (gramine for dimethyltryptamine and dihydroxybenzylamine for DA) *(see* **Subhead-**

ing 4.2.2.). The results taken together do not unequivocally point to a direct role for Phe (position 76 in rDAT), conserved among DAT and NET, and Asp (position 79 in rDAT), conserved among DAT, NET, and SERT, in substrate recognition.

Other residues listed in Table 8, potentially involved in substrate recognition/translocation/ion coupling, are: two serine residues in TM7, first examined by the group of Uhl *(97)* in positions 356 and 359 in rDAT, a phenylalanine (Phe^{364} in rDAT) in the same TM lying one helix turn above Ser^{359} (rDAT), which is the second serine of the pair mentioned previously, and a serine in TM11, studied as S545 in rSERT and S528 in hDAT. In the first two cases—the serines and the phenylalanine in TM 7—substrate uptake was reduced by replacement of the residues; where measured, K_m values were increased as well as K_i values for inhibition of radiolabeled inhibitor binding by substrate. It is therefore possible that defects in substrate recognition represent an underlying mechanism in all of these cases. However, decreased sensitivity of 5-HT uptake to Na$^+$ stimulation was also observed in some cases (S372A and F380C in rSERT), whereas double mutation of the same pair of TM 7 serine residues decreased V_{max} in DAT but not NET. Thus, differences between biogenic amine transporters, as well as similarities, exist for the function of a particular residue in the entire protein structure. This also applies to the serine residue in TM 11 listed in Table 8, which appears to play a role in cation dependence of substrate uptake in SERT but not DAT.

The remaining residues shown in Table 8 have been linked to structural features of biogenic amine transporters or to conformational changes occurring upon substrate interaction. Again, both differences and similarities between DAT, NET, and SERT are evident. The cysteine residue in EL1, C109 in rSERT and C90 in hDAT, played a different role in the effects of MTS treatment of SERT and DAT: substitution of Li$^+$ for Na$^+$ caused an appreciable increase in the rate of reaction of C90 in rSERT with MTS as measured by 5-HT uptake *(78)*, but Li$^+$ did not alter the rate of reaction of C90 in hDAT *(see* ref. *70)*. The isoleucine residue in TM 3, I179 in rSERT and I155 in rNET, appears to be conformationally sensitive, responding to ambient substrate concentration; however, for the SERT, a step subsequent to substrate binding needed to be considered in reactivation experiments (Table 8), whereas the NET data reinforced the postulate that substrate binding converted the transporter into an inward-facing state sequestering residue 155 from the external medium, making it inaccessible to MTS. The two cysteines in EL2, C200, and 209 in rSERT and C180 and 189 in rDAT, are believed to be connected by a disulfide bridge, which retains the transporter in a form amenable to proper targeting to the plasma membrane.

For inhibitors, chimera studies have pointed to the role of certain residues in their interactions with biogenic amine transporters (*see* **Subheading 5.2.**). For substrates, this has not yet been the case. An attempt to take advantage of the differences between catecholamine and 5-HT transporters in recognizing DA and NE on the one hand and 5-HT on the other hand, was only partly successful in a series of chimeras between hNET and rSERT (*145,146*; also *see* Table 4). Only the chimeras with switched amino or carboxy tails were functionally active, and no clues for the remaining portion of the protein in substrate recognition/translocation could be obtained. DAT-SERT chimeras have not been reported, and, if attempted, may also not be functional. Differences between DA and NE in interacting with catecholamine transporters have been explored in NET-DAT chimeras (Table 4). It is known that the cloned NET recognizes both NE and DA with high affinity (DA being more potent), whereas the cloned DAT is generally much more selective for DA vs NE; comparing the two substrates may therefore provide information about regions of the transporter specifically required for NE. As discussed previously, this approach led to the identification of the C-terminal portion starting with TM 9 (*22,123,136*) and TMs 1–3 (*123,136*) for providing information about substrate recognition/selectivity. Thus, when the last four TM domains of the NET are introduced into DAT, the NET profile of NE and DA uptake is found (*22*). It would be of interest to examine the residues that are different between DAT and NET in that region, and substitute into DAT, one by one, the corresponding NET residues to observe their effect on NE and DA uptake. A similar approach could be adopted for the TM 1–3 region. In future interspecies chimera studies, an advantage can be taken from more recently cloned transporters such as the fET (bullfrog epinephrine transporter) (*168*) and dDAT (*19*).

It is a formidable challenge to place all attempted mutations into a perspective, providing us with a description of an initial binding pocket for substrate and a series of subsequent interactions between transporter and substrate as substrate moves down the pore from the external to the internal side. Clearly, it is still too early for such a comprehensive picture. Most information with site-directed mutagenesis approaches has been collected for the DAT, and many of the residues studied have been summarized by Lin et al. (*108*). Mutations of seven residues scattered through TMs 1–4 were found to increase DA uptake K_m with their side chains oriented to a common crevice when helices were modeled according to Edvardsen and Dahl (*98*). The data together would suggest a "wall" of a DA translocation corridor extending along TMs 1–4, a slightly larger area of the DAT than suggested by the chimera studies of Buck and Amara (*123,136*). A remarkable feature of this pathway is that it extends all the way from the external side of the TMs into the low interior. Future experimental evidence will be needed to corroborate the spatial features suggested by modeling of the DAT structure, combined with knowledge about the dynamic movements of protein substructures when compounds interact. As far as the latter is concerned, an energy-minimizing model suggests changes in DAT structure upon DA interaction (compare Fig. 2 top left and bottom left in ref. *98*), whereas conformational changes upon substrate and blocker binding are suggested by approaches combining site-directed mutagenesis with thermodynamic analysis *(109)* and with measurement of susceptibility to MTS reagents *(70,110,113,132,133)*. As far as the modeling is concerned, the existence of oligomeric transporter complexes may need to be taken into account. For the SERT, evidence is building for the existence of dimers and tetramers *(158–160*; also *see* **Subheading 4.1.**). For the NET, evidence for oligomers has not yet been advanced, whereas there are some indications that the DAT may be oligomeric (see ref. 169).

5.2. Inhibitors

As shown in Table 8, many of the same residues conserved among DAT, NET, and SERT that appear to play a role in substrate interactions could be involved in inhibitor interactions. Thus, changes in inhibitor potency have been observed upon mutation of the phenylalanine and aspartate in TM 1 corresponding position 76 and 79 in rDAT, the serines in TM 7 corresponding to positions 356 and 359 in rDAT, and, five residues higher up, the phenylalanine in TM 7, Phe[364] (rDAT). In some cases, similar measures have been made allowing comparisons. Cocaine potency was reduced at both D98E rSERT and D79A rDAT, and reductions were seen both in nisoxetine-binding potency in S354-S357A hNET and CFT-binding potency in S356-S359A(G) rDAT. In most other cases, parallel measures are not available to infer similar trends between transporters. As indicated in Table 8, two conformationally sensitive residues distinctly differ between the biogenic amine transporters in terms of how they are impacted by the action of inhibitors. Thus, Li[+] did not enhance the reaction of C90 in hDAT with MTS as judged from CFT-binding assays as opposed to the enhanced reaction observed for C109 in rSERT measured by 5-HT uptake *(78)*; in addition, cocaine accelerated the reaction of C90 in hDAT with MTS *(70)* but not C109 in rSERT (G. Rudnick, personal communication to Ferrer and Javitch *[70]*). Furthermore, reaction of C179 in TM 3 in I179C rSERT with MTS was not influenced by cocaine, whereas the reaction between MTS and the corresponding C155 in I155C rNET was accelerated by cocaine (Table 8).

It is likely that different inhibitors will interact in different ways with the same transporter. For example, removing the amine function of a cocaine analog by substituting O for N in the tropane ring impacted the way their binding to hDAT was affected by mutating D79 and F530 (rDAT numbering) *(140)*. This suggests that an amine and a nonamine cocaine analog bind to the DAT differently. It is also possible that the same inhibitor will interact differently with two different transporters. Thus, Phe[586] in TM 12 of hSERT (rSERT numbering) confers increased affinity for tricyclics such as imipramine, but introducing phenylalanine at the corresponding position in hNET, normally Meth[566], did not lead to increased potency of tricyclics including imipramine *(150)*. Clearly, we are just at the beginning of our understanding of how inhibitors interact with biogenic amine transporters.

Chimera studies can be a powerful tool for focusing on transporter regions involved in inhibitor interactions. Thus, h- and d-SERT chimeras led to the identification of Tyr[95] in hSERT-TM 1, just three residues upstream from Asp[98] corresponding to Asp[79] in rDAT, as being involved in the interactions with mazindol and citalopram *(95)*, and h- and r-SERT chimeras resulted in experiments revealing the role of Phe[586] in hSERT-TM 12 in the interaction with tricyclics but not cocaine (150). hNET-hDAT chimeras have led to a recent elimination of residues involved in the interaction of tricyclics—i.e., Phe[316], Val[356], and Gly[400] in hNET in TM 6, 7, and 8, respectively *(149)*. Similar approaches have not resulted in residues involved in inhibitor interactions specific for the DAT. However, a wealth of information has been obtained from numerous mutations of multiple residues, Tryp, Phe, Pro, Tyr, Thre, Ser, Asp, Glu, and Glun, and this collection of mutations has been recently summarized by Lin et al. *(108)*. It appears that cocaine analog affinities are altered by the mutation of many different residues in virtually all TMs. Notably, these residues are found at many different depths of the transporter structure when side-viewed in the plasma membrane, and there are also a few residues located more extracellularly. It is likely that a fair number of all of these residues are not providing inhibitor contact points *per se*, but are involved in ion interactions impacting on cocaine binding or in conferring conformational properties to the protein enabling cocaine-analog binding. For example, we have found that mutating certain conserved residues altered Na^+-sensitivity of CFT binding (unpublished data) and it is known that cocaine-analog binding in wild-type DAT is strongly Na^+-dependent *(64,170,171)*. Furthermore, conformational changes in DAT upon binding of inhibitors have been inferred from thermodynamic data early on *(172)*, and have been studied more recently by the combination of site-directed mutagenesis with thermodynamic analysis *(109)* and with measurement of MTS-sensitivity *(70,110,113,132,133)*.

At present, a steady flow of results are being gathered from site-directed mutagenesis studies on biogenic amine transporters. The group of Rudnick

(G. Rudnick, personal communication) is designing a website for combined monitoring and presenting published mutations, in a manner facilitating the comparison of positions in the various transporters across transporters and species. In time, this will evolve into an invaluable resource for investigators in the structure-function transporter field, and such a database will make future reviews on interactions of compounds with biogenic amine transporters much easier to write and read.

ACKNOWLEDGMENTS

We would like to thank the National Institute on Drug Abuse (DA 08379, DA 11978, and DA 13261 to M. E. A. R.).

REFERENCES

1. Giros, B., Jaber, M., Jones, S., Wightman, R. M., and Caron, M. G. (1996) Hyperlocomotion and indifference to cocaine and amphetamine in mice lacking the dopamine transporter. *Nature* **379,** 606–612.
2. Jones, S. R., Gainetdinov, R. R., Jaber, M., Giros, B., Wightman, R. M., and Caron, M. G. (1998a) Profound neuronal plasticity in response to inactivation of the dopamine transporter. *Proc. Natl. Acad. Sci. USA* **95,** 4029–4034.
3. Whitby, L. G., Hertting, G., and Axelrod, J. (1960) Effect of cocaine on the disposition of noradrenaline labeled with tritium. *Nature* **187,** 604,605.
4. Coyle, J. T. and Snyder, S. H. (1969) Antiparkinsonian drugs: inhibition of dopamine uptake in the corpus striatum as a possible mechanism of action. *Science* **166,** 899–901.
5. Iversen, L. L. (1971) Role of transmitter uptake mechanisms in synaptic neurotransmission. *Br. J. Pharmacol.* **41,** 571–591.
6. Shaskan, E. G. and Snyder, S. H. (1970) Kinetics of serotonin accumulation into slices from rat brain: relationship to catecholamine uptake. *J. Pharmacol. Exp. Ther.* **175,** 404–418.
7. Zoli, M., Jansson, A., Sykova, E., Agnati, L. F., and Fuxe, K. (1999) Volume transmission in the CNS and its relevance for neuropsychopharmacology. *Trends. Pharmacol. Sci.* **20,** 142–150.
8. Descarries, L., Seguela, P., and Watkins, K. C. (1991) Nonjunctional relationships of monoamine axon terminals in the cerebral cortex of adult rat, in *Volume Transmission in the Brain: Novel Mechanisms for Neural Transmission.* (Fuxe, K. and Agnati, L. F., eds.), Raven Press, New York, pp. 53–62.
9. Carlsson, A., Jonason, J., Lindqvist, M., and Fuxe, K. (1969) Demonstration of extraneuronal 5-hydroxytryptamine accumulation in brain following membrane-pump blockade by chlorimipramine. *Brain Res.* **12,** 456–460.
10. Garris, P. A., Ciolkowski, E. L., Pastore, P., and Wightman, R. M. (1994) Efflux of dopamine from the synaptic cleft in the nucleus accumbens of the rat brain. *J. Neurosci.* **14,** 6084–6093.
11. Ciliax, B. J., Drash, G. W., Staley, J. K., Haber, S., Mobley, C. J., Miller, G. W., Mufson, E. J., Mash, D. C., and Levey, A. I. (1999) Immunocytochemi-

cal localization of the dopamine transporter in human brain. *J. Comp. Neurol.* **409,** 38–56.

12. Jones, S. R., Gainetdinov, R. R., Wightman, R. M., and Caron, M. G. (1998b) Mechanisms of amphetamine action revealed in mice lacking the dopamine transporter. *J. Neurosci.* **18,** 1979–1986.

13. Millier, G. W., Gainetdinov, R. R., Levey, A. I., and Caron, M. G. (1999) Dopamine trasnporters and neuronal injury. *TiPS* **20,** 424–429.

14. Levi, G. and Raiteri, M. (1993) Carrier-mediated release of neurotransmitters. *Trends. Neurosci.* **16,** 415–419.

15. Baldessarini, R. J. (1996) In *Goodman and Gilman's The Pharmacological Basis of Therapeutics.* (Hardman, J. G., Limbird, L. E., Molinoff, P. B., Ruddon, R. W., and Goodman Gilman, A., eds.) McGraw-Hill, New York, pp. 431–459.

16. Tatsumi, M., Groshan, K., Blakely, R. D., and Richelson, E. (1997) Pharmacological profile of antidepressants and related compounds at human monoamine transporters. *Eur. J. Pharmacol.* **340,** 249–258.

17. Amara, S. G. and Sonders, M. S. (1998) Neurotransmitter transporters as molecular targets for addictive drugs. *Drug Alcohol Depend.* **51,** 87–96.

18. Chen, N. and Reith, M. E. A. (2000) Structure and function of the dopamine transporter. *Eur. J. Pharmacol.* **405,** 329–339.

19. Pörzgen, P., Park, S. K., Hirsh, J., Sonders, M. S., and Amara, S. G. (2001) The antidepressant-sensitive dopamine transporter in drosophila melanogaster: a primordial carrier for catecholamines. *Mol. Pharmacol.* **59,** 83–95.

20. Vandenbergh, D. J., Persico, A. M., and Uhl, G. R. (1992) A human dopamine transporter cDNA predicts reduced glycosylation, display a novel repetitive element and provides racially-dimorphic Taq I RFLPs. *Mol. Brain Res.* **15,** 161–166.

21. Giros, B., Mestikawy, S. E., Godinot, N., Zheng, K., Han, H., Yang-Feng, T., and Caron, M. G. (1992) Cloning, pharmacological characterization, and chromosome assignment of the human dopamine transporter. *Mol. Pharmacol.* **42,** 383–390.

22. Giros, B., Wang, Y.-M., Suter, S., McLeskey, S. B., Pifl, C., and Caron, M. G. (1994) Delineation of discrete domains for substrate, cocaine, and tricyclic antidepressant interactions using chimeric dopamine-norepinephrine transporters. *J. Biol. Chem.* **269,** 15,985–15,988.

23. Pristupa, Z. B., Wilson, J. M., Hoffman, B. J., Kish, S. J., and Niznik, H. B. (1994) Pharmacological heterogeneity of the cloned and native human dopamine transporter: disassociation of [^{3}H]WIN 35, 428 and [^{3}H]GBR 12, 935 binding. *Mol. Pharmacol.* **45,** 125–135.

24. Eshleman, A. J., Neve, R. L., Janowsky, A., and Neve, K. A. (1995) Characterization of a recombinant human dopamine transporter in multiple cell lines. *J. Pharmacol. Exp. Ther.* **274,** 276–283.

25. Shimada, S., Kitayama, S., Lin, C-L., Patel, A., Nanthakumar, E., Gergor, P., Kuhar, M., and Uhl, G. (1991) Cloning and expression of a cocaine-sensitive dopamine transporter complementary DNA. *Science* **25,** 576–578.

26. Kilty, J., Lorang, D., and Amara, S. G. (1991) Cloning and expression of a cocaine-sensitive dopamine transporter. *Science* **254,** 578,579.

27. Giros, B., Mestikawy, S. E., Bertrand, L., and Caron, M. G. (1991) Cloning and functional characterization of a cocaine-sensitive dopamine transporter. *FEBS Lett.* **295,** 149–154.

28. Kitayama, S., Wang, J.-B., and Uhl, G. R. (1993) Dopamine transporter mutants selectively enhance MPP$^+$ transport. *Synapse* **15,** 58–62.

29. Kitayama, S., Mitsuhata, C., Davis, S., Wang, J.-B., Sato, T., Morita, K., Uhl, G. R., and Dohi, T. (1998) MPP$^+$ toxicity and plasma membrane dopamine transporter: study using cell lines expressing the wild-type and mutant rat dopamine transporters. *Biochim. Biophys. Acta.* **1404,** 305–313.

30. Usdin, T. B., Mezey, E., Chen, C., Brownstein, M. J., and Hoffman, B. J. (1991) Cloning of the cocaine-sensitive bovine dopamine transporter. *Proc. Natl. Acad. Sci. USA* **88,** 11,168–11,171.

31. Lee, S.-H., Rhee, J., Koh, J.-K., and Lee, Y.-S. (1996b) Species differences in functions of dopamine transporter: paucity of MPP$^+$ uptake and cocaine binding in bovine dopamine transporter. *Neurosci. Lett.* **214,** 199–201.

32. Chang, M.-Y., Lee, S-H., Kim, J.-H., Lee, K.-H., Kim, Y.-S., Son, H., and Lee Y.-S. (2001) Protein kinase C-mediated functional regulation of dopamine transporter is not achieved by direct phosphorylation on the dopamine transporter protein. *J. Neurochem.* **77,** 754–761.

33. Brüss, M., Wieland, A., and Bönisch, H. (1999) Molecular cloning and functional expression of the mouse dopamine transporter. *J. Neural. Transm.* **106,** 657–662.

34. Wu, X. and Gu, H. H. (1999) Molecular cloning of the mouse dopamine transporter and pharmacological comparison with the human homologue. *Gene* **233,** 163–170.

35. Jayanthi, L. D., Apparsundaram, S., Malone, M. D., Ward, E., Miller, D. M., Eppler, M., et al. (1998) The Caenorhabditis elegans gene T23G5. 5 encodes an antidepressant- and cocaine-sensitive dopamine transporter. *Mol. Pharmacol.* **54,** 601–609.

36. Sugamori, K. S., Lee, F. J. S., Pristupa, Z. B., and Niznik, H. B. (1999) A cognate dopamine transporter-like activity endogenously expressed in a COS-7 kidney-derived cell line. *FEBS Lett.* **451,** 169–174.

37. Liu, Q. R., Lopez-Corcuera, B., Mandiyan, S., Nelson, H., and Nelson, N. (1993) Cloning and expression of a spinal cord- and brain-specific glycine transporter with novel structural features. *J. Biol. Chem.* **268,** 22,802–22,808.

38. Shafqat, S., Velaz-Faircloth, M., Henzi, V. A., Whitney, K. D., Yang-Feng, T. L., Seldin, M. F., and Fremeau, R. T. (1995) Human brain-specific L-proline transporter: molecular cloning, functional expression, and chromosomal localization of the gene in human and mouse genomes. *Mol. Pharmacol.* **48,** 219–229.

39. Kim, K. M., Kingsmore, S. F., Han, H., Yang-Feng, T. L., Godinot, N., Seldin, M. F., et al. (1994) Cloning of the human glycine transporter type 1: molecular and pharmacological characterization of novel isoform variants

and chromosomal localization of the gene in the human and mouse genomes. *Mol. Pharmacol.* **45,** 608–617.

40. Nelson, H., Mandiyan, S., and Nelson, N. (1990) Cloning of the human brain GABA transporter. *FEBS Lett.* **269,** 181–184.

41. Sora, I., Richman, J., Santoro, G., Wei, H., Wang, Y., Vanderah, T., et al. (1994) The cloning and expression of a human creatine transporter. *Biochem. Biophys. Res. Commun.* **204,** 419–427.

42. Jhiang, S. M., Fithian, L., Smanik, P., McGill, J., Tong, Q., and Mazzaferri, E. L. (1993) Cloning of the human taurine transporter and characterization of taurine uptake in thyroid cells. *FEBS Lett.* **318,** 139–144.

43. Borden, L. A., Dhar, T. G., Smith, K. E., Branchek, T. A., Gluchowski, C., and Weinshank, R. L. (1994) Cloning of the human homologue of the GABA transporter GAT-3 and identification of a novel inhibitor with selectivity for this site. *Receptors. Channels* **2,** 207–213.

44. Borden, L. A., Smith, K. E., Hartig, P. R., Branchek, T. A., and Weinshank, R. L. (1992) Molecular heterogeneity of the gamma-aminobutyric acid (GABA) transport system. Cloning of two novel high affinity GABA transporters from rat brain. *J. Biol. Chem.* **267,** 21,098–21,104.

45. Lopez-Corcuera, B., Liu, Q. R., Mandiyan, S., Nelson, H., and Nelson, N. (1992) Expression of a mouse brain cDNA encoding novel gamma-aminobutyric acid transporter. *J. Biol. Chem.* **267,** 17,491–17,493.

46. Borden, L. A., Smith, K. E., Gustafson, E. L., Branchek, T. A., and Weinshank, R. L. (1995) Cloning and expression of a betaine/GABA transporter from human brain. *J. Neurochem.* **64,** 977–984.

47. Yamauchi, A., Uchida, S., Kwon, H. M., Preston, A. S., Robey, R. B., Garcia-Perez, A., et al. (1992) Cloning of a Na(+)- and Cl(-)-dependent betaine transporter that is regulated by hypertonicity. *J. Biol. Chem.* **267,** 649–652.

48. Ramamoorthy, S., Bauman, A. L., Moore, K. R., Han, H., Yang-Feng, T., Chang, A. S., Ganapathy, V., and Blakely, R. D. (1993) Antidepressant- and cocaine-sensitive human serotonin transporter: molecular cloning, expression, and chromosomal localization. *Proc. Natl. Acad. Sci. USA* **90,** 2542–2546.

49. Pacholczyk, T., Blakely, R. D., and Amara, S. G. (1991) Expression cloning of a cocaine- and antidepressant-sensitive human noradrenaline transporter. *Nature* **350,** 350–354.

50. Povlock, S. L. and Amara, S. G. (1997) The structure and function of norepinephrine, dopamine, and serotonin transporters, in: *Neurotransmitter Transporters, Structure, Function, and Regulation* (Reith, M. E. A., ed.), Humana Press, Totowa, NJ, pp. 1–28.

51. Chang, A. S., Chang, S. M., Starnes, D. M., Schroeter, S., Bauman, A. L., and Blakely, R. D. (1996) Cloning and expression of the mouse serotonin transporter. *Mol. Brain Res.* **43,** 185–192.

52. Brüss, M., Porzgen, P., Bryan-Lluka, L. J., and Bönisch, H. (1997) The rat norepinephrine transporter: molecular cloning from PC12 cells and functional expression. *Mol. Brain Res.* **52,** 257–262.

53. Lingen, B., Brüss, M., and Bönisch, H. (1994) Cloning and expression of the bovine sodium- and chloride-dependent noradrenaline transporter. *FEBS Lett.* **342,** 235–238.

54. Hoffman, B. J., Mezey, E., and Brownstein, M. J. (1991) Cloning of a serotonin transporter affected by antidepressants. *Science* **254,** 579,580.

55. Blakely, R. D., Berson, H. E., Fremeau, R. T., Jr., Caron, M. G., Peek, M. M., Prince, H. K., et al. (1991) Cloning and expression of a functional serotonin transporter from rat brain. *Nature* **354,** 66–70.

56. Demchyshyn, L. L., Pristupa, Z. B., Sugamori, K. S., Barker, E. L., Blakely, R. D., Wolfgang, W. J., et al. (1994) Cloning, expression, and localization of a chloride-facilitated, cocaine-sensitive serotonin transporter from Drosophila melanogaster. *Proc. Natl. Acad. Sci. USA* **91,** 5158–5162.

57. Corey, J. L., Quick, M. W., Davidson, N., Lester, H. A., and Guastella, J. (1994) A cocaine-sensitive Drosophila serotonin transporter: cloning, expression, and electrophysiological characterization. *Proc. Natl. Acad. Sci. USA* **91,** 1188–1192.

57a. Fritz, J. D., Jayanthi, L. D., Thoreson, M. A., and Blakely, R. D. (1998) Cloning and chromosomal mapping of the murine norepinephrine transporter. *J. Neurochem.* **70,** 2241–2251.

58. Mortensen, O. V., Kristensen, A. S., Rudnick, G., and Wiborg, O. (1999) Molecular cloning, expression and characterizatipon of a bovine serotonin transporter. *Mol. Brain Res.* **71,** 120–126.

58a. Wade, P. R., Chen, J., Jaffe, B., Kassem, I. S., Blakely, R. D. and Gershon, M. D. (1996) Localization and function of a 5-HT transporter in crypt epithelia of the gastrointestinal tract. *J. Neurosci.* **16,** 2352–2364.

59. Miller, G. M., Yatin, S. M., De La Garza II, R., Goulet, M., and Madras, B. K. (2001) Cloning of dopamine, norepinephrine and serotonin transporters from monkey brain: relevance to cocaine sensitivity. *Mol. Brain Res.* **87,** 124–143.

60. Olivier, B. and Soudijn, W., and Van Wijngaarden, I. (2000) Serotonin, dopamine and norepinephrine transporters in the central nervous system and their inhibitors. *Prog. Drug Res.* **54,** 59–119.

61. Sonders, M. S., Zhu, S.-J., Zahniser, N. R., Kavanaugh, M. P., and Amara, S. G. (1997) Multiple ionic conductances of human dopamine transporter: the actions of dopamine and psychostimulants. *J. Neurosci.* **17,** 960–947.

62. Chen, N., Trowbridge, C. C., and Justice, J. B. Jr. (1999) Cationic modulation of human dopamine transporter: dopamine uptake and inhibition of uptake. *J. Pharmacol. Exp. Ther.* **290,** 940–949.

63. Earles, C. and Schenk, J. O. (1999) A multisubstrate mechanism for the inward trasnport of dopamine by the human dopamine transporter expressed in HEK cells and its inhibition by cocaine. *Synapse* **33,** 230–238.

64. Li, L. B. and Reith, M. E. A. (1999) Modeling of the interaction of Na^+ and K^+ with the binding of dopamine and [^{3}H]WIN 35, 428 to the human dopamine transporter. *J. Neurochem.* **72,** 1095–1109.

65. Li, L. B. and Reith, M. E. A. (2000) Interaction of Na^+, K^+, and Cl^- with the binding of amphetamine, octopamine, and tyramine to the human dopamine transporter. *J. Neurochem.* **74,** 1538–1552.

66. Chen, J-G. and Liu-Chen, S., Rudnick, G. (1998) Determination of External loop topology in the serotonin transporter by site-directed chemical labeling. *J. Biol. Chem.* **273,** 12,675–12,681.

67. Nirenberg, M. J., Vaughen, R. A., Uhl, G. R., Kuhar, M. J., and Pickel, V. M. (1996) The dopamine transporter is localized to dendritic and axonal plasma membranes of nigrostriatal dopaminergic neurons. *J. Neurosci.* **16,** 436–447.

68. Hersch, S. M., Yi, H., Heilman, C. J., Edwards, R. H., and Levey, A. I. (1997) Subcellular localization and molecular topology of the dopamine transporter in the striatum and substantia nigra. *J. Comp. Neurol.* **388,** 211–227.

69. Vaughan, R. A. and Kuhar, M. J. (1996) Dopamine transporter ligand binding domains. *J. Biol. Chem.* **271,** 21,672–21,680.

70. Ferrer, J. V. and Javitch, J. A. (1998) Cocaine alters the accessibility of endogenous cysteines in putative extracellular and intracellular loops of the human dopamine transporter. *Proc. Natl. Acad. Sci. USA* **95,** 9238–9243.

71. Norregaard, L., Frederiksen, D., Nielsen, E. Ø., and Gether, U. (1998) Delineation of an endogenous zinc-binding site in the human dopamine transporter. *Eur. Mol. Biol. J.* **17,** 4266–4273.

72. Loland, C. J., Norregaard, L., and Gether, U. (1999) Defining proximity relationship in -the tertiary structure of the dopamine transporter. *J. Biol. Chem.* **274,** 36,928–36,924.

73. Vaughan, R. A., Brown, V. L., McCoy, M. T., and Kuhar, M. J. (1996) Species- and brain region-specific dopamine transporters: immunological and glycosylation characteristics. *J. Neurochem.* **66,** 2146–2152.

74. Lin, Z., Itokawa M., Wang, W., and Uhl, G. R. (1999) Selective influences on dopamine uptake and cocaine analog binding by alanine substitutions for dopamine transporter transmembrane domain residues. *Neurosci. Soc. Abstr.* **25,** 283.

75. Melikian, H. E., Ramamoorthy, S., Tate, C. G., and Blakely, R. D. (1996) Inability to N-glycosylate the human norepinephrine transporter reduces protein stability, surface trafficking, and transport activity but not ligand recognition. *Mol. Pharmacol.* **50,** 266–276.

76. Nguyen, T. and Amara, S. G. (1996) N-Linked oligosaccharides are required for cell surface expression of the norepinephrine transporter but do not influence substrate or inhibitor recognition. *J. Neurochem.* **67,** 645–655.

77. Wang, J. B., Moriwaki, A., and Uhl, G. R. (1995) Dopamine transporter cysteine mutants: second extracellular loop cysteines are required for transporter expression. *J. Neurochem.* **64,** 1416–1419.

78. Chen, J.-G., Liu-Chen, S., and Rudnick, G. (1997a) External cysteines in the serotonin transporter. *Biochemistry* **36,** 1479–1486.

79. Sur, C., Schloss, P., and Betz, H. (1997a) The rat serotonin transporter: identification of cysteine residues important for substrate transport. *Biochem. Biophys. Res. Commun.* **241,** 68–72.

80. Vrindavanam, N. S., Arnaud, P., Ma, J. X., Altman-Hamamdzic, S., Parratto, N. P., and Sallee, F. R. (1996) The effect of phosphorylation on the functional regulation of an expressed recombinant human dopamine transporter. *Neuroscience* **216,** 133–136.

81. Huff, R. A., Vaughan, R. A., Kuhar, M. J., and Uhl, G. R. (1997) Phorbol esters increase dopamine transporter phosphorylation and decrease transport V_{max}. *J. Neurochem.* **68**, 225–232.

82. Vaughan, R. A., Huff, R. A., Uhl, G. R., and Kuhar, M. J. (1997) Protein kinase C-mediated phosphorylation and functional regulation of dopamine transporters in striatal synaptosomes. *J. Biol. Chem.* **272**, 15,541–15,546.

83. Zhu, S.-J., Kavanaugh, M. P., Sonders, M. S., Amara, S. G., and Zahniser, N. R. (1997) Activation of protein kinase C inhibits uuptake, currents and binding associated with the human dopamine transporter expressed in Xenopus oocytes. *J. Pharmacol. Exp. Ther.* **282**, 1358–1365.

84. Pristupa, Z. B., McConkey, F., Liu, F., Man, H. Y., Lee, F. J. S., Wang, Y. T., and Niznik, H. B. (1998) Protein kinase-mediated bidirectional trafficking and functional regulation of the human dopamine transporter. *Synapse* **30**, 79–87.

85. Melikian, H. E. and Buckley, K. M. (1999) Membrane trafficking regulates the activity of the human dopamine transporter. *J. Neurosci.* **19**, 7699–7710.

86. Kitayama, S., Dohi, T., and Uhl, G. R. (1994) Phorbol esters alter functions of the expressed dopamine transporter. *Eur. J. Pharmacol.* **268**, 115–119.

87. Zhang, L., Coffey, L. L., and Reith, M. E. A. (1997) Regulation of the functional activity of the human dopamine transporter by protein kinase C. *Biochem. Pharmacol.* **53**, 677–688.

88. Vaughan, R. A. and Pananusorn, B. (2000) Dopamine transporters are phosphorylated in vivo primarily on the N-terminal tail. *Neurosci. Soc. Abs.* **26**, 25.

89. Quan, H., Wang, Y.-M., Pifl, C., and Caron, M. G. (1998) Ca2+-dependent phosphorylation of dopamine transporter upregulates its transporter activity. *Neurosci. Soc. Abs.* **24**, 608.

90. Ramamoorthy, S. and Blakely, R. D. (1999) Phosphorylation and sequestration of serotonin transporter differentially modulated by psychostimulants. *Science* **285**, 763–766.

91. Meiergerd S. M. and Schenk J. O. (1994) Striatal Transporter for dopamine: Catechol structure-activity studies and susceptibility to chemical modification. *J. Neurochem.* **62**, 998–1008.

92. Chen, N. and Justice, J. B. Jr. (2000) Differential effect of structural modification of human dopamine transporter on the inward and outward transport of dopamine. *Mol. Brain Res.* **75**, 208–215.

93. Eshleman, A. J., Carmolli, M., Cumbay, M., Martens, C. R., Neve, K. A., and Janowsky, A. (1999) Characteristics of drug interactions with recombinant biogenic amine transporters expressed in the same cell type. *J. Pharmacol. Exp. Ther.* **289**, 877–885.

94. Zhang, L., Elmer, L. W., and Little, K. Y. (1998) Expression and regulation of the human dopamine transporter in a neuronal cell line. *Mol. Brain Res.* **59**, 66–73.

95. Barker, E. L., Perlman, M. A., Adkins, E. M., Houlihan, W. J., Pristupa, Z. B., Niznik, H. B., and Blakely, R. D. (1998) High affinity recognition of serotonin transporter antagonists defined by species-scanning mutagenesis. An aromatic residue in transmembrane domain I dictates species-selective recognition of citalopram and mazindol. *J. Biol. Chem.* **273**, 19,459–19,468.

96. Berfield, J. L., Wang, L. C., and Reith, M. E. A. (1999) Which form of dopamine is the substrate for the human dopamine transporter: the cationic or the uncharged species? *J. Biol. Chem.* **274,** 4876–4882.

97. Kitayama, S., Shimada, S., Xu, H., Markham, L., Donovan, D. M., and Uhl, G. R. (1992. Dopamine transporter site-directed mutations differentially alter substrate transport and cocaine binding. *Proc. Natl. Acad. Sci. USA* **89,** 7782–7785.

98. Edvardsen, Ø. and Dahl, S. G. (1994) A putative model of the dopamine transporter. *Mol. Brain Res.* **27,** 265–274.

99. Bönisch, H. (1998) Transport and drug binding kinetics in membrane vesicle preparation, in *Neurotranamitter Transporters*, (Amara, S., ed.), Methods in Enzymology, Vol. 296. Academic Press, San Diego, CA, pp. 259–278.

100. Barker, E. L., Moore, K. R., Rakhshan, F., and Blakely, R. D. (1999) Transmembrane domain 1 contributes to the permeation pathway for serotonin and ions in the serotonin transporter. *J. Neurosci.* **19,** 4705–4717.

101. Wang, W., Sonders, M. S., Scott, H., Kloetzel, M. N., and Surratt, C. K. (2000b) Mutation of the dopamine transporter transmembrane 1 aspartic acid residue differentiates sites for dopamine uptake and for high affinity binding of a cocaine analog. *Neurosci. Soc. Abstr.* **26,** 1169.

102. Chen, N., Vaughan, R. A., and Reith, M. E. A. (2001) The role of conserved tryptophan and acidic residues in the human dopamine transporter as characterized by site-directed mutagenesis. *J. Neurochem.,* **77,** 1116–1127

103. Dougherty, D. A. (1996) Cation-pi interactions in chemistry and biology: a new view of benzene, phe, Tyr, and Trp. *Science* **271,** 163–168.

104. Burley, S. K. and Petsko, G. A. (1985) Aromatic-aromatic interactions in proteins. *Science* **229,** 23–28.

105. Lin, Z., Wang, W., Kopajtic, T., Revay, R. S., and Uhl, G. R. (1999) Dopamine transporter: transmembrane phenylalanine mutations can selectively influence dopamine uptake and cocaine analog recognition. *Mol. Pharmacol.* **56,** 434–447.

106. Lin, Z., Itokawa, M., and Uhl, G. R. (2000a) Dopamine transporter proline mutations influence dopamine uptake, cocaine analog recognition, and expression. *FASEB J.* **14,** 715–728.

107. Danek Burgess, K. S., and Justice, Jr., J. B. (1999) Effects of serine mutations in transmembrane domain 7 of the human norepinephrine transporter on substrate binding and transport. *J. Neurochem.* **73,** 656–664.

108. Lin, Z., Wang, W., and Uhl, G. R. (2000b) Dopamine transporter tryptophan mutants highlight candidate dopamine- and cocaine-selective domains. *Mol. Pharmacol.* **58,** 1581–1592.

109. Itokawa, M., Lin, Z., Cai, N., Wu, C., Kitayama, S., Wang, J. B., and Uhl, G. R. (2000) Dopamine transporter transmembrane polar mutants: ΔG and ΔG values implicate regional important for transporter functions. *Mol. Pharmacol.* **57,** 1093–1103.

110. Chen, N., Ferrer, J. V., Javitch, J. A., and Justice, J. B. Jr. (2000) Transport-dependent accessibility of a cytoplasmic loop cystine in the human dopamine transporter. *J. Biol. Chem.* **275,** 1608–1614.

111. Smicun, Y., Campbell, S. D., Chen, M. A., Gu, H., and Rudnick, G. (1999) The role of external loop regions in serotonin transport. *J. Biol. Chem.* **274,** 36,058–36,064.

112. Golovanevsky, V. and Kanner, B. I. (1999) The reactivity of the g-aminobutyric acid transporter GAT-1 toward sulfhydryl reagents is conformationally sensitive. *J. Biol. Chem.* **274,** 23,020–23,026.

113. Chen, J.-G. and Rudnick, G. (2000) Permeation and gating residues in serotonin transporter. *Proc. Natl. Acad. Sci. USA* **97,** 1044–1049.

114. Mitsuhata, C., Kitayama, S., Morita, K., Vandenbergh, D., Uhl, G. R., and Dohi, T. (1998b) Tyrosine-533 of rat dopamine transporter: involvement in interactions with 1-methyl-4-phenylpyridinium and cocaine. *Mol. Brain Res.* **56,** 84–88.

115. Mitsuhata, C., Lin, Z., Kitryama, S., Dohi, T., and Uhl, G. R. (1998a) Molecular mapping of the sites on rat dopamine transporter for specific recognition of 1-methyl-4-phenylpyridium. *Neruosci. Soc. Abs.* **24,** 606.

116. Pifl, C., Agneter, E., Drobny, H., Reither, H., and Singer, E. A. (1997) Induction by low Na+ or Cl- of cocaine sensitive carrier-mediated efflux of amines from cell transfected with the cloned human catecholamine transporters. *Br. J. Pharmacol.* **121,** 205–212.

117. Eshleman, A. J., Henningsen, R. A., Neve, K. A., and Janowsky, A. (1994) Release of dopamine via the human transporter. *Mol. Pharmacol.* **45,** 312–316.

118. Wall, S. C., Gu, H., and Rudnick, G. (1995) Biogenic amine flux mediated by cloned transporters stably expressed in cultured cell lines: amphetamine specificity for inhibition and efflux. *Mol. Pharmacol.* **47,** 544–550.

119. Johnson, R. A., Eshleman, A. J., Meyers, T., Neve, K. A., and Janowsky, A. (1998) [³H]Substrate- and cell-specific effects of uptake inhibitors on human dopamine and serotonin transporter-mediated efflux. *Synapse* **30,** 97–106.

120. Sitte, H. H., Huck, S., Reither, H., Boehm, S., Signer, E. A., and Pifl, C. (1998) Carrier-mediated release, transport rates, and charge transfer induced by amphetamine, tyramine, and dopamine in mammalian cells transfected with the human dopamine transporter. *J. Neurochem.* **71,** 1289–1297.

121. Chen, N. and Justice, J. B. Jr. (1998) Cocaine acts as an apparent competitive inhibitor at the outward-facing conformation of the human norepinephrine transporter: kinetic analysis of inward and outward transporter. *J. Neurosci.* **18,** 10,257–10,268.

122. Pifl, C. and Singer, E. A. (1999) Ion dependence of carrier-mediated release in dopamine or norepinephrine transporter-transfected cells questions the hypothesis of facilitated exchange diffusion. *Mol. Pharmacol.* **56,** 1047–1054.

123. Buck, K. J. and Amara, S. G. (1995) Structural domains of catecholamine transporter chimeras involved in selective inhibition by antidepressants and psychomotor stimulants. *Mol. Pharmacol.* **48,** 1030–1037.

124. Vaughan, R. A. (1995) Photoaffinity-labeled ligand binding domains on dopamine transporters identified by peptide mapping. *Mol. Pharmacol.* **47,** 956–964.

125. Vaughan, R. A., Agoston, G. E., Lever, J. R., and Newman, A. H. (1999) Differential binding of tropane-based photoaffinity ligands on the dopamine transporter. *J. Neurosci.* **19,** 630–636.

126. Vaughan, R. A., Gaffaney, J. D., Lever, J. R., Reith, M. E. A., and Dutta, A. K. (2001) Dual incorporation of photoaffinity ligands on dopamine transporters implicates proximity of labeled domains. *Mol. Pharmacol.* **59**, 1157–1164.

127. Johnson, K. M., Bergmann, J. S., and Kozikowski, A. P. (1992) Cocaine and dopamine differentially protect [3H]mazindol binding sites from alkylation by N-ethylmaleimide. *Eur. J. Pharmacol.* **227**, 411–415.

128. Heron, C., Costentin, J., and Bonnet, J. J. (1994) Evidence that pure uptake inhibitors including cocaine interact slowly with the dopamine neuronal carrier. *Eur. J. Pharmacol.* **264**, 391–398.

129. Saadouni, S., Refahi-Lyamani, F., Costentin, J., and Bonnet, J. J. (1994) Cocaine and GBR 12783 recognize nonidentical, overlapping binding domains on the dopamine neuronal carrier. *Eur. J. Pharmacol.* **268**, 187–197.

130. Reith, M. E. A., Xu, C., and Coffey, L. L. (1996) Binding domains for blockers and substrates on the cloned human dopamine transporter studied by protection against N-ethylmaleimide-induced reduction of 2 beta-carbomethoxy-3 beta-(4-fluorophenyl)[3H]tropane ([3H]WIN 35,428) binding. *Biochem. Pharmacol.* **52**, 1435–1446.

131. Xu, C., Coffey, L. L., and Reith, M. E. A. (1997) Binding domains for blockers and substrates on the dopamine transporter in rat striatal membranes studied by protection against N-ethylmaleimide-induced reduction of [3H]WIN 35,428 binding. *Naunyn Schmiedebergs Arch. Pharmacol.* **355**, 64–73.

132. Reith, M. E. A., Berfield, J. L., Ferrer, J. V., and Javitch, J. A. (2000a) Cocaine and other inhibitors of dopamine uptake promote different conformational changes in the dopamine transporter. *Winter Conference Brain Res. Abs.* **33**, 99.

133. Reith, M. E. A., Berfield, J. L., Wang, L. C., Ferrer, J. V., and Javitch, J. A. (2001) The uptake inhibitors cocaine and benztropine differentially alter the conformation of the human dopamine transporter. *J. Biol. Chem.* **276**, 29,012–29,018.

134. Xu, C., Coffey, L. L., and Reith, M. E. A. (1995) Translocation of dopamine and binding of 2b-cabomethoxy-b-(4-fluorophenyl) tropane (WIN 35,428) measured under identical conditions in rat striatal synaptosomal preparations. Inhibition by various blockers. *Biochem. Pharmacol.* **49**, 339–350.

135. Uhl, G., Lin, Z., Metzger, T., and Dar, D. E. (1998) Dopamine transporter mutants, small molecules, and approaches to cocaine antagonist/dopamine transporter disinhibitor development. *Meth. Enzymol.* **296**, 456–465.

136. Buck, K. J. and Amara, S. G. (1994) Chimeric dopamine-norepinephrine transporters delineate structural domains influencing selectivity for catecholamines and 1-methyl-4-phenylpyridinium. *Proc. Natl. Acad. Sci. USA* **91**, 12.584–12.588.

137. Lee, S.-H., Cho, H.-K., Son, H., and Lee, Y.-S. (1997) Substrate transport and cocaine binding of human dopamine transporter is reduced by substitution of carboxyl tail with that of bovine dopamine transporter. *Neuroreport* **8**, 2591–2594.

138. Lee, S.-H., Chang, M.-Y., Lee, K.-H., Park, B. S., Lee Y.-S., Chin, H. R., and Lee, Y.-S. (2000) Importance of valine at position 152 for the substrate transport and 2_-carbomethoxy-3 _-(4-fluorophenyl)tropane binding of dopamine transporter. *Mol. Pharmacol.* **57,** 883–889.

139. Chen, J.-G., Sachpatzidis, A., and Rudnick, G. (1997b) The third transmembrane domain of the serotonin trasnporter contains residues associated with substrate and cocaine binding. *J. Biol. Chem.* **272,** 28,321–28,327.

140. Miller, G. M., Gracz, L. M., Goulet, M., Yang, H., Panas, H., N., Meltzer, P. C., Hoffman, B. J., and Madras, B. K. (1999) Point mutations of the human dopamine transporter (hDAT) reveal differential binding of an amine and a non-amine ligand. *Neurosci. Soc. Abs.* **25,** 283.

141. Meltzer, P. C., Blundell, P., and Madras, B. K. (1998) Structure activity relationships of inhibition of the dopamine transporter by 3-arylbicyclo-[3,2,1]octanes. *Med. Chem. Res.* **8,** 12–34.

142. Lee, F. J. S., Pristupa, Z. B., Ciliax, B. J., and Levey, A. I. (1996a) The dopamine transporter carboxyl-terminal tail. *J. Biol. Chem.* **271,** 20,885–20,894.

143. Bönisch, H., Hammermann, R., and Brüss, M. (1998) Role of protein kinase C and second messengers in regulation of the norepinephrine transporter. *Adv. Pharmacol.* **42,** 183–186.

144. Snyder, S. H. and Coyle, J. T. (1969) Regional differences in H3-norepinephrine and H3-dopamine uptake into rat brain homogenates. *J. Pharmacol. Exp. Ther.* **165,** 78–86.

145. Moore, K. R. and Blakely, R. D. (1994) Restriction site-independent formation of chimeras from homologous neurotransmitter-transporter cDNAs. *Biotechniques* **17,** 130–135, 137.

146. Barker, E. L. and Blakely, R. D. (1998) Structural determinants of neurotransmitter transport using cross- species chimeras: studies on serotonin transporter. *Meth. Enzymol.* **296,** 475–498.

147. Syringas, M., Janin, F., Mezghanni, S., Giros, B., Costentin, J., and Bonnet, J.-J. (2000) Structural domains of chimeric dopamine-noradrenaline human transporters involved in the Na⁺- and Cl⁻-dependence of dopamine transport. *Mol. Pharmacol.* **58,** 1404–1411.

148. Blakely, R. D., Moore, K. R., and Qian, Y. (1993) Tails of serotonin and norepinephrine transporters: deletions and chimeras retain function, in *Molecular Biology and Function of Carrier Proteins.* (Reuss, L., Russell, J. M., and Jennings, M. L., eds.) Rockefeller University Press, New York, pp. 284–300.

149. Roubert, C., Cox, P. J., Brüss, M., Hamon, M., Bönisch, H., and Giros, B. (2001) Determination of residues in the norepinephrine transporter that are critical for tricyclic antidepressant affinity. *J. Biol. Chem.* **276,** 8254–8260.

149a. Runkel, F., Brüss, M., Nöthen, M. M., Stöber, G., Propping, P., and Bönisch, H. (2000) Pharmacological properties of naturally occurring variants of the human norepinephrine transporter. *Pharmacogenetics* **10,** 397–405.

150. Barker, E. L. and Blakely, R. D. (1996) Identification of a single amino acid, phenylalanine 586, that is responsible for high affinity interactions of tricyclic antidepressants with the human serotonin transporter. *Mol. Pharmacol.* **50,** 957–965.

151. Gu, H. H., Wall, S., and Rudnick, G. (1996) Ion coupling stoichiometry for the norepinephrine transporter in membrane vesicles from stably transfected cells. *J. Biol. Chem.* **271**, 6911–6916.

152. Strader, C. D., Fong, T. M., Tota, M. R., and Underwood, D. (1994) Structure and function of G protein-coupled receptors. *Annu. Rev. Biochem.* **63**, 101–132.

153. Wang, C. D., Buck, M. A., and Fraser, C. M. (1991) Site-directed mutagenesis of alpha 2A-adrenergic receptors: identification of amino acids involved in ligand binding and receptor activation by agonists. *Mol. Pharmacol.* **40**, 168–179.

154. Tate, C. G. and Blakely, R. D. (1994) The effect of N-linked glycosylation on activity of the Na(+)- and Cl(-)-dependent serotonin transporter expressed using recombinant baculovirus in insect cells. *J. Biol. Chem.* **269**, 26,303–26,310.

155. Qian, Y., Galli, A., Ramamoorthy, S., Risso, S., DeFelice, L. J., and Blakely, R. D. (1997) Protein kinase C activation regulates human serotonin transporter in HEK-293 cells via altered cell surface expression. *J. Neurosci.* **17**, 45–57.

156. Sakai, N., Sasaki, K., Nakashita, M., Honda, S., Ikegaki, N., and Saito, N. (1997) Modulation of serotonin transporter activity by a protein kinase C activator and an inhibitor of type 1 and 2A serine/threonine phosphatases. *J. Neurochem.* **68**, 2618–2624.

157. Stephan, M. M., Chen, M. A., Penado, K. M. Y., and Rudnick, G. (1997) An extracellular loop region of the serotonin transporter may be involved in the translocation mechanism. *Biochemistry* **36**, 1322–1328.

158. Jess, U., Betz, H., and Schloss, P. (1996) The membrane-bound rat serotonin transporter, SERT1, is an oligomeric protein. *FEBS Lett.* **394**, 44–46.

159. Chang, A. S., Starnes, D. M., and Chang, S. M. (1998) Possible existence of quaternary structure in the high-affinity serotonin transport complex. *Biochem. Biophys. Res. Commun.* **249**, 416–421.

160. Kilic, F. and Rudnick, G. (2000) Oligomerization of serotonin transporter and its functional consequences. *Proc. Natl. Acad. Sci. USA* **97**, 3106–3111.

161. Schmid, J. A., Scholze, P., Kudlacek, O., Freissmuth, M., Singer, E. A., and Sitte, H. H. (2001) Oligomerization of the human serotonin transporter and of the rat GABA transporter 1 visualized by fluorescence resonance energy transfer microscopy in living cells. *J. Biol. Chem.* **276**, 3805–3810.

162. Barker, E. L., Kimmel, H. L., and Blakely, R. D. (1994) Chimeric human and rat serotonin transporters reveal domains involved in recognition of transporter ligands. *Mol. Pharmacol.* **46**, 799–807.

163. Cao, Y., Li, M., Mager, S., and Lester, H. A. (1998) Amino acid residues that control pH modulation of transport-associated current in mammalian serotonin transporters. *J. Neurosci.* **18**, 7739–7749.

164. Lin, F., Lester, H. A., and Mager, S. (1996) Single-channel currents produced by the serotonin transporter and analysis of a mutation affecting ion permeation. *Biophys. J.* **71**, 3126–3135.

165. Penado, K. M., Rudnick, G., and Stephan, M. M. (1998) Critical amino acid

residues in transmembrane span 7 of the serotonin transporter identified by random mutagenesis. *J. Biol. Chem.* **273,** 28,098–28,106.

166. Sur, C., Betz, H., and Schloss, P. (1997b) A single serine residue controls the cation dependence of substrate transport by the rat serotonin transporter. *Proc. Natl. Acad. Sci. USA* **94,** 7639–7644.

166a. Giros, B. and Caron, M. G. (1993) Molecular characterization of the dopamine transporter. Trends Pharmacol. Sci. 14, 43–49.

167. Kamdar, G., Penado, K. M. Y., Rudnick, G., and Stephan, M. M. (2001) Functional role of critical stripe residues in transmembrane span 7 of the serotonin transporter. *J. Biol. Chem.* **276,** 4038–4045.

168. Apparsundaram, S., Moore, K. R., Malone, M. D., Hartzell, H. C., and Blakely, R. D. (1997) Molecular cloning and characterization of an L-epinephrine transporter from sympathetic ganglia of the bullfrog, Rana catesbiana. *J. Neurosci.* **17,** 2691–2702.

169. Wang, L. C., Berfield, J. L., Kuhar, M. J., Carroll, F. I., and Reith, M. E. A. (2000a) RTI-76, an isothiocyanate derivative of a phenyltropane cocaine analog, as a tool for irreversibly inactivating dopamine transporter function in vitro. *Naunyn Schmiedebergs Arch. Pharmacol.* **362,** 238–247.

170. Kennedy, L. T. and Hanbauer, I. (1983) Sodium-sensitive cocaine binding to rat striatal membrane: possible relationship to dopamine uptake sites. *J. Neurochem.* **41,** 172–178.

171. Reith, M. E. A. and Coffey, L. L. (1993) Cationic and anionic requirements for the binding of 2 beta- carbomethoxy-3 beta-(4-fluorophenyl)[3H] tropane to the dopamine uptake carrier. *J. Neurochem.* **61,** 167–177.

172. Bonnet, J. J., Benmansour, S., Costentin, J., Parker, E. M., and Cubeddu, L. X. (1990) Thermodynamic analyses of the binding of substrates and uptake inhibitors on the neuronal carrier of dopamine labeled with [3H]GBR 12783 or [3H]mazindol. *J. Pharmacol. Exp. Ther.* **253,** 1206–1214.

4

Gene Organization and Polymorphisms of Monoamine Transporters

Relationship to Psychiatric and Other Complex Diseases

Maureen K. Hahn and Randy D. Blakely

1. INTRODUCTION

The monoamine neurotransmitters norepinephrine (NE), dopamine (DA), and serotonin (5-HT), play important roles in mood, cognition, learning, motor activity, reward, sleep, appetite, and cardiovascular functions. The availability of extracellular NE, DA, and 5-HT is limited by presynaptically localized transporters—NET, DAT, and SERT, respectively. The MA transporters retrieve released neurotransmitter, thus limiting the spread and duration of synaptic excitability and allowing neurotransmitter to be repackaged into synaptic vesicles. NET and SERT are the targets for many antidepressants, such as the tricyclic agents and the selective serotonin reuptake inhibitors (SSRIs) and with DAT, are the targets for the psychostimulants amphetamine and cocaine (1).

NET, DAT and SERT form a subfamily within a larger gene family of 12 transmembrane domain (TMD)-containing, Na^+- and Cl^--coupled cotransporters which includes gamma-aminobutyric acid (GABA), glycine, proline, and taurine transporters. Amino-acid identity across the gene family and within the MA transporter subgroup is highest in the TMDs and boundaries—particularly TMD 1 and 2 and 4 through 8—and lowest in the amino and carboxy terminal regions and extracellular loops (ELs) (2). The large second EL contains consensus sites for glycosylation, which have been shown to play a role in transporter trafficking and stability (3,4) (*see* also Chapter 10). There are also potential sites for serine and threonine and tyrosine kinase phosphorylation in the amino- and carboxy-termini and cytoplasmic loops. Recent studies suggest an important role for protein kinases and phosphatases in the acute modulation of transporter trafficking and activity (5–8) (*see* also Chapters 1 and 10). Regions of highest sequence

From: *Contemporary Neuroscience:*
Neurotransmitter Transporters: Structure, Function, and Regulation, 2nd Edition
Edited by: M. E. A. Reith © Humana Press Inc., Totowa, NJ

homology may subserve functions common to all monamine transporters, such as ion and monoamine binding or translocation and psychostimulant action, whereas regions of sequence divergence may correspond to sites that produce the selectivity of substrate and antagonist recognition observed for each MA transporter.

Variations in human MA transporter gene sequences known as polymorphisms may alter transporter expression levels, activity, or regulation. Cloning of the MA transporter genes has provided the means for examining transporter genetic variation and its potential contribution to human disease. The MA transporters are each encoded by a single gene, as opposed to other members of the gene family, such as the glycine and GABA transporters. In addition, there is highly specific tissue localization of MA transporters, each demonstrating expression that is restricted to its own neurotransmitter system. This suggests a potentially large impact of MA-transporter polymorphisms resulting from a limited opportunity for compensation by other gene products. This is supported by transgenic mouse models, in which genetic disruption of the MA transporters results in marked alterations in a variety of biochemical and behavioral measures *(9–11)* *(see* also Chapter 5). It is therefore not surprising that much work is currently devoted to evaluating polymorphisms in human MA-transporter genes for their contribution to human behavior and response to drugs and their relationship to psychiatric and other complex diseases. We review the structure and polymorphisms of human NET, DAT, and SERT, and examine the current data relating genetic variation to human disease.

2. HUMAN NOREPINEPHRINE TRANSPORTER (hNET, SLC6A4)

2.1. Cloning and Gene Organization

The human norepinephrine transporter (hNET) was identified by expression cloning using COS cells transfected with pools of cDNA clones derived from an SK-N-SH neuroblastoma library. Positive clones were identified by measuring the autoradiographic signal of radiolabeled substrate transported into cells, and pools of positive clones were subdivided until a single clone imparting NE uptake was identified *(12)*. This was the first monoamine transporter to be cloned, and it showed 46% identity of the predicted amino-acid sequence to that of the cloned GABA transporter (GAT1), confirming the existence of a Na^+/Cl^--dependent family of neurotransmitter transporters *(see* also Chapter 3, dendogram Fig. 1). The hNET cDNA identified was 1983 bp with an 1851-bp open reading frame coding for a 617 amino-acid protein with a predicted mol wt of 69,000 Da. hNET mRNA of 3.6 kb and 5.8 kb have been identified in SK-N-SH and human placenta *(12,13)*. There are several polyadenylation sites in the 3' region of the gene that are consistent with the possibility that differential polyadenylation contributes to the range of sizes of mRNA observed *(12,14)*. Linkage analysis first placed hNET on chromosome 16 at 16q13–q21 *(15)* and Southern blotting and fluorescence *in situ* hybridization (FISH) identified hNET as a single-copy gene located at 16q12.2 *(16)*. The hNET gene was cloned from a human-lung fibroblast library and the intron-exon borders were determined, reporting a gene of ~45 kb with 14 exons and 13 introns *(17*; Fig. 1A). In general, each hNET exon encodes no more than a single TMD, with some exception, and this splice junction pattern is conserved among other gene family members.

Evidence exists for hNET mRNA heterogeneity in 5' noncoding elements. Comparison of the 5' region of the hNET mRNA to the gene revealed a novel exon in the 5' end of the gene, with a 476-bp intron located within the previously identified exon 1 *(18*, Fig. 1A). An alternative splice site within this intron was also reported that results in an additional 183 bp in the splice variant of the new exon 1 *(19)*. 5' RACE and primer extension were used to confirm the transcription of mRNAs with the new 227–259 5' untranslated leader sequence, indicating obligate excision of this intron. These studies also demonstrated two clusters of transcription start sites, one located 708–703 bp upstream of the start codon and another set 30 bp upstream of the first, although G-capped products were not generated for the latter *(18)*. There are two potential TATA boxes located approx 40 bp upstream of each cluster of transcription start sites, and the promoter region is very GC-rich. Transient transfection of promoter constructs demonstrated that the new intron is a strong tissue-specific enhancer of transcriptional activity of the 5' region *(18)*. Polymorphisms identified in this region could have an influence on cell-specific mRNA expression.

hNET exhibits alternative splicing in the 3' end of the gene that exists in several species, including man. A bovine NET splice variant was first identified screening a bovine adrenal medullary cDNA library *(20)*. Comparison of this sequence to hNET genomic clones identified homologous sequences downstream of exon 15, revealing a new hNET exon, 16 *(14*; Fig. 1A). RT-PCR of mRNA from SK-N-SH cells with primers including the novel exon 16 sequences revealed two alternatively spliced transcripts. The splice variants skip exon 15 and differ in usage of 3' splice acceptor sites and stop codons of exon 16, yielding 18 or three additional amino acids *(14*; Fig. 1A). The bovine splice variant codes for a protein that does not transport NE and appears to remain intracellular *(20)*. A splice variant has also been identified in rat NET that codes for a nonfunctional transporter *(21)*. The C-tail of hNET contains consensus sequences that bind PDZ domains, present in proteins involved in targeting. NET was colocalized with a PDZ domain-

A

Human Norepinephrine Transporter (hNET; SLC6A2)

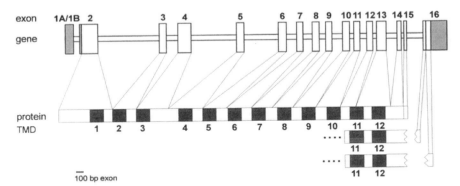

B

Human Dopamine Transporter (hDAT; SLC6A3)

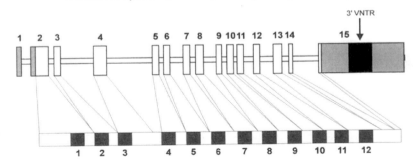

C

Human Serotonin Transporter (5-HTT; hSERT; SLC6A4)

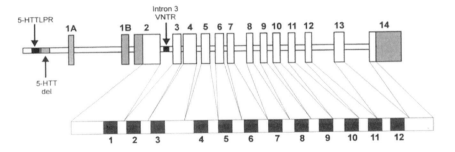

containing protein, PICK-1 by immunohistochemistry in the locus coeruleus *(22)*. Furthermore, removal of the terminal amino acid in hNET, part of the putative PDZ-binding domain, results in a decreased surface expression of mature, glycosylated hNET in transfected cells (P. Bauman, unpublished findings). Splice variants lacking exon 15 would lose this PDZ-domain-binding sequence, and thus may traffic to inappropriate cellular compartments. Potentially, polymorphisms that cause the selection of one splice pathway over the others could have great influence on hNET surface expression and thus affect the homeostatic regulation of NE neurotransmission and contribute to disease processes.

2.2. hNET as a Candidate Gene for Psychiatric and Cardiovascular Disease

There is evidence for the involvement of hNET in psychiatric and cardiovascular diseases. In the CNS, noradrenergic pathways subserve arousal, mood, attention, and the response to stress. It has long been hypothesized that depression involves the noradrenergic system, as evidenced by the mood-altering effects of reserpine, monoamine oxidase inhibitors, and tricyclic antidepressants—all compounds that impact catecholamine levels—and by studies of noradrenergic metabolite levels and receptor-binding sites in depression *(23–25)*. Animal studies have demonstrated that treatment with tricyclic antidepressants enhances the extracellular level of NE in terminal fields *(26,27)*. Furthermore, hNET levels measured by ^{3}H-nisoxetine binding are decreased in the brains of patients with major depression, the majority of whom died by suicide *(28)*.

The generation of a NET-knockout mouse supports the idea that NET is to involved in depression. In NET-knockout mice, the clearance rate of NE from the extracellular space, measured by fast-scan voltammetry, is decreased by sixfold, and extracellular levels of NE measured by in vivo microdialysis are increased twofold *(11)*. NET-knockout mice were compared to wild-type mice in the Porsolt forced-swim and tail-suspension tests. In these tests, the cessation of swimming or struggling is beleived to be a measure of behavioral despair, and antidepressants can reverse these effects on activity. NET-knockout mice showed an increase in activity compared to

---

Fig. 1. *(opposite page)* Exon/intron structure of MA-transporter genes showing the corresponding regions of the protein encoded by each exon. Untranslated regions of exons are shown in grey, and TMDs in protein are black boxes. **(A)** hNET; Broken boxes indicate alternative splice variants. **(B)** hDAT; 3' VNTR is shown in black. **(C)** 5-HTT/hSERT; VNTRs are shown in black and 5-HTT del is shown in grey. (Scale bar = 100 bp of exon sequence. Introns are not drawn to scale).

wild-type mice that was not further enhanced by antidepressant agents with actions at NET, DAT, or SERT. These data suggest that NET-knockout mice, possibly because of elevated extracellular NE levels, display behavior similar to that produced by antidepressant treatment *(11)*.

Noradrenergic systems and NET have been specifically implicated in diseases of the autonomic nervous system. NE is the major neurotransmitter in postganglionic sympathetic synapses, and it is estimated that NET can recapture approx 90% of NE released from sympathetic nerves terminals in the heart *(29,30)*. Additionally, as NET is dependent on the Na^+ and Cl^- gradients, efflux of NE through the transporter can be achieved if ion gradients are reversed or membranes depolarized, which can occur during cardiac ischemia *(30a)*. Decreases in NE uptake sites and activity are observed in cardiomyopathy and heart failure *(31–34)*. NE uptake is also impaired in a proportion of hypertensive patients *(35)*. The alterations of NET observed in human disease, and the impact of disruption of the NET gene in knockout mice, suggest that polymorphisms in the hNET gene that impact transporter expression may contribute to psychiatric and cardiovascular disorders.

2.3. hNET Taq1 RFLP

Gelernter and colleagues *(36)* identified the first polymorphic sites in hNET. A Taq1 restriction enzyme digestion, followed by a Southern blot probed with hNET cDNA, identified three alleles of fragment sizes 7.5, 6.8, and 4.6 kb, termed A1, A2, and A3, respectively. The frequencies of A1, A2, and A3 were determined in a Caucasian population to be 0.08, 0.82, and 0.1, respectively *(36)*. An analysis of six multiplex pedigrees with 45 individuals affected with bipolar disorder or major depression failed to demonstrate linkage with the Taq1 polymorphism *(37)*.

2.4. hNET Single Nucleotide Polymorphisms (SNPs)

2.4.1. Single Nucleotide Polymorphisms (SNPs)

Single nucleotide polymorphisms (SNPs) are common variants in the human genome that are believed to provide the genetic variability responsible for most traits and common diseases. As part of the Human Genome Project, there is currently a concerted effort to generate a map of SNPs in the human genome. A site is considered polymorphic if the frequency of the observed allele is >1%, and rare changes are termed mutations *(38)*. Coding SNPs (cSNPs) can be nonsynonymous, altering amino-acid sequence, or synonymous, leaving the sequence unchanged. Nonsynonymous cSNPs have the opportunity to alter protein structure and function, and thus are more likely to contribute to risk for functional disturbances in vivo. Thus, evaluating cSNPs—particularly those that result in nonconservative amino-acid substitutions in candidate gene studies—will increase the probability of marker association with the disease and by extension, the power of the analysis to detect an effect. However, nonsynonymous coding variants, particularly those with severe consequences for transporter function, are likely to be relatively rare events compared to cSNPs that result in conservative amino-acid changes or noncoding SNPs *(39)*. Polymorphisms with subtle effects will therefore be more frequent, and will potentially be responsible for variation in a greater percentage of the population *(40)*.

2.4.2 A457P and Orthostatic Intolerance

Our laboratory has attempted to refine the candidate gene analysis further, narrowing the phenotype definition to better encompass those populations likely to specifically harbor transporter mutations. In so doing, we have focused on autonomic disturbances in which NET dysfunction has been hypothesized. One such syndrome, Orthostatic Intolerance (OI) is a disorder characterized by an increase in standing heart rate of at least 30 beats per minute (bpm), which is not accompanied by hypotension *(41)*. OI may affect as many as 500,000 people in the United States. Many patients also suffer from palpitations, lightheadedness, altered mentation, headache, and fatigue when upright posture is assumed, and some patients demonstrate characteristics of a hyperadrenergic state and increased sympathetic outflow *(41)*. A proband with OI was identified who demonstrated standing-induced increased NE appearance in plasma (spillover), decreased NE disappearance from plasma (clearance), and decreased intraneuronal metabolism of NE as measured by decreased dihydroxyphenylglycol (DHPG) to NE ratios *(42)*. DHPG is the metabolite formed by monoamine oxidase in the cytoplasm of cells, and is therefore considered a correlative measure for NE that has been taken up by NET following release *(30)*. In addition, the proband demonstrated a blunted response to tyramine, a compound that causes an increase in plasma NE following uptake through NET *(42)*. Taken together, these findings suggested a selective involvement of an hNET dysfunction in this OI patient. Direct sequencing of the hNET gene revealed the proband to be a heterozygote for a G to C substitution at nucleotide 237 in exon 10 that results in a proline substitution for alanine at amino-acid position 457 (Table 1; *43*). This amino acid lies in a highly conserved region of TMD 9 (Fig. 2A). Family members of the proband were genotyped, identifying several heterozygotes for A457P (Fig 3). A high correlation was found between presence of the A457P polymorphism and standing-induced heart rate, plasma NE, and the plasma DHPG to NE ratio *(43)*. A457P was not identified in 254 unrelated normal subjects, patients with hypertension, or other OI patients *(43)*.

A Human Norepinephrine Transporter (hNET; SLC6A2)

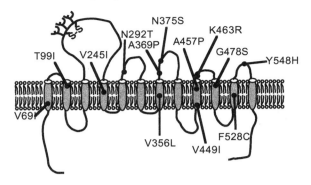

B Human Dopamine Transporter (hDAT; SLC6A3)

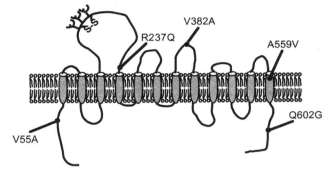

C Human Serotonin Transporter (hSERT, 5-HTT; SLC6A4)

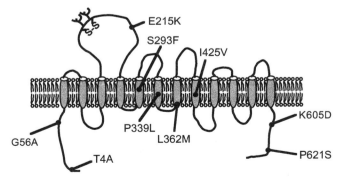

Fig. 2. Amino acid variants identified hNET (**A**), hDAT (**B**), and 5-HTT/hSERT (**C**). The transporters are shown as 12 transmembrane-domain-spanning proteins with intracellular N- and C-termini. The approximate location of the variant residues are shown. The number refers to the amino-acid position in the protein which is preceded by the single-letter code for the amino acid normally at that position, and is followed by the single-letter code for the variant.

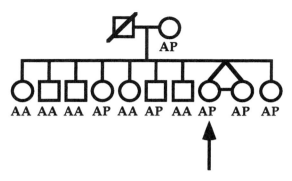

Fig. 3. Pedigree of the family of the proband with OI. *A* denotes alanine and *P* denotes proline, the variant amino acid. Female family members are shown as circles, and males as squares. A slash through the symbol indicates a deceased family member. The proband is identified by the arrow. (Adapted from ref. *43*).

The location of a nonconservative amino-acid substitution in a highly conserved TMD of hNET suggested that the A457P mutation could be highly disruptive to hNET structure and function. A proline substitution may cause a conformational change in the alpha-helical structure of TMD 9. Furthermore, a series of charged residues in the intracellular loop immediately preceding TMD 9 suggests a potential role in positioning the protein in the membrane, and therefore TMD 9 may be an important determinant of hNET structure and folding *(2)*. Transient transfection of A457P into heterologous expression systems revealed a protein that is devoid of transport activity *(43)*. In addition, A457P is not expressed in the mature, fully glycosylated form, and exhibits greatly diminished surface expression (Hahn et al., in preparation). Furthermore, A457P exerts a dominant-negative effect on wild-type hNET uptake activity (Hahn et al., in preparation). This finding suggests that transporters form multimeric complexes, and that individuals who are heterozygous for A457P or other transporter polymorphisms may be affected to a greater extent than predicted for harboring one mutant allele. This is the first demonstration of a coding mutation in a monoamine transporter gene with consequences for both in vitro transporter activity and neurotransmitter biochemistry and behavioral phenotype in the affected proband and family. The contribution of an hNET mutation to the tachycardia observed in OI may relate to the prominence of the reuptake mechanism in the heart, where synapses are 3× narrower than those in the vasculature *(44,45)*. NE is also the major neurotransmitter mediator of sympathetic nervous system regulation of heart rate. In contrast, the actions of NE in the brain occur within complex pathways utilizing multiple neurotransmitters. An hNET mutation in such an environment may have more subtle effects or a greater

Table 1
Single Nucleotide Polymorphisms in hNET Gene

Location	Nucleotide	Amino-acid variant	Ref.
exon 2	G254A	Val69Ile	46
exon 3	C233T	Thr99Ile	46
exon 5	G226A	Val245Ile	46
exon 6	C220G		50
exon 6	A258C	Asn292Thr	50
intron 6	A312G		46,50
exon 7	T181C		46
exon 8	G193C	Val356Leu	50
exon 8	G232C	Ala369Pro	50
exon 8	A251G	Asn375Ser	50
intron 8	C120A		46
exon 10	G155A		46,50
exon 10	G213A	Val449Ile	46
exon 10	G237C	Ala457Pro	43
exon 10	A256G	Lys463Arg	50
intron 10	G266A		46
exon 11	G743A	Gly478Ser	46
exon 12	T233G	Phe528Cys	50
intron 12	T263C		46
exon 13	T776C	Tyr548His	50
exon 14	G225A		46,50
intron 14	T342C		46
intron 14	G664C		50

Nucleotide positions are from Genbank Accession numbers X91117-X91127. Exon and intron numbers reflect the revised structure of the hNET gene following identification of a new exon 1 *(18)*. Additional SNPs in introns and flanking regions have been deposited in the Single Nucleotide Polymorphism database at NCBI (dbSNP) at http://www.ncbi.nlm.nih.gov/SNP.

opportunity for compensation. Thus, the elevated standing heart rate may be an endophenotype indicative of hNET dysfunction, which may successfully identify subgroups of subjects with functional hNET polymorphisms within populations with broader psychiatric diagnoses.

2.4.3 Other hNET SNPs

Stöber and colleagues were the first to identify SNPs in the coding region of the hNET gene, including several nonsynonymous SNPs (*46*; Table 1; Fig. 2A). Using Single-Strand Conformation Analysis (SSCA), they analyzed exons 2 through 15 in 46 control, 46 bipolar disorder, and 45 schizophrenic individuals. They identified thirteen variants—five nonsynonymous, three synonymous, and five in introns. The nonsynonymous polymorphisms were present in the population with frequencies of 0.01–0.02, and synonymous and noncoding SNPs were present at frequencies up to 0.40. An association study on new groups of 374 controls, 206 bipolar patients, and 456 schizophrenics revealed no association of any of the nonsynonymous cSNPs with schizophrenia or bipolar disorder *(46)*. Subsequently, this laboratory also demonstrated a lack of association of these SNPs with Tourette's Syndrome (TS) *(47)*. A lack of association for the common synonymous G1287A polymorphism with major depression or with suicidal ideation has also been reported *(48)*. The functional impact of the five identified nonsynonymous SNPs that cause amino-acid changes in hNET protein have been assayed in in vitro heterologous expression systems *(49)*. The single nucleotide substitutions were engineered by site-directed mutagenesis into expression vectors and transfected into COS-7 cells. The Gly478Ser variant demonstrated a fourfold increase in the K_m value for NE uptake *(49)*. This variant was previously reported solely in an individual from a control group with no evidence of disease *(46)*. The nonsynonymous cSNPs do not produce marked alterations in hNET function and have low frequencies of occurrence, qualities which may contribute to the inability to observe associations with disease states.

A recent study intended to determine the prevalence of SNPs in the human genome examined a group of genes that encode proteins linked to cardiovascular function. Subjects were taken from the top and bottom 2.5 percentile of blood-pressure measurements. Hybridization to a DNA microarray chip, which allows discrimination between a control and potential variant sequences, revealed seven nonsynonymous, three synonymous, and three noncoding SNPs in the hNET gene, with reported frequencies up to 0.1 (Table 1; Fig. 2A) *(50)*. Our laboratory has studied the effects of the nonsynonymous cSNPs on hNET expression and transport through transfection into heterologous cell culture systems *(51)*. These studies reveal both loss-of-function as well as gain-of-function hNET alleles, which now must be explored in the context of individual patient phenotypes. Nonetheless, the presence of these alleles in hyper- and hypotensive subjects, in addition to the mutation identified in OI, reinforces the potential of autonomic phenotypes to reveal functional hNET polymorphisms.

3. HUMAN DOPAMINE TRANSPORTER (HDAT, SLC6A3)

3.1. hDAT Cloning and Gene Organization

The cloning of the GAT and NET genes paved the way for cloning of other neurotransmitter transporters. Using degenerate oligonucleotide primers derived from the sequences of GAT and NET, a cDNA was amplified from rat substantia nigra (SN), and used to probe rat brain and SN cDNA libraries to clone the rat DAT (rDAT) *(52,53)*. Elucidation of the rDAT sequence provided the tools needed to isolate the human DAT (hDAT) cDNA from human brain stem and SN libraries *(54,55)*. The hDAT cDNA contained 5' and 3' untranslated sequence, and an open reading frame coding for a 620 amino acid protein with an estimated mol wt of approx 69,000 kDa and the 12 TMD predicted topology previously observed in GAT and NET *(54,55,* Fig. 1B). hDAT contains three potential N-linked glycosylation sites in the large EL between TMDs three and four *(see* also Chapter 10), and there is greater than 90% homology between rDAT and hDAT at the amino-acid level *(54,55)*. Northern blot analysis revealed a single 4.2-kb mRNA in human SN *(56)*. Southern blot analysis demonstrated that the hDAT gene (SLC6A3) is, like NET, a single copy gene, and additional studies have localized the gene to the distal short arm of chromosome 5 at 5p15.2–15.3 *(54,57)*. Cloning of hDAT revealed the gene to span more than 64 kb, with 15 exons and 14 introns *(56,58,59;* Fig. 1B). 5' RACE assays identified one transcriptional start site with no identifiable TATA- or CAAT-box sequences in proximity *(56,58)*. Possible sites for transcription-factor binding, including two E-box motifs and an Sp1 site, are contained in the first 1000 bp of 5' flanking region *(58)*.

3.2. hDAT as a Candidate Gene for Psychiatric Disorders

Dopamine (DA) systems are important mediators of motor function, mood, reward, and cognition *(60)*. DA systems are believed to play a role in many diseases including ADHD, schizophrenia, drug abuse, Parkinson's disease, and TS and hDAT gene variation could potentially contribute to many of these disorders *(61)*. An important role for DAT in DA physiology and behavior is supported by findings with DAT-knockout mice *(see* also Chapter 5). Clearance of DA from the extracellular space of the striatum of DAT-knockout mice is 300× slower than in intact animals, supporting the importance of the uptake mechanism, particularly in brain regions with a high density of DA neuron terminals *(10)*. Furthermore, DAT-knockout mice are hyperactive, demonstrate stereotypical behaviors, and do not respond to cocaine administration with a further increase in locomotion. The similarity of the DAT-knockout phenotype to symptomatology of disorders such as attention deficit/hyeractivity disorder (ADHD) and schizophrenia support the idea that polymorphisms in the hDAT gene could contribute to such illnesses *(12)*.

3.3. hDAT Taq1 Polymorphism

Taq1 restriction digestion of genomic DNA results in a variety of hDAT gene fragments, which is dependent on the probe employed. When using a probe in the 5' end of the gene, two bands were identified, 7 and 5.6 kb, termed Taq A1 and A2, respectively *(55)*. Frequencies of the Taq1 polymorphism differ between ethnic groups; the A1 allele has a frequency of 0.26 in whites and 0.42 in blacks. Another group showed that a Taq1 digestion of genomic DNA from 30 Caucasians, when probed with a more 3' probe, results in 16 unique fragments on a gel with frequencies ranging from 0.02 to 0.13 *(63)*.

3.4. hDAT 3' Variable Number of Tandem Repeats (VNTR)

A variable number of tandem repeats of a 40-bp unit has been identified in the hDAT gene in a region transcribed to the 3' untranslated region of the hDAT mRNA (Fig. 1B) *(57)*. Using PCR to amplify the repeat region from genomic DNA of 129 patients and controls, alleles containing 9-, 10-, or the rare3-, 5-,7-,8-, and 11-repeat units of the 40-bp element were observed. The 10-repeat allele was most common, present at a frequency of 0.70, followed by the 9-repeat allele at 0.24, and the remaining alleles had frequencies of approx 0.01–0.016. In this study, there were no significant differences in allele frequencies between American Caucasians and African Americans *(57)*. However, several studies have made it evident that great heterogeneity exists in allele frequencies across different ethnic groups. First, in a study of 107 unaffected Japanese individuals, the 10-repeat allele had a much higher frequency—0.93—than in the initial study described above, the 9-repeat allele was present at a frequency of 0.04, and the 7- and 11-repeat alleles were present at 0.009 and 0.02, respectively *(64)*. Another study that examined individuals from a variety of ethnic backgrounds found that allele frequencies differed among several of these populations *(65)*. For example, European Americans and African Americans demonstrated frequencies of the 9-repeat allele of 0.28 and 0.15, respectively, and the 3-, 7- ,and 8-repeat alleles were solely present in the African American population. Other variability included a Chinese population with allele frequencies of 0.06 and 0.93 for the 9- and 10-repeat alleles, respectively, and a Mbuti population from Africa with a high frequency of 0.34 for the 7-repeat allele, which is rare in most ethnic populations *(65)*. Kang and colleagues also performed an extensive survey of many populations of varying ancestry *(66)*. The 10-repeat allele was most common in most populations, with some populations of North/Central or South American ancestry demonstrating only the 10-repeat allele. In some African and Middle Eastern populations, the 10-repeat allele was as low as 0.35–0.50. In the African populations, the 7-repeat allele appeared with higher frequencies of 0.15–0.33. The 9-repeat allele ranged widely overall, from 0.06 to 0.5, and ranged from 0.1 to 0.3 in European populations. The latter is notable, as a Caucasian population is often used as a homogenous population in association studies. Another study included a population with Siberian ancestry, and revealed further heterogeneity that included both a relatively high frequency of the 7-repeat allele and the presence of a 13-repeat allele *(67)*. In summary, the 10-repeat allele occurs most frequently in most ethnic populations, followed by the 9-repeat allele, but there is wide diversity in the observed frequencies across ethnic groups. This variability is evident even among groups often considered as one population in association studies. This can result in positive results in case-control association studies because of ethnic differences, or stratification, between the control and affected groups. This suggests that stratification must be considered as a possible cause of significant results observed in many association studies, particularly when the differences in reported allele frequencies ascribed to diagnostic groupings are often smaller than some of the differences caused by the ethnic diversity.

3.5. hDAT Intron 8 VNTR

A VNTR has also been identified in intron 8, 2.6 kb 3' to exon 8 and 390 bp 5' to exon 9, consisting of five or six copies of a 30-bp repetitive element (*see* Fig. 1B). The frequency of the 6-repeat allele was 0.38 in Caucasians and 0.79 in African Americans *(59)*.

3.6. hDAT Single Nucleotide Polymorphisms (SNPs)

A large-scale SNP discovery study that examined 106 genes encoding proteins with functions relevant to cardiovascular and neurologic disorders was the first to identify SNPs in the hDAT gene *(39)*. Hybridization of human genomic DNA sequences to a high-density array of oligonucleotides and denaturing high-pressure liquid chromatography (HPLC)-(DHPLC) were both used to identify six synonymous SNPs and one nonsynonymous SNP (Table 2; Fig. 2B). Vandenbergh and colleagues recently screened with single-strand conformation polymorphism (SSCP), followed by sequencing the coding region and exon/intron boundaries of hDAT from Caucasian Tourette's Syndrome (TS) patients and controls *(59)*. Twelve variants were reported: ten synonymous SNPs, five of which were previously identified, two nonsynonymous cSNPs, and one variant in the 3' untranslated region (UTR) (Table 2). The two coding mutations, Val55Ala and Val382Ala, had frequencies of <0.01, the synonymous variants had frequencies of 0.05–0.33, and the noncoding region variant had a frequency of 0.26. Using a similar approach, another group examined 45 patients with bipolar disorder and 46 controls of German descent by SSCA and sequencing to identify SNPs *(68)*. They found ten SNPs: two novel nonsynonymous SNPs, three previously identified synonymous SNPs, and five novel SNPs in introns (Table 2). One

Table 2
Single Nucleotide Polymorphisms in hDAT Gene

Location	Nucleotide	Amino-acid variant	Ref.
5' flank	T-67A		*71b*
5' flank	G-660C		*71b*
5' flank	C-839T		*71b*
5' flank	C-1169G		*71b*
5' flank	T-1476G		*71b*
exon 2	C242T		*39,59,68*
exon 2	G278T		*39,59*
exon 2	C290T		*39*
exon 2	T292C	Val55Ala	*59*
exon 2	C299T		*59*
intron 3	C20A		*68*
exon 5	G838A	Arg237Gln	*39*
exon 6	C938T		*39,59*
exon 7	C1106T		*59*
exon 8	C1178A		*59*
exon 8	G1196A		*59*
exon 8	T1273C	Val382Ala	*59*
exon 9	A1343G		*39,59,68,*
			70,71,97
intron 9	G282T		*71*
intron 9	A1518G		*68*
intron 9	G1553A		*68,71,97*
intron 10	G1819A		*71*
intron 11	G2411A		*68*
exon 12	G1653A		*59*
exon 13	C1804T	Ala559Val	*68*
exon 13	C1859T		*39,59,68*
exon 14	A1933G	Glu602Gly	*68*
intron 14	629InsTC		*68*
exon 15/UTR	C2026T		*59,70*
exon 15/UTR	G2241C		*71*
exon 15/UTR	G2319A		*69,71*
exon 15/UTR	T3779C		*71*

Nucleotide positions in exons are from Genbank Accession number M95167. Nucleotide positions in 5' flanking region are 5' to the transcriptional start site in Kawarai et al., 1997; Genbank number D8856S01. Nucleotide positions in introns are from Genbank Accession numbers AF306558-AF306564, AF321320-AF321321. Additional SNPs in introns and flanking regions have been deposited in the Single Nucleotide Polymorphism database at NCBI (dbSNP) at http://www.ncbi.nlm.nih.gov/SNP.

group has reported a G2319A polymorphism, located in the 3' UTR of the hDAT gene in a Japanese population (Table 2) *(69)*. Direct sequencing and SSCP of exons 2 through 15 of the hDAT gene in Japanese populations yielded two previously identified SNPs (Table 2) *(70)*. Finally, complete sequencing of exons and substantial sequencing of introns, some in entirety, yielded 92 SNPs, including insertions and deletions in SNP *(71)*. All of these were noncoding, with the exception of one previously identified synonymous coding SNP. A subset of these SNPs used by this group in linkage analysis are listed in Table 2.

3.7. hDAT Polymorphisms: Linkage and Association Studies

3.7.1. Affective Disorders

The hDAT 3' VNTR has been examined for its role in affective disorders. Negative results have been obtained in association studies of both bipolar disorder and major depressive disorder (MDD) *(71a,173)*. However, family-based studies of linkage have yielded intriguing results. Within-family designs that provide a means for removing population stratification because of ethnic or socioeconomic influences have been pursued to evaluate the contribution of the hDAT 3' VNTR to bipolar disorder. The transmission disequilibrium test (TDT) is an example of this type of analysis, which determines if there is preferential allele transmission to the affected offspring by comparing parental alleles transmitted vs not transmitted. Using this approach, the 10-repeat allele of the 3' VNTR and the longer alleles of the Taq1 polymorphism were found to be in linkage disequilibrium with bipolar disorder *(72)*. A linkage study of Amish and Icelandic bipolar patient pedigrees examining several markers, including the 3' VNTR and Taq1 polymorphisms, failed to show significant logarithm of the odds (LOD) scores, although there was a trend toward linkage in individual families *(73)*. This group has recently extended these findings by increasing the number of markers in the hDAT gene used to determine linkage disequilibrium in bipolar disorder pedigrees, including the Amish group from the initial study *(71)*. Sequencing several individuals, this group identified novel polymorphic sites in hDAT including promoter and intron polymorphisms which could be used in linkage studies. They identified 92 polymorphisms including one in the coding region which resulted in a synonymous change. Fourteen of these SNPs were used in a linkage disequilibrium study revealing that haplotypes generated from markers in the 3' region of the gene were all in linkage disequilibrium with bipolar disorder. Thus, the 3' end of the hDAT gene appears to contain several informative elements that potentially account for the conflicting results obtained when using the 3' VNTR allelic variation as the sole marker for analysis. Another group failed to replicate this linkage in a study of 122 families with bipolar disorder, using the TDT test to compare only the transmission of the 9- vs 10-repeat allele *(74)*. Interestingly, two amino-acid substitutions, Ala559Val and Glu602Gly, previously identified by Grunhage and colleagues, appeared once each in bipolar patients in this study *(68)*. The relatively small sample and rare occurrence of these alleles diminishes the ability to observe differences between bipolar patients and controls. However, the Ala559Val and Glu602Gly variants were observed in families in which one parent was also affected, allowing some observations to be made regarding genotype and symptoms. In one family, both the affected patient and the parent with major depression shared the allele coding for glycine at position 602, whereas in the second family the affected individual did not receive the allele coding for the variant valine at position 559 from the affected parent. It will be important to determine the consequences of these amino-acid substitutions for transport function, and whether these variants are observed more frequently upon examination of a greater number of individuals with bipolar disorder or other phenotypes. In summary, the data are equivocal regarding a role for the hDAT 3' VNTR in affective disorder, but evidence of linkage of other sequences in the 3' region of the hDAT gene to bipolar disorder appears to be emerging. This finding highlights the power of using a dense SNP map in haplotype studies of disorders believed to be linked to hDAT for which positive evidence has been elusive. Indeed, this approach has also yielded positive results in other studies, including those of ADHD in which there has been conflicting evidence regarding involvement of the 3' VNTR.

3.7.2. Substance Abuse/Dependence

Animal studies have demonstrated that substances of abuse including cocaine, nicotine, opiates, and ethanol, all enhance the activity of mesocorticolimbic DA pathways and increase extracellular DA levels in the nucleus accumbens, and that the actions of DA in the nucleus accumbens are involved in mediating the acute reinforcing properties of cocaine and nicotine *(75)*. Cocaine and d-amphetamine elevate extracellular DA levels by blocking—and in the case of d-amphetamine, also reversing—DAT. Furthermore, in DAT-knockout mice, psychostimulants fail to increase extracellular levels of DA or enhance locomotor activity *(10)*. However, these mice can still be trained to self-administer cocaine *(76)*. Taken together, the evidence highlights the integral role of DAT in the actions of psychostimulants, emphasizes the complexity of both the effects of psychostimulants and potential compensatory mechanisms in DAT-knockout mice, and supports the theory that polymorphisms in the hDAT gene may contribute to substance abuse and dependence.

The 3' VNTR has been examined in a study of cocaine dependence. Cocaine-dependent patients were divided into groups with and without evidence of paranoia, and white and black patient groups were analyzed separately. A significant increase in the 9-repeat allele was seen in the cocaine-dependent group with paranoia, both for the total group and white patient subgroup *(77)*. Another study, using a group of over 200 U.S. subjects who completed a survey and were divided into moderate vs heavy drug use, found no association of drug use with the 9- or 10-repeat alleles or with the Taq1 polymorphism *(78)*. However, these researchers did not report the drugs used by subjects in the study. Since the action of cocaine at DAT to increase extracellular levels of DA is critical to reinforcing properties of this drug, additional studies are clearly needed to explore the contribution of the hDAT gene to cocaine abuse and dependence.

Several studies have investigated the relationship of the 3' VNTR to smoking. The 9-repeat allele was associated with a reduced likelihood of being a smoker and longer periods of smoking cessation for smokers *(79)*. A gene-gene interaction was found whereby this effect was greater if the 9-repeat allele individuals also carried the A2 allele of a polymorphic site in the DA D2 receptor gene. In a second study, those with the 9-repeat allele were 50% more likely to quit smoking. In this study, allele (and smoking cessation) were also associated with low novelty-seeking on the Temperament and Character Inventory *(80)*. However, in a large study of 861 Caucasians, the presence of the 9-repeat allele was not associated with current, former, or nonsmoker groups, and ratings on a personality scale were not associated with allele or smoking group *(81)*.

A series of studies have attempted to establish a relationship between the 3' VNTR and alcoholism. One group has demonstrated an increased frequency of the 9- vs 10-repeat allele in German alcoholics experiencing withdrawal seizures and/or delirium compared to controls or alcoholics negative for these symptoms *(82,83)*. These researchers have also recently demonstrated an increase in the 9-repeat allele in epilepsy, suggesting that the association of the 9-repeat allele may be to seizure susceptibility rather than to alcohol withdrawal *per se (84)*. However, in a family-based study of 87 alcohol-dependent patients, divided into groups with or without seizures and delirium, alcoholism in the total group or subdivided sample was not linked to the presence of the 9-repeat allele *(85)*. Case-control studies have also reported a lack of association of the DAT 3' VNTR with alcoholism *(86)*. In an association study using a Japanese population, where the 7-, 9-, and 10-repeat alleles had frequencies in controls of 0.013, 0.06, and 0.90, respectively, alcoholics with a mutation in the aldehyde dehydrogenase-2 gene had a 2.5-fold greater frequency of the 7-repeat allele *(87)*. This suggests that different allelic variants may contribute to a disease when in the context of a different genetic background. Examination of the single-nucleotide coding variants identified by Vandenbergh and colleagues demonstrated no differences in allele frequencies in alcohol-dependent patients vs controls *(59)*. One group has examined a G2319A polymorphism, located in the 3' UTR, and the DAT 3' VNTR in Japanese alcoholics *(69)*. There was a significant association of alcoholism with both the 2319A allele and A-containing genotypes. In addition, haplotype analysis revealed a significant association of the 2319A allele and the 10-repeat of the 3' VNTR with alcoholism. However, this may reflect solely the association with the A allele with alcoholism, as there may have been a large enough sample size of 10-repeat alleles, but not 9-repeat alleles, to see significance upon division of the sample into haplotypes. Overall, the link of the 9- or 7-repeat allele of DAT 3' VNTR with alcoholism is not yet strong, and although the potential link of the 9-repeat allele to seizure susceptibility is intriguing, it calls for further replication.

3.7.3. Attention Deficit/Hyperactivity Disorder

The proposed role of DA in ADHD, that DAT is a target for psychostimulant drugs used to treat ADHD, and the hyperactive phenotype of DAT-knockout mice all suggest that hDAT polymorphisms which may be identified in ADHD could contribute to this disorder *(88)*. A link between ADHD and the 3' VNTR 10-repeat allele has been demonstrated by several groups, using several tests of both linkage and association. Cook and colleagues used a haplotype-based haplotype relative risk (HHRR) approach to study 56 families, composed of ADHD-affected offspring with one or two parents, to demonstrate that the 10-repeat allele was preferentially transmitted to the child *(89)*. A similarly designed study, using HRR analysis, replicated the preferential transmission of the 10-repeat allele to ADHD affected offspring *(90)*. A comprehensive study of 117 families using several methods of analysis and examination of hyperactive-impulsive and inattentive subtypes also found evidence of higher risk associated with the 10-repeat allele *(91)*. In this study, ADHD patients and their siblings were rated on several scales by their parents and assigned one of three levels of intensity of symptoms. First, they used a between-family test to find a significant association of the number of 10-repeat alleles, designated as high-risk, with ADHD symptoms. Discordant sibling pairs with a greater number of the high-risk allele also demonstrated increased hyperactive-impulsive and inattentive symptoms. They also used within-family TDT, and found linkage disequilibrium between DAT and ADHD-combined subtypes, with the odds of transmitting a high-risk allele increasing with increasing symptom severity. Another within-family study assessed ADHD patients and their parents using HHRR and TDT for transmission of 10-repeat allele vs other alleles not transmitted *(92)*. In 19 cases for which parents were retrogradely diagnosed with ADHD, and were heterozygous for the 10-repeat allele, this allele was transmitted 15 times. Another TDT study also revealed a preferential transmission of the 10-repeat allele in a British families *(93)*. However, several studies have failed to replicate these positive findings. A study of 80 trios employing HRR and TDT did not reveal linkage or association of ADHD with the 10- vs 9-repeat *(88)*. Additionally, a study employing case-control association and TDT analysis also found no association of the hDAT 3' VNTR with ADHD *(94)*. Interestingly, the latter was a group of ADHD-combined type patients with a demonstrated clinical response to methylphenidate, and it is possible that patient selection affected the outcome. These authors suggest that the lack of association found in their study and others may stem from the use of structured diagnostic interviews in these studies to define ADHD compared to, for example, parental report. Another study also failed to detect an association in a Brazilian sample of the 10-repeat allele in subgroups defined by the dimensions of hyperactivity and inattention *(95)*. A longitudinal study of Australian children extending over 16 yr examined multiple aspects of temperament involved in ADHD, including hyperactivity, aggression, and inattention, and found no association between any measure of temperament and the 10-repeat allele *(96)*.

Vandenbergh and colleagues have assessed the relationship of the SNPs identified by their laboratory to ADHD *(59)*. Using the TDT test, they demonstrated a threefold greater transmission to the affected offspring of the 10-repeat allele of the 3' VNTR (previously shown in these same patients by Cook and colleagues) and an approx twofold greater transmission of C2026, located in the 3' UTR (Table 2) *(59,89)*. There was no transmission disequilibrium of Val55Ala or Val382Ala in the ADHD families. This group took further advantage of the increasing number of identified markers to perform an analysis of haplotypes which were composed of the 3' VNTR, an exon 9 SNP, and an intron 9 SNP A1343G and G1553A, respectively *(97;* Table 2). Using a TDT analysis, they found no preferential transmission of any of the alleles individually. However, a haplotype analysis revealed evidence of both biased transmission and nontransmission. Interestingly, haplotypes containing the 10-repeat allele and the A allele of the intron 9 polymorphism, but differing in the presence of the A or G allele of the exon 9 polymorphism, were preferentially transmitted or not transmitted, respectively. These data may explain the seemingly conflicting results generated from studies examining a link between the 10-repeat allele and ADHD. The 10-repeat allele may not be a marker or the sole marker for ADHD, but may signify the presence of another nearby marker in linkage disequilibrium.

The grouping of mixed haplotypes in an analysis of only the 10-repeat allele could contribute to the lack of significance observed in other studies. As more markers are identified to construct haplotypes and perform haplotype analyses, the region of the hDAT gene linked to ADHD and other diseases will become more defined.

The extent to which the VNTR or SNPs located in the 3' UTR affect hDAT expression remains to be determined. Interestingly, recent imaging studies have reported alterations in hDAT-binding sites in ADHD. Increases of 15–70% in hDAT-binding sites in the striatum in adult ADHD patients have been reported by several groups *(98–100)*. In two of these studies, elevated DAT-binding sites were decreased to less than control by methylphenidate treatment *(99,100)*. Although these findings support an involvement of hDAT in ADHD, these studies did not examine the potential contribution of genotype to DAT expression level. Nonetheless, evidence of both linkage and association of the 10-repeat allele of the hDAT 3' VNTR and 3' region haplotypes with ADHD, in combination with reports of altered hDAT expression observed in ADHD, makes this a promising area for pursuit of MA transporter genetic variation and disease.

3.7.4. *Schizophrenia*

The hypothetical role of DA systems in schizophrenia suggests that hDAT polymorphisms may contribute to this disorder. A linkage study of sixteen pedigrees with schizophrenia failed to reveal linkage between schizophrenia and either the hDAT Taq1 or 3' VNTR polymorphism *(101)*. Case-control association studies of various ethnic groups have mainly found a lack of association of the 9- and 10-repeat alleles with schizophrenia, schizophrenia subtypes, or specific symptoms. An association study of Italian patients with delusional disorder showed no association with the 3' VNTR 9- or 10-repeat allele or genotypes *(102)*. An association study of Japanese populations showed no significant differences in 3' VNTR allele frequencies for the total schizophrenia group, or when divided into subgroups based on age of onset, family history, or symptomatology, including the presence of delusions or negative symptoms *(103)*. One study of Italian schizophrenics compared to controls showed a significant increase in homozygote genotypes, with no change in the frequency of the 9- and 10-repeat alleles *(104)*. This association was not found when the schizophrenic group was divided into subtypes. A lack of association of the 9- or 10-repeat allele was also found in a study of Caucasians of European descent *(105)*. One comprehensive study used between-family linkage analysis and two within-family tests to determine the impact of the 3' VNTR or a microsatellite marker, D5 S406, on family members affected with schizophrenia or schizoaffective disorder *(106)*. No linkage was demonstrated using all families comprising 337 individuals, or using 75 sib pairs and their parents in an inheritance by descent (IBD) analysis. Finally, no association of alleles transmitted vs not transmitted was identified in an HRR analysis of these data. The G2319A polymorphism, located in the 3' UTR and the DAT 3' VNTR were not associated with schizophrenia in a Japanese population *(69)*. These data do not establish a relationship of the DAT 3' VNTR to schizophrenia, even with a division of patient populations into the symptom or diagnostic subgroup. Given the complex constellation of subtypes of schizophrenia and related diagnoses, further efforts to determine a narrow phenotype likely to involve hDAT are warranted.

3.7.5. Tourette's Syndrome

Involvement of the hDAT 3' VNTR in Tourette's Syndrome (TS) has been investigated. A linkage study in three families with TS excluded the DAT gene locus, including the 3' VNTR and Taq1 polymorphisms as linked to TS *(107)*. In an association study of a U.S. population of TS patients, their relatives, and unrelated controls, the TS patients and their relatives had an increased frequency of the 10-repeat allele compared to controls *(108)*. Interestingly, those TS cases with a 10/10 genotype vs those with other genotypes demonstrated a higher level of comorbidity with ADHD, somatization, depression, and alcohol abuse. Examination of the single nucleotide coding variants identified by Vandenbergh and colleagues demonstrated no differences in allele frequencies in TS patients vs controls *(59)*. In summary, there is no strong evidence for a genetic contribution of hDAT to TS, but the association of the 10-repeat allele with TS comorbid with ADHD emphasizes that ADHD is a promising phenotype for the involvement of hDAT genetic variability.

3.7.6. Parkinson's Disease

Parkinson's disease is characterized by severe motor dysfunction, including tremor, rigidity, bradykinesia, and an inability to initiate movements *(109)*. It is accompanied by a loss of midbrain dopaminergic neurons, and is treated primarily by L-DOPA to elevate DA levels *(109)*. The role of DAT in normal homeostasis of DA uptake, as well as being a conduit for neurotoxic substances, suggest that polymorphisms in hDAT may contribute to susceptibility to Parkinson's disease *(23)*. In a case-control association study of French Parkinson's disease patients and controls, there were no differences in the allele or genotype frequencies of the 3' VNTR and an analysis of variance revealed no effect of the 9- vs 10-repeat allele on age of onset of Parkinson's disease *(110)*. A study of Australian Caucasian Parkinson's disease patients found an increase in the rare 11-repeat allele in Parkinson's disease patients compared to controls, generating an odds ratio of 10, or a 10-fold increased risk, associated with the 11-repeat allele *(111)*. The latter study exhibits a large effect, albeit for a rare allele of the hDAT 3' VNTR, suggesting a contribution to only a minority of cases. The A1343G SNP was examined for association with Parkinson's disease, and the 1343G allele was found to be significantly higher in controls vs Parkinson's disease patients *(70)*. Given the highly selective pathology of dopaminergic neurons in Parkinson's disease, more work is clearly needed in this area.

3.8. hDAT 3' VNTR Genotype and hDAT Expression

It is not known how the hDAT 3' VNTR polymorphism—which is located in the 3' UTR and thus does not change the amino acid sequence—could affect hDAT protein expression. The 3' untranslated region of genes may serve a variety of functions, including sorting of mRNA to distinct compartments in the cell, regulating the rate of translation, and signaling stabilization or destabilization of mRNA to influence turnover rate *(112–114)*. The sequences or length disparity created by different repeat numbers in the hDAT 3' VNTR may impact the conformation of DNA and the ability of factors to bind effectively. Thus, polymorphisms in the 3' VNTR could result in differential regulation of hDAT mRNA polyadenylation, sorting, stability, and translational efficiency among individuals.

Few studies have examined the relationship of genotype of the hDAT 3' VNTR to hDAT-binding sites in the brain. In a study of 30 subjects of mostly European background, single positron emission computerized tomography (SPECT) analysis, using [^{3}H]β-CIT as ligand, was used to measure hDAT-binding sites. There was a small 13% decrease in binding to the transporter in the striatum in the 10/10 genotype group *(115)*. However, a study of 31 abstinent alcoholics and 23 controls demonstrated a 22% decrease in β-CIT binding in the putamen in those subjects harboring a 9-repeat allele *(86)*. In addition, there were no genotype differences between the control and alcoholic groups. The differences observed in these two studies are small and occur in the opposite direction. The study of Jacobsen and colleagues used control subjects alone, which may be the appropriate group for gaining an initial understanding of the relationship of genotype to binding sites. Indeed, a recent study of alcoholics admitted to detoxification treatment demonstrated decreased striatal [^{125}I]β-CIT binding compared to controls that inccreased to control levels following 4 wk of abstinence *(116)*. The findings of Heinz and colleagues, which indicate an increase in binding sites in those with a 10/10 genotype are consistent with both the association of the 10-repeat allele with ADHD and an increase in binding sites in ADHD that has been reported *(89,91)*. Three recent studies reported 15–70% increased DAT-binding sites in the striatum in adult ADHD patients which were decreased to less than control by methylphenidate treatment in two of the studies *(98–100)*. A recent study utilized SPECT scanning to examine the role of 3' VNTR genotype in DA availability in the striatum following a single amphetamine injection in both controls and schizophrenics *(117)*. A radiotracer selective for the dopamine D2 receptor was used to measure the displacement of DA from the D2 receptor as an index of DA release following amphetamine exposure. Amphetamine decreased the binding of radioligand to D2 receptors in both control and schizophrenic groups, but there was no significant effect of genotype. SPECT analysis of $[^{123}I]\beta$-CIT binding to hDAT revealed no difference in transporter levels across diagnosis or genotype. This work suggests that the hDAT 3' VNTR genotype does not give rise to a functional change in transport activity of hDAT in the striatum.

In summary, further studies, particularly ones using unaffected individuals, are required to understand the potential relationship of the hDAT 3' VNTR genotype to hDAT expression levels and function. Such studies would facilitate our understanding of the seemingly complex interaction of genotype, disease symptomatology and pharmaceutical treatments to impact hDAT expression levels and activity.

4. HUMAN SEROTONIN TRANSPORTER (SERT; 5-HTT; SLC6A4)

4.1. 5-HTT Cloning and Gene Organization

The human serotonin transporter (hSERT)* cDNA was initially cloned by Ramamoorthy and colleagues from a human placental cDNA library *(118)*. The cDNA encoded a 630 amino-acid protein and shared highest homology (44–48%) with NET and DAT when compared to Na^+/Cl^- cotransporter-gene-family members (*see* also Chapter 3, dendogram Fig. 1). Transfection of this cDNA into HeLa cells conferred Na^+ and Cl^--dependent, saturable 5-HT uptake that was potently antagonized by paroxetine and other selective serotonin reuptake inhibitors (SSRIs) *(118)*. Southern blotting of somatic cell hybrids and *in situ* hybridization localized the gene for SERT (5-HTT; SLC6A4) to chromosome 17 at 17q11.1–17q12 *(118,119)*. A cDNA encoding an identical hSERT protein was also isolated by polymerase chain reaction (PCR) from human dorsal raphe using oligonucleotide primers of rat SERT sequence *(120)*. Northern blot analysis re-

---

*We refer to serotonin transporter cDNA and protein with the acronym SERT and the cognate gene with the acronym 5-HTT, the latter proposed by Lesch and colleagues *(123)*.

vealed three major mRNA species, 6.8–8.2, 4.9, and 3.0–3.3 kb, in human dorsal raphe, platelets, lung and placenta, and *in situ* hybridization localized SERT mRNA to neurons in human dorsal raphe *(118,121,122)*. There are multiple, degenerate polyadenylation sequences that may result in alternate site usage *(122,123)*. 3' RACE from JAR cells or human whole blood identified two commonly used polyadenylation sites 567 and 690 bp downstream of the stop codon *(124)*. The gene for SERT (5-HTT; SLC6A4) was cloned by Heils and colleagues from a human genomic library, and intron-exon boundaries were identified by direct sequencing, revealing a 14-exon and 13-intron structure *(123*; Fig. 1C).

Subsequent studies have further elucidated the structure of the 5' region of the 5-HTT gene. 5' RACE suggested a single transcriptional start site located approx 14 kb upstream of translation initiation, with proximally located TATA-like box, as well as both an AP-1 and a CRE site *(123)*. A 1.4 kb region of the promoter confers tissue-specific activity to JAR cells (derived from placenta), but not SK-N-SH cells, and a repressor element lies within –1428 to –1185 bp *(123)*. Using 5' RACE and primer extension, Bradley and Blakely discovered an alternative splicing event in the 5' noncoding region of the gene, whereby an alternative exon, termed exon 1B, is joined to the 5' end of exon 2 and the 3' end of the previously identified exon 1, now termed exon 1A *(125*; Fig. 1C). A genomic clone was isolated, confirming exon 1A and 1B sequences, and their separation by a 736-bp intron. The splice variants, with the insertion or deletion of exon 1B, were expressed in equal proportions in JAR cells and human dorsal raphe *(22)*. An additional 381-bp sequence in the 5' region of the 5-HTT gene has recently been identified *(126–128*; Fig. 1C). The difficulty in identifying this region may stem from an instability of the DNA, making it prone to deletion during library construction. This suggests this region may also be subject to deletions in vivo that could affect expression.

4.2. 5-HTT as a Candidate Gene for Psychiatric Disorders

Serotonin plays a role in mood, aggression, appetite, sleep, cognition, and sexual and motor activity. Not surprisingly, 5-HT and SERT are implicated in some way in most psychiatric conditions. SSRIs are the most efficacious and commonly employed pharmaceutical tools in use. Alterations in the 5-HT system have been found in depression, including reduced levels of the 5-HT metabolite, 5-HIAA, increased relapse of SSRI-treated patients precipitated by a diet low in the 5-HT precursor tryptophan, and enhanced serotonergic transmission in response to antidepressant treatment *(129)*. In addition to its role as target for SSRIs, there is evidence that SERT-binding sites are reduced in both the platelets of depressed patients and the brains of depressed suicide victims *(129)*. 5-HT neurotransmission has also been associated with alcohol consumption. Increased 5-HT transmission has been associated with decreased alcohol self-administration, and 5-HT uptake into platelets was found to be increased in former alcoholics *(130,131)*. Animal stuies have shown an inverse relationship between levels of 5-HT and aggressive behavior. It has been suggested that 5-HT is involved in aggressive, impulsive behavior that may be present across a broad spectrum of psychiatric disorders, such as violent suicide *(131a)*. Reductions of 5-HIAA in brain stem and cerebrospinal fluid and SERT binding sites in several brain regions of suicides or attempted suicides have been reported *(131a)*. Overall, this evidence provides a good rationale for studies of the relationship of 5-HTT gene polymorphisms to psychiatric disorders.

4.3. 5-HTT Long/Short 5' VNTR: 5-HTTLPR

4.3.1. Identification of 5-HTTLPR and Relation to hSERT Expression Levels

Through PCR amplification of the 5' region of the 5-HTT gene, Heils and colleagues identified a repeat element with a 44-bp deletion polymorphism at position –1212 to –1255 *(133)* (Fig. 1C). The polymorphism, termed the 5-HTTLPR, generated PCR products of 484 or 528 bp, representing the short (s) or long (l) allele, respectively. In an analysis of 104 Caucasians, the s and l alleles had frequencies of 0.39 and 0.61, respectively. Transfection of constructs containing the l or s allele into the human placental choriocarcinoma cell line—which constitutively expresses 5-HTT—demonstrated that the l form had three-fold greater basal transcriptional activity than the 5 form *(133)*. Transfection of promoter constructs into lymphoblast cell lines demonstrated, for the l allele, two- to threefold greater basal transcriptional activity and increased stimulated activation through protein kinase C (PKC) and adenylate cyclase pathways *(134,135)*. Lymphoblast cell lines generated from individuals expressing the three genotypes—ss, sl, or ll—demonstrated that the ll genotypes, compared to the two genotypes with at least one s allele, had 1.4- to 1.7-fold greater mRNA levels, 30–40% more [^{125}I]RTI-55 membrane binding and 2× the 5-HT uptake *(134,135)*. That the ls and ss genotypes demonstrated similar characteristics, suggested that the s allele had a dominant effect on maintaining lower levels of hSERT expression *(134,135)*. A mechanism explaining this dominance has not yet been determined. The difference in promoter size or sequence of the s vs l alleles could affect the binding of transcription factors to the polymorphic region or to nearby segments of DNA.

The enhanced expression of hSERT associated with the 5-HTTLPR ll genotype has been replicated to some extent in a study of platelets. Platelet 5-HT uptake is considered a relevant measure because the SERTs expressed in brain and platelets have been shown to be identical *(136)*. In addition, 5-

HT uptake capacity is heritable, based on a study of twins, and there is evidence of altered platelet 5-HT uptake in depression *(129,137)*. In a study of 44 males, mostly Caucasian, who had not been taking psychotropic medications within the previous 2 mo, 5-HT uptake was 30% greater in platelets from subjects with the ll genotype compared to s-containing genotypes *(138)*. Interestingly, there was no difference in platelet membrane [^{3}H]paroxetine maximal binding or affinity or in platelet 5-HT content. However, it is probably an oversimplification to model brain SERT activity on that of platelets and more conflicting results have actually been observed when the correlation between genotype and brain SERT-binding site availability have been examined. For example, a recent SPECT study of 16 unaffected individuals revealed no 5HTT-LPR genotype-associated differences in [1^{23}I]β-CIT radioligand binding in several brain regions *(138a)*.

4.3.2. Population Frequencies of l and s Alleles of 5-HTTLPR

Several studies examining a variety of ethnic populations have expanded on the original observations of Lesch and colleagues regarding population frequencies of the 5-HTTLPR alleles. One such report found significant ethnic differences in l and s allele frequencies, with the l allele present at a frequency of 0.83 in Central Africans, 0.56 in North American Caucasians, and 0.33 in East Asians *(139)*. Another group also demonstrated a large range of allele frequencies in different populations; the s allele ranged from 0.25 in African Americans to 0.8 in the Japanese *(140)*. This group extended their work to additional ethnic groups of African, Asian, and European populations, finding large variations of the s and l alleles, with the s allele frequency ranging from 0.1 to 0.7 across different ethnic groups *(141)*. As noted previously, such a large variation could contribute to potential false-positive associations in studies using subject groups of mixed ethnic background, and should be considered when interpreting results from such reports.

These and other studies have also revealed several additional, although rare, repeat alleles. A novel 20-repeat allele has been identified *(139,142)*. This allele was termed extra-long (xl) and was observed in African Americans at a frequency of 0.02 and in a Congolese population at 0.06, but not in any North American or Western European Caucasians or East Asians *(139)*. Another study also found the xl allele, but only in the Japanese and African American populations *(140)*. Eighteen and 19-repeat alleles have also been reported *(140,142,143)*. Recently, Nakamura and colleagues reported novel alleles of 15- and 22-repeats, as well as novel alleles of s and l that differ in sequence composition of the individual repeat units *(144)*. The 15-, 19-, 20-, and 22-repeat alleles occurred rarely, and only in the Japanese but not Caucasian groups examined *(144)*.

4.4. 5-HTT Intron 3 VNTR

A VNTR has been identified in intron 3 (formerly designated intron 2 prior to the identification of a novel upstream intron) of the 5-HTT gene *(122;* Fig. 1C). The VNTR is composed of 17-bp repeats, and the most prevalent alleles are the 10- and 12-repeat *(122).* Examination of various ethnic groups has revealed wide variation in frequencies; the 10-repeat allele ranged from a frequency of 0.02 in the Japanese to 0.47 in European Americans *(140,141).* The 10-repeat allele was present at 0.26 in African Americans. This is worth noting, as this difference in frequencies between the European and African Americans is greater than the size of effects commonly observed in association studies, where these two ethnically diverse populations are often combined. A rare 9-repeat allele with a frequency of 0.01 and an 11-repeat allele with a frequency of 0.13 in a Mbuti population have also been identified *(140,141).*

Despite its location outside of the coding or promoter regions of the 5-HTT gene, recent evidence suggests that the intron 3 VNTR polymorphism may exert an influence on 5-HTT gene transcription. Transgenic embryos were generated that expressed either the 10- or 12-repeat of the intron 3 VNTR linked to a reporter gene. Both the 10- and 12-repeat sequences acted as strong transcriptional regulators. When cloned into an otherwise inactive LacZ promoter, the 10- and 12-repeat region imparted LacZ expression to the rostral hindbrain. The 12-repeat was strongly expressed in two additional brain-stem regions *(145).* Transient transfection of a luciferase reporter construct containing two copies of the 10- or 12-repeat into embryonic stem cells resulted in repressor and enhancer activity from both constructs, and the 12-repeat was a much stronger enhancer when cells were differentiated *(146).* The greater transcriptional activation from the 12-repeat could be the result of an increase in the number of transcriptional elements present in the larger 12-repeat construct, or caused by the size difference, which may allow a more favorable conformation of DNA for binding transcription factors. There were also sequence differences between the 10- and 12-repeat that could result in the ability of the 12-repeat to bind different factors than the 10-repeat. These findings support the theory that intron 3 of the 5-HTT gene plays a role in transcriptional regulation, and the polymorphic region may direct different levels of expression. As a variant with potential effects on transcriptional activity, the intron 3 VNTR holds promise for an association with clinical conditions.

4.4.1. 5-HTT 3' UTR Pst1 RFLP

A Pst1 RFLP has been identified in the 3' UTR of the 5-HTT gene *(147).* PCR amplification and restriction digestion with Pst1 revealed two alleles of 5.9 and 3.5 kb—with frequencies of 0.40 and 0.60, respectively—in a Caucasian population.

4.5. 5-HTT Deletion (5-HTT del)

As discussed previously, a novel 381-bp region of DNA was recently identified in genomic clones (*126,128*). The previous inability to isolate clones containing this region may be caused by the instability of this genomic sequence leading to deletion during the construction of libraries. This novel sequence is located between the 5-HTTLPR and the transcriptional start site, and contains AP1, AP2, CRE, and Sp1 consensus sites (*126,128;* Fig 1C). Transfection of promoter constructs containing this novel sequence into RN46A immortalized raphe cells and JAR cells enhances transcriptional expression compared to constructs lacking this region (*128*). The difficulty in cloning this site suggests that it may be unstable in vivo and thus polymorphic, demonstrating mosaicism in an individual, with deletion of the sequence in some cells but not others. Furthermore, the localization of this DNA to the 5' flanking region, the presence of consensus sites for *cis*-acting elements, and evidence of enhancer activity all suggest that polymorphisms at this site could be associated with altered expression. One group has found by PCR amplification assay that the deletion occurred in approx 50% of normal individuals examined (*127*). However, two other groups, including our laboratory, have found that all subjects examined contain the full sequence (*126,128*). The high rate of deletion observed by Lesch and colleagues could be caused by the use of PCR amplification as compared with direct Southern analysis. PCR does not allow rigorous quantitation, and thus the deletion may represent a rare event or be enhanced by PCR. Also, in judging relevance of this observation, it will be important to determine the extent to which this rearrangement occurs in tissues that express SERT, such as the dorsal raphe nucleus. It remains an interesting question as to whether this region represents a deletion carried by individuals that could contribute to genetic variability or an artifact of isolation and manipulation of genomic DNA or PCR amplification.

4.6. 5-HTT Single Nucleotide Polymorphisms (SNPs)

Initial studies of the 5-HTT gene structure revealed a limited number of SNPs. Lesch and colleagues isolated mRNA from platelets of 17 patients with MDD and 4 controls, used reverse-transcription to generate cDNA, and performed sequence analysis (*148*). One synonymous change in one bipolar patient was found (Table 3). Elucidation of the structure and sequence of the 5' region of the 5-HTT gene by Bradley and Blakely identified a SNP in exon 1B (*125;* Table 3). A SNP was also identified in one of the potential

Table 3
Single Nucleotide Polymorphisms in the 5-HTT Gene

Location	Nucleotide	Amino-acid variant	Ref.
exon 1B	A2631C		*39*
exon 1B	A2689T		*125*
exon 2	A483G	Thr4Ala	*149*
exon 2	G640C	Gly56Ala	*39,149*
exon 4	C1073T		*149*
exon 4	G1074A	Glu215Lys	*149*
exon 4	G1131A		*71b*
intron 4	G1226A		*149*
exon 5	C1931T		*149*
exon 6	C262T	Ser293Phe	*149*
exon 6	T308C		*148,149*
exon 7	C767T	Pro339Leu	*149*
intron 7	C229T		*149*
exon 8	C258A	Leu362Met	*149*
exon 8	C323T		*149*
intron 8	C220T		*149*
intron 8	C242T		*149*
exon 9	G302A		*149*
exon 9	A318G	Ile425Val	*149*
exon 9	G362A		*39,149*
intron 11	C518T		*149*
exon 13	A469C	Lys605Asp	*39,149*
exon 14	C274T	Pro621Ser	*149*
exon 14	G689T*		*124*

Nucleotide positions are from Genbank Accession numbers X76754-X76762, U79746. References: 6, dbSNP. *Nucleotide position is from *(124)*. Additional SNPs in introns and flanking regions have been deposited in the Single Nucleotide Polymorphism database at NCBI (dbSNP) at http://www.ncbi.nlm.nih.gov/SNP.

polyadenylation sites in the 3' UTR of the 5-HTT gene (*124*; Table 3). Recent assessments of larger numbers of individuals have yielded more fruitful results in the search for 5-HTT SNPs. A survey of the occurrence of SNPs in genes that encode proteins with functions relevant to cardiovascular and neurologic disorders, using hybridization to a high-density array of oligonucleotide DNA sequences or by denaturing HPLC, yielded four 5-HTT SNPs; two nonsynonymous, one synonymous, and one noncoding (*39*; Fig. 2C, Table 3). Utilizing DHPLC and direct sequencing of DNA from 450 individuals in the DNA Polymorphism Discovery Resource, additional low-frequency 5-HTT SNPs were revealed, 7 of which were novel amino acid variants (*149*; Fig. 2C, Table 3). Functional properties of all the 5-HTT variants have yet to be assessed.

4.7. 5-HTT Polymorphisms: Linkage and Association Studies

4.7.1. Anxiety/Personality Traits

In addition to identifying the 5-HTTLPR, Lesch and colleagues also demonstrated the first association of a polymorphism of the 5-HTT gene with a phenotype *(135)*. In their study, 505 subjects were genotyped, generating allele frequencies of 0.57 and 0.43 for the l and s alleles, respectively. The NEO personality inventory (NEO-PI-R), which measures neuroticism, based on anxiety and depressive symptoms, and estimates harm avoidance, based on anxiety symptoms, was administered to the subjects. The results revealed an association of s-containing genotypes with higher neuroticism scores, established by significantly higher ratings on the items for anxiety, anger, hostility, and impulsiveness. Furthermore, in 78 discordant sib pairs from within the total group, s genotypes were significantly linked to higher scores on neuroticism and harm avoidance *(135)*. Similar to the expression observed cell lines, the presence of the s allele exerted dominance on the observed in personality traits.

Many studies have attempted to replicate this finding, using different populations and a variety of personality inventories. Using the Temperament Character Inventory (TCI) and the NEO-PI, another group failed to replicate the association of the s allele with anxiety-related traits *(150)*. However, the l and s allele frequencies in the Japanese population in this study were 0.17 and 0.83, respectively. These are very different from the frequencies reported in the study of Lesch and colleagues. The ethnic differences, as well as the relatively small number of ll genotypes, may contribute to an inability to determine an association in the Japanese population. Treating neuroticism as a quantitative trait, another group took the upper and lower extremes of scores and compared the allele and genotype frequencies of the 5-HTTLPR, as well as the intron 3 VNTR *(151)*. No association with either polymorphism was identified. One danger in the use of this design is that the extremes of a quantitative trait may have different genetic determinants than the population in the central parts of the distribution. To diminish this problem, another study used the top and bottom 20% of scores for neuroticism on the NEO-FFI scale. There were no significant associations of personality trait with the 5-HTTLPR or intron 3 polymorphisms *(152)*. An analysis of an Israeli population employed the TPQ questionnaire, with the dimension of harm avoidance that incorporates aspects of anxiety and is correlated with NEO-PI neuroticism—the latter previously shown to be associated with s allele *(153)*. Despite the use of very similar personality assessment, no association was found. These studies can be criticized because their sample sizes are too limited to observe a small but real effect. However, a large case-control study of 759 Caucasians, with a reported power of 80% to detect an association accounting for 1% of the variance, also found no association of the 5-HTTLPR with anxiety-related traits, including neuroticism and depressive traits *(154)*. Furthermore, a study of 225 subjects, utilizing both the NEO-PI and TPQ inventories, failed to replicate the findings of Lesch *(155)*. One group has replicated the increase in scores of harm avoidance on the TPQ and of anxiety and depression subscales of neuroticism on the NEO-PI-R *(156)*. This group also found that the s allele in the presence of the homozygous genotypes of a COMT polymorphism was associated with higher scores on the RD2 scale of the TPQ questionnaire, which measures mainly a personality trait of persistence *(157)*. Finally, one study has reported that the presence of the s allele, although not associated with smoking *per se*, was associated with neuroticism scores in smokers and a correlation of neuroticism scores with smoking to heighten stimulation and decrease negative mood *(1058,159)*.

Recently, Lesch and colleagues returned to this issue and examined a population of 394 subjects, tmostly female, either alone or as a combined group with the population from the initial study *(160)*. As in the first study, the new group was composed of siblings, allowing family-based analyses. They found a highly significant association of s-containing genotypes with increased neuroticism, decreased agreeableness and anxiety. These associations were still strong when Caucasians alone were analyzed, indicating that population stratification was not responsible for the observed associations. Discordant sib-pair analysis also maintained the association of neuroticism and anxiety. In summary, reproducibility of the association of the s allele of the 5-HTTLPR with neuroticism has met with conflicting results, even when using large sample sizes and the same personality assessments, but the positive results suggest the presence of an associated trait at least in some populations. The definition of behavioral characteristics of this trait may need to be refined in order to gain reproducibility and a stronger association.

4.7.2. Affective Disorders

Efforts to identify an association of the 5-HTT intron 3 polymorphism with affective disorder have yielded mixed results. One group has found an association of the 9-repeat allele with unipolar depression in Scottish populations, in which the allele frequencies in controls were 0.01 for the 9-, 0.4 for the 10-, and 0.59 for the 12-repeat allele *(161,162)*. The total affective disorders group and the unipolar subgroup were both significantly different from controls in allele frequency. The difference was caused by an increase in the 9-repeat allele in patients, with a 4.4-fold increased risk associated with the presence of this allele. However, the 9-repeat allele is rare, comprising only 1% of the population alleles in controls and 3.6% in the affective disorders group. This finding suggests that the 9-repeat allele may contribute to affective disorder in only a small subset of patients. Furthermore, this has not been replicated in several studies. For example, a study of German unipolar and bipolar patients found similar allele frequencies compared to the control Scottish populations in the Harmar and Ogilvie studies, but no association of the 9-repeat allele with either depression diagnosis *(163)*. Furthermore, an association of the 12-repeat allele was reported in another study *(132)*. A study employing a meta-analysis design, which included the results of the studies described previously and had a calculated power of 80% to detect an odds ratio of <1.5, found no involvement of the intron 3 VNTR in unipolar or bipolar depression *(164)*. Another study has demonstrated an association of bipolar depression with the 12/12 genotype in a Japanese population *(142)*. However, the controls in this study had frequencies of 0.89 for the 12-repeat allele and 0.11 for the 10-repeat allele, which differ from the frequencies observed in Caucasians participating in studies reporting an association of the 9-repeat allele. Other recent work also demonstrated a lack of association of the intron 3 VNTR with bipolar disorder or MDD *(165,166)*. These conflicting results suggest that further replication is needed to support an involvement of the 5-HTT intron 3 polymorphism in affective disorder.

There have also been analyses of the 5-HTTLPR and affective disorder. Two family-based studies found no linkage of the 5-HTTLPR and bipolar disorder. In the first report, 223 bipolar-affected individuals and their parents and other relatives demonstrated no linkage of the s or l alleles in inheritance by descent (IBD) or TDT analysis *(167)*. The second study assessed, also with the TDT, 102 bipolar probands and their parents. They calculated a power of 90% to detect an increased risk of 1.9 to 4, but observed no significant linkage; the l and s were transmitted equally to affected individuals *(168)*. However, one family-based study of bipolar probands and their parents used TDT to find a preferential transmission of the s allele *(169)*. There is also some suggestion in case-control studies of association of the s allele and ss genotype with MDD, and unipolar or bipolar depression *(164,170,171)*. A significant association of the s allele with seasonal affective disorder (SAD) has also been shown *(172)*. However, the controls and patients in this study differed in gender and race. Furthermore, an equal number of reports support a lack of association of the 5-HTTLPR with depressive disorders *(142,165,173–175)*.

The relationship of SNPs in the 5-HTT gene to affective disorder has also been examined. A SNP identified in one of the potential polyadenylation sites in the 3' UTR of the 5-HTT gene was not associated with bipolar or unipolar depression *(124)*. Interestingly, a study using extended TDT (ETDT) to examine haplotype transmission from parents to their offspring with bipolar depression demonstrated a preferential transmission of the s allele of the 5-HTTLPR with the T allele of the 3' UTR polymorphism *(169)*. Overall, there is not strong support for the contribution of polymorphisms of the 5-HTT gene to affective disorder, and given that some of these studies employed large sample sizes or reported a high level of power to detect an effect, defining subtypes of affective disorder that suggest hSERT involvement may be important for future success.

The efficacy of SSRIs as antidepressant agents and their primary site of action as blockers of SERT activity indicate that polymorphisms in the 5-HTT gene could influence treatment response. Indeed, there are some intriguing recent reports of the impact of 5-HTTLPR genotype on antidepressant response. MDD patients with psychotic features treated with the SSRI fluvoxamine did not respond as well, as measured by scores on the Hamilton Depression Rating Scale (HDRS), if they were of the ss genotype *(176)*. Furthermore, this difference in response was eliminated when fluvoxamine treatment was supplemented with pindolol. Pindolol is hypothesized to hasten the ameliorative actions of SSRIs by antagonizing the 5-HT1A autoreceptors to alleviate the initial inhibitory effects of SSRIs on 5-HT neuronal firing. This group extended these findings to a group of MDD patients without psychotic features in which paroxetine-treated ll and ls genotypes improved significantly on HDRS scores at 2 and 4 wk compared to ss genotypes *(177)*. Furthermore, the association of the s-containing genotypes with poorer treatment response and pindolol augmentation appears to be independent of depression diagnostic subtype or the presence of delusional features *(178)*. These results were supported by another group showing that older depressed patients with the ll genotype responded more rapidly to the SSRI, paroxetine, compared to those patients with s-containing genotypes. Interestingly, there was no difference in the response of different genotype groups to nortriptyline, an uptake inhibitor acting more selectively at NET *(179)*. One group has examined 5-HTTLPR genotype in bipolar patients experiencing manic episodes during, and believed to be provoked by, antidepressant treatment *(168)*. Highly significant results were obtained for association of the s allele and ss genotype with patients who had experienced such episodes. In sumary, although there does not appear to be strong evidence for the 5-HTTLPR in affective disorder, the 5-HTTLPR genotype may have predictive power for SSRI treatment response. This presents an opportunity for a pharmacogenomic approach to the treatment of affective disorder whereby selection of therapeutics can be matched to genetic background.

4.7.3. Suicide

Findings of altered 5-HT metabolites and SERT-binding sites in suicide suggest a potential role for 5-HTT polymorphisms. One study examined alcohol-dependent patients who had previous suicide attempts. No association with alcohol dependence was found, but s-containing genotypes were increased in alcoholics with suicide attempts compared to controls *(180)*. Furthermore, dividing the suicide attempters into groups by number of attempts, resulted in an increased frequency of the s allele as the number of attempts increased. These results were not replicated in another study of 82 suicides and 132 nonsuicide controls. There were no significant associations of allele or genotype frequencies with depression or suicide, except for a higher rate of heterozygous genotypes in the depressed group *(181)*. The subjects in the latter study were a mixture of ethnicities that have demonstrated differences in 5-HTTLPR allele frequencies (whites, Hispanics, African Americans, and Asians) and the ethnic makeup differed between suicides and controls, mainly because there were less African Americans (9 vs 28%) in the depressed group. This difference may have limited their ability to observe an effect of genotype on suicide or depression. Additionally, although there was decreased [^{3}H]cyanoimipramine binding in the prefrontal cortex in depressed patients and suicide victims, there was no effect of genotype on binding sites. Therefore, this work does not support an association of the 5-HTTLPR with suicide or reports of lower uptake sites of s-containing genotypes described in cell lines and platelets. A third study also suggested only small involvement of the 5-HTTLPR in suicide. Forty-nine patients hospitalized for a suicide attempt and 49 controls were compared, and there were no 5-HTTLPR allele or genotype differences between these groups *(182)*. However, when these patients were grouped by genotype, the ll genotype had higher hopelessness scores than the grouped ls or ss scores. Another recent study has failed to replicate an association of the 5-HTTLPR with suicide *(183)*. Association of the intron 3 VNTR has also been examined in a study of 346 suicide attempters and 382 controls. The patients were admitted to the hospital following a suicide attempt, 94% of which were overdose of medication, as opposed to a violent attempt *(184)*. There were no allele or genotype differences between controls and patients, whereas there was an association between low scores for anxiety and the 12-repeat allele. Taken together, these findings fail to support a strong association of 5-HTT gene polymorphisms with suicide. Interestingly, research that focuses on suicide with elements of impulsive behavior has revealed an association with the 5-HTTLPR, emphasizing once again the importance of refining phenotype definition.

4.7.4. Substance Abuse and Dependence

Several studies have examined the potential association of the 5-HTTLPR polymorphism with substance abuse and dependence. Several groups have found an excess of the s allele and/or ss genotype in alcohol-dependent subjects (185–188). In one study, there was a trend for an increase in ss genotypes accompanying increasing severity of withdrawal symptoms, including seizures and delusions (187). Another study reported that the s allele and ss genotype were significantly higher in young adults with high alcohol tolerance compared to controls (189). However, one study associated the presence of the ll genotype with alcohol dependence. In a longitudinal study of alcohol dependence, 41 subjects were genotyped, and it was found that subjects with ll genotypes had, in the initial phase of the study, significantly lower levels of response to alcohol, which is thought to be a predisposing factor to alcoholism (190). Additionally, ll genotypes were more likely to have received a diagnosis of alcohol abuse or dependence during the study. Finally, no association of the 5-HTTLPR or intron 3 VNTR with alcohol dependence was found in a large European American population (140). Overall, there is support for an association of the s allele of the 5-HTTLPR with alcohol dependence, and continued success in this area may depend on the designation of subgroups, such as symptom severity or personality traits.

Two reports have examined the 5-HTTLPR as well as SERT-binding sites in the brain in alcohol-dependent subjects and cocaine users. In these reports, there were no significant genotype differences between the controls and any of the drug-use groups, but there were interesting findings regarding the relationship between genotype and binding-site availability. In one report, the sl genotypes were accompanied by less post-mortem $[^{125}I]\beta$-CIT binding and SERT mRNA in the dorsal raphe in a group combining cocaine users and controls compared to ll genotypes, but there were too few ss genotypes to make conclusions about this group (191). However, this relationship was not observed in the alcohol-use group, as $[^{125}I]\beta$-CIT binding was increased in alcoholics with s-containing genotypes compared to cocaine user and control groups (191). This apparent interaction of genotype with alcoholism on SERT-binding sites could be the result of a different array of genetic contributions in alcoholics, or the effects of alcohol and many other environmental factors associated with chronic alcohol use. Therefore, studies correlating genotype with binding sites that use clinical populations as part of the sample should be interpreted with caution. In the second study, examining controls only, the ll genotype demonstrated twofold greater $[^{123}I]\beta$-CIT binding in the dorsal brain stem than ss-carrier controls (192).

The increased binding in the controls with the ll genotype supports the data from cell-culture and platelet studies that this genotype directs higher levels of expression. However, alcohol-dependent subjects with the ll genotype demonstrated less $[^{123}I]\beta$-CIT binding than ll controls. The lack of an association of genotype with binding sites in alcohol-dependent patients in both of the previous studies suggests a more complex interaction that may involve genotype as well as environmental influences—particularly chronic alcohol use, which merits further study. These results indicate that more comprehensive studies in groups free from psychiatric diagnoses and exposure to medications and drugs of abuse are required to understand the relationship between genotype and expression levels.

4.7.5. Impulsivity

There have been efforts to establish an association of 5-HTT gene polymorphisms with clinical conditions that involve elements of impulsivity, such as alcohol dependence. Early onset of alcoholism, with dissocial and impulsive-aggressive behavior, describes the Type 2 subgroup of alcoholics *(193)*. Type 2 alcoholics exhibit alterations in aspects of serotonergic system function and high novelty-seeking and low harm-avoidance scores on the TPQ personality assessment *(193)*. One study examined the relationship of 5-HTTLPR genotype to alcohol-dependent subjects with dissocial personality *(194)*. They found a trend for an excess of ss genotypes in the dissocial alcoholic group, as well as a trend for higher novelty-seeking scores and lower harm-avoidance scores in the dissocial alcoholics. Furthermore, personality-trait scores were significantly different when dissocial alcoholics with ss genotypes were compared to dissocial alcoholics with other genotypes, suggesting an interaction effect of ss genotype on dissocial alcoholics. In another study, impulsive, violent, early-onset alcoholics were compared to late-onset alcoholics and controls for frequency of the 5-HTTLPR polymorphism. The odds ratio for ss vs ll was 3.9 when impulsive, violent, early-onset alcoholics were compared to controls, and the s allele frequency was also significantly higher *(195)*. However, this finding was not replicated in German alcohol-dependent patients, divided into high- and low-impulsivity groups, compared to controls *(196)*. An association has also been made between the 5-HTTLPR and suicidal behavior with elements of violence. One compelling study reported that 91% of suicides but only 67% of controls had an s-containing genotype, and the s allele frequency was 0.41 for controls and 0.59 for suicides *(197)*. The authors suggest an association with violent suicide, as most suicides in this study were committed by violent means. Indeed, of the five suicides not committed by violent means, three of the individuals were homozygous for ll genotypes and none were homozygous ss genotypes. This was recently replicated by another group *(198)*. Taken together, the data suggest an association of the s allele of the 5-HTTLPR with impulsive violent behavior, that may be alternately expressed as violence comorbid with alcohol dependence or suicide attempts. These findings advocate extracting a common phenotype from a cluster of characteristics of complex disorders that may more accurately reflect the traits to which the gene is contributing.

4.7.6. Schizophrenia

Most studies have found no association of the 5-HTTLPR with schizophrenia *(199–201)*. One report found a subgroup association; when neuroleptic-free for 2 wk, patients with the ll genotype had significantly higher scores on the thought disturbance scale, because of increased hallucinations *(202)*. There is also limited evidence for an association of the intron 3 VNTR. One study demonstrated a highly significant preferential transmission of the 12-repeat allele of the intron 3 VNTR in a study of 133 trios *(203)*. However, another report observed no differences in allele frequencies between controls and schizophrenics *(204)*.

4.7.7. OCD

Obsessive-compulsive disorder (OCD) is uniquely treated by SERT-preferring antidepressants, suggesting that SERT function contributes to this disorder. Analyses of the 5-HTTLPR and OCD have yielded evidence of both linkage disequilibrium of the l allele and association of the ll genotype with OCD *(205,206)*. However, other studies have failed to find a relationship *(207,208)*. Additionally, one group has sequenced 22 OCD patients and found no polymorphisms *(209)*. Based on results of a genome-wide linkage analysis in seven OCD pedigrees that revealed a LOD score of 2.06 for markers near the 5-HTT locus, our laboratory assessed these OCD patients for SNPs in the 5-HTT gene *(210)*. This revealed that the A2631C SNP in exon 1B was expressed at similar frequencies in OCD patients and controls (S. Belous, unpublished observation). Although the latter findings suggest that common coding variants do not precipitate OCD, it will be important to evaluate this further, in larger sample sizes or in subgroups of larger samples demonstrating specific phenotypes.

4.7.8. Autism

Elevated platelet 5-HT levels have been reported in autistic patients, suggesting a potential contribution of SERT to this disorder *(211)*. Cook and colleagues were the first to demonstrate an involvement of 5-HTT gene polymorphisms in autism *(212)*. Eighty-six trios, each consisting of a proband with autism and parents, were assessed for the 5-HTTLPR and intron 3

VNTR polymorphisms. The majority were Caucasian—with African-, Hispanic-, and Asian-Americans comprising the remaining subjects—and almost all probands were male. TDT analysis showed an increased transmission of the s allele to the proband *(212)*. However, another family based TDT of German families showed linkage of the l allele in a combined group that included autistic probands that had no language delay *(213)*. These two reports agree in finding no preferential transmission of the intron 3 VNTR. The l allele was also shown to be in excess in, and preferentially transmitted to, the autistic probands in another family-based study *(214)*. Employing severity subgroups, an enhanced transmission of the l allele to affected probands with mild social interaction and communication symptoms was identified *(215)*. A genome-wide linkage analysis of 152 sib pairs revealed significant LOD scores for chromosome 17 *(216)*. Notably, a LOD score of 2.3 was generated for a marker located specifically within intron 2 of the 5-HTT gene. Furthermore, there appeared to be a parent-of-origin effect at this marker with increased paternal transmission. Case-control studies have been unsuccessful in demonstrating an association of the 5-HTTLPR with autism *(217)*. In order to establish a link between the 5-HTT gene and autism, additional studies utilizing comparable diagnostic criteria and an increased number of markers may yield more definitive results.

4.7.9. Other Diseases

A limited amount of data is available on the role of 5-HTT polymorphisms in other neuropsychiatric diseases. An association of the s allele and ss genotype was found in both a Brazilian and a Caucasian population of patients with Alzheimer's disease *(218,219)*. The 5-HTTLPR association with Alzheimer's disease was not replicated in German and Japanese populations *(220)*. One case-control study has shown no association of the 5-HTTLPR with panic disorder *(221)*. Finally, two studies failed to show a relationship between the 5-HTTLPR and anorexia nervosa *(222,223)*. However, one report has suggested an association of bulimia with an excess of s-containing genotypes *(224)*.

5. SUMMARY AND FUTURE DIRECTIONS

The cloning of the MA transporter genes over the last 10 yr has provided gene structure information that can be used to begin to understand transporter expression and function. There is much yet to learn about the role of the 5' promoter region, 3' region, and introns in MA transporter gene expression, the role of amino-acid sequence in transporter structure and function, and how polymorphisms in any of these regions may influence behavior and personality and affect predisposition to illness. Indeed, SNP discovery has only recently begun in earnest, and this search should yield variants that contribute to the genetic diversity of MA transporters. It also remains to be determined to what extent alternative splicing of hNET, hSERT, and potentially hDAT, occurs in the human brain, whether splice variants alter transporter expression and function, and if there are polymorphisms that favor one splice pattern over another.

Our expanding knowledge of complex disorders has allowed some hypotheses of which diseases are predicted to exhibit genetic variability in MA transporters. Armed with these tools, researchers have met with some success in identifying genetic variants in the MA-transporter genes, their association with disease, and their potential impact on transporter function. The link between the hDAT 3' VNTR 10-repeat allele and ADHD serves as a useful example of how promising inroads are being established while many questions remain. The selection of hDAT as a candidate gene was strongly hypothesis-driven, given the importance of DA systems in motor activity and attentional processes, both disrupted in ADHD and evidence of altered hDAT-binding sites in the brains of ADHD patients. This supported hDAT involvement in ADHD and the potential of finding a hDAT association with the disorder. However, the evidence of a correlation of the 10- vs 9-repeat of the 3' VNTR with hDAT expression in normals is unclear. This is troubling, as most scientists will search not only for an association of a polymorphism with a trait, but some understanding of the mechanism by which a polymorphism exerts its effects. It is possible, in the case of 10-repeat allele of the hDAT 3' VNTR, that the associated polymorphism may reflect a linkage disequilibrium with another polymorphism or that variation in other genes, in combination with the hDAT gene polymorphism, determine hDAT expression. The former is suggested by the exciting studies of haplotypes that examine the interaction of several polymorphic sites as a single unit potentially linked to disease, or signify in a more refined way than one marker the localization of relevant gene sequences. Furthermore, the polymorphism could alter an aspect of hDAT function not yet assayed, such as transcriptional activation under conditions of stress. Such issues will continually resurface, and must be addressed as novel MA transporter polymorphisms and interactions with disease continue to emerge. Finding an association of a polymorphism with disease is clearly the first of many steps toward understanding how genetic differences shape us.

Constructing hypotheses of disease etiology and designating candidate genes enhances the power of genetic studies, but proving the association of complex diseases influenced by many genes that contribute small individual effects requires a large sample size, and even then may prove elusive. It may be possible to combat this limitation by applying phenotype definitions that more narrowly define the disorder and more accurately model the underlying genetic composition. Our laboratory has successfully used such an approach. Designating alterations in DHPG levels and tyramine responsiveness as indicative of hNET dysfunction, the identification of the nonfunctional polymorphism, A457P, was achieved. This provides a rational basis for assessing these biochemical measures in other potential subjects for hNET polymorphism discovery. The information obtained by identifying just one individual with a nonfunctional hNET is invaluable. Elevated standing heart rate may be a highly predictive endophenotype, particularly for those polymorphisms that are highly disruptive to hNET function. This would provide a simple method to identify a subgroup within an affected group of interest. For example, although ADHD has been studied for hDAT polymorphisms, the role of the noradrenergic system in cognition and attention through actions in the prefrontal cortex suggests an involvement in ADHD *(225)*. Additionally, TCAs such as desipramine, and the experimental selective norepinephrine reuptake inhibitor tamoxetine, have been effective in ADHD *(225)*. Perhaps a subset of ADHD involves the NE system to a greater extent than the DA system, and a subset of these patients have hNET polymorphisms. Selecting a subset of ADHD patents with elevated standing heart rate may greatly increase the chances of finding an association of hNET with this disorder.

Similar refinements can be applied to the search for hDAT and 5-HTT polymorphisms. For example, our laboratory is currently attempting to identify 5-HTT polymorphisms in psychiatric patients whose diagnoses include gastrointestinal complaints. 5-HT has excitatory actions on myenteric neurons, and SERT is expressed in the crypt epithelial cells of the intestine *(226,227)*, Furthermore, gastrointestinal complaints are often comorbid with mood and anxiety disorders *(228)*. 5-HTT gene polymorphisms may generate phenotypes of combined psychiatric and gastrointestinal abnormalities. Similarly, DAT is also expressed in the bowel *(141)*, and may play a role in gastrointestinal function that will also provide a new phenotype to explore. For example, ADHD patients with bowel disorders could harbor hDAT polymorphisms.

The rapid evolution of technology including microarray chips, DHPLC, and microsphere-based addressing and sorting of SNPs will also aid in the identification of MA transporter polymorphisms in disease. These techniques allow increased throughput for both SNP discovery and genotyping of identified SNPs. Advantages include the ability to identify SNPs and larger-length variants without sequencing the entire region (DHPLC), and the researcher need only examine nucleotide positions identified as variable by the analysis software (microarray chips). In addition, multiplexing opportunities are feasible with all of these techniques. Higher throughput methodologies will aid in assaying the increasing number of markers that are being identified and can be used in more complex haplotype analyses. Utilizing highly selective phenotype definitions in combination with multiple informative markers should enhance our ability to clarify how genetic variation in transporters poses risk to complex human disease.

REFERENCES

1. Barker, E. L. and Blakely, R. D. (1995) Norepinephrine and serotonin transporters: molecular targets of antidepressant drugs, in *Psychopharmacology: The Fourth Generation of Progress*. (Bloom, F. E. and Kupfer, D. J., eds.), Raven Press, New York, NY.
2. Amara, S. G. and Kuhar, M. J. (1993) Neurotransmitter transporters: recent progress. *Annu. Rev. Neurosci.* **16**, 73–93.
3. Melikian, H. E., McDonald, J. K, Gu, H., Rudnick, G., Moore, K. R., and Blakely, R. D. (1994) Human norepinephrine transporter. Biosynthetic studies using a site- directed polyclonal antibody. *J. Biol. Chem.* **269**, 12,290–12,297.
4. Melikian, H. E., Ramamoorthy, S., Tate, C. G., and Blakely, R. D. (1996) Inability to N-glycosylate the human norepinephrine transporter reduces protein stability, surface trafficking, and transport activity but not ligand recognition. *Mol. Pharmacol.* **50**, 266–276.
5. Apparsundaram, S., Galli, A., DeFelice, L. J., Hartzell, H. C., and Blakely, R. D. (1998) Acute regulation of norepinephrine transport: I. protein kinase C- linked muscarinic receptors influence transport capacity and transporter density in SK-N-SH cells. *J. Pharmacol. Exp. Ther.* **287**, 733–743.
6. Apparsundaram, S., Schroeter, S., Giovanetti, E., and Blakely, R. D. (1998) Acute regulation of norepinephrine transport: II. PKC-modulated surface expression of human norepinephrine transporter proteins. *J. Pharmacol. Exp. Ther.* **287**, 744–751.
7. Bauman, A. L., Apparsundaram, S., Ramamoorthy, S., Wadzinski, B. E., Vaughan, R. A., and Blakely, R. D. (2000) Cocaine and antidepressant-sensitive biogenic amine transporters exist in regulated complexes with protein phosphatase 2A. *J. Neurosci.* **20**, 7571–7578.
8. Ramamoorthy, S. and Blakely, R. D. (1999) Phosphorylation and sequestration of serotonin transporters differentially modulated by psychostimulants. *Science* **285**, 763–766.
9. Bengel, D., Murphy, D. L., Andrews, A. M., Wichems, C. H., Feltner, D., Heils, A., et al. (1998) Altered brain serotonin homeostasis and locomotor insensitivity to 3, 4-methylenedioxymethamphetamine ("Ecstasy") in serotonin transporter-deficient mice. *Mol. Pharmacol.* **53**, 649–655.
10. Giros, B., Jaber, M., Jones, S. R., Wightman, R. M., and Caron, M. G. (1996) Hyperlocomotion and indifference to cocaine and amphetamine in mice lacking the dopamine transporter. *Nature* **379**, 606–612.

11. Xu, F., Gainetdinov, R. R., Wetsel, W. C., Jones, S. R., Bohn, L. M., Miller, G. W., et al. (2000) Mice lacking the norepinephrine transporter are supersensitive to psychostimulants. *Nature Neurosci.* **3,** 465–471.
12. Pacholczyk, T., Blakely, R. D., and Amara, S. G. (1991) Expression cloning of a cocaine- and antidepressant-sensitive human noradrenaline transporter. *Nature* **350,** 350–354.
13. Ramamoorthy, S., Prasad, P. D., Kulanthaivel, P., Leibach, F. H., Blakely, R. D., and Ganapathy, V. (1993) Expression of a cocaine-sensitive norepinephrine transporter in the human placental syncytiotrophoblast. *Biochemistry* **32,** 1346–1353.
14. Pörzgen, P., Bönisch, H., Hammermann, R., and Brüss, M. (1998) The human noradrenaline transporter gene contains multiple polyadenylation sites and two alternatively spliced C-terminal exons. *Biochim. Biophys. Acta* **1398,** 365–370.
15. Gelernter, J., Kruger, S., Pakstis, A. J., Pacholczyk, T., Sparkes, R. S., Kidd, K. K., et al. (1993) Assignment of the norepinephrine transporter protein (NET1) locus to chromosome 16. *Genomics* **18,** 690–692.
16. Brüss, M., Kunz, J., Lingen, B., and Bönisch, H. (1993) Chromosomal mapping of the human gene for the tricyclic antidepressant-sensitive noradrenaline transporter. *Human Genet.* **91,** 278–280.
17. Pörzgen, P., Bönisch, H., and Brüss, M. (1995) Molecular cloning and organization of the coding region of the human norepinephrine transporter gene [published erratum appears in *Biochem. Biophys. Res. Commun.* (1996) **227(2),** 642,643]. *Biochem. Biophys. Res. Commun.* **215,** 1145–1150.
18. Kim, C. H., Kim, H. S., Cubells, J. F., and Kim, K. S. (1999) A previously undescribed intron and extensive 5' upstream sequence, but not Phox2a-mediated transactivation, are necessary for high level cell type-specific expression of the human norepinephrine transporter gene. *J. Biol. Chem.* **274,** 6507–6518.transporter. *Hum. Genet.* **91,** 278–280.
19. Meyer, J., Wiedemann, P., Okladnova, O., Bruss, M., Staab, T., Stober, G., et al. (1998) Cloning and functional characterization of the human norepinephrine transporter gene promoter. *J. Neural Transm.* **105,** 1341–1350.
20. Burton, L. D., Kippenberger, A. G., Lingen, B., Bruss, M., Bonisch, H. and Christie, D. L. (1998) A variant of the bovine noradrenaline transporter reveals the importance of the C-terminal region for correct targeting to the membrane and functional expression. *Biochem. J.* **330,** 909–914.
21. Kitayama, S., Ikeda, T., Mitsuhata, C., Sato, T., Morita, K., and Dohi, T. (1999) Dominant negative isoform of rat norepinephrine transporter produced by alternative RNA splicing. *J. Biol. Chem.* **274,** 10,731–10,736.
22. Torres, G. E., Tao, W. D., Mohn, A. R., Quan, H., Kim, K. M., Levey, A. L., et. al. (2001) Functional interaction between monoamine plasma membrane transporters and the synaptic PD2 domain-containing protein PICK1. *Neuron.* **30,** 121–134.
23. Leonard, B. E. (1997) The role of noradrenaline in depression: a review. *J. Psychopharmacol.* **11,** S39–S47.

24. Ressler, K. J. and Nemeroff, C. B. (1999) Role of norepinephrine in the pathophysiology and treatment of mood disorders. *Biol. Psychiatry* **46,** 1219–1233.
25. Schildkraut, J. J. (1965) The catecholamine hypothesis of affective disorders: a review of supporting evidence. *Am. J. Psychiatry* **122,** 509–522.
26. Brady, L. S. (1994) Stress, antidepressant drugs, and the locus coeruleus. *Brain Res. Bull.* **35,** 545–556.
27. Linner, L., Arborelius, L., Nomikos, G. G., Bertilsson, L., and Svensson, T. H. (1999) Locus coeruleus neuronal activity and noradrenaline availability in the frontal cortex of rats chronically treated with imipramine: effect of alpha 2-adrenoceptor blockade. *Biol. Psychiatry* **46,** 766–774.
28. Klimek, V., Stockmeier, C., Overholser, J., Meltzer, H. Y., Kalka, S., Dilley, G., and Ordway, G. A. (1997) Reduced levels of norepinephrine transporters in the locus coeruleus in major depression. *J. Neurosci.* **17,** 8451–8458.
29. Eisenhofer, G., Cox, H. S., and Esler, M. D. (1990) Parallel increases in noradrenaline reuptake and release into plasma during activation of the sympathetic nervous system in rabbits. *Naunyn Schmiedebergs Arch. Pharmacol.* **342,** 328–335.
30. Eisenhofer, G., Esler, M. D., Meredith, I. T., Dart, A., Cannon, R. O., Quyyumi, A. A., et al. (1992) Sympathetic nervous function in human heart as assessed by cardiac spillovers of dihydroxyphenylglycol and norepinephrine. *Circulation* **85,** 1775–1785.
30a. Imamura, M., Lander, H. H., and Levi, R. (1996) Activation of histamine H-3 receptor inhibits carrier-mediated norepinephrine release during protracted myocardialeschemia. Comparison with adenosine A1-receptors and alpha-2 adrenoceptors, *Circ. Res.* **78,** 475–481.
31. Bohm, M., La Rosee, K., Schwinger, R. H., and Erdmann, E. (1995) Evidence for reduction of norepinephrine uptake sites in the failing human heart. *J. Am. Coll. Cardiol.* **25,** 146–153.
32. Liang, C. S., Fan, T. H., Sullebarger, J. T., and Sakamoto, S. (1989) Decreased adrenergic neuronal uptake activity in experimental right heart failure. A chamber-specific contributor to beta-adrenoceptor downregulation. *J. Clin. Invest.* **84,** 1267–1275.
33. Merlet, P., Dubois-Rande, J. L., Adnot, S., Bourguignon, M. H., Benvenuti, C., Loisance, D., et al. (1992) Myocardial beta-adrenergic desensitization and neuronal norepinephrine uptake function in idiopathic dilated cardiomyopathy. *J. Cardiovasc. Pharmacol.* **19,** 10–16.
34. Schafers, M., Dutka, D., Rhodes, C. G., Lammertsma, A. A., Hermansen, F., Schober, O., et al. (1998) Myocardial presynaptic and postsynaptic autonomic dysfunction in hypertrophic cardiomyopathy. *Circ. Res.* **82,** 57–62.
35. Esler, M., Jackman, G., Bobik, A., Leonard, P., Kelleher, D., Skews, H., et al. (1981) Norepinephrine kinetics in essential hypertension. Defective neuronal uptake of norepinephrine in some patients. *Hypertension* **3,** 149–156.
36. Gelernter, J., Kruger, S., Kidd, K. K., and Amara, S. (1993) TaqI RFLP at norepinephrine transporter protein (NET) locus. *Human Mol. Genet.* **2,** 820.
37. Hadley, D., Hoff, M., Holik, J., Reimherr, F., Wender, P., Coon, H., et al.

(1995) Manic-depression and the norepinephrine transporter gene. *Human Hered.* **45**, 165–168.

38. Roses, A. D. (2000) Pharmacogenetics and the practice of medicine. *Nature* **405**, 857–865.
39. Cargill, M., Altshuler, D., Ireland, J., Sklar, P., Ardlie, K., Patil, N., et al. (1999) Characterization of single-nucleotide polymorphisms in coding regions of human genes. *Nat. Genet.* **22**, 231–238.
40. Weiss, K. M. and Terwilliger, J. D. (2000) How many diseases does it take to map a gene with SNPs? *Nature Genet.* **26**, 151–157.
41. Robertson, D. (1999) The epidemic of orthostatic tachycardia and orthostatic intolerance. *Am. J. Med. Sci.* **317**, 75–77.
42. Jacob, G., Shannon, J. R., Costa, F., Furlan, R., Biaggioni, I., Mosqueda-Garcia, R., et al. (1999) Abnormal norepinephrine clearance and adrenergic receptor sensitivity in idiopathic orthostatic intolerance. *Circulation* **99**, 1706–1712.
43. Shannon, J. R., Flattem, N. L., Jordan, J., Jacob, G., Black, B. K., Biaggioni, I., et al. (2000) Clues to the origin of orthostatic intolerance: a genetic defect in the cocaine- and antidepressant sensitive norepinephrine transporter. *N. Engl. J. Med.* **342**, 541–549.
44. Goldstein, D. S., Brush, Jr., J. E., Eisenhofer, G., Stull, R., and Esler, M. (1988) In vivo measurement of neuronal uptake of norepinephrine in the human heart. *Circulation* **78**, 41–48.
45. Novi, A. M. (1968) An electron microscopic study of the innervation of papillary muscles in the rat. *Anat. Rec.* **160**, 123–141.
46. Stöber, G., Nothen, M. M., Porzgen, P., Bruss, M., Bonisch, H., Knapp, M., et al. (1996) Systematic search for variation in the human norepinephrine transporter gene: identification of five naturally occurring missense mutations and study of association with major psychiatric disorders. *Am. J. Med. Genet.* **67**, 523–532.
47. Stöber, G., Hebebrand, J., Cichon, S., Bruss, M., Bonisch, H., Lehmkuhl, G., et al. (1999) Tourette syndrome and the norepinephrine transporter gene: results of a systematic mutation screening. *Am. J. Med. Genet.* **88**, 158–163.
48. Owen, D., Du, L., Bakish,D., Lapierre, Y. D., and Hrdina, P. D. (1999) Norepinephrine transporter gene polymorphism is not associated with susceptibility to major depression. *Psychiatry Res.* **87**, 1–5.
49. Runkel, F., Bruss, M., Nothen, M. M., Stober, G., Propping, P., and Bonisch, H. (2000) Pharmacological properties of naturally occurring variants of the human norepinephrine transporter. *Pharmacogenetics* **10**, 397–405.
50. Halushka, M. K., Fan, J. B., Bentley, K., Hsie, L., Shen, N., Weder, A., et al. (1999) Patterns of single-nucleotide polymorphisms in candidate genes for blood-pressure homeostasis. *Nature Genet.* **22**, 239–247.
51. Hahn, M. K., Mazei, M. S., Robertson, D., and Blakely, R. D. (2000) Role of human norepinephrine transporter gene single nucleotide polymorphisms in cardiovascular disease. *Am. J. Med. Genet.* **67**, 369.
52. Giros, B., el Mestikawy, S., Bertrand, L., and Caron, M. G. (1991) Cloning and functional characterization of a cocaine-sensitive dopamine transporter. *FEBS Lett.* **295**, 149–154.

53. Kilty, J. E., Lorang, D., and Amara, S. G. (1991) Cloning and expression of a cocaine-sensitive rat dopamine transporter. *Science* **254**, 578,579.

54. Giros, B., el Mestikawy, S., Godinot, N., Zheng, K., Han, H., Yang-Feng, T., and Caron, M. G. (1992) Cloning, pharmacological characterization, and chromosome assignment of the human dopamine transporter. *Mol. Pharmacol.* **42**, 383–390.

55. Vandenbergh, D. J., Persico, A. M., and Uhl, G. R. (1992) A human dopamine transporter cDNA predicts reduced glycosylation, displays a novel repetitive element and provides racially-dimorphic TaqI RFLPs. *Brain Res. Mol. Brain Res.* **15**, 161–166.

56. Donovan, D. M., Vandenbergh, D. J., Perry, M. P., Bird, G. S., Ingersoll, R., Nanthakumar, E., et al. (1995) Human and mouse dopamine transporter genes: conservation of 5'-flanking sequence elements and gene structures. *Brain Res. Mol. Brain Res.* **30**, 327–335.

57. Vandenbergh, D. J., Persico, A. M., Hawkins, A. L., Griffin, C. A., Li, X., Jabs, E. W., et al. (1992) Human dopamine transporter gene (DAT1) maps to chromosome 5p15.3 and displays a VNTR. *Genomics* **14**, 1104–1106.

58. Kawarai, T., Kawakami, H., Yamamura, Y., and Nakamura, S. (1997) Structure and organization of the gene encoding human dopamine transporter. *Gene* **195**, 11–18.

59. Vandenbergh, D. J., Thompson, M. D., Cook, E. H., Bendahhou, E., Nguyen, T., Krasowski, M. D., et al. (2000) Human dopamine transporter gene: coding region conservation among normal, Tourette's disorder, alcohol dependence and attention-deficit hyperactivity disorder populations. *Mol. Psychiatry* **5**, 283–292.

60. Carlsson, A. (1987) Perspectives on the discovery of central monoaminergic neurotransmission. *Annu. Rev. Neurosci.* **10**, 19–40.

61. Bannon, M. J., Granneman, J. G., and Kapatos, G. (1995) The dopamine transporter. Potential involvement in neuropsychiatric disorders, in *Psychopharmacology: The Fourth Generation of Progress*. (Bloom, F. E. and Kupfer, D. J., eds.), Raven Press, New York, NY, pp. 179–187.

62. Gainetdinov, R. R., Jones, S. R., and Caron, M. G. (1999) Functional hyperdopaminergia in dopamine transporter knock-out mice. *Biol. Psychiatry* **46**, 303–311.

63. Byerley, W., Hoff, M., Holik, J., Caron, M. G., and Giros, B. (1993) VNTR polymorphism for the human dopamine transporter gene (DAT1). *Hum. Mol. Genet.* **2**, 335.

64. Sano, A., Kondoh, K., Kakimoto, Y., and Kondo, I. (1993) A 40-nucleotide repeat polymorphism in the human dopamine transporter gene. *Human Genet.* **91**, 405,406.

65. Gelernter, J., Kranzler, H., and Lacobelle, J. (1998) Population studies of polymorphisms at loci of neuropsychiatric interest (tryptophan hydroxylase [TPH], dopamine transporter protein [SLC6A3], D3 dopamine receptor [DRD3], apolipoprotein E [APOE], mu opioid receptor [OPRM1], and ciliary neurotrophic factor [CNTF]). *Genomics* **52**, 289–297.

66. Kang, A. M., Palmatier, M. A., and Kidd, M. A. (1999) Global variation of a 40-bp VNTR in the 3'-untranslated region of the dopamine transporter gene (SLC6A3). *Biol. Psychiatry* **46,** 151–160.

67. Mitchell, R. J., Howlett, S., Earl, L., White, N. G., McComb, J., Schanfield, M. S., et al. (2000) Distribution of the 3' VNTR polymorphism in the human dopamine transporter gene in world populations. *Human Biol.* **72,** 295–304.

68. Grunhage, F., Schulze, T. G., Muller, D. J., Lanczik, M., Franzek, E., Albus, M., et al. (2000) Systematic screening for DNA sequence variation in the coding region of the human dopamine transporter gene (DAT1). *Mol. Psychiatry* **5,** 275–282.

69. Ueno, S., Nakamura, M., Mikami, M., Kondoh, K., Ishiguro, H., Arinami, T., et al. (1999) Identification of a novel polymorphism of the human dopamine transporter (DAT1) gene and the significant association with alcoholism. *Mol. Psychiatry* **4,** 552–557.

70. Morino, H., Kawarai, T., Izumi, Y., Kazuta, T., Oda, M., Komure, O., et al. (2000) A single nucleotide polymorphism of dopamine transporter gene is associated with Parkinson's disease. *Ann. Neurol.* **47,** 528–531.

71. Greenwood, T. A., Alexander, M., Keck, P. E., McElroy, S., Sadovnick, A. D., Remick, R. A., et al. (2001) Evidence for linkage disequilibrium between the dopamine transporter and bipolar disorder. *Am. J. Med. Genet.* **105,** 145–151.

71a. Manki, H., Kanba, S., Muramatsu, T., Higuchi, S., Suzuki, E., Matsushita, S., et al. (1996) Dopamine D2, D3 and D4 receptor and transporter gene polymorphisms and mood disorders. *J. Affect. Disord.* **40,** 7–13.

71b. Rubic, C., Schmidt, F., Knapp, M., Spandel, J., Wiegand, C., Meyer, J., et al. (2001) The human dopamine transporter gene: the 5'-flanking region reveals five diallelic polymorphic sites in a caucasian population sample. *Neurosci. Lett.* **297,** 125–128.

72. Waldman, I. D., Robinson, B. F., and Feigon, S. A. (1997) Linkage disequilibrium between the dopamine transporter gene (DAT1) and bipolar disorder: extending the transmission disequilibrium test (TDT) to examine genetic heterogeneity. *Genet. Epidemiol.* **14,** 699–704.

73. Kelsoe, J. R., Sadovnick, A. D., Kristbjarnarson, H., Bergesch, P., Mroczkowski-Parker, Z., Drennan, M., et al. (1996) Possible locus for bipolar disorder near the dopamine transporter on chromosome 5. *Am. J. Med. Genet.* **67,** 533–540.

74. Kirov, G., Jones, I., McCandless, F., Craddock, N., and Owen, M. J. (1999) Family-based association studies of bipolar disorder with candidate genes involved in dopamine neurotransmission: DBH, DAT1, COMT, DRD2, DRD3 and DRD5. *Mol. Psychiatry* **4,** 558–565.

75. Koob, G. F., Sanna, P. P., and Bloom, F. E. (1998) Neuroscience of addiction. *Neuron* **21,** 467–476.

76. Rocha, B. A., Fumagalli, F., Gainetdinov, R. R., Jones, S. R., Ator, R., Giros, B., et al. (1998) Cocaine self-administration in dopamine-transporter knockout mice [see comments] [published erratum appears in *Nature Neurosci.* (1998) **1(4),** 330]. *Nature Neurosci.* **1,** 132–137.

77. Gelernter, J., Kranzler, H. R., Satel, S. L., and Rao, P. A. (1994) Genetic

association between dopamine transporter protein alleles and cocaine-induced paranoia. *Neuropsychopharmacology* **11,** 195–200.

78. Persico, A. M., Vandenbergh, D. J., Smith, S. S., and Uhl, G. R. (1993) Dopamine transporter gene polymorphisms are not associated with polysubstance abuse. *Biol. Psychiatry* **34,** 265–267.

79. Lerman, C., Caporaso, N. E., Audrain, J., Main, D., Bowman, E. D., Lockshin, B., et al. (1999) Evidence suggesting the role of specific genetic factors in cigarette smoking. *Health Psychol.* **18,** 14–20.

80. Sabol, S. Z., Nelson, M. L., Fisher, C., Gunzerath, L., Brody, C. L., Hu, S., et al. (1999) A genetic association for cigarette smoking behavior [comment]. *Health Psychol.* **18,** 7–13.

81. Jorm, A. F., Henderson, A. S., Jacomb, P. A., Christensen, H., Korten, A. E., Rodgers, B., et al. (2000) Association of smoking and personality with a polymorphism of the dopamine transporter gene: results from a community survey. *Am. J. Med. Genet.* **96,** 331–334.

82. Sander, T., Harms, H., Podschus, J., Finckh, U., Nickel, B., Rolfs, A., et al. (1997) Allelic association of a dopamine transporter gene polymorphism in alcohol dependence with withdrawal seizures or delirium. *Biol. Psychiatry* **41,** 299–304.

83. Schmidt, L. G., Harms, H., Kuhn, S., Rommelspacher, H., and Sander, T. (1998) Modification of alcohol withdrawal by the A9 allele of the dopamine transporter gene. *Am. J. Psychiatry* **155,** 474–478.

84. Sander, T., Berlin, W., Ostapowicz, A., Samochowiec, J., Gscheidel, N., and Hoehe, M. R. (2000) Variation of the genes encoding the human glutamate EAAT2, serotonin and dopamine transporters and Susceptibility to idiopathic generalized epilepsy. *Epilepsy Res.* **41,** (2000) 75–81.

85. Franke, P., Schwab, S. G., Knapp, M., Gansicke, M., Delmo, C., Zill, P., Trixler, M., et al. (1999) DAT1 gene polymorphism in alcoholism: a family-based association study. *Biol. Psychiatry* **45,** 652–654.

86. Heinz, A., Goldman, D., Jones, D. W., Palmour, R., Hommer, D., Gorey, J. G., et al. (2000) Genotype influences in vivo dopamine transporter availability in human striatum. *Neuropsychopharmacology* **22,** 133–139.

87. Muramatsu, T. and Higuchi, S. (1995) Dopamine transporter gene polymorphism and alcoholism. *Biochem. Biophys. Res. Commun.* **211,** 28–32.

88. Swanson, J. M., Flodman, P., Kennedy, J., Spence, M. A., Moyzis, R., Schuck, S., et al. (2000) Dopamine genes and ADHD. *Neurosci. Biobehav. Rev.* **24,** 21–25.

89. Cook, Jr., E. H., Stein, M. A., Krasowski, M. D., Cox, N. J., Olkon, D. M., Kieffer, J. E., et al. (1995) Association of attention-deficit disorder and the dopamine transporter gene. *Am. J. Hum. Genet.* **56,** 993–998.

90. Gill, M., Daly, G., Heron, S., Hawi, Z., and Fitzgerald, M. (1997) Confirmation of association between attention deficit hyperactivity disorder and a dopamine transporter polymorphism. *Mol. Psychiatry* **2,** 311–313.

91. Waldman, I. D., Rowe, D. C., Abramowitz, A., Kozel, S. T., Mohr, J. H., Sherman, S. L., et al. (1998) Association and linkage of the dopamine transporter gene and attention-deficit hyperactivity disorder in children: heterogeneity owing to diagnostic subtype and severity. *Am. J. Human Genet.* **63,** 1767–1776.

92. Daly, G., Hawi, Z., Fitzgerald, M., and Gill, M. (1999) Mapping susceptibility loci in attention deficit hyperactivity disorder: preferential transmission of parental alleles at DAT1, DBH and DRD5 to affected children. *Mol.
Psychiatry* **4,** 192–196.

93. Curran, S., Mill, J., Tahir, E., Kent, L., Richards, S., Gould, A., et al. (2001)
Association study of a dopamine transporter polymorphism and attention
deficit hyperactivity disorder in UK and Turkish samples. *Mol. Psychiatry*
6, 425–428.

94. Holmes, J., Payton, A., Barrett, J. H., Hever, T., Fitzpatrick, H., Trumper,
A. L., et al. (2000) A family-based and case-control association study of the
dopamine D4 receptor gene and dopamine transporter gene in attention deficit hyperactivity disorder. *Mol. Psychiatry* **5,** 523–530.

95. Roman, T., Schmitz, M., Polanczyk, G., Eizirik, M., Rohde, L. A., and Hutz,
M. H. (2001) Attention-deficit hyperactivity disorder: a study of association with both the dopamine transporter gene and the dopamine D4 receptor
gene. *Am. J. Med. Genet.* **105,** 471–478.

96. Jorm, A. F., Prior, M., Sanson, A., Smart, D., Zhang, Y., and Easteal, S. (2001)
Association of a polymorphism of the dopamine transporter gene with exter-
nalizing behavior problems and associated temperament traits: a longitudinal
study from infancy to the mid-teens. *Am. J. Med. Genet.* **105,** 346–350.

97. Barr, C. L., Xu, C., Kroft, J., Feng, Y., Wigg, K., Zai, G., et al. (2001)
Haplotype study of three polymorphisms at the dopamine transporter locus
confirm linkage to attention-deficit/hyperactivity disorder. *Biol. Psychiatry*
49, 333–339.

98. Dougherty, D. D., Bonab, A. A., Spencer, T. J., Rauch, S. L., Madras, B. K.,
and Fischman, A. J. (1999) Dopamine transporter density in patients with
attention deficit hyperactivity disorder. *Lancet* **354,** 2132,2133.

99. Dresel, S., Krause, J., Krause, K. H., LaFougere, C., Brinkbaumer, K., Kung,
H. F., et al. (2000) Attention deficit hyperactivity disorder: binding of
[99mTc]TRODAT-1 to the dopamine transporter before and after methylphenidate treatment. *Eur. J. Nucl. Med.* **27,** 1518–1524.

100. Krause, K. H., Dresel, S. H., Krause, J., Kung, H. F., and Tatsch, K. (2000)
Increased striatal dopamine transporter in adult patients with attention deficit hyperactivity disorder: effects of methylphenidate as measured by single
photon emission computed tomography. *Neurosci. Lett.* **285,** 107–110.

101. Persico, A. M., Wang, Z. W., Black, D. W., Andreasen, N. C., Uhl, G. R.,
and Crowe, R. R. (1995) Exclusion of close linkage of the dopamine transporter gene with schizophrenia spectrum disorders. *Am. J. Psychiatry* **152,**
134–136.

102. Persico, A. M. and Catalano, M. (1998) Lack of association between dopamine transporter gene polymorphisms and delusional disorder. *Am. J. Med.
Genet.* **81,** 163–165.

103. Inada, T., Sugita, T., Dobashi, I., Inagaki, A., Kitao, Y., Matsuda, G., et al.
(1996) Dopamine transporter gene polymorphism and psychiatric symptoms
seen in schizophrenic patients at their first episode. *Am. J. Med. Genet.* **67,**
406–408.

104. Persico, A. M. and Macciardi, F. (1997) Genotypic association between dopamine transporter gene polymorphisms and schizophrenia. *Am. J. Med. Genet.* **74,** 53–57.

105. Daniels, J., Williams, J., Asherson, P., McGuffin, P., and Owen, M. (1995) No association between schizophrenia and polymorphisms within the genes for debrisoquine 4-hydroxylase (CYP2D6) and the dopamine transporter (DAT). *Am. J. Med. Genet.* **60,** 85–87.

106. Maier, W., Minges, J., Eckstein, N., Brodski, C., Albus, M., Lerer, B., et al. (1996) Genetic relationship between dopamine transporter gene and schizophrenia: linkage and association. *Schizophr. Res.* **20,** 175–180.

107. Gelernter, J., Vandenbergh, D., Kruger, S. D., Pauls, D. L., Kurlan, R., Pakstis, A. J., et al. (1995) The dopamine transporter protein gene (SLC6A3): primary linkage mapping and linkage studies in Tourette syndrome. *Genomics* **30,** 459–463.

108. Comings, D. E., Wu, S., Chiu, C., Ring, R. H., Gade, R., Ahn, C., et al. (1996) Polygenic inheritance of Tourette syndrome, stuttering, attention deficit hyperactivity, conduct, and oppositional defiant disorder: the additive and subtractive effect of the three dopaminergic genes—DRD2, D beta H, and DAT1. *Am. J. Med. Genet.* **67,** 264–288.

109. Hagan, J. J., Middlemiss, D. N., Sharpe, P. C., and Poste, G. H. (1997) Parkinson's disease: prospects for improved drug therapy. *Trends Pharmacol. Sci.* **18,** 156–163.

110. Mercier, G., Turpin, J. C., and Lucotte, G. (1999) Variable number tandem repeat dopamine transporter gene polymorphism and Parkinson's disease: no association found. *J. Neurol.* **246,** 45–47.

111. Le Couteur, D. G., Leighton, P. W., McCann, S. J., and Pond, S. (1997) Association of a polymorphism in the dopamine-transporter gene with Parkinson's disease. *Mov. Disord.* **12,** 760–763.

112. Dix, D. J., Lin, P. N., McKenzie, A. R., Walden, W. E., and Theil, E. C. (1993) The influence of the base-paired flanking region on structure and function of the ferritin mRNA iron regulatory element. *J. Mol. Biol.* **231,** 230–240.

113. Kislauskis, E. H., Li, Z., Singer, R. H., and Taneja, K. L. (1993) Isoform-specific 3'-untranslated sequences sort alpha-cardiac and beta-cytoplasmic actin messenger RNAs to different cytoplasmic compartments [published erratum appears in J. Cell Biol. (1993) **123(6 Pt 2),** following 1907]. *J. Cell Biol.* **123,** 165–172.

114. Schiavi, S. C., Wellington, C. L., Shyu, A. B., Chen, C. Y., Greenberg, M. E., and Belasco, J. G. (1994) Multiple elements in the c-fos protein-coding region facilitate mRNA deadenylation and decay by a mechanism coupled to translation. *J. Biol. Chem.* **269,** 3441–3448.

115. Jacobsen, L. K., Staley, J. K., Zoghbi, S. S., Seibyl, J. P., Kosten, T. R., Innis, R. B., et al. (2000) Prediction of dopamine transporter binding availability by genotype: a preliminary report . *Am. J. Psychiatry* **157,** 1700–1703.

116. Laine, T. P., Ahonen, A., Torniainen, P., Heikkila, J., Pyhtinen, J., Rasanen, P., et al. (1999) Dopamine transporters increase in human brain after alcohol withdrawal. *Mol. Psychiatry* **4,** 189–191.

117. Martinez, D., Gelernter, J., Abi-Dargham, A., van Dyck, C. H., Kegeles, L., Innis, R. B., et al. (2001) The variable number of tandem repeats polymorphism of the dopamine transporter gene is not associated with significant change in dopamine transporter phenotype in humans. *Neuropsychopharmacology* **24,** 553–560.
118. Ramamoorthy, S., Bauman, A. L., Moore, K. R., Han, H., Yang-Feng, T., Chang, A. S., et al. (1993) Antidepressant- and cocaine-sensitive human serotonin transporter: molecular cloning, expression, and chromosomal localization. *Proc. Natl. Acad. Sci. USA* **90,** 2542–2546.
119. Gelernter, J., Pakstis, A. J., and Kidd, K. K. (1995) Linkage mapping of serotonin transporter protein gene SLC6A4 on chromosome 17. *Human Genet.* **95,** 677–680.
120. Lesch, K. P., Wolozin, B. L., Estler, H. C., Murphy, D. L., and Riederer, P. (1993) Isolation of a cDNA encoding the human brain serotonin transporter. *J. Neural Transm. Gen. Sect.* **91,** 67–72.
121. Austin, M. C., Bradley, C. C., Mann, J. J., and Blakely, R. D. (1994) Expression of serotonin transporter messenger RNA in the human brain. *J. Neurochem.* **62,** 2362–2367.
122. Lesch, K. P., Balling, U., Gross, J., Strauss, K., Wolozin, B. L., Murphy, D. L., et al. (1994) Organization of the human serotonin transporter gene. *J. Neural Transm. Gen. Sect.* **95,** 157–162.
123. Heils, A., Teufel, A., Petri, S., Seemann, M., Bengel, D., Balling, U., Riederer, P., et al. (1995) Functional promoter and polyadenylation site mapping of the human serotonin (5-HT) transporter gene. *J. Neural Transm. Gen. Sect.* **102,** 247–254.
124. Battersby, S., Ogilvie, A. D., Blackwood, D. H., Shen, S., Muqit, M. M., Muir, W. J., et al. (1999) Presence of multiple functional polyadenylation signals and a single nucleotide polymorphism in the 3' untranslated region of the human serotonin transporter gene. *J. Neurochem.* **72,** 1384–1388.
125. Bradley, C. C. and Blakely, R. D. (1997) Alternative splicing of the human serotonin transporter gene. *J. Neurochem.* **69,** 1356–1367.
126. Flattem, N. L. and Blakely, R. D. (2000) Modified structure of the human serotonin transporter promoter. *Mol. Psychiatry* **5,** 110–115.
127. Lesch, K. P., Jatzke, S., Meyer, J., Stober, G., Okladnova, O., Mossner, R., et al. Mosaicism for a serotonin transporter gene promoter-associated deletion: decreased recombination in depression. *J. Neural Transm.* **106,** 1223–1230.
128. Mortensen, O. V., Thomassen, M., Larsen, M. B., Whittemore, S. R., and Wiborg, O. (1999) Functional analysis of a novel human serotonin transporter gene promoter in immortalized raphe cells. *Brain Res. Mol. Brain Res.* **68,** 141–148.
129. Owens, M. J. and Nemeroff, C. B. (1994) Role of serotonin in the pathophysiology of depression: focus on the serotonin transporter. *Clin. Chem.* **40,** 288–295.
130. Ernouf, D., Compagnon, P., Lothion, P., Narcisse, G., Benard, J. Y., and Daoust, M. (1993) Platelets 3H 5-HT uptake in descendants from alcoholic patients: a potential risk factor for alcohol dependence? *Life Sci.* **52,** 989–995.

131. Sellers, E. M., Higgins, G. A., and Sobell, M. B. (1992) 5-HT and alcohol abuse. *Trends Pharmacol. Sci.* **13**, 69–75.

131a. Coccaro, E. F., (1989) Central serotoni and impulsive aggression. *Br. J. Psychiatry Suppl.* 52–62.

132. Collier, D. A., Arranz, M. J., Sham, P., Battersby, S., Vallada, H., Gill, P., et al. (1996) The serotonin transporter is a potential susceptibility factor for bipolar affective disorder. *Neuroreport* **7**, 1675–1679.

133. Heils, A., Teufel, A., Petri, S., Stober, G., Riederer, P., Bengel, D., et al. (1996) Allelic variation of human serotonin transporter gene expression. *J. Neurochem.* **66**, 2621–2624.

134. Heils, A., Mossner, R., and Lesch, K. P. (1997) The human serotonin transporter gene polymorphism—basic research and clinical implications. *J. Neural Transm.* **104**, 1005–1014.

135. Lesch, K. P., Bengel, D., Heils, A., Sabol, S. Z., Greenberg, B. D., Petri, S., et al. (1996) Association of anxiety-related traits with a polymorphism in the serotonin transporter gene regulatory region . *Science* **274**, 1527–1531.

136. Lesch, K. P., Wolozin, B. L., Murphy, D. L., and Reiderer, P. (1993) Primary structure of the human platelet serotonin uptake site: identity with the brain serotonin transporter. *J. Neurochem.* **60**, 2319–2322.

137. Meltzer, H. Y. and Arora, R. C. (1988) Genetic control of serotonin uptake in blood platelets: a twin study. *Psychiatry Res.* **24**, 263–269.

138. Greenberg, B. D., Tolliver, T. J., Huang, S. J., Li, Q., Bengel, D., and Murphy, D. L. (1999) Genetic variation in the serotonin transporter promoter region affects serotonin uptake in human blood platelets. *Am. J. Med. Genet.* **88**, 83–87.

138a. Willeit, M., Stastny, J., Pivker, W., Prascha-Kieder, N., Nevmeister, A., Asenbaum, S., et al. (2001) No evidence for in vivo regulation of mid-brain serotonin transporter availablility by serotonin transporter promoter gene polymorphism. *Biol. Psychiatry* **50**, 8–12.

139. Delbruck, S. J., Wendel, B., Grunewald, I., Sander, T., Morris-Rosendahl, D., Crocq, M. A., et al. (1997) A novel allelic variant of the human serotonin transporter gene regulatory polymorphism. *Cytogenet. Cell Genet.* **79**, 214–220.

140. Gelernter, J., Kranzler, H., and Cubells, J. F. (1997) Serotonin transporter protein (SLC6A4) allele and haplotype frequencies and linkage disequilibria in African- and European-American and Japanese populations and in alcohol-dependent subjects. *Human Genet.* **101**, 243–246.

141. Gelernter, J., Cubells, J. F., Kidd, J. R., Pakstis, A. J., and Kidd, K. K. (1999) Population studies of polymorphisms of the serotonin transporter protein gene. *Am. J. Med. Genet.* **88**, 61–66.

142. Kunugi, H., Hattori, M., Kato, T., Tatsumi, M., Sakai, T., Sasaki, T., et al. (1997) Serotonin transporter gene polymorphisms: ethnic difference and possible association with bipolar affective disorder. *Mol. Psychiatry* **2**, 457–462.

143. Michaelovsky, E., Frisch, A., Rockah, R., Peleg, L., Magal, N., Shohat, M., et al. (1999) A novel allele in the promoter region of the human serotonin transporter gene. *Mol. Psychiatry* **4**, 97–99.

144. Nakamura, M., Ueno, S., Sano, A., and Tanabe, H. (2000) The human serotonin transporter gene linked polymorphism (5-HTTLPR) shows ten novel allelic variants. *Mol. Psychiatry* **5,** 32–38.

145. MacKenzie A. and Quinn, J. (1999) A serotonin transporter gene intron 2 polymorphic region, correlated with affective disorders, has allele-dependent differential enhancer-like properties in the mouse embryo. *Proc. Natl. Acad. Sci. USA* **96,** 15,251–15,255.

146. Fiskerstrand, C. E., Lovejoy, E. A., and Quinn, J. P. (1999) An intronic polymorphic domain often associated with susceptibility to affective disorders has allele dependent differential enhancer activity in embryonic stem cells. *FEBS Lett.* **458,** 171–174.

147. Gelernter, J. and Freimer, M. (1994) PstI RFLP at the SERT locus. *Human Mol. Genet.* **3,** 383.

148. Lesch, K. P., Gross, J., Franzek, E., Wolozin, B. L., Riederer, P., and Murphy, D. L. (1995) Primary structure of the serotonin transporter in unipolar depression and bipolar disorder. *Biol. Psychiatry* **37,** 215–223.

149. Glatt, C. E., DeYoung, J. A., Delgado, S., Service, S. K., Giacomini, K. M., Edwards, R. H., et al. (2001) Screening a large reference sample to identify very low frequency sequence variants: comparisons between two genes. *Nature Genet.* **27,** 435–438.

150. Nakamura, T., Muramatsu, T., Ono, Y., Matsushita, S., Higuchi, S., Mizushima, H., et al. (1997) Serotonin transporter gene regulatory region polymorphism and anxiety-related traits in the Japanese. *Am. J. Med. Genet.* **74,** 544,545.

151. Ball, D., Hill, L., Freeman, B., Eley, T. C., Strelau, J., Riemann, R., et al. (1997) The serotonin transporter gene and peer-rated neuroticism. *Neuroreport* **8,** 1301–1304.

152. Deary, I. J., Battersby, S., Whiteman, M. C., Connor, J. M., Fowkes, F. G., and Harmar, A. (1999) Neuroticism and polymorphisms in the serotonin transporter gene. *Psychol. Med.* **29,** 735–739.

153. Ebstein, R. P., Gritsenko, I., Nemanov, L., Frisch, A., Osher, Y., and Belmaker, R. H. (1997) No association between the serotonin transporter gene regulatory region polymorphism and the Tridimensional Personality Questionnaire (TPQ) temperament of harm avoidance. *Mol. Psychiatry* **2,** 224–226.

154. Jorm, A. F., Henderson, A. S., Jacomb, P. A., Christensen, H., Korten, A. E., Rodgers, B., Tan, X., (1998) An association study of a functional polymorphism of the serotonin transporter gene with personality and psychiatric symptoms. *Mol. Psychiatry* **3,** 449–451.

155. Flory, J. D., Manuck, S. B., Ferrell, R. E., Dent, K. M., Peters, D. G., and Muldoon, M. F. (1999) Neuroticism is not associated with the serotonin transporter (5-HTTLPR) polymorphism. *Mol. Psychiatry* **4,** 93–96.

156. Osher, Y., Hamer, D., and Benjamin, J. (2000) Association and linkage of anxiety-related traits with a functional polymorphism of the serotonin transporter gene regulatory region in Israeli sibling pairs. *Mol. Psychiatry* **5,** 216–219.

157. Benjamin, J., Osher, Y., Lichtenberg, P., Bachner-Melman, R., Gritsenko,

I., Kotler, M., et al. (2000) An interaction between the catechol O-methyltransferase and serotonin transporter promoter region polymorphisms contributes to tridimensional personality questionnaire persistence scores in normal subjects. *Neuropsychobiology* **41**, 48–53.

158. Hu, S., Brody, C. L., Fisher, C., Gunzerath, L., Nelson, M. L., Sabol, S. Z., Sirota, L. A., et al. (2000) Interaction between the serotonin transporter gene and neuroticism in cigarette smoking behavior. *Mol. Psychiatry* **5**, 181–188.

159. Lerman, C., Caporaso, N. E., Audrain, J., Main, D., Boyd, N. R., and Shields, P. G. (2000) Interacting effects of the serotonin transporter gene and neuroticism in smoking practices and nicotine dependence. *Mol. Psychiatry* **5**, 189–192.

160. Greenberg, B. D., Li, Q., Lucas, F. R., Hu, S., Sirota, L. A., Benjamin, J., et al. (2000) Association between the serotonin transporter promoter polymorphism and personality traits in a primarily female population sample. *Am. J. Med. Genet.* **96**, 202–216.

161. Harmar, A. J., Ogilvie, A. D., Battersby, S., Smith, C. A., Blackwood, D. H., Muir, W. J., et al. (1996) The serotonin transporter gene and affective disorder. *Cold Spring Harb. Symp. Quant. Biol.* **61**, 791–795.

162. Ogilvie, A. D., Battersby, S., Bubb, V. J., Fink, G., Harmar, A. J., Goodwim, G. M., et al. (1996) Polymorphism in serotonin transporter gene associated with susceptibility to major depression. *Lancet* **347**, 731–733.

163. Stöber, G., Heils, A., and Lesch, K. P. (1996) Serotonin transporter gene polymorphism and affective disorder. *Lancet* **347**, 1340,1341.

164. Furlong, R. A., Ho, L., Walsh, C., Rubinsztein, J. S., Jain, S., Paykel, E. S., et al. (1998) Analysis and meta-analysis of two serotonin transporter gene polymorphisms in bipolar and unipolar affective disorders. *Am. J. Med. Genet.* **81**, 58–63.

165. Hoehe, M. R., Wendel, B., Grunewald, I., Chiaroni, P., Levy, N., Morris-Rosendahl, D., et al. (1998) Serotonin transporter (5-HTT) gene polymorphisms are not associated with susceptibility to mood disorders. *Am. J. Med. Genet.* **81**, 1–3.

166. Saleem, Q., Ganesh, S., Vijaykumar, M., Reddy, Y. C., Brahmachari, S. K., et al. (2000) Association analysis of 5HT transporter gene in bipolar disorder in the Indian population. *Am. J. Med. Genet.* **96**, 170–172.

167. Esterling, L. E., Yoshikawa, T., Turner, G., Badner, J. A., Bengel, D., Gershon, E. S., et al. (1998) Serotonin transporter (5-HTT) gene and bipolar affective disorder. *Am. J. Med. Genet.* **81**, 37–40.

168. Mundo, E., Walker, M., Tims, H., Macciardi, F., and Kennedy, J. L. (2000) Lack of linkage disequilibrium between serotonin transporter protein gene (SLC6A4) and bipolar disorder. *Am. J. Med. Genet.* **96**, 379–383.

169. Mynett-Johnson, L., Kealey, C., Claffey, E., Curtis, D., Bouchier-Hayes, L., Powell, C., et al. (2000) Multimarkerhaplotypes within the serotonin transporter gene suggest evidence of an association with bipolar disorder. *Am. J. Med. Genet.* **96**, 845–849.

170. Collier, D. A., Stober, G., Li, T., Heils, A., Catalano, M., Di Bella, D., et al. (1996) A novel functional polymorphism within the promoter of the serotonin transporter gene: possible role in susceptibility to affective disorders. *Mol. Psychiatry* **1**, 453–460.

171. Gutierrez, B., Pintor, L., Gasto, C., Rosa, A., Bertranpetit, J., Vieta, E., and Fananas, L. (1998) Variability in the serotonin transporter gene and increased risk for major depression with melancholia. *Human Genet.* **103**, 319–322.

172. Rosenthal, N. E., Mazzanti, C. M., Barnett, R. L., Hardin, T. A., Turner, E. H., Lam, G. K., et al. (1998) Role of serotonin transporter promoter repeat length polymorphism (5-HTTLPR) in seasonality and seasonal affective disorder. *Mol. Psychiatry* **3**, 175–177.

173. Frisch, A., Postilnick, D., Rockah, R., Michaelovsky, E., Postilnick, S., Birman, E., et al. (1999) Association of unipolar major depressive disorder with genes of the serotonergic and dopaminergic pathways. *Mol. Psychiatry* **4**, 389–392.

174. Minov, C., Baghai, T. C., Schule, C., Zwanzger, P., Schwarz, M. J., Zill, P., et al. (2001) Serotonin-2A-receptor and -transporter polymorphisms: lack of association in patients with major depression. *Neurosci. Lett.* **303**, 119–122.

175. Seretti, A., Cusin, C., Lattuada, E., Di Bella, D., Catalano, M., and Smeraldi, E. (1999) Serotonin transporter gene (5-HTTLPR) is not associated with depressive symptomatology in mood disorders. *Mol. Psychiatry* **4**, 280–283.

176. Smeraldi, E., Zanardi, R., Benedetti, F., Di Bella, D., Perez, J., and Catalano, M. (1998) Polymorphism within the promoter of the serotonin transporter gene and antidepressant efficacy of fluvoxamine. *Mol. Psychiatry* **3**, 508–511.

177. Zanardi, R., Benedetti, F., Di Bella, D., Catalano, M., and Smeraldi, E. (2000) Efficacy of paroxetine in depression is influenced by a functional polymorphism within the promoter of the serotonin transporter gene. *J. Clin. Psychopharmacol.* **20**, 105–107.

178. Zanardi, R., Serretti, A., Rossini, D., Franchini, L., Cusin, C., Lattuada, E., et al. (2001) Factors affecting fluvoxamine antidepressant activity: influence of pindolol and 5-httlpr in delusional and nondelusional depression. *Biol. Psychiatry* **50**, 323–330.

179. Pollock, B. G., Ferrell, R. E., Mulsant, B. H., Mazumdar, S., Miller, M., Sweet, R. A., et al. (2000) Allelic variation in the serotonin transporter promoter affects onset of paroxetine treatment response in late-life depression. *Neuropsychopharmacology* **23**, 587–590.

180. Gorwood, P., Batel, P., Ades, J., Hamon, M., and Boni, C. (2000) Serotonin transporter gene polymorphisms, alcoholism, and suicidal behavior. *Biol. Psychiatry* **48**, 259–264.

181. Mann, J. J., Huang, Y. Y., Underwood, M. D., Kassir, S. A., Oppenheim, S., et al. (2000) A serotonin transporter gene promoter polymorphism (5-HTTLPR) and prefrontal cortical binding in major depression and suicide. *Arch. Gen. Psychiatry* **57**, 729–738.

182. Russ, M. J., Lachman, H. M., Kashdan, T., Saito, T., and Bajmakovic-Kacila, S. (2000) Analysis of catechol-O-methyltransferase and 5-hydroxytryptamine transporter polymorphisms in patients at risk for suicide. *Psychiatry Res.* **93**, 73–78.

183. Fitch, D., Lesage, A., Seguin, M., Trousignant, M., Bankelfat, C., Rouleau, G. A., et al. (2001) Suicide and the serotonin transporter gene. *Mol. Psychiatry* **6,** 127,128.

184. Evans, J., Battersby, S., Ogilvie, A. D., Smith, C. A., Harmar, A. J., Nutt, D. J., and Goodwin, G. M. (1997) Association of short alleles of a VNTR of the serotonin transporter gene with anxiety symptoms in patients presenting after deliberate self harm. *Neuropharmacology* **36,** 439–443.

185. Hammoumi, S., Payen, A., Favre, J. D., Balmes, J. L., Benard, J. Y., Husson, M., et al. (1999) Does the short variant of the serotonin transporter linked polymorphic region constitute a marker of alcohol dependence? *Alcohol* **17,** 107–112.

186. Matsushita, S., Yoshino, A., Murayama, M., Kimura, M., Muramatsu, T., and Higuchi, S. (2001) Association study of serotonin transporter gene regulatory region polymorphism and alcoholism. *Am. J. Med. Genet.* **105,** 446–450.

187. Sander, T., Harms, H., Lesch, K. P., Dufeu, P., Kuhn, S., Hoehe, M., et al. (1997) Association analysis of a regulatory variation of the serotonin transporter gene with severe alcohol dependence. *Alcohol Clin. Exp. Res.* **21,** 1356–1359.

188. Thompson, M. D., Gonzalez, N., Nguyen, T., Comings, D. E., George, S. R., and BF, O. D. (2000) Serotonin transporter gene polymorphisms in alcohol dependence. *Alcohol* **22,** 61–67.

189. Turker, T., Sodmann, R., Goebel, U., Jatzke, S., Knapp, M., Lesch, K. P., et al. (1998) High ethanol tolerance in young adults is associated with the low-activity variant of the promoter of the human serotonin transporter gene. *Neurosci. Lett.* **248,** 147–150.

190. Schuckit, M. A., Mazzanti, C., Smith, T. L., Ahmed, U., Radel, M., Iwata, N., et al. (1999) Selective genotyping for the role of 5-HT2A, 5-HT2C, and GABA alpha 6 receptors and the serotonin transporter in the level of response to alcohol: a pilot study. *Biol. Psychiatry* **45,** 647–651.

191. Little, K. Y., McLaughlin, D. P., Zhang, L., Livermore, C. S., Dalack, G. W., McFinton, P. R., et al. (1998) Cocaine, ethanol, and genotype effects on human midbrain serotonin transporter binding sites and mRNA levels. *Am. J. Psychiatry* **155,** 207–213.

192. Heinz, A., Jones, D. W., Mazzanti, C., Goldman, D., Ragan, P., Hommer, D., et al. (2000) A relationship between serotonin transporter genotype and in vivo protein expression and alcohol neurotoxicity. *Biol. Psychiatry* **47,** 643–649.

193. Cloninger, C. R. (1987) Neurogenetic adaptive mechanisms in alcoholism. *Science* **236,** 410–416.

194. Sander, T., Harms, H, Dufeu, P., Kuhn, S., Hoehe, M., Lesch, K. P., et al. (1998) Serotonin transporter gene variants in alcohol-dependent subjects with dissocial personality disorder. *Biol. Psychiatry* **43,** 908–912.

195. Hallikainen, T., Saito, T., Lachman, H. M., Volavka, J., Pohjalainen, T., Ryynanen, O. P., et al. (1999) Association between low activity serotonin transporter promoter genotype and early onset alcoholism with habitual impulsive violent behavior. *Mol. Psychiatry* **4,** 385–388.

196. Preuss, U. W., Soyka, M., Bahlmann, M., Wenzel, K., Behrens, S., de Jonge, S., et al. (2000) Serotonin transporter gene regulatory region polymorphism (5-HTTLPR), [3H]paroxetine binding in healthy control subjects and alcohol-dependent patients and their relationships to impulsivity. *Psychiatry Res.* **96**, 51–61.

197. Bondy, B., Erfurth, A., de Jonge, S., Kruger, M., and Meyer, H. (2000) Possible association of the short allele of the serotonin transporter promoter gene polymorphism (5-HTTLPR) with violent suicide. *Mol. Psychiatry* **5**, 193–195.

198. Courtet, P., Baud, P., Abbar, M., Boulenger, J. P., Castelnau, D., Mouthon, D., et al. (2001) Association between violent suicidal behavior and the low activity allele of the serotonin transporter gene. *Mol. Psychiatry* **6**, 338–341.

199. Mendes de Oliveira, J. R., Otto, P. A., Vallada, H., Lauriano, V., Elkis, H., Lafer, B., et al. (1998) Analysis of a novel functional polymorphism within the promoter region of the serotonin transporter gene (5-HTT) in Brazilian patients affected by bipolar disorder and schizophrenia. *Am. J. Med. Genet.* **81**, 225–227.

200. Serretti, A., Catalano, M., and Smeraldi, E. (1999) Serotonin transporter gene is not associated with symptomatology of schizophrenia. *Schizophr. Res.* **35**, 33–39.

201. Tsai, S. J., Hong, C. J., Yu, Y. W., Lin, C. H., Song, H. L., Lai, H. C., et al. (2000) Association study of a functional serotonin transporter gene polymorphism with schizophrenia, psychopathology and clozapine response. *Schizophr. Res.* **44**, 177–181.

202. Malhotra, A. K., Goldman, D., Mazzanti, C., Clifton, A., Breier, A., and Pickar, D. (1998) A functional serotonin transporter (5-HTT) polymorphism is associated with psychosis in neuroleptic-free schizophrenics. *Mol. Psychiatry* **3**, 328–332.

203. Hranilovic, D., Schwab, S. G., Jernej, B., Knapp, M., Lerer, B., Albus, M., et al. (2000) Serotonin transporter gene and schizophrenia: evidence for association/linkage disequilibrium in families with affected siblings. *Mol. Psychiatry* **5**, 91–95.

204. Bonnet-Brilhault, F., Laurent, C., Thibaut, F., Campion, D., Chavand, O., Samolyk, D., et al. (1997) Serotonin transporter gene polymorphism and schizophrenia: an association study. *Biol. Psychiatry* **42**, 634–636.

205. Bengel, D., Greenberg, B. D., Cora-Locatelli, G., Altemus, M., Heils, A., Li, Q., and Murphy, D. L. (1999) Association of the serotonin transporter promoter regulatory region polymorphism and obsessive-compulsive disorder. *Mol. Psychiatry* **4**, 463–466.

206. McDougle, C. J., Epperson, C. N., Price, L. H., and Gelernter, J. (1998) Evidence for linkage disequilibrium between serotonin transporter protein gene (SLC6A4) and obsessive compulsive disorder. *Mol. Psychiatry* **3**, 270–273.

207. Billett, E. A., Richter, M. A., King, N., Heils, A., Lesch, K. P., and Kennedy, J. L. (1997) Obsessive compulsive disorder, response to serotonin reuptake inhibitors and the serotonin transporter gene. *Mol. Psychiatry* **2**, 403–406.

208. Hanna, G. L., Himle, J. A., Curtis, G. C., Koram, D. Q., Veenstra-VanderWeele, J., Leventhal, B. L., et al. (1998) Serotonin transporter and seasonal variation in blood serotonin in families with obsessive-compulsive disorder. *Neuropsychopharmacology* **18**, 102–111.
209. Altemus, M., Murphy, D. L., Greenberg, B., and Lesch, K. P. (1996) Intact coding region of the serotonin transporter gene in obsessive-compulsive disorder. *Am. J. Med. Genet.* **67**, 409–411.
210. Hanna, G. L., Veenstra-Vander Weele, J., Cox, N. J., Boehnke, M., Himle, J. A., Curtis, G. C., et al. (1999) Genome scan of early-onset obsessive-compulsive disorder. *Annual Meeting of the American Academy of Child and Adolescent Psychiatry*.
211. Anderson, G. M., Freedman, D. X., Cohen, D. J., Volkmar, F. R., Hoder, E. L., McPhedran, P., Minderaa, R. B., et al. (1987) Whole blood serotonin in autistic and normal subjects. *J. Child Psychol. Psychiatry* **28**, 885–900.
212. Cook, Jr., E. H., Courchesne, R., Lord, C., Cox, N. J., Yan, S., Lincoln, A., et al. (1997) Evidence of linkage between the serotonin transporter and autistic disorder. *Mol. Psychiatry* **2**, 247–250.
213. Klauck, S. M., Poustka, F., Benner, A., Lesch, K. P., and Poustka, A. (1997) Serotonin transporter (5-HTT) gene variants associated with autism? *Human Mol. Genet.* **6**, 2233–2238.
214. Yirmiya, N., Pilowsky, T., Nemanov, L., Arbelle, S., Feinsilver, T., Fried, I., et al. (2001) Evidence for an association with the serotonin transporter promoter region polymorphism and autism. *Am. J. Med. Genet.* **105**, 381–386.
215. Tordjman, S., Gutknecht, L., Carlier, M., Spitz, E., Antoine, C., Slama, F., et al. (2001) Role of the serotonin transporter gene in the behavioral expression of autism. *Mol. Psychiatry* **6**, 434–439.
216. M. G. S. o. A. C. (IMGSAC) (2001) A genomewide screen for autism: strong evidence for linkage to chromosomes 2q, 7q, and 16p. *Am. J. Hum. Genet.* **69**, 570–581.
217. Zhong, N., Ye, L., Ju, W., Brown, W. T., Tsiouris, J., and Cohen, I. (1999) 5-HTTLPR variants not associated with autistic spectrum disorders. *Neurogenetics* **2**, 129–131.
218. Hu, M., Retz, W., Baader, M., Pesold, B., Adler, G., Henn, F. A., et al. (2000) Promoter polymorphism of the 5-HT transporter and Alzheimer's disease . *Neurosci. Lett.* **294**, 63–65.
219. Oliveira, J. R., Gallindo, R. M., Maia, L. G., Brito-Marques, P. R., Otto, P. A., Passos-Bueno, M. R., Morais, Jr., M. A., et al. (1998) The short variant of the polymorphism within the promoter region of the serotonin transporter gene is a risk factor for late onset Alzheimer's disease. *Mol. Psychiatry* **3**, 438–441.
220. Zill, P., Padberg, F., de Jonge, S., Hampel, H., Burger, K., Stubner, S., et al. (2000) Serotonin transporter (5-HT) gene polymorphism in psychogeriatric patients. *Neurosci. Lett.* **284**, 113–115.
221. Matsushita, S., Muramatsu, T., Kimura, M., Shirakawa, O., Mita, T., Nakai, T., et al. (1997) Serotonin transporter gene regulatory region polymorphism and panic disorder. *Mol. Psychiatry* **2**, 390–392.

222. Hinney, A., Barth, N., Ziegler, A., von Prittwitz, S., Hamann, A., Hennighausen, K., et al. (1997) Serotonin transporter gene-linked polymorphic region: allele distributions in relationship to body weight and in anorexia nervosa. *Life Sci.* **61,** PL 295–303.

223. Sundaramurthy, D., Pieri, L. F., Gape, H., Markham, A. F., and Campbell, D. A. (2000) Analysis of the serotonin transporter gene linked polymorphism (5-HTTLPR) in anorexia nervosa. *Am. J. Med. Genet.* **96,** 53–55.

224. Di Bella, D. D., Catalano, M., Cavallini, M. C., Riboldi, C., and Bellodi, L. (2000) Serotonin transporter linked polymorphic region in anorexia nervosa and bulimia nervosa [letter]. *Mol. Psychiatry* **5,** 233,234.

225. Biederman, J. and Spencer, T. (1999) Attention-deficit/hyperactivity disorder (ADHD) as a noradrenergic disorder. *Biol. Psychiatry* **46,** 1234–1242.

226. Wade, P. R., Chen, J., Jaffe, B., Kassem, I. S., Blakely, R. D., and Gershon, M. D. (1996) Localization and function of a 5-HT transporter in crypt epithelia of the gastrointestinal tract. *J. Neurosci.* **16,** 2352–2364.

227. Wade, P. R., Tamir, H., Kirchgessner, A. L., and Gershon, M. D. (1994) Analysis of the role of 5-HT in the enteric nervous system using anti-idiotopic antibodies to 5-HT receptors. *Am. J. Physiol.* **266,** G403–G416.

228. O'Malley, P. G., Wong, P. W., Kroenke, K., Roy, M. J., and Wong, R. K. (1998) The value of screening for psychiatric disorders prior to upper endoscopy. *J. Psychosom. Res.* **44,** 279–287.

5

Monoamine Transporters

Their Role in Maintaining Neuronal Homeostasis

Raul R. Gainetdinov and Marc G. Caron

1. INTRODUCTION

Central monoaminergic transmission is controlled by several critical processes. A complex balance between the amount of neurotransmitter synthesized, stored, released, metabolized, and recaptured determines the intensity of monoaminergic signaling *(1)*. Released monoamines undergo enzymatic degradation and dilution by diffusion; however, a major mechanism in the control of extracellular monoamine dynamics is selective uptake by presynaptic neurons via plasma-membrane monoamine transporters *(2–7)*. Monoamine transporters, such as that for dopamine (DAT), norepinephrine (NET) and serotonin (SERT), remove neurotransmitter from outside cells and recycle it back into the releasing neurons *(2,3,5)*. Accordingly, the drugs that interfere with the activity of these transporters, such as the psychostimulants, cocaine and amphetamine, produce elevated levels of monoamines in the extracellular spaces of the brain *(8)*. Several studies have revealed that certain chemicals can produce neurotoxic reactions by entering the neurons through plasma-membrane monoamine transporters thereby revealing another functional role of transporters, such as a molecular gateway for neurotoxins *(9)*. Thus, until recently, the major roles of the monoamine transporters are commonly believed to be limited to the control of the extracellular lifetime of monoamines, and to serve as a gateway and/or target for neurotoxins, antidepressants, and psychostimulants.

Relatively little attention has been given to another important aspect of transporter function. Several lines of pharmacological and physiological evidence suggest that neurotransmitter transporters may be critically involved in regulating presynaptic neuronal homeostasis. However, until genetic animal models with targeted disruption of these transporters became available, this role of plasma-membrane transporters was not fully appreci-

From: *Contemporary Neuroscience:*
Neurotransmitter Transporters: Structure, Function, and Regulation, 2nd Edition
Edited by: M. E. A. Reith © Humana Press Inc., Totowa, NJ

ated. In this chapter, recent findings on DAT-, SERT-, and NET-knockout mice highlighting the homeostatic role of monoamine transporters are discussed.

2. DAT-KO MICE

The DAT-knockout (DAT-KO) mice were generated through genetic deletion of the DAT by homologous recombination (10). The removal of the DAT leads to a distinct biochemical and behavioral phenotype. DAT-KO mice are hyperactive *(10–12)*, dwarf *(13)*, and display numerous cognitive disturbances *(11)*, disrupted sensorimotor gating *(14)*, and sleep dysregulation *(15)*. Despite normal social interactions in adult animals *(12)*, females lacking the DAT show an impaired capacity to care for their offspring *(10)*. Anterior pituitary hypoplasia and alterations in the parameters of the hypothalamo-pituitary axis, highlighting the role of hypothalamic dopamine (DA) reuptake in the neurohormonal regulation, were detected in DAT-KO mice *(13)*. Importantly, the enhanced locomotion of mutant mice is dependent on intact dopaminergic transmission, since the inhibition of tyrosine hydroxylase (TH) by α-methyl-para-tyrosine (α-MPT) and blockade of dopamine receptors by haloperidol completely immobilize them *(11)*. As might be expected, cocaine and amphetamine do not produce further increases in locomotion in DAT-KO mice *(10)*, but, actually inhibit hyperactivity of these mice through serotonergic mechanisms *(11)*. At the same time, cocaine was still rewarding *(16)*, and morphine had increased rewarding properties in DAT-KO mice *(17)*. In addition, 1-methyl-4-phenyl-1,2,3,6-tetrahydropyridine (MPTP) *(18,19)* and methamphetamine *(20)* treatment failed to produce dopaminergic neurotoxicity in these mice, and the wakefulness stimulant modafinil did not exert a wake-promoting effect in these mice *(15)*.

2.1. Extracellular DA Dynamics

Pharmacological evidence has indicated that elimination of the DAT would result in elevation of the extracellular portion of DA. Accordingly, hyperactivity of central dopaminergic transmission in DAT-KO mice was demonstrated in cyclic voltammetry experiments in mouse striatal slices, which showed a 300-fold increase in the amount of time DA remains in the extracellular space *(10,21)*. Moreover, cyclic voltammetry studies have revealed that the rate of DA elimination, over the time it takes to clear DA released by a single-pulse stimulation, was not affected by inhibitors of the serotonin and norepinephrine transporters or by selective inhibitors of the DA degradative enzymes monoamine oxidase (MAO) and catechol-O-methyl transferase (COMT) (pargyline and tolcapone) *(21)*. Both cocaine and amphetamine were found to be unable to affect clearance and extracellular DA levels in the striatum of DAT-KO mice *(10,16,22)*. Thus, it has been suggested that diffusion plays major role in removing DA from the extracellular space in DAT-KO mice *(21)*. Generally, the same conclusions were reached in subsequent experiments with DAT-KO mice with a complementary technique to study the amplitude of DA release and the kinetics of DA elimination—carbon fiber amperometry *(23)*. In these in vivo experiments, striatal DA release was evoked by electrical stimulation of the medial forebrain bundle. DA half-life was estimated to be at least two orders of magnitude higher in these mice. Similarly to studies in brain slices, inhibition of COMT by tolcapone did not affect DA clearance. However, inhibition of MAO by pargyline modestly slowed down DA elimination in knockout mice, suggesting that metabolism of DA by MAO may play a role in addition to uptake and diffusion in the clearance of DA released over the course of multiple stimulations. It has been demonstrated that in mice lacking the DAT, low-frequency firing resulted in consistently high extracellular DA levels, which could not be distinguished from the DA levels achieved by high-frequency firing. Based on this observation, it has been suggested that in DAT-KO mice the burst-firing activity cannot be specifically translated into phasic changes in extracellular DA *(23)*.

Thus, both voltammetric and amperometric studies convincingly demonstrated remarkably prolonged clearance of striatal DA in DAT-KO mice. To directly prove that this prolonged clearance can result in alterations in extracellular DA concentrations, an alternative approach to assess extracellular DA dynamics—a quantitative "no net flux" microdialysis technique *(24)* in freely moving mice—was used *(21,25)*. These studies revealed a fivefold elevation in steady-state extracellular DA in DAT-KO mice in comparison to wild-type mice. Together, these neurochemical data establish the DAT-KO mice as a genetic model of persistent functional hyperdopaminergia.

From the initial neurochemical characterization, it has become clear that in DAT-KO mice there are remarkable changes in extracellular DA dynamics, as well as numerous alterations in both pre-and postsynaptic components of dopaminergic transmission. However, alterations directly related to consequences of disruption of DAT-mediated transport (which presumably would result in mostly presynaptic changes) should be distinguished from indirect alterations induced by persistently elevated dopaminergic tone (which may be relevant to both pre- and postsynaptic plasticity).

For example, the remarkable prolongation in striatal DA clearance rate detected in both voltammetry and amperometry experiments was associated with a significantly decreased amount of DA molecules (75% in voltammetric experiments, 93% in aperometric studies) released in response to stimulation *(21,23)*. Similarly, the response to high K$^+$ stimulation in microdialysis experiments was also profoundly reduced. Infusion of 100 m*M* K$^+$ through the dialysis probe showed that less than 10% of the amount of DA released from wild-type mice was measured in DAT-KO mice *(26)*. These findings suggest that the actual amount of releasable DA in the DAT-KO mice is decreased. Although a slow clearance rate may be inferred from the well-known effects of DAT inhibitors, the decrease in release parameters was quite unexpected. It is reasonable to suggest that this fact reflects a greatly reduced amount of DA available for release in DAT-KO mice, which may be a direct consequence of the lack of DAT-mediated inward transport rather than indirect consequence of hyperdopaminergia. These observations may provide an explanation for the fact that the greatly attenuated clearance rate (>300×) in mice lacking the DAT brings extracellular DA levels to only about fivefold higher than in wild-type mice. It is important to also note that in microdialysis experiments DA extracellular levels in DAT-KO mice were impulse flow-dependent and Ca^{2+}- sensitive. Infusion of tetrodotoxin gradually reduced levels of extracellular DA to undetectable levels. Similarly, infusion of Ca^{2+}-free artificial media markedly attenuated the levels of extracellular DA *(26)*. These data suggest that these elevated extracellular levels of DA in DAT-KO mice are still reflective of depolarization-dependent vesicular exocytosis *(27)*. The divergent consequences of DAT removal on different parameters of extraneuronal DA dynamics are consistent with several hypotheses on the complex regulation of dopaminergic transmission *(28)*.

2.2 Presynaptic Homeostasis

Another example of the direct consequences of the DAT deletion is the fact that total striatal-tissue DA levels in the striatum, which mostly reflect the intraneuronal vesicular storage pool of DA, are drastically reduced in DAT-KO mice *(21)*. Tissue levels of DA in the DAT-KO mice were only 1/20th of that in wild-type mice. Importantly, these low levels of DA in the striatum were extremely sensitive to the inhibition of TH by α-MPT, suggesting that these levels may represent mainly a newly synthesized pool of dopamine.

Because reductions in DA levels may be potentially explained as a consequence of abnormal development or degeneration of DA neurons, additional DA neuron markers were examined. The striatal levels of tyrosine hydroxylase (TH), the rate-limiting enzyme in DA synthesis, were markedly decreased *(21,29)*. However TH-positive neurons in the substantia nigra (SN) were only slightly decreased, with no modification in the ratio of TH mRNA levels per neuron, which cannot account for the dramatic decreases in the levels of TH and dopamine in the striatum. These results highlight the complex mechanisms involved in TH regulation at the level of mRNA expression, protein synthesis, activity, and distribution (29). Although the exact mechanism responsible for such dramatic alterations in TH regulation is unclear it is likely that a decrease in intracellular dopamine is primarily responsible for this effect. In addition, dihydroxyphenylalanine (DOPA) decarboxylase levels were not modified, and the neuronal vesicular monoamine transporter (VMAT2) levels were decreased only marginally, which suggests that the decrease of TH labeling in the striatum is not caused by a loss of dopaminergic projections *(29)*. Remarkably, comparable reductions in striatal DA have been observed in normal animals following depletion of the DA storage pools by the inhibitors of vesicular transport—reserpine and tetrabenazine *(30)*. However, in DAT-KO mice no significant changes were found in mRNA for VMAT2 in the SN, tetrabenazine binding, VMAT2 protein levels, or functional uptake of DA in striatal vesicular preparations, showing that the depletion in striatal DA cannot be explained either by destruction of DA neurons or dysfunction of VMAT2-mediated vesicular uptake mechanisms *(21,25)*. These findings strongly suggest that depletion of the DA storage pools and decreased DA release in DAT-KO mice is directly caused by the absence of inward transport of DA through the DAT. Consequently, in a normal situation, a tight dependence of DA storage on recycled DA must exist *(21,25)*.

In contrast to DA levels, the tissue content of DA metabolites were unaltered (DOPAC), or elevated (HVA) in DAT-KO mice. Strikingly, depletion of DA reserves occurs although the DA synthesis rate, measured in vivo by accumulation of L-DOPA following inhibition of DOPA decarboxylase by 3-hydroxybenzylhydrazine (NSD-1015), was found to be significantly elevated (approx 200% of control) *(21)*. This indicates that both DA synthesis and turnover are extremely high in mutant animals, despite the low levels of striatal TH protein levels *(21,25)*. This seemingly paradoxical observation may be explained by the disinhibition of TH, which under normal conditions is the subject of tonic inhibition by both intraneuronal and extraneuronal DA *(31)*. In the DAT-KO mice, intraneuronal DA is greatly reduced, and could result in a disinhibition of TH and a significant increase in the DA synthetic rate. Alternatively, activation of TH may be explained by a loss of autoreceptor function caused by pronounced extracellular DA concentrations *(1,32,33)*. Indeed, D2 autoreceptor mRNA and binding were found to be decreased by 50% in the SN and ventral tegmental area of the DAT-KO mice *(10)*. In addition, functional studies revealed marked desensitization in the major autoreceptor functions: regulation of neuronal firing rate, nerve-terminal DA release, and synthesis *(34)*. The firing rate of DA neurons in the ventral midbrain was found to be markedly elevated and only slightly sensitive to DA-agonist application. Striatal nerve-terminal release-regulating autoreceptors were also found to be desensitized. Quinpirole, an autoreceptor agonist, elicited only a slight decrease in striatal DA release in the knockout mice, as measured both by voltammetry in striatal slices and by microdialysis in freely moving mice. To study the sensitivity of terminal DA autoreceptors controlling dopamine synthesis, measurement of the effect of quinpirole on L-DOPA levels in the striatum of freely moving mice during infusion of the DOPA decarboxylase inhibitor NSD-1015 under cessation of dopaminergic impulse flow by gamma-butyrolactone (GBL) was performed *(35)*. Both GBL and quinpirole failed to produce significant alterations in the DA biosynthesis rate measured in vivo in DAT-KO mice. Thus, striatal autoreceptors controlling DA synthesis were also found to be essentially nonfunctional *(34)*. Together, these data, which demonstrate a profound neurochemical plasticity of dopaminergic neurons, illustrate the critical role of DAT in the maintenance of presynaptic functions.

2.3. Postsynaptic Receptor Adaptations

Another consequence of remarkably altered extracellular dopamine dynamics is dysregulation in the responsiveness of postsynaptic DA receptors. It was found that protein and mRNA levels of the two major postsynaptic DA receptors, D1 and D2, are downregulated by approx 50% in the striatum of DAT-KO mice *(10)*. Surprisingly, however, in the DAT-KO mice certain postsynaptic DA receptors appear to be supersensitive. In quantitative *in situ* hybridization studies, decreased mRNA levels for both D1 (caudate putamen, –34%; nucleus accumbens, –45%) and D2 receptors (caudate putamen, –36%; nucleus accumbens, –33%) but increased mRNA levels encoding the D3 receptor (caudate putamen, +60–85%; nucleus accumbens, +40–107%) were found *(36)*. Moreover, an increased density of preproenkephalin A-negative neurons that express the D3 receptor mRNA were found in the nucleus accumbens (+35–46%) of DAT-KO mice *(36)*. In addition, preliminary examinations of the firing rate of DA-responsive neurons in the nucleus accumbens of DAT-KO mice have revealed unchanged responsiveness of postsynaptic receptors to a microiontophoretically applied D1-receptor agonist (SKF 81297), and the normal inhibitory effect of a D2 receptor agonist (quinpirole) was completely replaced by an excitatory effect *(37)*, despite the marked decrease in receptor numbers. Further confirmation of an elevated sensitivity of postsynaptic DA receptors was gained using an in vivo approach *(26)*. In DA-depleted DAT-KO mice, apomorphine induced a more pronounced activation in comparison to wild-type controls. Thus, it appears that different populations of postsynaptic receptors have followed divergent paths in their response to the inactivation of DAT, in often unexpected directions—some are downregulated and others become supersensitive.

2.4. Gene-Dose Effect of DAT Deletion

Importantly, most of the alterations listed in **Subheadings 2.1. and 2.2.** display a clear gene-dose effect (21,34). In most of the neurochemical studies, mice that are heterozygous for DAT deletion displayed an intermediate phenotype. For example, in these mice, tissue levels of DA are decreased by 30%, DA synthesis is modestly elevated, a-MPT produces faster depletion of DA, autoreceptor regulation is partially diminished, DA clearance is prolonged by about twofold, and extracellular DA is twofold higher in comparison to wild-type mice. Similar results were found in mice expressing only about 10% of DAT *(38)*. It is worth mentioning , however, that the magnitude of these changes was not directly proportional to the level of expression of DAT. For example, the alterations in DA neurochemical parameters detected in mice expressing only 10% of the DAT *(38)* were greater than in DAT-heterozygous mice *(21)*, but the magnitude of the changes was less than the results expected from a directly proportional relationship. This observation may be akin to the situation in humans where greater than 90–95% of dopaminergic cell loss is necessary to develop overt Parkinsonism-like phenotype.

2.5. DAT as a Determinant of the Mode of DA Neurotransmission

Striatal DA transmission is an ideal model for neurochemical studies on DA physiology. The highest levels of DA in the brain are reported in nigrostriatal neurons, which contain the highest density of the DAT *(1,6,39)*. Importantly, the alterations described in the striatum of DAT-KO mice are less evident in other brain areas of these mice. For example, minimal alterations in DA and metabolite concentrations are found in the frontal cortex in DAT-KO mice (Gainetdinov and Caron, unpublished results). Moreover, in DAT-KO mice, the mode of striatal DA-neuron transmission resembles that described previously for frontal-cortex DA neurons in normal animals. Several characteristic regulatory properties for mesocortical dopaminergic neurons have been described, which differ markedly from nigrostriatal neurons *(1)*. First, there are relatively few DA transporters *(6)*, and a low DA uptake rate *(40)*. The firing rate of mesocortical DA neurons is elevated, possibly indicating less activity of impulse-regulating autoreceptors *(41)*. There are few DA synthesis-modulating autoreceptors in the frontal cortex *(42)*. Tissue DA content is disproportionally low when compared to either the amount of basal extracellular DA *(43)* or stimulation-evoked DA release *(40)*. The tissue level of DA in the frontal cortex is tightly dependent on ongoing synthesis, as shown by an increased responsiveness to TH inhibition by α-MPT *(1)*. The fact that all of the hallmarks of DA-neuron homeostasis in the frontal cortex are now observed in the striatum of DAT-knockout mice leads to the theory that the relative efficiency of DA uptake may play an important role in determining these features. All these findings suggest that the DAT should be considered to be an important component in terminating extracellular DA signals, as well as a primary determinant of DA-system homeostasis *(21,25)*. Thus, the density of DAT per neuron may be an important determinant of dopaminergic neuron signaling, controlling both the mode of transmission from more synaptically limited to "volume"-like *(44)* or "nonsynaptic" *(45)* transmission and the profile of presynaptic dopaminergic machinery. Dopaminergic groups of cells expressing various transporter levels may have dramatically different profiles of transmission. In addition, factors affecting DAT regulation such as development, aging, and exposure to pharmacological agents may induce a substantial shift between these modalities.

3. SERT-KO MICE

The serotonin transporter (SERT) plays a key role in regulating the intensity of serotonin (5-hydroxytryptamine, 5-HT) transmission, and is the primary target for the selective serotonin reuptake inhibitors (SSRIs) as well as for substituted amphetamines such as (+)-3,4-methylenedioxymethamphetamine (MDMA). To evaluate the functions of the serotonin transporter (SERT), the gene was disrupted in mice by homologous recombination *(46)*. Despite evidence that excess extracellular 5-HT during embryonic development may disrupt several critical processes of embryogenesis, no obvious developmental phenotype was reported in the SERT-KO mice. Although high doses of d-amphetamine induced hyperactivity similarly in both mutant and wild-type mice, the locomotor stimulation induced by MDMA was completely disrupted in SERT-KO mice *(46)*. In addition, these mice demonstrated pronounced and increased cocaine-conditioned place preference in comparison to wild-type mice, suggesting that the rewarding properties of cocaine may be enhanced in these mutants *(47)*.

3.1. Extracellular 5-HT Dynamics and Presynaptic Homeostasis

Neurochemical studies on serotonergic neuron homeostasis performed in these mice generally recapitulate the observations of DAT-KO mice with respect to dopaminergic transmission. Initial studies demonstrated that high-affinity [3H]5-HT uptake was completely absent in the brain synaptosomes of SERT-KO mice *(46)*, yet a recent study using primary neuronal cultures of embryonic wild-type and SERT-KO mice has demonstrated plasticity at the level of serotonin uptake *(48)*. In particular, [3H]5-HT uptake of SERT-KO neuronal cultures was observed—although very weak—suggesting that in mice that do not express SERT, 5-HT may be taken up through other transporters. Nevertheless, disrupted uptake of 5-HT in SERT-KO mice was shown to result in a substantial increase in extracellular levels of 5-HT. In microdialysis studies, an approximately sixfold elevation in extracellular 5-HT was found *(49, 49a)*. Similar to observations in DAT-KO mice *(21,25)*, in adult SERT-KO mice, marked reductions (60–80%) in 5-HT tissue levels were measured in several brain regions *(46,49)*, suggesting deficient intraneuronal storage of this monoamine. Analysis of the 5-HT synthesis revealed an increased synthesis rate in these mice, but a modestly decreased synthesis rate in the brain stem and striatum was also reported *(49)*. These observations, and the studies of DAT-KO mice *(21,25)* support the theory that storage of monoamines is tightly controlled by the amount of monoamines recaptured, rather than synthesized. More detailed studies of possible adaptive changes in tryptophan hydroxylase (TPH) levels, regulation, and activity, as well as additional neurochemical studies—particularly electrochemical carbon-fiber measurements of stimulated 5-HT release and clearance—would certainly clarify the regulatory role of SERT in 5-HT neuronal homeostasis.

3.2. Pre- and Postsynaptic 5-HT-Receptor Responsiveness

The responsiveness of pre- and postsynaptic receptors in response to elevated serotonergic tone in mice lacking SERT has been extensively investigated. Numerous studies have demonstrated the remarkable plasticity of both pre- and postsynaptic receptor sensitivity following chronic SSRI treatments *(50)*. In agreement with these studies, expected changes in 5-HT-receptor regulation were found in SERT-KO mice.

In vivo electrophysiological studies in anesthetized animals have shown that the spontaneous firing rate of the dorsal raphe nucleus (DRN) 5-HT neurons was significantly decreased in SERT-KO mice by 66% and in heterozygous mice by 36%, as compared with wild-type littermates *(50)*. Importantly, the cell-body 5-HT1A autoreceptors were remarkably desensitized in both SERT-KO and heterozygous mice. In addition, the recovery time of the firing rate of hippocampal (CA-3) pyramidal neurons following iontophoretic applications of 5-HT was also significantly prolonged, but only in SERT-KO mice. Thus, it has been suggested that in SERT-KO mice a marked desensitization of both pre- and postsynaptic 5-HT1A receptors occurs, while only presynaptic receptors are affected in the heterozygous mice *(50)*.

In another study, 5-HT1A-binding sites in the hypothalamus and in the DRN were found to be significantly decreased in SERT-KO mice. In addition, significantly altered hypothermic and neuroendocrine responses to the 5-HT1A agonist 8-OH-DPAT were noted *(51)*. In the SERT-KO mice, the hypothermic response to 8-OH-DPAT was completely abolished. Furthermore, mutant mice had significantly attenuated plasma oxytocin and corticosterone responses to 8-OH-DPAT. No significant alterations in the hypothermic or hormonal responses to the agonist were reported in heterozygous mice. These results confirm that disruption of the SERT is associated with a functional desensitization of 5-HT1A-receptor responses. Histochemical studies have also showed that 5-HT1A receptor protein and mRNA levels were significantly decreased in the DRN, but increased in the hippocampus and unchanged in other forebrain areas of SERT-KO vs wild-type mice (49). Similar regional differences were also found with respect to 5-HT1B receptors. For example, a decrease in the density of these receptors was found in the SN (–30%), but not in the globus pallidus of mutant mice. Intermediate changes were found in heterozygous mice. In addition, quantification of [^{35}S]GTP-γ-S binding evoked by potent 5-HT1A- and 5-HT1B-receptor agonists revealed that a decrease in receptor coupling occurs in the DRN (–66%) and the SN (–30%), but not other brain areas in SERT-KO mice. In functional neurochemical studies, a decrease in the brain 5-HT turnover rate after administration of ipsapirone (a 5-HT1A agonist), and an increased 5-HT release in the SN by GR 127935 (a 5-HT1B/1D antagonist) were disrupted in SERT-KO mice, further confirming the marked alterations in 5-HT1A- and 5-HT1B-autoreceptor functions in these animals (49).

Another population of postsynaptic serotonin receptors has also been downregulated (52). Adaptive changes in brain 5-HT2A receptors in SERT-KO mice were assessed by autoradiographic labeling of these receptors with the selective antagonist [^{3}H]MDL 100,907. These studies revealed an expected downregulation in the density of 5-HT2A receptors in SERT mutants in some brain areas (–30–40% in the claustrum, cerebral cortex, and lateral striatum). These data convincingly demonstrate that autoreceptor function is remarkably desensitized in SERT-KO mice, yet postsynaptic receptor regulation may be more complex, depending on the brain area and the receptor subtype.

Together, these data illustrate that the presence of a functional SERT is essential for proper 5-HT neuronal homeostasis and receptor regulation.

4. NET-KO MICE

Rapid termination of norepinephrine (NE) neurotransmission is mediated by reuptake of the catecholamine into presynaptic terminals via the plasma-membrane NE transporter (NET) (3). This process regulates a diverse assortment of central and sympathetic nervous system functions that include learning and memory, mood, sleep-wakefulness cycles, blood flow, and metabolism. NET is a primary target of antidepressants such as the tricyclic desipramine, which inhibits reuptake of NE and increases extracellular NE concentrations. Using homologous recombination, mice lacking the NET (NET-KO) have recently been generated (53,54). The homozygous NET-KO mice are viable to adulthood, but display a slightly lower body weight. Similar to wild-type mice treated with desipramine, the NET-KO mice exhibit prolonged escape attempts in either an automated tail suspension test or the forced swim test, two widely popular tests used as behavioral screens for antidepressants. NET-KO mice also habituate faster in a novel environment and are somewhat less active than wild-type mice *(54)*. In addition, NET-KO mice display lower body temperature in a cold-tolerance test, an impairment consistent with the regulatory role played by NE thermoregulation (Xu and Caron, unpublished observations). In the warm-water tail-flick assay, a greater morphine analgesia was found in NET-KO mice, the effect apparently mediated by enhanced NE stimulation of α2–adrenoreceptors *(55)*. Interestingly, in the NET-KO mice, desipramine did not further enhance analgesia, but was still able to produce inhibitory effects on the locomotor activity of these mutants, suggesting that not all of the effects of this drug are exclusively mediated through interactions with the NET *(55)*.

The locomotor responses to cocaine and amphetamine were significantly elevated in the NET-KO mice compared to controls. In a cocaine-sensitization paradigm, chronic cocaine treatment of NET-KO mice produced no significant enhancement of the responsiveness to cocaine beyond what was already established in the mice, and the sensitivity of the NET-KO was equal to that observed in wild-type mice after chronic treatment with the psychostimulant *(54)*. Moreover, when NET-KO mice were tested in a place-preference conditioning paradigm where the rewarding property of cocaine was assessed, NET-KO mice showed an increased preference for the chamber paired with cocaine. These enhanced responses to psychostimulants have been correlated with secondary perturbations in midbrain dopaminergic function that are accompanied by postsynaptic D2/D3 DA-receptor supersensitivity *(54)*.

4.1. Extracellular NE Dynamics, Presynaptic Homeostasis, and Adrenergic-Receptor Responsiveness

Similar to observations in DAT-KO and SERT-KO mice, NET-KO mice demonstrated profound alterations in both presynaptic neuron homeostasis and postsynaptic-receptor responsiveness *(54)*. To test whether disruption of the NET results in alterations in extracellular NE dynamics, fast-scan cyclic voltametry was used to measure NE release and clearance. The bed nucleus of the stria terminalis, pars ventralis, was chosen because it is rich in NE and very low in DA and serotonin *(56)*. Release of NE in response to electrical stimulation is reduced by 60%, and the rates of clearance following stimulation are at least sixfold slower in the NET-KO mice compared to their wild-type controls. These altered dynamics result in approximately a twofold elevation of extracellular levels of NE, as evidenced by quantitative "low perfusion rate" in vivo microdialysis. NE-enriched brain regions were examined for NE content, which mostly reflects neuronal storage of neurotransmitter *(21)*. In the prefrontal cortex, hippocampus, cerebellum, and spinal cord of NET-KO mice, tissue concentrations of NE were 55–70% lower than in wild-type animals *(54,55)*. To assess whether NE synthesis was altered, the hippocampus was further examined. Following inhibition of L-aromatic amino-acid decarboxylase by NSD-1015, accumulation of L-DOPA was augmented about 1.7-fold over that of wild-type mice, indicative of disinhibition of monamine synthesis in neuronal systems without active transport. Since the hippocampus is enriched in NE and contains little DA, this observation suggests that NE synthesis is increased in NET-KO mice. These results again indicate that neuronal storage of monoamines are primarily controlled by reuptake rather than by synthesis. Adaptive changes to elevated extracellular levels of NE is revealed by a significant decrease (30%) in postsynaptic α1 adrenergic-receptor binding in the hippocampus of NET-KO mice, as determined by saturation binding with ^{3}H-prazosin *(54)*. Detailed assessments of adrenergic autoreceptor regulation in NET-KO mice are not currently available. In the binding assessment of α2 adrenergic-receptor density in the spinal cord, no significant alterations were found in these mutants *(55)*.

Taken together, these observations indicate that the NET regulates the extracellular lifetime of NE, and is critically involved in pre- and postsynaptic NE system homeostasis. These results extend the observations made in the DAT-KO and SERT-KO mice, for which the disruption of transporters is found to exert similar changes in neuronal functions.

5. CONCLUSIONS

From the initial studies on monoamine-transporter-knockout mice, it has become obvious that elimination of the active transport process results in a fundamental shift in the mode of neuronal functioning. Several critical neuronal functions are dramatically altered in these mice (Table 1), and a proper understanding of the causal links of these alterations may shed light on some previously unappreciated points in monoamine neuronal homeostasis.

5.1. Disrupted Clearance and Elevated Extracellular Levels of Monoamines in Transporter-Knockout Mice

First of all, there is the expected disruption of extracellular monoamine clearance, which results in a remarkably potentiated extracellular lifetime of monoamines. The most remarkable (300-fold) prolongation of extracellular clearance was noted in DAT-KO mice *(21)*. Currently, data on 5-HT clear-

Table 1
Alterations in Monoamine Homeostasis in DAT-KO, SERT-KO, and NET-KO Mice

Neurochemical parameters	DAT-KO mice	SERT-KO mice	NET-KO mice
Monoamine clearance (electrochemistry)	Prolonged (300-fold)	Not tested	Prolonged (sixfold)
Stimulated release (electrochemistry)	Decreased by 75–90%	Not tested	Decreased by 57%
Extracellular levels (microdialysis)	Elevated fivefold	Elevated sixfold	Elevated twofold
Tissue content (storage)	Decreased by 95%	Decreased by 65–80%	Decreased by 55–75%
Monoamine synthesis	Elevated twofold	Elevated or decreased as reported	Elevated by 70%
Autoreceptor function	Disrupted	Disrupted	Not tested
Postsynaptic receptor responsiveness	Downregulated, but a certain population is supersensitive	Downregulated, but not uniformly	Downregulated

ance in the SERT-KO mice are not available to make a direct comparison. However, interestingly, in voltammetry experiments with the NET-KO mice *(54)*, the clearance rate of NE is prolonged by only sixfold compared to that in wild-type animals. These results are in contrast to the changes in clearance rate of DA in DAT-KO mice. Importantly, these differences are essentially caused by alterations in the intrinsic clearance rates of these neurotransmitters in the wild-type controls, since the clearance of NE and DA in the two respective mutant strains of mice are identical. It is important to underscore that assessment of DA clearance was performed in the striatum of DAT-KO mice, where high levels of DAT are expressed in normal mice, whereas measurements of NE dynamics were performed in the bed nucleus stria terminalis, one of the many areas expressing modest levels of NET. Thus, regional differences in transporter expression per neuron may be an important determinant of the distance monoamines can reach in extracellular space beyond their site of release. The concept of a diffusion-mediated extrasynaptic mode of neurotransmission, termed "volume" or "nonsynaptic" transmission *(44,45)*, may be a suitable framework to describe the extracellular fate of monoamines in transporter-knockout mice. This mode of transmission has been postulated for neuropeptides *(44)*, and also for classical neurotransmitters, including dopaminergic neurotransmission in the prefrontal cortex *(40,56)* and serotonin transmission in most regions of the brain *(57)*. The characteristics of volume transmission sites have been described in the context of the synaptic organization of the anatomical area, including the proximity of release sites and receptors *(44)*. The possibility that membrane transporters function as a determinant of volume or nonsynaptic transmission sites should also be considered *(21,45)*.

As a result of the protracted clearance, the extracellular levels of the respective monoamines are elevated in all three mutant strains. Interestingly, the degree of elevation is highly variable. For example, a fivefold increase was observed in DAT-KO mice *(21)*, but only a twofold elevation in extracellular NE levels was found in NET-KO mice *(54)*. Several points should be made here. In both DAT-KO and NET-KO mice, the rate of clearance of the monoamine approximated that for clearance by diffusion *(21,54)*. However, these extracellular monoamine levels are much lower than those predicted for clearance by diffusion alone. A contributing factor for these findings is that the actual amount of monoamine released per pulse was found to be decreased for both DA and NE *(21,54)*, and these parameters appear to vary in different regions of the brain. In addition, the contribution of metabolic enzymes and the potential role of additional transporter systems to the clearance of monoamines could vary for each neurotransmitter in each particular region. Together, these genetic studies confirm general principles of extracellular monoamine dynamics inferred from pharmacological studies with the use of monoamine-transporter blockers.

5.2. Adaptive Changes in Pre- and Postsynaptic-Receptor Responsiveness in Transporter-Knockout Mice

The pharmacological models of chronic psychostimulant or antidepressant exposure which results in a sustained elevation of extracellular levels of monoamines has provided a fairly comprehensive picture of adaptive alterations in pre- and postsynaptic receptors. However, the broad spectrum of effects of most pharmacological agents has prevented the definitive clarification of whether elevated monoamine concentrations *per se* or other effects of the drugs are responsible for alterations in neuronal function. In this respect, some interesting generalizations can be made from the studies on mutant mice. Thus, it has been convincingly shown that persistently elevated monoamine levels induce marked downregulation and functional desensitization of presynaptic autoreceptors in both DAT-KO *(34)* and SERT-KO *(49–51)* mice. With regard to postsynaptic receptors, a more complex picture seems to emerge. In general, all three mutant mice depict a downregulation of major postsynaptic receptors *(10,49–52,54)*. However this downregulation was not uniform, and certain postsynaptic receptors were found to be actually upregulated *(36,49)*, supersensitive *(26)*, or not altered *(49)*. Although this nonuniform pattern in postsynaptic receptor responsiveness is not well understood and deserves closer future examination, it probably reflects a consequence of remarkably altered extracellular neurotransmitter dynamics. It should be noted that the actual amount of neurotransmitter released per stimulation is decreased in mutant mice, thereby suggesting that the normal steep gradient in the concentrations of monoamine (extremely high in the synapse; decreased with increasing distance from release site; low nanomolar in extrasynaptic space) has been changed to a more uniform one (lower concentrations close to release site as compared with wild-type, but relatively elevated concentrations in extrasynaptic compartments). Thus, a different outcome of receptor regulation, depending on the relative localization of these postsynaptic receptors (synaptic vs extrasynaptic), may be expected.

5.3. Disrupted Storage of Monoamines in Transporter-Knockout Mice

One of the most striking observations made in mice lacking monoamine transporters is the depletion of intraneuronal storage of monoamines. The most remarkable decrease in monoamine tissue content was found in DAT-KO *(21)* mice (95%), and more modest decreases occurred in SERT-KO *(46)* mice (60–80%) and NET-KO *(54)* mice (55–75%). These observations seem to indicate a tight dependence of the intraneuronal storage pool on the monoamine uptake system *(25)*. The fact that these depletions occur independently of levels of monoamine synthesis in these mice, which is in fact disinhibited (in NET-KO and DAT-KO mice), strongly suggest that the contribution of newly synthesized monoamines to maintenance of storage is negligible. It is reasonable to speculate that these remarkable decreases in monoamine storage pools may in turn change several critical intracellular processes—and, for example, may account for the altered regulation of monoamine synthesis *(21,25)* in the mutant mice.

5.4. Homeostatic Role of Monoamine Transporters

Together, these observations, while still incomplete, strongly suggest that protracted clearance, elevated extracellular levels, depletion of intraneuronal stores of transmitter, and disinhibition of neuronal amine synthesis can be hallmarks of neuronal systems without active monoamine transport. This postulate may have important implications. It is important to note that the decrease in tissue content was not uniform, and there are some brain areas (for example, frontal cortex in DAT-KO mice or hypothalamus in NET-KO mice (Gainetdinov and Caron, unpublished observations) where alterations in monoamine tissue contents were less pronounced. With regard to the DA system, the highest levels of DAT expression and transporter protein are found in nigrostriatal and mesolimbic DA neurons, significantly less in frontal cortex and hypothalamus, and low levels in the olfactory bulb and pituitary *(6,39,58)*. The pattern of DAT distribution, which does not completely overlap with the density of DA neurons shows that the level of DAT expression can vary among various DA cell groups *(6,58)*. It is also important to note that DA release is well-established before mature rates of transport are reached during development *(59)*. In addition, DAT is subject to substantial structural and functional maturation postnatally, and specific developmentally determined modes of DA transmission may occur in younger animals *(60)*. An age-dependent decrease in DAT levels has also been described *(61)*. These observations suggest that differential expression of DAT in various populations of DA neurons may determine the type of DA signaling which occurs in any given anatomical area, with some areas utilizing a classical synaptic type of signaling and others a more paracrine or volume transmission type of signaling *(44,45,57)*. Similar regulatory mechanisms may apply also to NE and 5-HT systems.

5.5. Alterations in Monoamine Homeostasis by Transporter Inhibitors

Another potential implication concerns the effects of monoamine-transporter inhibitors. In general, depletions in monoamine content found after chronic administration of drugs that interact with monoamine transporters, such as amphetamines, were usually interpreted as a consequence of neurotoxicity induced by these drugs *(8)*. However, several observations have shown that these depletions do not necessarily reflect damage or loss of DA neurons *(62–64)*. Specifically, it has been shown that chronic methamphetamine abuse by humans results in reductions in striatal DA, TH, and DAT levels, but not VMAT2 or DOPA decarboxylase levels, demonstrating that DA depletion following methamphetamine in these subjects does not reflect the destruction of DA neurons, but may be caused by a chronically diminished DA reuptake process *(63)*. Modest decreases in 5-HT content are found in brain tissues following chronic treatment with SERT inhibitors or MDMA *(65–67)*. Similar decreases in brain NE levels are observed following chronic treatment with NET inhibitors *(68,69)*. These observations, along with our findings, suggest that caution should be taken in the interpretation of monoamine depletion produced by high doses or prolonged treatment with amphetamines or other drugs that interfere with monoamine transporters. Direct depletion of monoamine storage caused by diminished monoamine-transporter function may also account for these effects. Relative inefficiency of most current monoamine-transporter inhibitors to significantly affect monoamine storage may be caused by the relatively low potency or short-term duration of action of these drugs. It may be proposed that depletion of monoamines by monoamine-transporter inhibitors may take place only following long-term effective blockade of reuptake. The studies employing new generations of extremely potent monoamine transporter inhibitors such as Research Triangle Institute compounds *(70)* or novel cocaine analog, 2 beta-propanoyl-3beta-(4-tolyl)-tropane (PTT) *(71)* could potentially resolve this issue.

Overall, these observations, which reveal previously unappreciated points in the regulation of monoamine homeostasis, may provide new insights in understanding the basic mechanisms underlying the etiology and pathology of several neuropsychiatric conditions. Particularly, alterations in monoamine homeostasis induced by chronic monoamine-transporter inhibition should be considered when attempts are made to understand neurotoxic, addictive, or therapeutic effects of the drugs interacting with these transporters. In addition, numerous other factors regulating monoamine-transporter function, which may cause secondary alterations in monoamine homeostasis, should be considered as potential contributing factors in the pathogenesis of some neuropsychiatric disorders. These could include developmental abnormalities *(60)*, aging *(61)*, or pharmacological and environmental interventions *(8,9)*. It would also be interesting to explore the possibility that the regulatory control which the monoamine transporters exerts over presynaptic monoamine neuronal function may extend to all neurotransmitter systems with transporters.

ACKNOWLEDGMENTS

We wish to thank Dr. Laura Bohn for help with editing of the manuscript. MGC is an Investigator of the Howard Hughes Medical Institute. RRG is a visiting scientist from the Institute of Pharmacology, Russian Academy of Medical Sciences, Baltiyskaya 8, 125315, Moscow, Russia.

REFERENCES

1. Roth, R. H. and Elsworth, J. D. (1995) Biochemical pharmacology of midbrain dopamine neurons, in: *Psychopharmacology: The Fourth Generation of Progress*, (Bloom F. E. and Kupfer D. J., eds.), Raven, New York, pp. 227–243.
2. Amara, S. G. and Kuhar, M. J. (1993) Neurotransmitter transporters: recent progress. *Annu. Rev. Neurosci.* **16,** 73–93.
3. Blakely, R. D., De Felice, L. J., and Hartzell, H. C. (1994) Molecular physiology of norepinephrine and serotonin transporters. *J. Exp. Biol.* **196,** 263–281.
4. Garris, P. A. and Wightman, R. M. (1994) Different kinetics govern dopaminergic transmission in the amygdala, prefrontal cortex, and striatum: an in vivo voltammetric study. *J. Neurosci.* **14,** 442–450.

5. Giros, B. and Caron, M. G. (1993) Molecular characterization of the dopamine transporter, *Trends Pharmacol. Sci.* **14,** 43–49.

6. Shimada, S., Kitayama S., Walther, D., and Uhl, G. (1992) Dopamine transporter mRNA: dense expression in ventral midbrain neurons. *Mol. Brain. Res.* **13,** 359–362.

7. Reith, M. E. A., Xu, C., and Chen, N. -H. (1997) Pharmacology and regulation of the neuronal dopamine transporter. *Eur. J. Pharmacol.* **324,** 1–10.

8. Seiden, L. S. and Sabol, K. E. (1995) Neurotoxicity of methamphetamine-related drugs and cocaine, in: *Handbook of Neurotoxicology*, (Chang L. W. and Dyer R. S., eds.), Marcell Dekker, New York, pp. 825–843.

9. Miller, G. W., Gainetdinov, R. R., Levey, A. I., and Caron, M. G. (1999) Dopamine transporters and neuronal injury. *Trends Pharmacol. Sci.* **20,** 424–429.

10. Giros, B., Jaber, M. Jones, S. R., Wightman, R. M., and Caron, M. G. (1996) Hyperlocomotion and indifference to cocaine and amphetamine in mice lacking the dopamine transporter. *Nature* **379,** 606–612.

11. Gainetdinov, R. R., Wetsel, W. C., Jones, S. R., Levin, E. D., Jaber, M., and Caron, M. G. (1999) Role of serotonin in the paradoxical calming effect of psychostimulants on hyperactivity. *Science* **283,** 397–401.

12. Spielewoy, C. Roubert, C., Hamon, M., Nosten-Bertrand, M., Betancur, C., and Giros, B. (2000) Behavioural disturbances associated with hyperdopaminergia in dopamine-transporter knockout mice. *Behav. Pharmacol.* **11,** 279–290.

13. Bosse, R., Fumagalli, F., Jaber, M., Giros, B., Gainetdinov, R. R., Wetsel, W. C., Missale, C., and Caron, M. G. (1997) Anterior pituitary hypoplasia and dwarfism in mice lacking the dopamine transporter. *Neuron* **19,** 127–138.

14. Ralph, R. J., Paulus, M. P., Fumagalli, F., Caron, M. G., and Geyer, M. A. (2001) Prepulse inhibition deficits and perseverative motor patterns in dopamine transporter knock-out mice: differential effects of D1 and D2 receptor antagonists. *J. Neurosci.* **21,** 305–313.

15. Wisor, J. P., Nishino, S., Sora, I., Uhl, G. H., Mignot, E., and Edgar, D. M. (2001) Dopaminergic role in stimulant-induced wakefulness. *J. Neurosci.* **21,** 1787–1794.

16. Rocha, B. A., Fumagalli, F., Gainetdinov, R. R., Jones, S. R., Ator, R., Giros, B., et al. (1998) Cocaine self-administration in dopamine transporter knockout mice. *Nat. Neurosci.* **1,** 132–137.

17. Spielewoy, C., Gonon, F., Roubert, C., Fauchey, V., Jaber, M., Caron, M. G., et al. (2000) Increased rewarding properties of morphine in dopamine-transporter knockout mice. *Eur. J. Neurosci.* **12,** 1827–1837.

18. Gainetdinov, R. R., Fumagalli, F., Jones S. R., and Caron, M. G. (1997) Dopamine transporter is required for in vivo MPTP neurotoxicity: evidence from mice lacking the transporter. *J. Neurochem.* **69,** 1322–1325.

19. Bezard, E., Gross, C. E., Fournier, M. C., Dovero, S., Bloch, B., and Jaber, M. (1999) Absence of MPTP-induced neuronal death in mice lacking the dopamine transporter. *Exp. Neurol.* **155,** 268–273.

20. Fumagalli, F., Gainetdinov, R. R., Valenzano, K. J., and Caron, M. G. (1998) Role of dopamine transporter in methamphetamine-induced neurotoxicity: evidence from mice lacking the transporter. *J. Neurosci.* **18,** 4861–4869.
21. Jones, S. R., Gainetdinov, R. R., Wightman, R. M., and Caron, M. G. (1998) Profound neuronal plasticity in response to inactivation of the dopamine transporter. *Proc. Natl. Acad. Sci. USA* **95,** 4029–4034.
22. Jones, S. R., Gainetdinov, R. R., Wightman R. M., and Caron, M. G. (1998) Mechanisms of amphetamine action revealed in mice lacking the dopamine transporter. *J. Neurosci.* **18,** 1979–1986.
23. Benoit-Marand, M., Jaber, M., and Gonon, F. (2000) Release and elimination of dopamine in vivo in mice lacking the dopamine transporter: functional consequences. *Eur. J. Neurosci.* **12,** 2985–2992.
24. Justice, J. B. Jr. (1993) Quantitative microdialysis of neurotransmitters. *J. Neurosci. Meth.* **48,** 263–276.
25. Gainetdinov, R. R., Jones, S. R., Fumagalli, F., Wightman, R. M., and Caron, M. G. (1998) Re-evaluation of the role of the dopamine transporter in dopamine system homeostasis. *Brain Res. Rev.* **26,** 148–153.
26. Gainetdinov, R. R., Jones, S. R., and Caron M. G. (1999) Functional hyperdopaminergia in dopamine transporter knock-out mice. *Biol. Psychiatry* **46,** 303–311.
27. Westerink, B. H., Damsma, G., Rollema, H., De Vries, J. B., and Horn, A. S. (1987) Scope and limitations of in vivo brain dialysis: a comparison of its application to various neurotransmitter systems. *Life Sci.* **41,** 1763–1776.
28. Grace, A. A. (1991) Phasic versus tonic dopamine release and the modulation of dopamine system responsivity: a hypothesis for the etiology of schizophrenia. *Neuroscience* **41,** 1–24.
29. Jaber, M., Dumartin, B., Sagne, C., Haycock, J,W., Roubert, C., Giros, B., Bloch, B., and Caron, M. G. (1999) Differential regulation of tyrosine hydroxylase in the basal ganglia of mice lacking the dopamine transporter. *Eur. J. Neurosci.* **11,** 3499–3511.
30. Carlsson, A. (1987) Perspectives on the discovery of central monoaminergic neurotransmission. *Annu. Rev. Neurosci.* **10,** 19–40.
31. Seeman, P. (1981) Brain dopamine receptors. *Pharmacol. Rev.* **32,** 229–313.
32. Henry, D. J. and White, F. J. (1995) The persistence of behavioral sensitization to cocaine parallels enhanced inhibition of nucleus accumbens neurons. *J. Neurosci.* **15,** 6287–6299.
33. Woolverton, W. L. and Johnson, K. M. (1992) Neurobiology of cocaine abuse. *Trends Pharmacol. Sci.* **13,** 193–200.
34. Jones, S. R., Gainetdinov, R. R., Hu, X.-T., Cooper, D. C, Wightman, R. M., White, F. J., and Caron, M. G. (1999) Loss of autoreceptor functions in mice lacking the dopamine transporter. *Nat. Neurosci.* **2,** 649–655.
35. Walters, J. R. and Roth, R. H. (1976) Dopaminergic neurons: an in vivo system for measuring drug interactions with presynaptic receptors. *Naunyn-Schmiedeberg's Arch. Pharmacol.* **296,** 5–14. *Monogr.* **105,** 147–153.
36. Fauchey, V., Jaber, M., Caron, M. G., Bloch, B., and Le Moine, C. (2000) Differential regulation of the dopamine D1, D2 and D3 receptor gene expression and changes in the phenotype of the striatal neurons in mice lacking the dopamine transporter. *Eur. J. Neurosci.* **12,** 19–26.

37. Cooper, D. C., Hu, X.-T., Jones, S. R., Giros, B., Caron, M. G., and White, F. J. (1997) In vivo neurophysiological assessment of mesoaccumbens dopamine function in dopamine transporter knockout mice. *Soc. Neurosci. Abstr.* **23,** 1210.
38. Zhuang, X., Oosting, R. S., Jones, S. R. Gainetdinov, R. R., Miller, G. W., et al. (2001) Hyperactivity and impaired response habituation in hyperdopaminergic mice. *Proc. Natl. Acad. Sci. USA* **98,** 1982–1987.
39. Hersch, S. M., Yi, H., Heilman, C. J., Edwards, R. H., and Levey, A. I. (1997) Subcellular localization and molecular topology of the dopamine transporter in the striatum and substantia nigra. *J. Comp. Neurol.* **388,** 211–227.
40. Garris, P. A., Collins, L. B., Jones, S. R., and Wightman, R. M. (1993) Evoked extracellular dopamine in vivo in the medial prefrontal cortex. *J. Neurochem.* **61,** 637–647.
41. Bannon, M. J., Freeman, A. S., Chiodo, L. A., Bunney, B. S., and Roth R. H. (1987) The pharmacology and electrophysiology of mesolimbic dopamine neurons, in *Handbook of Psychopharmacology*, Vol. 19, (Iversen L. L., ed.), Plenum Press, New York, pp. 329–374.
42. Kilts, C. D., Anderson C. M., Ely, T. D., and Nishita, J. K. (1987) Absence of synthesis modulating nerve terminal autoreceptors on mesoamygdaloid and other mesolimbic dopamine neuronal populations. *J. Neurosci.* **7,** 3961–3975.
43. Moghaddam, B., Roth, R. H., and Bunney, B. S. (1990) Characterization of dopamine release in the rat medial prefrontal cortex as assessed by in vivo microdialysis: comparison to the striatum. *Neuroscience* **36,** 669–676.
44. Agnati, L. F., Zoli, M., Stromberg, I., and Fuxe, K. (1995) Intracellular communication in the brain: wiring versus volume transmission. *Neuroscience* **69,** 711–726.
45. Vizi, E. S. (2000) Role of high-affinity receptors and membrane transporters in nonsynaptic communication and drug action in the central nervous system. *Pharmacol. Rev.* **52,** 63–89.
46. Bengel, D., Murphy, D. L., Andrews, A. M., Wichems, C. H., Feltner, D., Heils, A., et al. (1998) Altered brain serotonin homeostasis and locomotor insensitivity to 3, 4-methylenedioxymethamphetamine ("Ecstasy") in serotonin transporter-deficient mice. *Mol. Pharmacol.* **53,** 649–655.
47. Sora, I., Wichems, C., Takahashi, N., Li, X. F., Zeng, Z., Revay, R., et al. (1998) Cocaine reward models: conditioned place preference can be established in dopamine- and in serotonin-transporter knockout mice. *Proc. Natl. Acad. Sci. USA* **95,** 7699–7704.
48. Pan, Y., Gembom, E., Peng, W., Lesch, K., Mossner, R., and Simantov, R. (2001) Plasticity in serotonin uptake in primary neuronal cultures of serotonin transporter knockout mice. *Brain Res. Dev. Brain Res.* **126,** 125–129.
49. Fabre, V., Beaufour, C., Evrard, A., Rioux, A., Hanoun, N., Lesch, K. P., et al. (2000) Altered expression and functions of serotonin 5-HT1A and 5-HT1B receptors in knock-out mice lacking the 5-HT transporter. *Eur. J. Neurosci.* **12,** 2299–2310.
49a. Murphy, D. L., Wickems, C., Andrews, A. M., Li, Q., Hamer, D., and Greenberg, B. D. (1999) Consequences of engineered and spontaneous genetic alterations of the 5-HT ransporter in mice, men, and women. *Behav. Pharmacol.* **10,** 565.

50. Gobbi, G., Murphy, D. L., Lesch , K. P., and Blier, P. (2001) Modifications of the serotonergic system in mice lacking serotonin transporters: an in vivo electrophysiological study. *J. Pharmacol. Exp. Ther.* **296,** 987–995.

51. Li, Q., Wichems, C., Heils, A., Van De Kar, L. D., Lesch, K. P., and Murphy, D. L. (1999) Reduction of 5-hydroxytryptamine (5-HT)(1A)-mediated temperature and neuroendocrine responses and 5-HT(1A) binding sites in 5-HT transporter knockout mice. *J. Pharmacol. Exp. Ther.* **291,** 999–1007.

52. Rioux, A., Fabre, V., Lesch, K. P., Moessner, R., Murphy, D. L., Lanfumey, L., Hamon, M., and Martres, M. P. (1999) Adaptive changes of serotonin 5-HT2A receptors in mice lacking the serotonin transporter. *Neurosci. Lett.* **262,** 113–116.

53. Wang, Y. M., Xu, F., Gainetdinov, R. R., and Caron, M. G. (1999) Genetic approaches to studying norepinephrine function: knockout of the mouse norepinephrine transporter gene. *Biol. Psychiatry* **46,** 1124–1130.

54. Xu, F., Gainetdinov, R. R., Wetsel, W. C., Jones, S. R., Bohn, L. M., Miller, G. W., et al. (2000) Mice lacking the norepinephrine transporter are supersensitive to psychostimulants. *Nat. Neurosci.* **3,** 465–471.

55. Bohn, L. M., Xu, F., Gainetdinov, R. R., and Caron, M. G. (2000) Potentiated opioid analgesia in norepinephrine transporter knock-out mice. *J. Neurosci.* **20,** 9040–9045.

56. Garris, P. A., Ciolkowski, E. L., Pastore, P., and Wightman, R. M. (1994) Efflux of dopamine from the synaptic cleft in the nucleus accumbens of the rat brain. *J. Neurosci.* **14,** 6084–6093.

56a. Palij, P. and Stamford, J. A. (1994) Real-time monitoring of endogenous noradrenaline release in rat brain slices using fast cyclic voltammetry: 3. Selective detection of noradrenaline efflux in the locus coeruleus. *Brain Res.* **634,** 275–282.

57. Bunin, M. A. and Wightman, R. M. (1999) Paracrine neurotransmission in the CNS: involvement of 5-HT. *Trends Neurosci.* **22,** 377–382.

58. Nirenberg, M. J., Chan, J., Pohorille, A., Vaughan, R. A., Uhl, G. R., Kuhar, M. J., and Pickel, V. M. (1997) The dopamine transporter: comparative ultrastructure of dopaminergic axons in limbic and motor compartments of the nucleus accumbens. *J. Neurosci.* **17,** 6899–6907.

59. Jones, S. R., Bowman, B. P., Kuhn, C. M., and Wightman, R. M. (1996) Development of dopamine neurotransmission and uptake inhibition in the caudate nucleus as measured by fast-cyclic voltammetry. *Synapse* **24,** 305–307.

60. Patel, A. P., Cerruti, C., Vaughan, R. A., and Kuhar, M. J. (1994) Developmentally regulated glycosylation of dopamine transporter. *Brain Res. Dev. Brain. Res.* **83,** 53–58.

61. Bannon, M. J., Poosch, M. S., Xia, Y., Goebel, D. J., and Cassin, B. (1992) Dopamine transporter mRNA content in human substantia nigra decreases precipitously with age. *Proc. Natl. Acad. Sci. USA* **89,** 7095–7099.

62. Wilson, J. M., Levey A. I., Bergeron, C., Kalasinsky, K., Ang, L., Peretti, F., et al. (1996a) Striatal dopamine, dopamine transporter, and vesicular monoamine transporter in chronic cocaine users. *Ann. Neurol.* **40,** 428–439.

63. Wilson, J. M., Kalasinsky, K. S., Levey, A. I., Bergeron, C., Reiber, G., Anthony, R. M., et al. (1996b) Striatal dopamine nerve terminal markers in human, chronic methamphetamine users. *Nat. Med.* **2,** 699–703.

64. Xia, Y., Goebel, D. J., Kapatos, G., and Bannon, M. J. (1992) Quantification of rat dopamine transporter mRNA: effects of cocaine treatment and withdrawal. *J. Neurochem.* **59,** 1179–1182.
65. Cabrera-Vera, T. M., Garcia, F., Pinto, W., and Battaglia, G. (1997) Effect of prenatal fluoxetine (Prozac) exposure on brain serotonin neurons in prepubescent and adult male rat offspring. *J. Pharmacol. Exp. Ther.* **280,** 138–145.
66. Feenstra, M. G., Van Galen, H., Te Riele, P. J., Botterblom, M. H., and Mirmiran, M. (1996) Decreased hypothalamic serotonin levels in adult rats treated neonatally with clomipramine. *Pharmacol. Biochem. Behav.* **55,** 647–652.
67. Rattray, M. (1991) Ecstasy: towards an understanding of the biochemical basis of the actions of MDMA. *Essays Biochem.* **26,** 77–87.
68. Avni, J., Gerson, S., Draskoczy, P. R., and Schildkraut, J. J. (1975) Norepinephrine content of various rat organs after chronic administration of desmethylimipramine. *Arch. Int. Pharmacodyn. Ther.* **218,** 106–109.
69. Pugsley, T. A. and Lippmann, W. (1979) Effect of acute and chronic treatment of tandamine, a new heterocyclic antidepressant, on biogenic amine metabolism and related activities. *Naunyn Schmiedebergs Arch. Pharmacol.* **308,** 239–247.
70. Carroll, F. I., Rahman, M. A., Philip, A., Lewin, A. H., Boja, J. W., and Kuhar, M. J. (1991) Synthesis and receptor binding of cocaine analogs. *NIDA Res.* **105,** 147–153.
71. Freeman, W. M., Yohrling, G. J., Daunais, J. B., Gioia, L., Hart, S. L., Porrino, L. J., et al. (2000) A cocaine analog, 2beta-propanoyl-3beta-(4-tolyl)-tropane (PTT), reduces tyrosine hydroxylase in the mesolimbic dopamine pathway. *Drug Alcohol Depend.* **61,** 15–21.

6

Family of Sodium-Coupled Transporters for GABA, Glycine, Proline, Betaine, Taurine, and Creatine

Pharmacology, Physiology, and Regulation

Scott L. Deken, Robert T. Fremeau, Jr., and Michael W. Quick

1. INTRODUCTION

For many decades it has been recognized that a variety of endogenous molecules are taken up by central-nervous-system tissue through structurally specific, high-affinity, sodium-dependent, plasma-membrane transport processes. Only recently, however, through the development of modern molecular cloning techniques, has it been realized that despite significant variability in the chemical and biochemical nature of these substances, there is remarkable similarity between the proteins and mechanisms responsible for their transport. High-affinity transport proteins have been cloned from the brain, spinal cord, and non-neural tissues for the biogenic amine neurotransmitters as described in Chapters 3 and 4, as well as the inhibitory neurotransmitters γ-amino butyric acid (GABA) and glycine. In addition, transport proteins have been cloned for several other small mol wt compounds, including proline, betaine, taurine, and creatine.

This chapter discusses recent advances in our understanding of the molecular properties, pharmacology, localization, regulation, physiological function, and clinical relevance of the cloned, nonbiogenic amine members of this transporter family, which we refer to as the GABA subfamily of Na⁺-dependent plasma-membrane transporters (Fig. 1). Discussion of the biogenic amine transporters, as well as the details of structure-function relationships for the GABA subfamily of transporters, are reserved for other chapters (*see* Chapters 3, 4, and 7). The primary goal of this chapter is to provide specific information about each transporter in the GABA subfamily, with an emphasis on their known and/or potential roles in central ner-

From: *Contemporary Neuroscience:*
Neurotransmitter Transporters: Structure, Function, and Regulation, 2nd Edition
Edited by: M. E. A. Reith © Humana Press Inc., Totowa, NJ

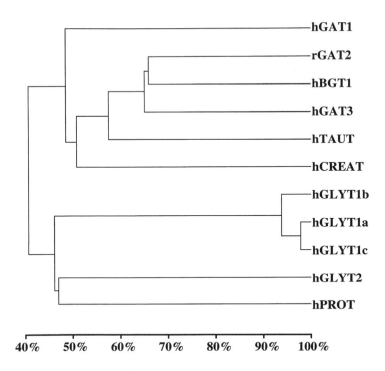

Fig. 1. Dendrogram of the amino-acid sequence relationships between members of the Na$^+$- (and Cl$^-$-) dependent transporter gene family. The percent amino-acid sequence identity between pairs of transporters at convergence points is shown at the bottom. The human homologs for each transporter are shown, except in the case of GAT2, for which no human sequence has been published. As shown in the dendrogram, two distinct subfamilies can be resolved based on amino-acid sequence identities: the subfamily of GABA, betaine, taurine, and creatine transporters and the subfamily of amino-acid (glycine and proline) transporters. hGAT1, human GABA transporter 1; mGAT2, mouse GABA transporter 2; hBGT1, human betaine transporter; hGAT3, human GABA transporter 3; hTAUT, human taurine transporter; hCREAT, human creatine transporter; hGLYT1b, human glycine transporter 1b; hGLYT1a, human glycine transporter 1a; hGLYT1c, human glycine transporter 1c; hGLYT2, human glycine transporter 2; and hPROT, human proline transporter.

vous system (CNS) function. It will become clear, however, that the similarities and differences among these transporters provide a broader view of how nature has utilized an effective mechanism, Na$^+$- (and Cl$^-$-) dependent translocation, and adapted it for varied, yet specific functions with both common and wide-ranging pharmacological and physiological characteristics. These functions include, but are not limited to, the termination of neural signals. Consequently, these transporters are recognized as potential targets for therapeutic and pathological alterations of both neurological and non-neurological physiological functions.

Before examining each transporter in detail, it may be helpful to examine the common themes among this subfamily of transporters. These transporters share >40% primary sequence identity, which establishes them as a distinct gene family (Fig. 2). Although no high-resolution information exists on transporter structure, based upon hydropathy analysis and other experiments, each of these transporters is believed to be composed of 12 transmembrane α-helical domains, cytoplasmic amino and carboxyl termini, and a large glycosylated extracellular loop separating putative transmembrane domains (TMDs) 3 and 4 (Fig. 3). Each of these transporters has an absolute requirement for extracellular Na^+ and at least a partial requirement for extracellular Cl^- in order to translocate substrates. Usually two or three sodium molecules are cotransported with each substrate molecule, allowing for the accumulation of substrate against its concentration gradient and for the setting of fairly low levels of extracellular substrate. In addition, each of these transporters has a rather high affinity for its substrate (typically in the low micromolar range), and transports substrates rather slowly (on the order of a few tens of cycles per second). These transporters, under certain physiological (or pathophysiological) conditions, can operate "in reverse" and move substrate from the cytoplasm to the extracellular space. In addition, each transporter has been shown to undergo functionally important post-translational modifications via a number of different mechanisms.

2. GABA TRANSPORTERS

The accumulation of exogenously administered GABA into mammalian brain tissue was first described over 40 years ago *(1)*. A decade later, Iversen and colleagues *(2)* provided detailed characteristics of this process: GABA uptake in rat cortical slices was found to be structurally specific, Na^+- and temperature-dependent, and saturable, with an experimentally determined Km (22 mM) indicative of high-affinity transport. These studies provided convincing evidence that a plasma membrane transport protein existed in mammalian brain that was specific for GABA. Moreover, subsequent experiments indicated that there were two GABA-transporter subtypes, which could be distinguished by their differential sensitivities to the GABA uptake inhibitors *cis*-1,3-aminocyclohexane carboxylic acid (ACHC) *(3)* and β-alanine *(2)*. It was further suggested that these inhibitor sensitivities indicated localization of these transport proteins to specific cell types within the brain (i.e., the ACHC-sensitive subtype was believed to be expressed in neuronal cells, whereas the β-alanine-sensitive subtype was believed to be

```
Key:  Identity = *    Similarity = +      Predicted Transmembrane Region = ^

hGAT1    MATNGS-----KVADGQISTEVSEAPVANDKPKTLV--------------------------                                    31
rGAT2    M----------DNRVSGTTSNG--ETKPV-CP-----------------------------                                    19
hGAT3    MTAEKA-----LPLGNGKAAEEARES--EAPGGGCS-------------------------                                    29
hGLYT1a  M-----------VGKGA-----------------------------------------KGML-                                  10
hGLYT1b  M-----------AAAHGPVAPSSPEQVTLLPVQRSFFLPP----------FSGATPSTSLAESV-----LKVWHGAYNSGLLFQL--           59
hGLYT1c  M-----------AAAHGPVASSPEQ-----------------------------------                                     15
hGLYT2   MDCSAPKEMNKLPANSPEAAAAQGHPDGPCAPRTSPEQELPAAAAPPPRVPRSASTGAQTFQSADARACEAERPGVGSCKLSSPRAQAASAALRDLREA 100
hPROT    M-----------KKLQGA-------------------------------------DL--                                     17
hBGT1    M-----------DGKVAVQERGP--PAVSW-VP--------HLRKPVTP------------                                    19
hTAUT    MATKEK-----LQCLKDFHKDMVKPS--PGKSPGTR-------------------------                                    29
hCREAT   MAKKSA-----ENGIYSVSGDEKKGPLIAPGPDGAP-------------------------                                    29
         *

hGAT1    -----------------------------VKVQ--------------------------KKAADLPDRDTWKGRFDF                    53
rGAT2    ------------------------------V--MEKVE----------------------EDGTLEREQWTNKMEF                    41
hGAT3    ------------------------------SGGAAPARHPR-------------------VKRDKAVHERGHWNNKVEF                  59
hGLYT1a  ------------------------------------------------------NG-----AVPSEATKRDQNLK-RGNWGNQIEF           36
hGLYT1b  --------------------------------------MAQHSLAMAQNG-----NG-----AVPSEATKRDQNLK-RGNWGNQIEF           95
hGLYT1c  ------------------------------------------------------NG-----AVPSEATKRDQNLK-RGNWGNQIEF           41
hGLYT2   QGAQASPPPGSSGPGNALHCKIPSLRGPEGDANVSVGKGTLERNNTPVVGWVNMSQSTVVLGTDGITSVLPGSVATVATQEDEQGDENKARGNWMSSKLDF 200
hPROT    --------------------------LMTPSDQGDVDLD--------------------RGNWTGKLDF                            46
hBGT1    --------------------------EEGEKLDQED----------------------EDQVKDRGQWTNKMEF                       45
hTAUT    --------------------------PEDEAEGKPP----------------------QREKWSSKIDF                            50
hCREAT   --------------------------AKGDGPVGLGT--------------------PGGRLAVPPRETWTRQMDF                     61
                                                                  *  *   +   + +*
                                                                                      <

hGAT1    LMSCVGYAIGLGNVWRFPYLCGKNGGGAFLIPYFLTLIFAGVPLFLLECSLGQYTSIGGLGVW-KLAPMFKGVGLAAAVLSFWLNIYYIVIVISWAIYYLY 152
rGAT2    VLSVAGEIIGLGNVWRFPYLCYKNGGGAFFIPYLIFLFTCGIPVFFLETALGQYTNQGGITAWRKICPIFEGIGYASQMIVSLLNVYYIVVLAWALFYLF   141
hGAT3    VLSVAGEIIGLGNVWRFPYLCYKNGGGAFLIPYVVFFICCGIPVFFLETALGQFTSEGGITCWRKVCPLFEGIGYATQVIERAHLNVYYIILAWAIFYLS   159
hGLYT1a  VLTSVGYAVGLGNVWRFPYLCYRNGGGAFMFPYFIMLIFCGIPLFFMELSFGPFASQGCLGVWR-ISPMFKGVGYGMMVVSTYIGIYYNVICIAFYYFF   135
hGLYT1b  VLTSVGYAVGLGNVWRFPYLCYRNGGGAFMFPYFIMLIFCGIPLFFMELSFGFASQGCLGVWR-ISPMFKGVGYGMMVVSTYIGIYYNVICIAFYYFF    194
hGLYT1c  VLTSVGYAVGLGNVWRFPYLCYRNGGGAFMFPYFIMLIFCGIPLFFMELSFGFASQGCLGVWR-ISPMFKGVGYGMMVVSTYIGIYYNVICIAFYYFF    140
hGLYT2   ILSMCVGYAVGLGNVWRFPYLAFQNGGGAFLIPYLMMLALAGLPIFFLEVSLGQFASQGPVSVWK-AIPALQGCGIAMLIISVLIALYYNVICYLFYLF   299
hPROT    LLSCIGYCVGLGNVWRFPYLCYRNGGGAFLIPYFIFLFGCGIPLFFLELSLGQFASQGSVSVWK-AIPALQGCGIAMLIISVLIAIYYNVICYLFYLF    145
hBGT1    VLSVAGEIIGLGNVWRFPYLCYKNGGGAFFIPYFIFFVCGIPYFFLEVALGQYTSQGSVTAWRKICPLFQGIGLASVVIESYLNVYYIIILAWALFYLF   145
hTAUT    VLSVAGGFVGLGNVWRFPYLCYKNGGGAFLIPYFIFLFGSGLPVFFLEIIIGQYTSEGGITCWEKICPLFSGIGYASVVIVSLLNVYYIVILAWATYYLF   150
hCREAT   IMSCVGFAVGLGNVWRFPYLCYKNGGGVFLIPYLVLIALVGGIPIFFLEISLGQFMKAGSINVWN-ICPLFKGLGYASMVIVFYCNTYYIMVLAWGFYYLV 160
         +++  *    **********     *****+++ ** +     *+*+*+++    ++++ + + + **   * * +  * + +   ++   +++  +*+
         ^^^^^^^TM1^^^^^^^^       ^^^^^^^^TM2^^^^^^^                          ^^^^^^^^^^TM3^^^^^^^^
```

```
hGAT1    NSFTTTLPWKQCDNPWNTDRCFSNY----------SMVNTT--------------------------NMTSAVVEFWERNMHQM--TDGLDKPGQIRWPLA  215
rGAT2    SSFTTDLPWGSCSHEWNTENCVE-FQ--------KTNNSL-NVTS-E--------------------NATSPVIEFWERRVLKI--SDGIQHLGSLRWELV  209
hGAT3    NCFTTTLPWATCGHEWNTENCVE-F---------QKLNVSNYSHVSLQ------------------NATSPVIEFWEHRVLAI--SDGIEHIGNLRWELA  229
hGLYT1a  SSMTHVLPWAYCNNPWNTHDCAG--------VLDASNLTNGSRPA--------ALPS-NLSHLLNHSLQR-TSPSEEYWRLYVLKL--SDDIGNFGEVRLPLL  218
hGLYT1b  SSMTHVLPWAYCNNPWNTHDCAG--------VLDASNLTNGSRPA--------ALPS-NLSHLLNHSLQR-TSPSEEYWRLYVLKL--SDDIGNFGEVRLPLL  277
hGLYT1c  SSMTHVLPWAYCNNPWNTHDCAG--------VLDASNLTNGSRPA--------ALPS-NLSHLLNHSLQR-TSPSEEYWRLYVLKL--SDDIGNFGEVRLPLL  223
hGLYT2   ASFVSVLPWGSCNNPWNTPECKDKTKLLLDSCVISDHPKIQIKNSTFCMTAYPNVTMVNFTSQANKTFVSGSEYFKYFVLKI--SAGIEYPGEIRWPLA  397
hPROT    ASLTSDLPWEHCGNWWNTELCLEHR------------VSKDGNGALPLNLTCT-----------------VSPSEEYWSRYVLHIQGSQGIGSPGEIRWNLC  218
hBGT1    SSFTSELPWTTCNNFWNTEHCTD-F-------------LNHSGAG-TVTPFE-------------------NFTSPVMEFWERRVLGI--TSGIHDLGSLRWELA  214
hTAUT    QSFQKELPWAHCNHSWNTPHCMEDT--------------MRKNKSVWITISST-------------------NFTSPVIEFWERRVLSL--SPGIDHPGSLKWDLA  221
hCREAT   KSFTTTLPWATCGHTWNTPDCVEIF----------RHEDCANASLANLTC--------------------DQLADRRSPVIEFWENKVLRL--SGGLEVPGALNWEVT  236
            ++   ***   *++ ***   *              *              *       +      * ++ +     + + +++   *  ++  +
                                                                                                          ^^^^

hGAT1    ITLAIAWILVYFCIWKGVGWTGKVVYFSATYPYIMLLILFFRGVTLPGAKEGILFYITPNFRKLSDSEVWLDAATQIFFSYGLGLGSLIALGSYNSFHNN  315
rGAT2    LCLLLAWIICYFCIWKGVKSTGKVVYFTATFPYIMLVLLIRGVTLPGAAQGIQFYLYPNITRLWDPQVWMDAGTQIFFSFAICLGCLTALGSYNKYHNN  309
hGAT3    LCLLAAWTICYFCIWKGTKSTGKVVYVTATFPYIMLLLLLIRGVTLPGASEGIKFYLYPDLSRLSDPQVWVDAGTQIFFSYAICLGCLTALGSYNNYNNN  329
hGLYT1a  GCLGVSWLVVFLCLLIRGVKSSGKVVYFTATFPYVVLTILFVRGVTLEGAFDGIMYLTPQWDKILEAKVWGDAASQIFYSLACAWGGLITMASYNKFHNN  318
hGLYT1b  GCLGVSWLVVFLCLLIRGVKSSGKVVYFTATFPYVVLTILFVRGVTLEGAFDGIMYLTPQWDKILEAKVWGDAASQIFYSLACAWGGLITMASYNKFHNN  377
hGLYT1c  GCLGVSWLVVFLCLLIRGVKSSGKVVYFTATFPYVVLTILFVRGVTLEGAFDGIMYLTPQWDKILEAKVWGDAASQIFYSLACAWGGLITMASYNKFHNN  323
hGLYT2   LCLFLLAWVIVYASLAKGIKTSGKVVYFTATFPYVVLIILLMLLVRGVTLPGAGAGIWYFITPKWEKLTDATVWKDAATQIFFSLSAAWGGLITLSSYNKFHNN  497
hPROT    LCLLLAWVIVFLCILKGVKSSGKVVYFTATFPYVYYILILLMLLVRGVTLPGAWKGIQFYLTPQFHHILSSKVWIEAALQIFYSLGVGFGGLLTFASYNTFHQN  318
hBGT1    LCLLLAWVICYFCICKGVRSTGKVVYFTATFPYIMLVLLIRGVTLPGAYQGIIYYLKPDLFRLKDPQVWMDAGTQIFFSFAICQGCLTALGSYNKYHNN  314
hTAUT    LCLLLVWLVCFFCICKGVRSTGKVVYFTATFPYVVLVVLLVRGLTLPGAGRGIKFYLYPDITRLEDPQWIDAGTQIFFSYAICLGAMTSLGSYNKYKYN  321
hCREAT   LCLLACWVLVYFCVWKGVKSTGKIVYFTATFPYVVLVVLLVRGVLLRGVLLPGALDGIIYYLKPDWSKLGSPQVWIDAGTQIFFSYAIGLGALTALGSYNRFNNN  336
          *  * + + ++ +++     +++***  +++*** *   +  ++ ++  *  *   ++ ++   *   *   +++ +++** ++ * + + ++++** ++ *
         ^^^^TM4^^^^^^^                                        ^^^^^TM5^^^^^^^            ^^^^^^TM6^^^^^^^

hGAT1    VYRDSIIVCCINSCTSMFAGFVIFSIVGFMAHVTKRSIADVAASGPGLAFLAYPRAVTQLPISPLWAILFFSMLLMLGIDSQFCTIVEGFITALVDEYPRL  415
rGAT2    CYRDCVALCILNSSTSFVAGFAIFSILGFMSQEQGVPISEVAESGPGLAFIAYPRAVVMLPFSPIWACCFFMVVLLGLDSQFVCVESLVTALVDMYPRV  409
hGAT3    CYRDCIMLCCLNSGTSFVAGFAIFSVLGFMAYEQGVPIAEVAESGPGLAFIAYPKAVTMPLSPLWATLFFMLIFLGLDSQFVCVESLVTAVVDMYPKV  429
hGLYT1a  CYRDSVIISITNCATSVYAGFVIFSILGFMANHLGVDVSRVADHGPGLAFVAYPEALTLLPISPLWSLLFFFMLIILLGLGTQFCLLETLVTAIVDEVGNE  418
hGLYT1b  CYRDSVIISITNCATSVYAGFVIFSILGFMANHLGVDVSRVADHGPGLAFVAYPEALTLLPISPLWSLLFFFMLIILLGLGTQFCLLETLVTAIVDEVGNE  477
hGLYT1c  CYRDSVIISITNCATSVYAGFVIFSILGFMANHLGVDVSRVADQGPGLAFVAYPEALTLLPISPLWSLLFFFMLIILLGLGTQFCLLETLVTAIVDEVGNE  423
hGLYT2   CYRDILVTCTNSATSIFAGFVIFSVLGFMANEREKVNIENVADQGPGIAFVYPEALTRLPLSPFWAIIFFLMLLTIGLDTMFATIETEITVTSISDEFPK-  596
hPROT    IYRDTFIVTLGNAITSILAGFAIFSVLGYMSQELGVPVDQVAKAGPGLAFVVYPQAMTMLPLSPFWSFLFFFMLLTIGLDSQFAFLETIVTAVTDEFPY-  417
hBGT1    CYKDCIALCFLNSATSFVAGFVVFSILGFMSQEQGVPISEVAESGPGLAFIAFPRAVTMPLSQLWSCLFFIMLIFIGLDSQFVCVECLVTASIDMFPRQ  414
hTAUT    SYRDCMLLGCLNSGTSFVSGFAIFSILGFMAQEQGVDIADVAESGPGLAFIAYPKAVTMPLPTFWSIIFFIMLLLGLDSQFVEVEGQITSLVDLYPSF  421
hCREAT   CYKDAIILALINSGTSFFAGFVISLGFMAAEQGVHISKVAESGPGLAFIAYPRAVTMLMPVAPIWAALFFFMLLLLGLDSQFVGVEGFITGLLDLLPAS  436
          *+*  +   * **  *+   ***+*+** +++ * ++ +**  *+++ ++ **** ++++ * ***+** ++ * + +*+
         ^^^^^^^TM7^^^^^^^^^                                                       ^^^^^^^^TM8^^^^^^^
```

Fig. 2.

```
hGAT1    LRN--RRELFIAAVCIISYLIGLSNITQGGIYVFKLFDYYSASGMSLLFLVFFECVSISWFYGVNRFYDNIQEMVGSRPCIWWKLCWSFFTPIIVAGVFI   513
rGAT1    FRKKNRREILILIVSVSFFIGLIMLTEGGMYVFQLFDYYAASGMCLLFVAIFESLCVAMVYGASRFYDNIEDMIGYKPWPLIKYCWLFFTPAVCLATFL    509
hGAT3    FRRGYRRELLILALSVISYFLGLVMLTEGGMYIFQLFDSYAASGMCLLFVAIFECICIGWVYGSNRFYDNIEDMIGYRPPSLIKWCWIMTPGICAGIFI   529
hGLYT1a  WILQ--KKTYVTLGVAVAGFLLGIPLTSQAGIYWLL-MDNYAAS-FSLVVISCIMCVAIMYIYGHRNYFQDIQMMLGFPPPLFFQICWRFVSPAIIFFILV  515
hGLYT1b  WILQ--KKTYVTLGVAVAGFLLGIPLTSQAGIYWLLLMDNYAAS-FSLVVISCIMCVAIMYIYGHRNYFQDIQMMLGFPPPLFFQICWRFVSPAIIFFILV  575
hGLYT1c  WILQ--KKTYVTLGVAVAGFLLGIPLTSQAGIYWLLLMDNYAAS-FSLVVISCIMCVAIMYIYGHRNYFQDIQMMLGFPPPLFFQICWRFVSPAIIFFILV  521
hGLYT2   YLRT-HKPVFTLGCCICFFIMGFPMITQGGIYMFQLVDTYAAS-YALVIIAIFELVGISYVYGLQRFCEDIEMMIGFQPNIFWKVCWAFVTPTILJTFILC  694
hPROT    YLRP-KKAVFSGLICVAMYLMGLLITDGGMYWLVLLDDYSAS-FGLMVVITTCLAVTRVYGIQRFCRDIHMMLGFKPGLYFRACWLFLSPATLLALLV     515
hBGT1    LRKSGRRELLILTIAVMCYLIGLFLVTEGGMYIFQLFDYYASSGICLLFLSLFEVVCISWVYGADRFYDNIEDMIGYRPWPLVKISWLFLTPGLCLATFL   514
hTAUT    LRKGYRREIFIAFVCSISYLLGLMVTTEGGMYVFQLFDYYAASGVCLLWVAFFECFVIAWIYGGDNLYDGIEDMIGYRPGPWMKYSWVI-TPVLVCGCFI   520
hCREAT   YYFRFQREISVALCCALCFVIDLSMVTDGGMYVFQLFDYYASGTTLLWQAFWECVVVAWNYGADRFMDDIACMIGYRPCPWMKWCWSFFTPLVCMGIFI    536
              ++  ++++  +++*++* +  *  *++*  +    +  +  ** *   ++  *   ++  *** *  + +** + +*
         ^^^^^^^^TM9^^^^^^^^^                                                            ^^^^^^^^TM11^^^^


hGAT1    FSAVQMTPLTMGN-YVFPKWGQGVGWLMAL--SSMVLIPGYMAYMFLALKG-SLKQRIQVMVQPSED----TVRPENGPEHAQAGSS-----------    592
rGAT2    FSLIKYTPLTVNKKYTYPWWGDALGWLLAL--SSMVCIPAWSIYKLRTLKG-PLRERLRQLVCPAED-----LPQ-KS---QPELTSPA------TPMTSL  592
hGAT3    FPLIKYKPLKYNNIVTYPAWGYGIGWLMAL--SSMLCIPLWICITVWKTEG-TLPEKLQKLTTPSTD----LKMRGKLGVSPRMVTVNDCDAKLKSDGTI   622
hGLYT1a  FTVIQYQPITVNH-YQYPGWAVAIGFLMAL--SSVLCIPLYAMFRLCRTDGDTLLQRLKNATKPSRDWGPALLEHRTGRYAPTIAPSPEDGFEVQSLH--   610
hGLYT1b  FTVIQYQPITVNH-YQYPGWAVAIGFLMAL--SSVLCIPLYAMFRLCRTDGDTLLQRLKNATKPSRDWGPALLEHRTGRYAPTIAPSPEDGFEVQSLH--   670
hGLYT1c  FTVIQYQPITVNH-YQYPGWAVAIGFLMAL--SSVLCIPLYAMFRLCRTDGDTLLQRLKNATKPSRDWGPALLEHRTGRYAPTIAPSPEDGFEVQSLH--   616
hGLYT2   FSFYQWEPMTYGS-YRYPNWSMVLGWLM--LACSVIWIPIMFVIKMHLAPGR-FIERLKLVCSQPDWGPFLAQHRGERY-KNMIDPLGTSSLGLKLPV-    788
hPROT    YSIVKYQPSEYGS-YRFPPWAELLGLMGLLSCLM--IPAGMLVAVLREEGS-LWERLQQASRMGYVATLAGSQSPKPLMVHMRKY                 611
hBGT1    FSLSKYTPLKYNNVYVYPPWGYSIGWLFLAL--SSMVCVPLFVVITLLKTRG-PFRKRLRHVITPDSS----LPQPKQ---HPCLDGSAGRNFGPSPTREG  604
hTAUT    FSLVKYVPLTVNKTVSPTWAIGLGWSLAL--SSMLCVPLVIVIRLCQTEG-PFLVRVKYLLTP-------REPNRWAVEREGATPYNSRTVMNGALVK     609
hCREAT   FNVVYYEPLVYNNTYVYPWGEAMGWAFAL--SSMLCVPLHLLGCLLRAKG-TMAERWQHLTQPIWG-----LHHLEYRAQDADVRGL----TTLTPVSES   625
          +    *    +     + +  *  * *++           * +  +  *       *      *    +  + + +
         ^^^^^^^^TM12^^^^^^^^^


hGAT1    ---TSK--------------EAYI  599
rGAT1    LRLTEL--------------ESNC  602
hGAT3    AAITEK--------------ETHF  632
hGLYT1a  ---PDKAQIPIVGSNGSSRLQDSRI  632
hGLYT1b  ---PDKAQIPIVGSNGSSRLQDSRI  692
hGLYT1c  ---PDKAQIPIVGSNGSSRLQDSRI  638
hGLYT2   -------KDLELGTQC-------  797
hPROT    GGITSFENTAIEVDREIAEEEESMM  636
hBGT1    LIAGEK--------------ETHL  614
hTAUT    PTHIIV--------------ETMM  619
hCREAT   SKVVVV--------------ESVM  635
          +     * * * *   +  + ++
         ^^^^
```

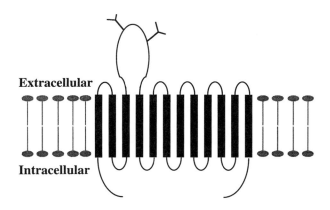

Fig. 3. Schematic representation of the putative transmembrane topology of the Na^+- (and Cl^--) dependent transporter(s). The proposed model of these transporters includes 12 transmembrane α-helical domains, cytoplasmic amino and carboxyl termini, and a glycosylated EL separating TMDs 3 and 4.

expressed in glial cells). We now recognize that there are at least four distinct high-affinity GABA-transporter subtypes that exhibit distinct pharmacologies and unique temporal and spatial patterns of expression. We have also learned much about their microscopic transport properties, their ability to be functionally regulated, and their physiological and pathophysiological roles.

2.1. GABA-Transporter Subtypes

The molecular and pharmacological characterization of GABA-transport proteins began with the pioneering work of Kanner and colleagues, who purified the first neurotransmitter transporter: the rat-brain GABA transporter GAT1 *(4)*. GAT1 cDNA predicted an open reading frame of 1797 nucleotides, encoding a 599 amino-acid protein with a mol wt of 67 kDa. On expression in *Xenopus* oocytes, GAT1 was found to be highly selective for GABA, with a Km of 7 μM. An absolute dependence on extracellular Na^+ ions for transport was demonstrated, consistent with the findings in rat cortical slices. A dependence on extracellular Cl^- ions was also shown. In addi-

---

Fig. 2. (196–198) Alignment of the predicted amino acid sequences encoding the subfamily of GABA, glycine, proline, betaine, taurine, and creatine Na^+- (and Cl^--) dependent transporter(s). Asterisks are used for amino acids identically conserved for all members of the subfamily, and plus signs are used for amino acids similarly conserved for all members of the subfamily. Predicted TMDs are indicated with carats below the appropriate amino-acid residues. For abbreviations, *see* Fig. 1.

tion, GAT1 was found to be strongly inhibited by the GABA uptake inhibitors ACHC, nipecotic acid, and 2,4-diaminobutyric acid (DABA), but only weakly inhibited by β-alanine. GAT1 transporters are localized to both pre- and postsynaptic neurons *(5,6)*, as well as to glial cells *(7)*. This latter finding dismissed the traditional distinction of the GAT1 transporter as "neuronal," as was initially suggested by uptake-inhibitor sensitivities and by immunolocalization *(8)*.

After the cloning and characterization of the second member of the neurotransmitter transporter family, the norepinephrine transporter *NET (9)*, it became possible to use polymerase chain reaction (PCR)-based homology screening to search for other family members. Two additional GABA transporters were soon cloned from the rat brain *(10)*, and denoted GAT2 and GAT3. These cDNAs encoded proteins of 602 and 627 amino acids, respectively, and shared 52% amino-acid identity with GAT1, and 67% amino-acid identity with each other. When heterologously expressed in mammalian cells, both GAT2 (Km = 17 μM) and GAT3 (Km = 33 μM) showed high affinity GABA transport. However, they both differed pharmacologically from GAT1 because of lower affinities for nipecotic acid and ACHC, and higher affinities for β-alanine. These two transporters also showed less inhibition of transport than GAT1 following the removal of extracellular Cl⁻.

Many other GABA transporters from many different species were cloned almost simultaneously, leading to some confusion over GABA-transporter nomenclature. There are probably four GABA-transporter subtypes: GAT1, GAT2, GAT3, and BGT1. How these present names relate to the earlier cloning literature (names placed in quotes) are as follows: In 1992, Lopez-Corcuera et al. *(11)* isolated a novel high-affinity GABA transporter from mouse, which they called "GAT2," which exhibited 49% identity with the GAT1 GABA transporter. When expressed in *Xenopus* oocytes, mouse "GAT2" exhibited distinct pharmacological properties. It had an approx 10-fold lower affinity for GABA (Km, 79 μM) compared to GAT1 (Km 7 μM), and it exhibited different sensitivities to the GABA uptake inhibitors nipecotic acid and β-alanine. Moreover, mouse "GAT2" was shown to be significantly inhibited by the osmolyte and methyl donor betaine. Mouse "GAT2" was found to have 88% sequence identity with a transporter (BGT1) that had recently been cloned from canine kidney cells, and had been shown to transport both GABA and betaine with relatively high affinity *(12)*. Thus, mouse "GAT2" is likely BGT1. This topic is discussed in more detail in **Subheading 4.2.** on betaine transporters. At about the same time, Clark et al. *(13)* described the cloning and expression of "GAT-B" from the rat midbrain. "GAT-B" was found to have 50% sequence identity with GAT1 and 63% identity with BGT1. On transient expression in HeLa cells, "GAT-B" transported GABA with a Km of 2.3 μM. The transport was strongly inhibited by β-alanine (IC50 = 6.7 μM) and nipecotic acid (lC50 = 53 μM), and was weakly inhibited by DABA, THPO, taurine, and ACHC. Sequence and functional analysis suggested that "GAT-B" is GAT3. In the meantime, Liu et al. *(14)* cloned four high-affinity GABA transporters from the mouse brain that were designated "GAT1"–"GAT4." Molecular, pharmacological, and localization studies suggest that "GAT1" is GAT1, "GAT2" is BGT1, "GAT3" is GAT2, and "GAT4" is GAT3.

2.2. *In Vivo, Cellular, and Subcellular Transporter Distribution*

Northern blot or *in situ* hybridization analysis has shown that the three GABA transporters (excluding BGT1) have differential tissue distributions, with significant expression outside the nervous system. GAT1 is found throughout the brain and other tissues, but is not found in the liver *(4)*. GAT2 is found in the brain, retina, liver, and kidney *(10)*. GAT 3 is found almost exclusively in the nervous system *(13)*. The differences in expression, along with the differences in pharmacological profiles, suggest different physiological functions for these transporters.

In the brain, GAT1 is the most prominent and widely expressed GABA-transporter subtype *(15)*. GAT1 is found throughout the neocortex, hippocampal formation, basal ganglia, thalamus, brain stem, spinal cord, cerebellum, olfactory bulb, and retina. The distribution of GAT1 corresponds well with GABAergic markers such as glutamic acid decarboxylase and GABA *(15,16)*. In contrast, GAT3 has a much less widespread distribution. Expression is high in basal ganglia structures, and in the olfactory bulb and retina *(13)*, with minimal expression in the neocortex, hippocampus, and spinal cord. GAT2 has an even more interesting distribution in the brain. GAT2 staining is intense in meninges, ependyma, and choroid plexus *(17)*, providing a good match with the mRNA localization *(15)*. Some GAT2 expression is also evident in cells within the neocortex *(17)* and the retina *(16)*. Thus, all three GABA-transporter subtypes are expressed in the retina; however, their distribution patterns are distinct, suggesting different functions within retinal circuits.

To participate in the termination of synaptic signaling, or to perform any specific function in the brain, a transporter presumably must be localized at or near a nerve terminal with a relatively high local density. Little is known, however, about the mechanisms by which transporters are sorted to functionally relevant subcellular locales (i.e., neuronal and glial plasma membranes). The cloning of the Na^+- (and Cl^--) dependent transporters provides the opportunity to determine structural motifs within the proteins that are important for subcellular targeting. Although not neuronal in origin, a good model for this purpose may be the Madin-Darby canine kidney (MDCK) cell line. It was from these cells that the BGT1 transporter was initially cloned *(12)*. Under appropriate cell-culture conditions, MDCK cells form a polarized bilayer consisting of basolateral and apical membranes. Pietrini et al. *(18)* observed that BGT1 expressed in MDCK cells accumulated primarily at the basolateral membrane. Upon stable transfection in MDCK cells, GAT1 accumulated primarily in the apical membrane. Thus, BGT1 and GAT1 were targeted to different subcellular locations within this cell type.

These results were the basis of a study by Ahn et al. *(19)* that explored the expression of GAT transporters in MDCK cells and cultured hippocampal neurons. In MDCK cells, the GAT2 transporter was found to localize to the basolateral membrane, like BGT1, whereas GAT3 was found to localize to the apical membrane, like GAT1. Similar sorting phenomena were observed in hippocampal neurons microinjected with the GABA transporter DNA; BGT1 was found only in somatodendritic membranes, whereas GAT3 was found in axons as well as cell bodies and dendrites. The subcellular distributions of GAT1 and GAT2 within hippocampal neurons were not determined, but based on the MDCK data it was predicted that their expressions would follow those observed for BGT1 and GAT3, respectively.

More recently, the molecular mechanisms underlying the differential sorting of GABA-transporter subtypes has been examined. Using chimeric molecules constructed from GAT1 and BGT1, it has been shown that the C-terminal tail of BGT1 contains a stretch of basic amino-acid residues which provide the sorting information necessary to direct transporters to the basolateral surface in polarized epithelial cells *(20)*. A similar story exists for chimeric GAT2 and GAT3 molecules. Sorting information in the C-terminal tail of GAT2 is necessary to direct it to the basolateral surface in MDCK cells *(21)*. Interestingly, BGT1 contains a motif in its C-terminal domain that mediates protein-protein interactions with the protein LIN-7, a member of the PDZ family of localization proteins. This domain is not essential to target BGT1 to the basolateral surface, but is necessary to maintain BGT1 on the basolateral surface *(20)*. A similar PDZ-binding motif exists in GAT3, which suggests a role for PDZ-binding proteins in the retention of GAT3 at the apical cell surface *(21)*.

2.3. Permeation Properties

Initial investigations into GABA transport came from biochemical studies of rat-brain preparations. These studies suggested Na^+ and Cl^- dependence on transport as well as a stoichiometry of $Na^+:Cl^-:GABA$ of 2:1:1 *(22)*. From the time that the four GAT clones were isolated, the majority of studies examining the details of GABA-transporter permeation have been performed using GAT1. Electrophysiological studies of GAT1 expressed in oocytes *(23)* confirmed these earlier data and demonstrated a net inward current associated with GABA uptake (because of a net 1 positive charge entering the cell with each transport cycle). Voltage-jump experiments performed in the absence of GABA, thus permitting only partial reactions, showed that Na^+ binding could still occur, and that the rate of the translocation process was on the order of tens of cycles per second *(24)*. These slow rates of transport suggested that if transporters were to have a significant impact on signaling in the brain, they would have to exist in large numbers near these signaling proteins.

As mentioned in other chapters (*see* Chapters 2, 3, and 8), an interesting question in the field of transporter permeation is the extent to which transporters resemble ion channels—is each cycle of the transport process stoichiometric, and does it permit loading and unloading of substrate in a defined manner (the alternating access model), or is the process more akin to ions moving through a pore in a less defined manner (channel model)? Fluctuation analysis shows that uptake resembles movement of ions through an occluded pore, as seen in ion-channel permeation *(25)*. Although this may suggest an ion-channel model for GAT1, recent studies on the transporter using giant membrane patches suggests that an alternating access model can account for all the various behaviors of GAT1 *(26)*.

The structural domains of GABA transporters that underlie their functional activities are discussed in detail in Chapter 7. However, we have recently identified the N-terminal cytoplasmic domain of GAT1 as an important factor in substrate translocation *(27)*. Removal of the N-terminal tail, or mutagenesis of three aspartic residues in this tail, results in a transporter that shows decreased rates of transport. Co-expression of the N-terminal tail fragment along with these other mutants in *Xenopus* oocytes rescues wild-type activities. We have postulated that the N-terminal tail interacts with the permeation path to permit normal substrate permeation. Manipulations that disrupt this interaction decrease translocation rates. How this may be achieved mechanistically is currently under investigation.

2.4. Physiological Roles

One physiological role for GABA transporters is in terminating cell signaling. In the brain, there are two general classes of receptors which mediate GABA signaling both presynaptically and postsynaptically. GABA-A and GABA-C receptors are ligand-gated ion channels that act on a time-scale of milliseconds; GABA-B receptors are G-protein-coupled receptors that mediate their actions on a time-scale of seconds. In a variety of experimental preparations, application of GABA-uptake inhibitors slows the decay rate of signaling mediated by both of these receptor types *(28,29)*, suggesting that GABA transporters remove transmitter from the vicinity of these receptors. Because of the slower response time of GABA-B receptors, GABA-transporter blockade also decreases peak responses mediated through this receptor type. The mechanism underlying the GABA transporter's regulation of synaptic signaling has not yet been determined. One possibility is that transport *per se* limits receptor signaling. However, as discussed previously, the translocation rate of a single transporter is fairly slow, suggesting that transport may play a role only in regulating GABA-B signals. The other possibility is that the transporter acts as a binding site for GABA, sequestering it away from receptors.

Given the stoichiometry of the transport process, under most conditions the concentration of cotransported ions, the concentration of GABA, and the cell's membrane potential will result in uptake of transmitter into the cell cytoplasm. However, under certain conditions the transporter will operate "in reverse" *(30)* and efflux transmitter. In some retinal cells, GABA efflux through the transporter caused by depolarization may be the principal release mechanism *(31)*. Efflux may also be a cellular response to prolonged hyperexcitability. Cell swelling and increases in extracellular GABA following acute depolarization-induced excitotoxicity are both ameliorated by GABA transport inhibitors *(32)*.

2.5. Regulation

If appropriate transmitter levels are needed for normal brain function, then the regulation of transport would be one mechanism that cells could use for the short- or long-term regulation of transmitter levels. This was our initial hypothesis when we began to examine the regulation of GAT1. Our initial studies focused on the observation that the sequence of the GAT1 protein predicts several sites of phosphorylation by protein kinase C (PKC). Thus, we used a variety of PKC activators and inhibitors, and protein phosphatases, on GAT1 expressed in *Xenopus* oocytes *(33)*. We showed that PKC could regulate GAT1 function twofold on the time-scale of minutes. Interestingly, this regulation occurred even when the three consensus PKC sites were removed by site-directed mutagenesis, suggesting either phosphorylation of nonconsensus sites or indirect action of PKC *(34)*. This effect of PKC and phosphatases was also shown for GABA transport in a variety of systems, although the magnitude and direction of the modulation appears to be system- and condition-dependent *(35,36)*. PKC regulation of transport has been shown to be a common theme throughout this subfamily of transporters.

We found that the action of PKC on the transporter was to alter the maximal transport capacity of GAT1, with minimal effect on Km. These data were consistent with a regulatory mechanism associated with alterations in the number of functional transporters on the membrane. Subcellular fractionation experiments showed that GAT1 protein was being redistributed between plasma-membrane fractions and cytosolic fractions in the presence or absence of PKC *(33)*. We also used electrophysiological procedures to count the number of functional GABA transporters by estimating the number of transporter-binding sites; this showed that PKC altered the number of functional GABA transporters on the plasma membrane *(37)*.

More recently, we have documented that protein tyrosine kinases directly regulate GAT1 expressed endogenously in hippocampal neurons and are expressed heterologously in Chinese Hamster Ovary (CHO) cells *(38)*. Inhibitors of tyrosine kinases decreased GABA uptake; inhibitors of tyrosine phosphatases increased GABA uptake. The decrease in uptake in the presence of tyrosine kinase inhibitors correlated with a decrease in tyrosine phosphorylation of GAT1 and resulted in a redistribution of the transporter from the cell surface to intracellular locations. A mutant GAT1 construct in which all putative tyrosine kinase sites were eliminated was refractory to tyrosine phosphorylation and could not be regulated by tyrosine kinase inhibitors. PKC was additive to the effects of tyrosine kinase inhibitors, suggesting that multiple signaling pathways control transporter redistribution. These data are consistent with direct phosphorylation of GAT1 by tyrosine kinases, and suggest that transporter expression and function are controlled by the interplay of multiple cell-signaling cascades.

A related question concerns the initial triggers responsible for the regulation of GAT1. Neurotransmitter receptors that activate these second messengers are one probable trigger. In support of this, we have shown that the application of brain-derived neurotrophic factor (BDNF), which activates receptor tyrosine kinases, upregulates GAT1 function *(38)*. In addition, we have shown that specific agonists of G-protein-coupled acetylcholine, glutamate, and serotonin receptors downregulate GAT1 function *(39)*. The inhibition is dose-dependent, mimicked by PKC activators, and prevented by specific receptor antagonists and PKC inhibitors. The inhibition correlates with a redistribution of GAT1 from the plasma membrane to intracellular locations. These results suggest that a number of receptor systems can alter GAT1 function, and raise the possibility that some effects of G-protein-mediated alterations in synaptic signaling may occur through changes in the number of transporters expressed on the plasma membrane and subsequent effects on synaptic neurotransmitter levels. Serotonin may also regulate GAT2 and GAT3 function *(40)*.

If maintaining appropriate transmitter levels in the cleft represents a critical role for GABA transporters, then transporters may be regulated directly by those transmitter levels. We have shown that extracellular GABA induces chronic changes in GABA transport that occur in a dose-dependent and time-dependent manner *(41)*. In addition to GABA, ACHC and nipecotic acid, both substrates of GAT1, upregulate transport; GAT1 transport inhibitors that are not transporter substrates downregulate transport. These changes occur in the presence of blockers of GABA receptors and protein synthesis inhibitors, and are not influenced by intracellular GABA. This increase in surface expression is caused by a net slowing of GAT1 internalization in the presence of extracellular GABA. These data suggest that the GABA transporter fine-tunes its function in response to extracellular GABA, and would act to maintain a constant level of neurotransmitter at the synaptic cleft.

Most recently, we have been investigating the role of protein-protein interactions in the regulation of GAT1. We have shown that syntaxin 1A, a component of the docking and fusion apparatus involved in neurotransmitter release, regulates GAT1 function both in heterologous systems *(37)* and endogenously *(34)*. Syntaxin 1A binds the cytoplasmic N-terminal domain of GAT1, perhaps through electrostatic interactions with aspartic-acid residues in the N-terminal tail. This interaction results in an upregulation of GAT1 expression, consistent with the evidence that syntaxin 1A is a positive regulator of vesicle fusion; however, the interaction also results in a downregulation of function, because of a decrease in the substrate translocation rate *(27)*. The physiological importance of syntaxin 1A as a positive modulator of GAT1 expression but a negative regulator of GAT1 transport is unknown. One theory is that neurons use syntaxin to increase GAT1 surface expression while maintaining the transporter functionally suppressed until a time when removal of GABA from the synaptic cleft is a priority.

2.6. GABA-Transporter Pathophysiology and Therapeutic Targeting

GABA is the major inhibitory neurotransmitter in the mammalian CNS, and alterations in GABAergic neurotransmission have been implicated in the pathophysiology of several CNS disorders, particularly epilepsy *(42)*. A simple hypothesis is that there is a lack of GABA in the synaptic cleft, leading to hyperexcitability. What is the role of the GABA transporter in this process? One possibility is that there may be too many transporters in this situation. However, this is not the case. In fact, there is a decrease in transporters in temporal-lobe epilepsy *(43)*. The decrease in transporters in epilepsy has led to the hypothesis that transport reversal may be compromised in these individuals during excitability, leading to a seizure state *(44)*. However, whether the decrease in the number of GABA transporters is a causative factor, or whether it reflects a down-regulation of the transporter resulting from reduced GABA levels has not yet been determined.

In addition to epilepsy, there is evidence for a role of GABA transporters in several clinically important conditions. First, there are data indicating that some anesthetics can inhibit the rate of uptake of GABA, and that this effect may play a role in the mechanism of anesthesia of these agents *(45)*. Second, postmortem analysis of schizophrenic brains has revealed a decreased number of GABA uptake sites in subcortical regions, suggesting that GABAergic mechanisms are abnormal in schizophrenia *(46)*. Third, advanced liver disease can impair high-affinity GABA uptake into liver, thus increasing circulating GABA levels *(47)*. This has been postulated as a causative factor in hepatic encephalopathy and hypotension during advanced liver disease. Also, high circulating levels of GABA caused by GABA-synthesizing bacterial pathogens have been implicated in the decreased levels of consciousness observed during bacterial sepsis *(47)*. Fourth, the rabies virus has been shown to increase GABA release and to decrease GABA uptake in primary cortical neuronal cultures, suggesting the involvement of GABA in the neuropathological sequelae of rabies virus infection *(48)*. Lastly, the nonsteroidal anti-inflammatory drug indomethacin has been shown to be a noncompetitive inhibitor of mouse cortical synaptosomal GABA uptake *(49)*. This has been postulated to be responsible for some CNS side effects of indomethacin, including the impairment of psychomotor functions.

One potentially effective strategy for the treatment of GABA-related disorders has been the development of orally active inhibitors of GABA transport. Such inhibitors would cross the blood-brain barrier and enhance GABAergic neurotransmission by increasing the time that GABA remains in the synapse. Inhibitors such as nipecotic acid are not particularly lipid-soluble, do not cross the blood-brain barrier, and thus have very little effect in vivo. However, the addition of lipophilic side chains to the nitrogen atom of some of the known GABA transport inhibitors has resulted in the identification of several compounds that are active in vivo. These include: SKF 89976A, CI-966, NNC-711, and the most widely tested compound tiagabine, among others *(50)*. Some or all of these drugs have been shown to act as anticonvulsants in laboratory animals, to increase extracellular levels of GABA in the brain *(51,)* and to be highly specific for the GAT1-transporter subtype *(50)*. A novel GAT inhibitor known as (S)-SNAP-5114 has been identified, which apparently is highly specific for human GAT3 *(52)*. Two other anticonvulsant compounds, NNC 05-2045 and NNC 05-2090, have also been shown to act preferentially on non-GAT1 GABA uptake sites *(53)*.

Clearly, further study is required to elucidate the physiological relationships between GABA-uptake and pathological conditions. Perhaps the most intriguing possibility is that the four GABA transporter subtypes may be differentially related to these conditions. This theory implies that pharmacological agents may be developed that target specific subtypes and consequently have highly specific primary effects and fewer side effects.

3. GLYCINE TRANSPORTERS

The elucidation of the properties of glycine transport closely mirror that of the GABA transporters. First, experimental evidence has indicated the existence of more than one glycine-transporter subtype with distinctive pharmacological properties *(54,55)*. Second, a potent inhibitor of glycine uptake, sarcosine (N-methylglycine), was identified *(56)*, which later proved to differentially affect the glycine-transporter subtypes. Then, the isolation of molecular clones revealed that there were actually four glycine-transporter subtypes with distinct pharmacologies and unique temporal and spatial patterns of expression, as observed for GABA transporters. In addition, glycine-transporter subtypes were found to be expressed in both the CNS and peripheral tissues. Since then, many studies have focused on understanding the structure-function, regulation, and physiological and pathophysiological roles of glycine transporters.

3.1. Cloning and Characterization of Subtypes

In 1992, Smith et al. *(57)* reported the first molecular cloning and characterization of a high-affinity glycine transporter. Low-stringency screening of a rat-brain cDNA library with probes derived from the rat GAT1 transporter led to the isolation of complementary DNA, which exhibited 45% sequence identity with rat GAT1. Pharmacological analysis of this clone after transient expression in COS cells revealed that it specifically transported glycine with a Km of 120 μM. Glycine transport in this system was dependent on both extracellular Na^+ and Cl^-. No other neurotransmitters or amino acids, including closely related alanine, were found to be transported. The screening of potential uptake blockers revealed that only glycine derivatives were potent inhibitors, with sarcosine the most significant (IC50 approx 50 μM).

A short time later, similar probing efforts led to the independent cloning and characterization of Na^+- and Cl^--dependent glycine transporters from mouse and rat brains by Liu et al. *(58)* and Guastella et al. *(59)*, respectively. These transporters were expressed in *Xenopus* oocytes, and were found to exhibit somewhat different kinetics than the transporter isolated by Smith et al. *(57)*. Specifically, mean Km values for these transporters were four- to fivefold lower (20 and 33 μM for mouse and rat, respectively). However, the three transporters shared similar sensitivities to sarcosine. Subsequent comparison of the primary sequences of these isoforms showed that they differed only in the amino terminus; the protein encoded by Smith et al. had an initial 15-residue sequence that was distinct from the initial 10-residue sequence found in the Liu et al. *(58)* and Guastella et al. *(59)* transporters. The nomenclature for these transporters was later defined, with those of Liu et al. *(58)* and Guastella et al. *(59)* assigned the name GLYT1a, and that of Smith et al. *(57)* assigned GLYT1b *(60)*.

In 1994, Kim et al. *(60)* reported the cloning of a human version of GLYT1b, as well as a third subtype termed GLYT1c. These isoforms were found, after transient expression in COS cells, to transport glycine with Km values of 72 and 90 μM, respectively. Sarcosine was found to inhibit GLYT1b, as had been shown for the rat version *(57)*, but its effect on GLYT1c was not determined. Comparison of the primary sequences of the three isoforms again showed differences exclusive to the amino termini; GLYT1c had an initial 15-residue sequence that was equivalent to that of GLYT1b, which was followed by a unique 54-residue sequence that was not found in the other two isoforms. Except for these amino-terminal differences, the three isoforms shared a high degree of sequence identity, which suggested that they were all products of the same genetic locus. This idea was supported by the efforts of Adams et al. *(61)*, who determined that the gene encoding the GLYT1 transporters was created by alternative splicing and/or separate promoters which gave rise to the distinct isoforms. The GLYT1 gene contains 15 exons, with the first two specific for GLYT1a and the third specific for GLYT1b *(62)*. Kim et al. *(60)* localized the GLYT1 gene to human chromosome 1p31.3-p32 and mouse chromosome 4.

As the isoforms of GLYT1 were being cloned and characterized, a fourth Na+- and Cl−-dependent high-affinity glycine transporter was isolated from the rat brain (GLYT2) by Liu et al. *(58)*. This subtype exhibited only 48% identity with mouse GLYTla. Upon expression in *Xenopus* oocytes, GLYT2 was found to transport glycine with a Km of 17 μM. Unlike the GLYT1 isoforms, GLYT2 was not sensitive to inhibition by sarcosine. Therefore, GLYT2 was considered to be pharmacologically distinct or to be a product of a separate genetic locus. The human GLYT2 gene, located on human chromosome 11p15.1-15.2, shows similar functional characteristics to that of rat GLYT2 in terms of glycine affinity and the inability to be inhibited by sarcosine *(63)*.

3.2. Distribution

The four cloned GLYT subtypes show differences in primary structure and pharmacological characteristics as well as differential expression patterns in various brain regions and peripheral tissues. This has been extensively studied, using both Northern analyses and *in situ* hybridization. GLYT1a mRNA from the rat has been localized to various brain regions, as well as to the pancreas, uterus, stomach, spleen, liver, and lung *(64)*. However, rat GLYT1b and human GLYT1c are found only in the brain and CNS *(60,64)*. Interestingly, the use of probes designed to detect mRNA for all three GLYT1 subtypes has detected expression in the mouse and/or human brain, kidney, pancreas, lung, placenta, liver, heart, and muscle *(60,61)*. These findings stand in contrast to the data from the rat, in which the heart and kidney show no mRNA for GLYT1a, -1b, or 1c *(57,60,64)*. Whether this is the result of a failure to detect signal in the rat, an indication of yet another isoform, or differences between the species is unknown.

Within the CNS, transcripts arising from the GLYT1 gene have been shown to be expressed throughout the brain and spinal cord. As for cellular differences in expression, in most brain regions GLYT1 is expressed in glial cells *(61)*. Unlike GLYT1, GLYT2 appears to be expressed solely in neurons, with the notable exception of Golgi cells of the granular cell layer in the cerebellum *(65)*. In the retina, GLYT1 protein is found in many different cell types, including amacrine cells *(65)*, and transiently during development in ganglion, horizontal, and photoreceptor cells *(66)*. Interestingly— GLYT2—but not GLYT1, transcription is regulated by auditory afferents. The amount of GLYT2 transcription is directly correlated with the amount of activity in this system *(67)*.

The developmental expression patterns of GLYT1 and GLYT2 are also distinct. Both types have been shown to be expressed in prenatal embryos. GLYT1 mRNA in the mouse has been seen early in development, with a peak at embryonic stages E13 to E18 *(61)*. In the rat, maximal GLYT1 mRNA levels are not observed until postnatal day 14–21, with protein levels lagging slightly behind *(68)*. Peak GLYT2 mRNA expression has been observed early in postnatal life, but then decreases after 21 days. The protein levels peak similarly, but do not show a substantial decrease after postnatal day 21.

3.3. Structure-Function

A brief discussion of glycine transporter structure-function is included here because much is being learned regarding GLYT function based upon detailed analysis of GLYT structure. The amino and carboxyl termini of GLYT appear to be important for subcellular localization. Olivares et al. *(69)* transiently expressed amino- and carboxyl-terminal rat GLYT1b deletion mutants in COS cells. They found that most of the amino terminus and a significant portion of the carboxyl terminus are not required for transport. However, more extensive deletions in the carboxyl terminus yielded inactive transport because of decreased protein targeting to the plasma membrane. When expressed in polarized epithelial cells, GLYT1b localizes to the basolateral surface *(70)*. Site-directed mutagenesis of residues in the amino terminus and of two di-leucine motifs in the carboxyl terminus revealed them to be crucial in mediating this localization.

GLYT cDNA encodes for seven potential N-linked glycosylation sites, with four of these sites presumed to be extracellular. These four sites are present in the large extracellular domain between TMDs 3 and 4. To establish the functional role of these N-glycosylation sites, Aragon and colleagues *(71,72)* examined the effects of deglycosylation on transporter function using glycosidase treatments and site-directed mutagenesis. Deglycosylation assays on a purified glycine transporter confirmed the presence of N-linked carbohydrates, but did not reveal any O-linked carbohydrate modifications *(71)*. Reconstitution of the purified protein in liposomes followed by treatment with deglycosylating enzymes led to reduced uptake, with only 25% of the activity of the fully glycosylated transporter. Treatment of GLYT1 expressed in COS cells with the N-glycosylation inhibitor tunicamycin produced a protein of 46 kDa *(72)*. Site-directed mutagenesis of the four putative extracellular asparagine glycosylation sites also reduced the GLYT1 protein to a 46-kDa mobility. Successively higher mobilities were observed for mutations of 3, 2, 1, and 0 asparagines, respectively, indicating the presence of as many as four N-linked sugars in the native protein. Immunofluorescence studies of the GLYT1 mutant lacking all four glycosylation sites showed abnormal targeting to the plasma membrane, thus establishing the importance of N-linked sugars to the proper subcellular sorting of this transporter *(11)*. Because all Na^+- (and Cl^--) dependent transporters contain potential glycosylation sites in the same extracellular region, the importance of N-glycosylation on protein targeting and sorting may apply to the entire transporter family.

As with GAT1, the hydrophobicity-derived transmembrane topology for glycine transporters has recently come into question *(73)*. Using a strategy based upon introducing unnatural glycosylation sites along various portions of the GLYT1 protein, it appears that the N-terminal third of GLYT 1 may have a different structure from that proposed in earlier models. Although there may still be 12 TMDs, the first TMD may not fully cross the membrane. This analysis places the loop between hydrophobic domains 2 and 3 on the extracellular side.

3.4. Regulation

As with GABA transporters, studies of the regulation of glycine transporters have focused upon the triggers of the regulation and protein-protein interactions that may control the function of the transporter. Two studies on the effect of PKC modulators *(74,75)* showed that activation of PKC inhibits the transport of glycine by GLYT1. In HEK 293 cells expressing mouse GLYT1B, PMA caused a 60% reduction in [^{3}H]-glycine uptake. The effect of PMA was on the maximal transport capacity. PKC inhibitors were also found to block the effect of PMA on uptake. Four consensus intracellular PKC phosphorylation sites were identified, and were mutated individually and together (75). PMA treatment of HEK 293 cells expressing these mutants continued to show reduced uptake, indicating that PKC activation was occurring at a cryptic nonconsensus phosphorylation site or was entirely unrelated to phosphorylation. Activation of PKC by phorbol ester in glia causes a reduction in glycine uptake similar to that seen in expression systems (75).

As mentioned previously, GLYT1 and GLYT2 localize predominantly to glial cells and neurons, respectively. Interestingly, when glial cells are cultured in the absence of neurons, GLYT1 is not expressed (76). In mixed neuronal and glial cultures, GLYT1 can be found in a number of different glial types. This expression is lost following the elimination of neurons in these cultures. The neuronal factor required for glial GLYT expression, the signal transduction process that mediates the process, and the physiological relevance of this effect have not yet been elucidated.

GLYT subtypes also associate with several different cellular proteins, suggesting that transporter function can be regulated through protein-protein interactions. As with GABA-transporter GAT1, both GLYT1 and GLYT2 physically interact with syntaxin 1A both in brain tissue and when heterologously expressed in COS cells. In COS cells, the presence of syntaxin results in a 40% decrease in glycine transport. The physical interaction can be disrupted by Munc18, a binding partner of syntaxin 1A, and the reduction correlates with a decrease in GLYT protein on the cell surface (77). These data link glycine transporters to the regulatory machinery involved in vesicle trafficking and neurotransmitter release. In the retina, a novel fourth C-terminal GLYT1 variant interacts with the GABA-C subtype of GABA-activated Cl$^-$ channels (78). Although the physiological relevance of this interaction is presently unknown, these data suggest that inhibitory signaling in the retina may be under the convergent control of both glycine and GABA.

3.5. Physiological Relevance

Detailed localization of the GLYT subtypes within the brain has suggested specific functional roles. The expression pattern of GLYT2 within the hindbrain and spinal cord has been found to overlap [^{3}H]strychnine-binding sites, colocalizing GLYT2 with strychnine-sensitive glycine receptors (79). Furthermore, high levels of GLYT2 protein have been detected in putative glycinergic nerve terminals *(65)*. This finding strongly suggests that GLYT2 functions to terminate glycinergic neurotransmission. In addition, it has been suggested that GLYT1 is involved in both glycinergic and N-methyl-D-aspartate (NMDA) receptor-mediated glutamatergic pathways. Glycine is a required co-agonist at NMDA receptors *(80)*, and GLYT1 expression is observed within areas not associated with glycinergic pathways *(65)*. However, a precise overlap with glutamatergic pathways does not occur. Moreover, high levels of GLYT1 have been detected in glial cells surrounding putative glycinergic synapses *(65)*. A specific isoform of GLYT1 may be associated with NMDA-receptor-expressing pathways, whereas other variants are associated with glycinergic pathways.

Physiological experiments confirm that GLYT plays an important role in NMDA-mediated synaptic transmission. NMDA-induced currents are reduced when GLYT1B is co-expressed in *Xenopus* oocytes *(81)*. The reduction is prevented when extracellular Na^+ is eliminated, or the NMDA receptor is activated by serine (which is not transported by GLYT1B) rather than glycine. These results suggest that GLYT1 is effective at controlling the local levels of glycine around NMDA receptors. Furthermore, this effect occurs in cells that endogenously express NMDA receptors and GLYT1. NMDA-mediated currents are increased in hippocampal neurons in which GLYT1 antagonists are used, suggesting that GLYT1 maintains sub-saturating glycine concentrations under physiological conditions *(82)*.

The possibility that extracellular levels of glycine may be determined in part by efflux through GLYT, either during physiological or pathological situations, must also be considered. Recently, the stoichiometry of the GLYT isoforms has been evaluated using a combination of electrophysiological and radiolabeled techniques *(83)*. The glial GLYT1 transports with a Na^+/Cl^-/glycine stoichiometry of 2:1:1, and the neuronal GLYT2 transports with a stoichiometry of 3:1:1. Depending on the exact physiological conditions, it is possible that glial glycine transporters may undergo reverse transport, whereas neuronal glycine transport is likely to be inward under all physiological conditions.

4. OTHER TRANSPORTERS IN THE FAMILY

4.1. Proline Transporters

Screening experiments isolated human- *(84)* and rat- *(85)* brain cDNA clones that encoded a high-affinity, Na^+ (and Cl^-) -dependent L-proline transporter (PROT). Mammalian brain PROT exhibits significant amino-acid sequence identity (42~50%) with the rest of the gene family of Na^+-(and Cl^--) dependent plasma-membrane transport proteins, and is most closely related to glycine transporters (~50% amino-acid sequence identity with the three isoforms of human GLYT and rat GLYT2). In transiently transfected HeLa cells, PROT transports L-proline with a Km of 5–10 μM *(84,85)*, which is within the range of values observed for L-proline uptake by rat hippocampal synaptosomes *(86)*. Interestingly, mammalian-brain PROT exhibits a narrow range of substrate specificity with striking selectivity for L-proline *(84,85)*. Sarcosine (N-methylglycine) competes with a high apparent affinity for PROT-induced L-proline uptake (Ki approx 24 μM) *(84,85)*. In addition, leu- and met-enkephalin, and their des-tyrosyl derivatives, potently and selectively inhibit high-affinity L-proline uptake in HeLa cells transfected with the recombinant PROT cDNA and in rat hippocampal synaptosomes (IC50 = 0.26 μM) *(87)*. This latter result confirms that PROT is a separate entity from other proline carriers, and identifies a compound that may be a useful tool for elucidating the structure-function properties and physiological roles of PROT.

As with other members of this family of transporters, PROT shows ionic conductances associated with transport and regulation by second messengers. Expression of PROT in human embryonic kidney cells induces a Na^+-dependent inward current that is activated by L-proline or its six-member ring cogener L-pipecolate. The whole-cell currents are voltage-dependent, suggesting an electrogenic uptake process, and fluctuation analysis reveals "channel-like" fluxes in the range of 20–60 fA at negative holding potentials *(88)*. Like other Na^+- and Cl^--dependent transporters, PKC downregulates PROT function *(89)*. Intracellular calcium levels also play a role in PROT function *(89)*, although the effect is more complex. Thapsigargin, which increases intracellular calcium levels by inhibiting the Ca^{2+}-ATPase, induces an immediate increase in PROT function. However, after several minutes of thapsigargin treatment, PROT function is reduced. The Ca^{2+}-mediated reduction can be prevented by application of a Ca^{2+}/calmodulin-dependent kinase II inhibitor, and the Ca^{2+} effect on PROT is additive with the PKC effect. This latter result suggests that the alterations in PROT function by these second messengers proceed via different signal transduction cascades.

Northern blot analysis reveals brain-specific expression of PROT mRNA in human tissues. mRNA transcript is heterogeneously distributed in different regions of human brain, but is not detected in the human heart, placenta, lung, liver, kidney, skeletal muscle, or pancreas. Similarly, a 68-kDa hPROT immunoreactive protein is found in crude synaptosomal membranes from various regions of human brain, but not in membranes prepared from human liver, kidney, or heart *(84)*. This apparent brain-specific expression of PROT is interesting because other members of the Na$^+$- (and Cl$^-$-) dependent transporter family do not show this degree of brain specificity.

Although little is known about the putative physiological roles of high-affinity L-proline uptake in nervous tissue, several lines of evidence support the hypothesis that such uptake modulates excitatory synaptic transmission in specific glutamatergic nerve terminals. Functional autoradiographic studies localize high-affinity L-proline uptake to only a subset of hippocampal glutamatergic pathways *(90)*. Terminal fields of the Schaffer collateral, commissural, and ipsilateral associational fibers in areas CA1, CA3, and the lateral perforant path, and associational commissural projections in the dentate molecular layer exhibit prominent high-affinity L-proline uptake; the medial perforant path and mossy fiber pathway, which also use L-glutamate as their excitatory transmitter, exhibit little or no high-affinity L-proline uptake.

Interestingly, mammalian-brain PROT protein is localized to synaptic vesicles in specific glutamatergic nerve terminals *(91)*. Electron microscopic ultrastructural studies show that immunogold-silver particles are predominantly localized over synaptic vesicles in labeled excitatory nerve terminals in both the caudate nucleus and in the stratum oriens of the CA1 region of the hippocampus. Subcellular fractionation studies reveal that PROT is substantially enriched in a highly purified synaptic vesicle fraction compared to markers for the plasma membrane, such as the NMDA receptor. Approximately 80–90% of the immunoreactive PROT protein is vesicular, whereas the remaining 10–20% is on the presynaptic plasma membrane. The topology of PROT on synaptic vesicles is reversed compared to its orientation on the plasma membrane, consistent with the theory that they are part of a recycling pathway to and from the plasma membrane. Whether these vesicles are different from small synaptic vesicles containing glutamate is unknown. Also unknown are the triggers that may redistribute the transporter to various subcellular compartments.

It is generally accepted that L-glutamate is the transmitter used by the vast majority of excitatory pathways in the mammalian brain. Since PROT has been localized to specific excitatory nerve terminals, high-affinity L-proline uptake may modulate some aspect of excitatory synaptic transmission in specific glutamatergic nerve terminals. In fact, glutamatergic synaptic transmission in the hippocampus is potentiated in the presence of proline *(92)*. Glutamatergic synapses have been implicated in diverse physiological processes, including synapse formation, synaptic plasticity, memory and learning, and neuroendocrine regulation *(93)*. Furthermore, abnormalities of glutamatergic transmission have been implicated in the pathophysiology of neurologic disorders in which excitotoxic processes are believed to play a role, including cerebral strokes, epilepsy, and head trauma

(93). Therefore, an understanding of the functional role of PROT in specific glutamatergic nerve terminals may provide new insights into presynaptic regulatory mechanisms involved in synaptic plasticity and excitotoxic nerve-cell damage.

4.2. Betaine Transporters

The first high-affinity, Na^+- (and Cl^--) dependent betaine transporter (BGT1) was isolated from MDCK cells by expression cloning *(12)*. Upon expression in *Xenopus* oocytes, BGT1 was found to transport betaine with a Km of 398 μM. This substrate affinity was similar to that calculated for betaine uptake into native MDCK cells (480 μM). However, betaine was not the only high-affinity substrate for BGT1, because Yamauchi et al. found that it also transported GABA. In fact, the Km for GABA transport in *Xenopus* oocytes (98 μM) and native MDCK cells (~20 μM) was significantly lower than for betaine. Despite this marked difference in substrate affinity, it was concluded that BGT1 was primarily a betaine transporter based on two observations: First, tissue localization through Northern analysis detected BGT1 mRNA in canine kidney, but not brain. Second, significant levels of betaine, but not GABA, were found to accumulate in the kidney because of a large difference in plasma concentrations (approx 180 μM and <1 μM for betaine and GABA, respectively) *(12).* Yamauchi et al. also found that BGT1 was relatively insensitive to the GABA uptake inhibitors nipecotic acid, β-alanine, and DABA (IC50 values > 1 μM). These observations, along with a higher Km for GABA, pharmacologically distinguished BGT1 from the previously cloned rat-brain GAT1 transporter *(23)*. Nonetheless, when primary structures were compared, BGT1 surprisingly shared significant sequence identity with GAT1 (53%). BGT1 also shared significant identity with the previously cloned human-brain NET *(9)*. These observations established BGT1 as a novel member of the Na^+-(and Cl^--) dependent transporter family.

Subsequently, the GABA transporter GAT2 was isolated from the mouse brain *(11)*. It exhibited 88% sequence identity with canine BGT1 and had similar pharmacological properties. As described in **Subheading 2.1.**, GABA transport by mouse GAT2 had a Km (79 μM) that was similar to BGT1 (98 μM). In addition, GABA uptake by mouse GAT2 was inhibited by betaine, a finding that was assumed to mean that betaine was a substrate for this transporter. These observations, along with the localization by Northern analysis of mouse GAT2 to kidney (in addition to brain), suggested that the two transporters were products of the same gene. However, this conclusion was questioned *(11)* because of the following observations: First, the 88% identity between mouse GAT2 and canine BGT1 was high, but not as high as the >97% identity shared by the human, mouse, and rat GAT1 transporters. Second, Northern analysis yielded a single mRNA transcript of 5 kb that hybridized with mouse GAT2, whereas a doublet of 3 and 2.4 kb had been previously reported for canine BGT1 *(12)*. Third, no homology was observed in the 3'-nontranslated cDNA sequences between the transporters. Based on these findings, it was concluded that mouse GAT2 and canine BGT1 were distinct gene products *(11)*.

BGT1 transporters were then cloned from the human brain *(94)* and kidney *(95)* which exhibited >87% sequence identity with mouse GAT2 and canine BGT1, and shared similar pharmacological properties. Based on these findings, it appears likely that the human and canine BGT1 proteins and the mouse GAT2 protein are products of the same gene *(94,95)*. Detailed investigation of the functional properties of BGT1, expressed in oocytes, suggests both similarities and differences with members of the GAT subfamily of GABA transporters *(96)*. Both betaine and GABA elicit similar voltage-dependent ionic currents. Substrate uptake and ionic fluxes are completely dependent on extracellular Na^+, but only partially on extracellular Cl^-. Stoichiometry experiments suggest that Na^+/Cl^-/substrate flux is 3:1:1 or 3:2:1, and data are consistent with sequential ordered binding of substrate first, followed by Na^+. In addition, at least for a betaine transporter endogenously expressed in squid motor neurons, betaine activates a Na^+-dependent Cl^- current, suggesting that the betaine transporter may have a channel-like mode of operation *(97)*.

Because of the localization of the putative betaine transporter to both brain and peripheral tissues *(95)*, speculation has centered on its physiological function. In the renal medulla, betaine serves as one of the major osmolytes used for volume regulation, and the betaine transporter is beleived to be a central component of this function. Indeed, BGT1 mRNA is induced by hypertonicity. A small 5'-flanking region upstream of the BGT1 promotor harbors a hypertonicity responsive enhancer element *(98)*. Thus, it appears that BGT1 plays a significant role in volume homeostasis. Volume regulation may be the function of the betaine transporter in other organs as well. However, in the brain, the betaine transporter may also function as a GABA transporter and consequently may play a role in modulation of GABAergic transmission *(94,95)*. It must be noted, however, that the distribution of human BGT1 in the brain does not match the localization pattern of GABAergic pathways *(94)*. This transporter may sequester GABA that has diffused away from synaptic regions *(94)*. Alternatively, it may not function to transport GABA at all. Finally, the brain localization may be indicative of an unrecognized role for betaine in neurotransmission.

4.3. Taurine Transporters

In 1992, several species homologs of a high-affinity, Na^+- (and Cl^--) dependent taurine transporter (TAUT) were cloned and characterized, with the first isolated by Smith et al. *(99)* from the rat brain. On transient expression in COS cells, TAUT transported taurine with a Km of 43 μM. Transport was strongly inhibited by hypotaurine and β-alanine (IC50 values between 10 and 100 μM), and weakly inhibited by GABA (IC50 approx 1 μM). Northern analysis indicated that TAUT was expressed in a variety of tissues, including the brain, retina, liver, kidney, heart, spleen, and pancreas.

Several different endogenous and heterologous systems have been exploited to identify factors that regulate taurine uptake. These factors include PKC, calmodulin, and taurine itself. As with many other members of the GABA-transporter family, pharmacological activation of PKC causes inhibition of TAUT in placental-derived cell lines *(100)*, intestinal-derived cell lines *(101)*, and *Xenopus* oocytes *(102)*. This effect of PMA is blocked by staurosporine, a potent inhibitor of PKC. Consistent with PKC modulation is the presence of six potential PKC recognition sites within the primary sequence of the transporter *(103)*. Site-directed mutagenesis of serine 322 in this segment creates a mutant transporter that is refractory to regulation by PKC *(102)*.

Similar studies have found that taurine transport is decreased by calmodulin inhibitors *(104)*. Its effect was shown not to involve transcriptional processes, but whether it involves changes in protein synthesis, protein targeting, substrate affinity, and/or substrate translocation is unknown. TAUT is also regulated by the level of extracellular taurine *(105)*. The amount of TAUT mRNA and protein is inversely related to the amount of extracellular taurine present in the culture medium, suggesting that intracellular taurine levels must be closely regulated by these cells.

The physiological functions of high-affinity TAUT are not well understood, although it probably plays a role in osmoregulation. Like BGT1, the expression of TAUT is regulated by hypertonicity. Culturing intestinal epithelial cell lines in hypertonic medium increases the intracellular accumulation of taurine. This hypertonicity-induced increase in taurine is correlated with an upregulation of mRNA for TAUT *(106)*. Taurine is also involved in a number of other important physiological processes, including bile-acid conjugation in hepatocytes, detoxification, membrane stabilization, and modulation of calcium flux and neural excitability *(107)*. Also indicative of potential functions for TAUT are the observed effects of taurine deficiency, which include abnormalities in immune, cardiac, and reproductive functions *(107)*. Taurine deficiency is also related to certain forms of retinal-related blindness. Of interest in this regard is the evidence that both glucose and insulin enhance retinal taurine uptake by altering the transport capacity of the transporter *(108)*. In addition, the TAUT gene has been mapped to human chromosome 3p24-p25 *(103)*, a region associated with mental retardation and other neurological defects. An interesting potential role for taurine transporters in the CNS is in ischemia. In the hippocampus, Na^+-dependent taurine efflux may exert a neuroprotective function in glutamate-induced excitotoxicity that accompanies ischemia. This efflux is regulated by nitric oxide-generating compounds, suggesting that the transduction process involves retrograde nitric oxide signals from NMDA receptors on postsynaptic neurons onto presynaptic neurons containing TAUT *(109)*.

4.4. Creatine Transporters

Creatine is a critical molecule in the maintenance of ATP homeostasis. Creatine kinase uses ATP to mediate the conversion of cellular creatine to phosphocreatine. In localized regions of high energy consumption—for example, muscle fibers during sustained contraction—creatine kinase uses phosphocreatine to phosphorylate ADP and regenerate ATP. Therefore, creatine is an important energy reservoir. For this reason, there is much recent interest in creatine as a nutritional supplement, especially among athletes *(110)*. Clinically, creatine may ameliorate the symptoms associated with neuromuscular diseases, and creatine decreases are correlated with cardiomyopathy *(111)*. Thus, there is interest in how creatine is taken into cells, and how creatine levels can be controlled.

In 1993, Guimbal et al. *(112)* isolated and characterized a high-affinity, Na^+- (and Cl^--) dependent creatine transporter from rabbit brain and muscle. The creatine transporter shows a Km for creatine of 20–50 μM when expressed in a variety of cell systems. In oocytes, ion-dependency experiments suggest a stoichiometry of 2 Na^+ and 1 Cl^- *(113)*. High-affinity inhibitors of creatine transport include β-guanidinopropionic acid, cyclocreatine, and 2-amino-1-imidazolidineacetic acid. Whether these are true inhibitors or also substrates of the transporter is unknown.

The creatine-transporter protein is widely expressed in a variety of mammalian tissues *(112)*. Creatine-transporter expression is also observed in a variety of brain regions, including the retina, cerebellum, and hippocampus *(114)*. This diverse tissue expression is consistent with a general metabolic role for the transporter, and its localization in brain corresponds well with the CNS distribution of creatine kinase *(114)*. To date, there is no strong evidence that creatine participates in neurotransmission.

The primary sequence identity between the creatine-transporter species homologs is very high (98–99%), whereas identity with BGT1 and TAUT is approx 50%. Because BGT1 and TAUT do not transport creatine, comparisons of the primary sequences should suggest which amino-acid domains are involved in substrate binding. In this regard, Nash et al. *(115)* proposed that TMDs 9, 10, and 11 are likely to be responsible for the observed differences in substrate affinities. Moreover, a creatine transporter has been cloned by Guimbal and Kilimann *(116)* from *T. marmorata* which has 64% identity with the mammalian creatine transporter. Inclusion of this species homolog in the comparison of primary structures suggests that only a few residues are candidates for involvement in creatine substrate specificity *(116)*. It is anticipated that site-directed mutagenesis will be employed to answer these predictions.

A potential role for the creatine transporter in genetic disease is suggested by its mapping to human chromosome Xq28 *(115)*. This genetic locus has been linked to several skeletoneuromuscular disorders. Additionally, creatine-transporter protein is reduced in humans with cardiomyopathy, and in animals with experimental heart failure *(111)*. Because of its similarity to other members of this gene family, regulation of the creatine transporter would be considered a likely mechanism for the regulation of creatine levels in cells. Like many other transporters of this family, creatine-transporter function is acutely decreased by PKC, at least in expression systems *(113)*, and chronic creatine treatment downregulates the in vivo expression of the transporter *(110)*. Whether this regulation has physiological relevance, and the mechanisms underlying these forms of regulation, are likely to be the subjects of future experiments.

4.5. Orphan Transporters

The identification of Na$^+$- and Cl$^-$-dependent transporters as members of the same gene family suggested cloning approaches based upon homology screening. Such experiments have led to the identification of a number of gene products that have approx 40% amino-acid identity to the family of GABA transporters; however, the substrates for these gene products have not yet been identified. Thus, these proteins have been termed "orphan" transporters (Fig. 4). Members of the orphan family of transporters include Ntt4 *(117)* (also known as Rxt1), rB21 *(118)*, ROSIT *(119)*, and v7-3 *(120)*. Structurally, the orphan transporters differ somewhat from other family members: they are predicted to have approx 700 amino acids, larger fourth and sixth extracellular loops, and a longer C-terminal cytoplasmic tail (Fig. 5).

Complicating the identification of substrates for these orphans is their widespread distribution in the brain and other tissues. *In situ* hybridization and antibody labeling often reveal staining patterns that do not match up well with the distribution of any known neurotransmitter or neuropeptide. Ntt4/Rxt1 appears to colocalize with NMDA receptors, suggesting that it

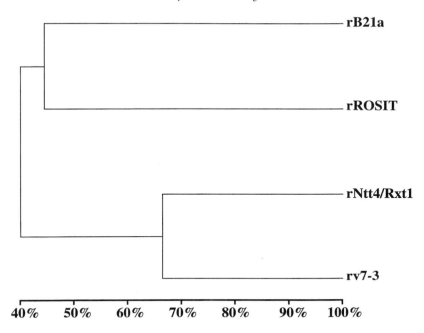

Fig. 4. Dendrogram of the amino-acid sequence relationships between members of the "orphan" Na^+- (and Cl^--) dependent transporter gene family. The percent amino-acid sequence identity between pairs of transporters at convergence points is shown at the bottom. The rat homologs for each transporter are shown. rROSIT, rat renal osmotic stress-induced transporter.

may play a role in glutamatergic transmission *(121)*. Interestingly, Ntt4/Rxt1 has been found to be localized to small synaptic vesicles, similar to H^+-dependent vesicular transporters, and little Ntt4/Rxt1 staining is ever observed on cellular plasma membranes *(122,123)*. This finding suggests that several of these transporters may have unique and specialized roles in nervous-system function unlike that of other family members.

The reasons why substrates for these transporters have remained a mystery are many. It may be that auxiliary subunits or heteromultimerization are necessary to confer substrate transport. If the transporters localize to vacuoles, or to vesicles as in the case of Ntt4/Rxt1, particular experimental conditions may not be appropriate for examining these specialized transport processes. The transporter may require particular cotransported ions that are unknown, or may be a transporter of ions only. Lastly, these transporters may be directing us toward the identification of unknown neurotransmitters. In any case, the study of this subfamily of orphans will likely lead to novel insights into the function of transporters in the brain.

```
ORPHAN TRANSPORTER ALIGNMENT
Key:  Identity = *   Similarity = +
rB21      MRLA--IKRRASRGQ------------------------RPGPDEKRARD--MEKARPQWGNPLQFVFACISYAVGLGNVWRFPYLCQMYGGGSF       67
rROSIT    M-----------AQ------------------------ASGMD--PLVD--IEDERPKWDNKLQYLLSCIGFAVGLGNIWRFPYLCHTHGGGAF       55
rNt4/Rxt1 MPKNSKVTQREHSNEHVTESVADLLALEEFVD--YKQSVLNVAGETGGKQKVAEEELDAEDRPAWNSKLQYILAQIGFSVGLGNIWRFPYLCQKNGGGAY       98
rv7-3     MPKNSKVVKRDL-DDDVIESVKDLLSNEDSVEDVSKKSELIVDVQEEKDTDAEDGSEVDDERPAWNSKLQYILAQVGFSVGLGNVWRFPYLCQKNGGGAY       99
          *                     +        +          +++++  *++  **+++ ++++++++++*+++++++++++++++        *****++

rB21      LVPYLIMLIVEGMPLLYLELAVGQRMRQGSIGAWRTISPYLSGVGVASVVSFFLSMYYNVINAWGFWYLFHSFQDPLPWSVCPL--NSNRTGYDEECEK      165
rROSIT    LIPYFIALVFEGIPLFYIELAIGQRLRRGSIGVWKTISPYLGGVGLGCFSVSFLVSLYYNTILLMVLWFFLNSFQHPLPWSTCPL--DLNRTGFVQECQS      153
rNt4/Rxt1 LVPYLVLLIIIGIPLFFLELAVGQRIRRGSIGVWHYVCPRLGGIGFSSCIVCLFVGLYYNVIIGWSVFYFFKSFQYPLPWSECPVIRNGTVAVVEPECEK      198
rv7-3     LLPYLILLLVIGIPLFFLELSVGQRIRRGSIGVWNYIISPKLGGIGFASCVVCYFVALYYNVIIGWTLFYFYSQSFQQPLPWDQCPLVKNASHTYIEPECEK      199
          *+++++ *+  *+*++++*++*+*********+++  *+  ++*   *+   ++  *+ +++++****+*     *     *+++   +  + +  *++++

rB21      ASSTQYFWYRKTLNISPSIQENGGVQWEPALCLTLAWLMYLCILRGTESTGKVVFTALMPYCVLIIVLVRGLTLHGATNGLMYMFTPKIEQLANPKAW      265
rROSIT    SGTVSYFWYRQTLNITSDITIQMKLFLCLVACWTTVYLCVIRGIESTGKVIYFTALFPYLVLTIFLIRGLTLPGATEGLTYLFTPNMKILQNSRVW      253
rNt4/Rxt1 SSATTYFWYREALDISNSISESGGLNWKMTVCLLVAWSIVGMAVVKGIQSSGKVMFSSLFPYVVLACFIVRGLLLRGAVDGILHMFTPKLDKMLDPQVW      298
rv7-3     SSATTYYWYREALAISSSISESGGLNWKMTGCLLAAWVMVCLAMIKGIQSSGKIMFSSLFPYVVLICFLIRSLLLNGSIDGIRHMFTPKLEMMLEPKVW      299
          ++++ *+****+++*++ +* ++*******++  ** *   * + ++++++*+++**++** **  +*+++*  * +++++  ++******  + ++++*

rB21      INAATQIFFSLGLGFGSLIAFASYNEPSNDCQKHAVIVSVINSSTSIFASIVTFSIYGFKATFNYENCLNKV--------ILLLTNSFDLEDGFLTA---------      354
rROSIT    LDAATQIFFSLSLAFGGHIAFASYNQPRNNCEKDAVTIALVNSMTSLYASITIFSIMGFKASNDYGRCLDRN---ILSLINEFDFPELSISR---------      342
rNt4/Rxt1 RERATQVFFALGLGFGGVIAFSSYNKQDNNCHFDAALVSFINFTSVLATLVFAVLGFKANIMNEKCVVENAEKILGYLNSNVLSRDLIPPHVNFSHLT      398
rv7-3     RERATQVFFALGLGFGGVIAFSSYNKDNNCHFDAVLVSFINFTSVLATLVFAVLGFKANIVNEKCISQNSEMILKLLKTGNVSWDVIPRHINLSAVT      399
          +*****++*+++*+*++** ***+*+****+ *+*++ ++*  + *  +**++****++ ++*++++++      ++++               ++

rB21      -SNLEEVKDYLASTYPNKYSEVFPHIRNCSLESELNTAVQGTGLAFIVYAEAIKNMEVSQLWSVLYFFMLLMLGMGSMLGNTAAILTPLTDSKVISSYLP      453
rROSIT    -DEYPSVLMYLNATQPERVARL--PLKTCHLEDFLDKSASGPGLAFIVFTEAVLHMPGASVWSVLFFGMLFTLGLSSMFGNMEGVITPLFDMGILPKGVP      439
rNt4/Rxt1 TKDYSEMYNVIMTVKEKQFSAL--GLDPCLLEDELDKSVQGTGLAFIAFTEAMTHPASPPWSVMFFLMLINLGLGSMIGTMAGITTPIID----TFKVP      492
rv7-3     AEDYHVVYDIIQKVKEEFAVL--HLKACQIEDELNKAVQGTGLAFIAFTEAMTHPASPFWSVMFFLMLINLGLGSMFGTIEGIITPVVD----TFKVR      493
          ++  +  + +++ +  + +*+      + *  *+*  *+* ***+  * *++ +*** ** ****+***+ +++ *** **  *   + +   + ++

rB21      KEAISGLVCLINCAVGMV-FTMEAGNYWFDIFNDYAATLSLLLIVLVETIAVCYVYGLRRFESDLRAMTGRPLNWYWKAMWAFVSPLLIIGLFIFYLSDY      552
rROSIT    KETMTGVVCFI-CFLSAICFTLQSGSYWLEIFDSFAASLNLIIFAFMEVVGVIHVGIKRFCDDIEWMTGRRPSLYWQVTWRVVSPMLLFGIFLSYIV-L      537
rNt4/Rxt1 KEMFTVGCCVFAFFVGLL-FVQRSGNYFVTMFDDYSATLPLTVILENIAVAWIYGTKRFKFMQELTEMLGFRPYRFYFYMWKFVSPLCMAVLTTASIIQ-      590
rv7-3     KEILTVICCLLAFCIGLM-FVQRSGNYFVTMFDDYSATLPLLIVILENIAVSFVYGIDKFLEDLTDMLGFAPSKYYYMWKYISPLMLVTLLIASIVN-      591
          ** ** *    ++ +   ++ + +**+ + +*** +++ *   *+     +++ ++   *      ++ *  * ***++ +  + +*++ + + +
```

```
rB21       ILTGTLQYQAW--DATQGQLVTKD---YPPHALAVIGLLVASSTMCIPLV----------------ALGTFI----------------------  603
rROSIT     LAQSSPSYKAW--NPQYEHFPSREEKLYPGWVQVTCVLLSFLPSLWVPGI--------------ALAQLLFQ----------------------  593
rNtt4/Rxt1 LGVSPPGYSAWIKEEAAERYLY-----FPNWAMALLITLIAVATLPIPVVFLLRHFHLLSDGSNTL-SVSYKKGRMMKDISNLEENDETRFTLSKVPSEA  684
rv7-3      MGLSPPGYNAWIKEKASEEFLS-----YPMWGMVVCFSLMVLAILPVPVFVIRRCNLIDDSSGNLASVTYKRGRVLKEPVNLDG-DDASLIHGKIPSEM  685
           +  ++  * *  +             +*    *   + + ++* +                                                +*

rB21       ---------RNRLK--------RGGSSPVA-------------  616
rROSIT     --------YRQRWKNTHLESALKPQESRGC------------  615
rNTT4/RXT1 PSPMPTHRSYLGPGSTSPLESSSHPNGRYGSGYLLASTP---ESEL  727
rv7-3      SSPNFGKNIYRKQSGSPTLDTA--PNGRYGIGYLMADMPDMPESDL  729
           +                                       ++
```

Fig. 5. Alignment of the predicted amino-acid sequences encoding the subfamily of "orphan" Na^+- (and Cl^--) dependent transporter(s). Asterisks are used for amino acids identically conserved for all members of the subfamily, and plus signs are used for amino acids similarly conserved for all members for the subfamily. For abbreviations, *see* Fig. 4.

223

5. FUTURE DIRECTIONS

It has been only a decade since the first Na^+- and Cl^--dependent plasma-membrane neurotransmitter transporter, GAT1, was cloned. Since then, many excellent studies have begun to elucidate the properties of this interesting family of proteins. Although we have learned a lot, a significant number of fundamental questions must be addressed.

(1) What are the physiological roles of these transporters? In the brain, we know that transporters are capable of regulating neural signaling. However, how much of this regulation is caused by binding of substrate, and how much is the result of actual substrate transport per se? We know that the stoichiometries, transport rates, and numbers of transporters will play a role in the time-course and extent to which substrate is concentrated within the cell. Is reuptake for reuse necessary for normal function, or a nonessential by-product of limiting extracellular substrate levels? We know that these transporters, based upon concentrations of cotransported ions and upon membrane potential, can operate "in reverse," so that substrates are transported to the extracellular space. Is this a physiological process or a pathophysiological process? We know that many of these transporters are found on both neurons and glia. Does this affect their primary physiological function? We also know that these transporters have unique distributions throughout nervous and non-nervous tissue. Are the roles of these transporters in non-neural tissues the same or different than that seen in the brain? High-resolution electrophysiological tools and modern genetic approaches have already begun to address these questions.

(2) What forms of transporter regulation are critical for in situ transporter function? We recognize many different signal transduction pathways that can regulate transporters—for example, PKC, phosphatases, and protein-protein interactions. But which of these pathways do alter transporter function in native tissues? We know that many different triggers initiate transporter regulation. Once again, what are the endogenous cell triggers, and which signaling pathways converge for the synergistic control of transporter function? And over what time-scale do these modulators exert their effects? We know that transporters can be regulated both by redistributing the transporter to and from the plasma membrane and by altering unitary properties of the transporter. How much of a role do each of these mechanisms play in the physiological function of these transporters?

(3) What are the unique cell-biological aspects of this gene family? We know that these transporters have a precise localization on the plasma membrane in a variety of cell types, and that this localization differs for various members of the family. What are the signals that control this targeting? We know that these transporters can be moved to and from the plasma membrane in response to modulators. Where do they go when they are inside of cells, what are the pathways that they use, and do the same transporters return to the plasma membrane? We know that the transporter participates in a variety of cellular functions. Does its localization dictate the function that it serves?

(4) What are the microscopic properties that underlie substrate transport? For many members of this family, we know the stoichiometry of the transport process. But what are the transition states that govern substrate binding and translocation? We know that biogenic amine transporters (discussed in Chapters 2 and 3) show a variety of conductance states that resemble ion channels. Is this true for GABA and its related transporters? Are there differences within this subfamily? As discussed in Chapters 2, 3, and 7, we are beginning to identify amino-acid residues that alter transport properties. Are these residues part of the permeation path? And what is the structure of this pathway and of these transporters in general?

(5) What are the clinical foci for transporter investigations? We know that multiple isoforms exist for many members of the GABA-transporter subfamily, suggesting that they play different roles in various brain regions. Can these functional isoform differences be exploited for the creation of therapeutics with high specificity? As discussed previously, transporters are regulated by many different signaling cascades. Can novel reagents be created that target these pathways to alter transporter distribution and/or function? Mutations in transporters from other neurotransmitter-transporter subfamilies are associated with particular disease states. Is this also true for the GABA subfamily? The completion of the human genome should help to answer these questions.

Of course, these are just a few of the many intriguing questions than can be addressed about this gene family. If the last decade of research on these proteins is any indication, the next decade will not only answer many of these questions, but will also leave us with many more interesting and unanswered questions.

REFERENCES

1. Elliot, K. A. C. and van Gelder, N. M. (1958) Occlusion and metabolism of γ-aminobutyric acid by brain tissue. *J. Neurochem.* **3,** 28–40.
2. Iversen, L. L. and Kelly, J. S. (1975) Uptake and metabolism of γ-aminobutyric acid by neurones and glial cells. *Biochem. Pharmacol.* **24,** 933–938.
3. Bowery, N. G., Jones, G. P., and Neal, M. J. (1976) Selective inhibition of neuronal GABA uptake by cis-1,3-aminocyclohexane carboxylic acid. *Nature* **264,** 281–284.
4. Guastella, J., Nelson, N., Nelson, H., Czyzyk, L., Keynan, S., Miedel, M. C., Davidson, N., Lester, H. A., and Kanner, B. I. (1990) Cloning and expression of a rat brain GABA transporter. *Science* **249,** 1303–1306.
5. Snow, H., Lowne, M. B., and Bennett, J. P. (1992) A postsynaptic GABA transporter in rat spinal motor neurons. *Neurosci. Lett.* **143,** 119–122.
6. Swan, M., Najierahim, A., Watson, R. F. B., and Bennett, J. P. (1994) Distribution of mRNA for the GABA transporter GAT-1 in the rat brain: evidence that GABA uptake is not limited to presynaptic neurons. *J. Anat.* **185,** 315–323.
7. Minelli, A., Brecha, N. C., Karschin, C., DeBiasi, S., and Conti, F. (1995) GAT-1, a high affinity GABA plasma membrane transporter, is localized to neurons and astroglia in the cerebral cortex. *J. Neurosci.* **15,** 7734–7746.

8. Mabjeesh, N. J., Frese, M., Rauen, T., Jeserich, G., and Kanner, B. I. (1992) Neuronal and glial γ-aminobutyric acid transporters are distinct proteins. *FEBS Lett.* **299,** 99–102.

9. Pacholczyk, T., Blakely, R. D., and Amara, S. G. (1991) Expression cloning of a cocaine-and antidepressant-sensitive human noradrenaline transporter. *Nature* **350,** 350–354.

10. Borden, L. A., Smith, K. F., Hartig, P.R., Branchek, T. A., and Weinshank, R. L. (1992) Molecular heterogeneity of the γ-aminobutyric acid (GABA) transport system: cloning of two novel high affinity GABA transporters from rat brain. *J. Biol. Chem.* **267,** 21,098–21,104.

11. Lopez-Corcuera, B., Liu, Q.-R., Mandiyan, S., Nelson, H., and Nelson, N. (1992) Expression of a mouse brain cDNA encoding novel γ-aminobutyric acid transporter. *J. Biol. Chem.* **267,** 17,491–17,493.

12. Yamauchi, A., Uchida, S., Kwon, H. M., Preston, A. S., Robey, R. B., Garcia-Perez, A., Berg, M. B., and Handler, J. S. (1992) Cloning of a Na+- and Cl—dependent betaine transporter that is regulated by hypertonicity. *J. Biol. Chem.* **267,** 649–652.

13. Clark, I. A., Deutch, A. Y., Gallipoli, P. Z., and Amara, S. G. (1992) Functional expression and CNS distributuon of a β-alanine-sensitive neuronal GABA transporter. *Neuron* **9,** 337–348.

14. Liu, Q.-R., Lopez-Corcuera, B., Mandiyan, S., Nelson, H., and Nelson, N. (1993) Molecular characterization of four pharmacologically distinct γ-aminobutyric acid transporters in mouse brain. *J. Biol. Chem.* **268,** 2106–2112.

15. Durkin, M. M., Smith, K. E., Borden, L. A., Weinshank, R. L., Branchek, T. A., Gustafson, E. L. (1995) Localization of messenger RNAs encoding three GABA transporters in rat brain: an in situ hybridization study. *Brain Res.* **33,** 7–21.

16. Johnson, J., Chen T. K., Rickman D.W., Evans, C., and Brecha, N. C. (1996) Multiple γ-aminobutyric acid plasma membrane transporters (GAT-1, GAT-2, GAT-3) in the rat retina. *J. Comp. Neurol.* **375,** 212–224.

17. Conti, F., Zuccarello, L. V., Barbaresi, P., Minelli, A., Brecha, N. C., and Melone, M. (1999) Neuronal, glial, and epithelial localization of gamma-aminobutyric acid transporter 2, a high-affinity γ-aminobutyric acid plasma membrane transporter, in the cerebral cortex and neighboring structures. *J. Comp. Neurol.* **409,** 482–494.

18. Pietrini, G., Suh, Y. J., Edelmann, L., Rudnick, G., and Caplan, M. J. (1994) The axonal γ-aminobutyric acid transporter GAT4 is sorted to the apical membranes of polarized epithelial cells. *J. Biol. Chem.* **269,** 4668–4674.

19. Ahn, I., Mundigl, O., Muth, T. R., Rudnick, G., and Caplan, M. J. (1996) Polarized expression of GABA transporters in Madin-Darby canine kidney cells and cultured hippocampal neurons. *J. Biol. Chem.* **271,** 6917–6924.

20. Perego, C., Bulbarelli, A., Longhi, R., Caimi, M., Villa, A., Caplan, M. J., and Pietrini, G. (1997) Sorting of two polytopic proteins, the γ-aminobutyric acid and betaine transporters, in polarized epithelial cells. *J. Biol. Chem.* **272,** 6584–6592.

21. Muth, T. R., Ahn, J., and Caplan, M. J. (1998) Identification of sorting de-

terminants in the C-terminal cytoplasmic tails of the γ-aminobutyric acid transporters GAT-2 and GAT-3. *J. Biol. Chem.* **273,** 25,616–25,627.

22. Kenyan, S. and Kanner, B. I. (1988) Gamma-aminobutyric acid transport in reconstituted preparations from rat brain: coupled sodium and chloride fluxes. *Biochemistry* **27,** 12–17.

23. Kavanaugh, M. P., Arriza, J. L., North, R. A., and Amara, S. G. (1992) Electrogenic uptake of gamma-aminobutyric acid by a cloned transporter expressed in Xenopus oocytes. *J. Biol. Chem.* **267,** 22,007–22,009.

24. Mager, S., Corey-Naeve, J., Quick, M. W., Labarca, C., Davidson, N., and Lester, H. A. (1993) Steady states, charge movements, and rates for a cloned GABA transporter expressed in Xenopus oocytes. *Neuron* **10,** 177–188.

25. Cammack, J. N., and Schwartz, E. A. (1996) Channel behavior in a gamma-aminobutyrate transporter. *Proc. Natl. Acad. Sci. USA* **93,** 723–727.

26. Hilgemann, D. W. and Lu C. C. (1999) GAT1 (GABA:Na+:Cl-) cotransport function. Database reconstruction with an alternating access model. *J. Gen. Physiol.* **114,** 459–475.

27. Deken, S., Beckman, M., Boos, L., and Quick, M. W. (2000) Transport rates of GABA transporters: regulation by the N-terminal domain and syntaxin 1A. *Nature Neurosci.* **3,** 998–1003.

28. Dingledine, R. and Korn, S. J. (1985) Gamma aminobutyric acid uptake and the termination of inhibitory synaptic potentials in the rat hippocampal slice. *J. Physiol (Lond.)* **366,** 387–409.

29. Solis, J. M. and Nicoll, R. A. (1992) Pharmacological characterization of GABAB-mediated responses in the CA1 region of the rat hippocampal slice. *J. Neurosci.* **12,** 3466–3472.

30. Keynan, S., Suh, Y.-J., Kanner, B. I., and Rudnick, G. (1992) Expression of a cloned γ-aminobutyric acid transporter in mammalian cells. *Biochemistry* **31,** 1974–1979.

31. Schwartz, E. A. (1982) Calcium-independent release of GABA from isolated horizontal cells of the toad retina. *J. Physiol. (Lond.)* **323,** 211–227.

32. Zeevalk, G. D. and Nicklas, W. J. (1996) Attenuation of excitotoxic cell swelling and GABA release by the GABA transport inhibitor SKF89976A. *Mol. Chem. Neuropath.* **29,** 27–36.

33. Corey, J. L., Davidson, N., Lester, H. A. Brecha, N., and Quick, M. W. (1994) Protein kinase C modulates the activity of a cloned γ-aminobutyric acid transporter expressed in Xenopus oocytes via regulated subcellular redistribution of the transporter. *J. Biol. Chem.* **269,** 14,759–14,767.

34. Beckman, M. L., Bernstein, E. M., and Quick, M. W. (1998) Protein kinase C regulates the interaction between a GABA transporter and syntaxin 1A. *J. Neurosci.* **18,** 6103–6112.

35. Gomeza, J., Casado, M., Gimenez, C., and Aragon, C. (1991) Inhibition of high-affinity γ-aminobutyric acid uptake in primary astrocyte cultures by phorbol esters and phospholipase C. *Biochem. J.* **275,** 435–439.

36. Osawa, I., Saito, N., Koga, T., and Tanaka, C. (1994) Phorbol ester-induced inhibition of GABA uptake by synaptosomes and by Xenopus oocytes expressing GABA transporter (GAT-1). *Neurosci. Res.* **19,** 287–293.

37. Quick, M. W. Corey, J. L., Davidson, N., and Lester, H. A. (1997) Second messengers, trafficking-related proteins, and amino acid residues that contribute to the functional regulation of the rat brain GABA transporter GAT1. *J. Neurosci.* **17,** 2967–2979.

38. Law, R., Stafford, A., and Quick, M. W. (2000) Functional regulation of GABA transporters by direct tyrosine phosphorylation. *J. Biol. Chem.* **275,** 23,986–23,991.

39. Beckman, M. L., Bernstein, E. M., and Quick M. W. (1999) Multiple G protein-coupled receptors initiate protein kinase C redistribution of GABA transporters in hippocampal neurons. *J. Neurosci.* **19,** RC9 1–6.

40. Voutsinos, B., Dutuit, M., Reboul, A., Fevre-Montange, M., Bernard, A., Trouillas, P., Akaoka, H., Belin, M. F., and Didier-Bazes, M. (1998) Serotoninergic control of the activity and expression of glial GABA transporters in the rat cerebellum. *Glia* **23,** 45–60.

41. Bernstein, E. M. and Quick, M. W. (1999) Regulation of gamma-aminobutyric acid (GABA) transporters by extracellular GABA. *J. Biol. Chem.* **274,** 889–895.

42. Krogsgaard-Larsen, P., Falch, E., Larsson, O. M., and Schousboe, A. (1987) GABA uptake inhibitors: relevance to antiepileptic drug research. *Epil. Res.* **1,** 77–93.

43. During, M. J., Ryder, K. M., and Spencer, D. D. (1995) Hippocampal GABA transporter function in temporal-lobe epilepsy. *Nature* **376,** 174–177.

44. Richerson, G. B. and Gaspary, H. L. (1997) Carrier-mediated GABA release: is there a functional role? *Neuroscientist* **3,** 151–157.

45. Mantz, J., Lecharny, J.-B., Laudenbach, V., Henzel, D., Peytavin, G., and Desmonts, J.-M. (1995) Anesthetics affect the uptake but not the depolarization-evoked release of GABA in rat striatal synaptosomes. *Anesthesiology* **82,** 502–511.

46. Simpson, M. D. C., Slater, P., Royston, M. C., and Deakin, J. F. W. (1992) Regionally selective deficits in uptake sites for glutamate and gamma-aminobutyric acid in the basal ganglia in schizophrenia. *Psychiatry Res.* **42,** 273–282.

47. Minuk, G. Y. (1993) Gamma-aminobutyric acid and the liver. *Diges. Disord.* **11,** 45–54.

48. Ladogana, A., Bouzamondo, E., Pocchiari, M., and Tsiang, H. (1994) Modification of tritiated γ-amino-n-butyric acid transport in rabies virus-infected primary cortical cultures. *J. Gen. Virol.* **75,** 623–627.

49. Wong, P. T.-H. (1993) Interactions of indomethacin with central GABA systems. *Arch. Internat. Pharmacodynam.* **324,** 5–16.

50. Borden, L. A., Dhar, T.G.M., Smith, K. F., Weinshank, R. L., Branchek, T. A., and Gluchowski, C. (1994) Tiagabine, SK&F 89976-A, CT-966, and NNC-711 are selective for the cloned GABA transporter GAT-1. *Eur. J. Pharmacol.-Mol. Pharmacol. Sec.* **269,** 219–224.

51. Suzdak, P. D. and Jansen, J. A. (1995) A review of the preclinical pharmacology of tiagabine: a potent and selective anticonvulsant GABA uptake inhibitor. *Epilepsia* **36,** 612–626.

52. Borden, L. A., Dhar, T.G.M., Smith, K. F., Branchek, T. A., Gluchowski, C., and Weinshank, R. L. (1994) Cloning of the human homologue of the GABA transporter GAT-3 and identification of a novel inhibitor with selectivity for this site. *Recept. Channels* **2,** 207–213.
53. Dalby, N. O., Thomsen, C., Fink-Jensen, A., Lundbeck, J., Sokilde, B., Man, C. M., Sorensen, P. O., and Meldrum, B. (1997) Anticonvulsant properties of two GABA uptake inhibitors NNC 05-2045 and NNC 05-2090, not acting preferentially on GAT-1. *Epilepsy Res.* **28,** 51–61.
54. Johnston, G. A. R. and Iversen, L. L. (1971) Glycine uptake in rat central nervous system slices and homogenates: evidence for different uptake systems in spinal cord and cerebral cortex. *J. Neurochem.* **18,** 1951–1961.
55. Logan, W. J. and Snyder, S. H. (1972) High affinity uptake systems for glycine, glutamic and aspartic acids in synaptosomes of rat central nervous tissues. *Brain Res.* **42,** 413–431.
56. Zafra, F. and Gimenez, C. (1989) Characteristics and adaptive regulation of glycine transport in cultured glial cells. *Biochem. J.* **258,** 403–408.
57. Smith, K. E., Borden, L. A., Hartig, P. R., Branchek, T., and Weinshank, R. L. (1992) Cloning and expression of a glycine transporter reveal colocalization with NMDA receptors. *Neuron* **8,** 927–935.
58. Liu, Q.-R., Nelson, H., Mandiyan, S., Lopez-Corcuera, B., and Nelson, N. (1992) Cloning and expression of a glycine transporter from mouse brain. *FEBS Lett.* **305,** 110–114.
59. Guastella, J., Brecha, N., Weigmann, C., Lester, H. A., and Davidson, N. (1992) Cloning, expression, and localization of a rat brain high-affinity glycine transporter. *Proc. Natl. Acad. Sci. USA* **89,** 7189–7193.
60. Kim, K.-M., Kingsmore, S. F., Han, H., Yang-Feng, T. L., Godinot, N., Seldin, M. F., Caron, M. G., and Giros, B. (1995) Cloning of the human glycine transporter type 1: molecular and pharmacological characterization of novel isoform variants and chromosomal localization of the gene in the human and mouse genomes. *Mol. Pharmacol.* **45,** 608–617.
61. Adams, R. H., Sato, K., Shimada, S., Tohyama, M., Püschel, A. W., and Betz, H. (1995) Gene structure and glial expression of the glycine transporter GLYTi in embryonic and adult rodents. *J. Neurosci.* **15,** 2524–2532.
62. Borowsky, B. and Hoffman, B. J. (1998) Analysis of a gene encoding two glycine transporter variants reveals alternative promoter usage and a novel gene structure. *J. Biol. Chem.* **273,** 29,077–29,085.
63. Morrow, J. A., Collie, I. T., Dunbar, D. R., Walker, G. B., Shahid, M., and Hill, D. R. (1998) Molecular cloning and functional expression of the human glycine transporter GlyT2 and chromosomal localisation of the gene in the human genome. *FEBS Lett.* **439,** 334–340.
64. Borowsky, B., Mezey, E., and Hoffman, B. J. (1993) Two glycine transporter variants with distinct localization in the CNS and peripheral tissues are encoded by a common gene. *Neuron* **10,** 851–863.
65. Zafra, F., Aragon, C., Olivares, L., Danbolt, N. C., Gimenez, C., and Storm-Mathisen, J. (1995) Glycine transporters are differentially expressed among CNS cells. *J. Neurosci.* **15,** 3952–3969.

66. Pow, D. V. and Hendrickson, A. E. (2000) Expression of glycine and the glycine transporter glyt-1 in the developing rat retina. *Vis. Neurosci.* **17,** 1–9.
67. Barmack, N. H., Guo, H., Kim, H. J., Qian, H., and Qian, Z. (1999) Neuronally modulated transcription of a glycine transporter in rat dorsal cochlear nucleus and nucleus of the medial trapezoid body. *J. Comp. Neurol.* **415,** 175–188.
68. Zafra, F., Gomeza, J., Olivares, L., Aragon, C., and Gimenez, C. (1995) Regional distribution and developmental variation of the glycine transporters GLYT1 and GLYT2 in the rat CNS. *Eur. J. Neurosci.* **7,** 1342–1352.
69. Olivares, L., Aragon, C., Gimenez, C., and Zafra, F. (1995) Carboxyl terminus of the glycine transporter GLYT1 is necessary for correct processing of the protein. *J. Biol. Chem.* **269,** 28,400–28,404.
70. Poyatos, I., Ruberti, F., Martinez-Maza, R., Gimenez, C., Dotti, C. G., and Zafra, F. (2000) Polarized distribution of glycine transporter isoforms in epithelial and neuronal cells. *Mol. Cell. Neurosci.* **15,** 99–111.
71. Nunez, I. F. and Aragon, C. (1994) Structural analysis and functional role of the carbohydrate component of glycine transporter. *J. Biol. Chem.* **269,** 16,920–16,924.
72. Olivares, L., Aragon, C., Gimenez, C., and Zafra, F. (1995) The role of N-glycosylation in the targeting and activity of the GLYT1 glycine transporter. *J. Biol. Chem.* **270,** 9437–9442.
73. Olivares, L., Aragon, C., Gimenez, C., and Zafra, F. (1997) Analysis of the transmembrane topology of the glycine transporter GLYT1. *J. Biol. Chem.* **272,** 1211–1217.
74. Gomeza, J., Zafra, F., Olivares, L., Gimenez, C., and Aragon C. (1995) Regulation by phorbol esters of the glycine transporter (GLYT1) in glioblastoma cells. *Biochim Biophys Acta* **1233,** 41–46.
75. Sato, K., Adams, R., Betz, H., and Schloss, P. (1995) Modulation of a recombinant glycine transporter (GLYT1b) by activation of protein kinase C. *J. Neurochem.* **65,** 1967–1973.
76. Zafra, F., Poyatos, I., and Gimenez, C. (1997) Neuronal dependency of the glycine transporter GLYT1 expression in glial cells. *Glia* **20,** 155–162.
77. Geerlings, A., Lopez-Corcuera, B., and Aragon, C. (2000) Characterization of the interactions between the glycine transporters GLYT1 and GLYT2 and the SNARE protein syntaxin 1A. *FEBS Lett.* **470,** 51–54.
78. Hanley, J. G., Jones, E. M., and Moss, S. J. (2000) GABA receptor rho1 subunit interacts with a novel splice variant of the glycine transporter, GLYT-1. *J. Biol. Chem.* **275,** 840–846.
79. Luque, J. M., Nelson, N., and Richards, J. G. (1995) Cellular expression of glycine transporter 2 messenger RNA exclusively in rat hindbrain and spinal cord. *Neuroscience* **64,** 525–535.
80. Johnson, J. W. and Ascher, P. (1987) Glycine potentiates the NMDA response in cultured mouse brain neurons. *Nature* **325,** 529–531.
81. Supplisson, S. and Bergman, C. (1997) Control of NMDA receptor activation by a glycine transporter co-expressed in Xenopus oocytes. *J. Neurosci.* **17,** 4580–4590.

82. Bergeron, R., Meyer, T. M., Coyle, J. T., and Greene R. W. (1998) Modulation of N-methyl-D-aspartate receptor function by glycine transport. *Proc. Natl. Acad. Sci. USA* **95,** 15,730–15,734.

83. Roux, M. J. and Supplisson, S. (2000) Neuronal and glial glycine transporters have different stoichiometries. *Neuron* **25,** 373–383.

84. Shafqat, S., Velaz-Faircloth, M., Henzi, V., Yang-Feng, T., Seldin, M., and Fremeau, R. T., Jr. (1995) Human brain-specific L-proline transporter: molecular cloning, pharmacological characterization, and chromosomal localization of the gene in human and mouse genomes. *Mol. Pharmacol.* **48,** 219–229.

85. Fremeau, R. T., Jr., Caron, M. G., and Blakely, R. D. (1992) Molecular cloning and expression of a high affinity L-proline transporter expressed in putative glutamatergic pathways of rat brain. *Neuron* **8,** 915–926.

86. Nadler, J. V. (1987) Sodium-dependent proline uptake in the rat hippocampal formation: association with ipsilateral commissural projections of CA3 pyramidal cells. *J. Neurochem.* **49,** 1155–1160.

87. Fremeau, R. T. Jr., Velaz-Faircloth, M., Miller, J. W., Henzi, V. A., Cohen, S. M., Nadler, J. V., Shafqat, S., Blakely, R. D., and Domin, B. (1996) A novel nonopioid action of enkephalins: competitive inhibition of the mammalian brain high affinity L-proline transporter. *Mol. Pharmacol.* **49,** 1033–1041.

88. Galli, A., Jayanthi, L. D., Ramsey, I. S., Miller, J. W., Fremeau, R. T. Jr., and DeFelice, L. J. (1999) L-proline and L-pipecolate induce enkephalin-sensitive currents in human embryonic kidney 293 cells transfected with the high-affinity mammalian brain L-proline transporter. *J. Neurosci.* **19,** 6290–6297.

89. Jayanthi, L. D., Wilson, J. J., Montalvo,. J., and DeFelice, L. J. (2000) Differential regulation of mammalian brain-specific proline transporter by calcium and calcium-dependent protein kinases. *Brit. J. Pharmacol.* **129,** 465–470.

90. Nadler, J. V., Bray, S.D., and Evenson, D. A. (1992) Autoradiographic localization of proline uptake in excitatory hippocampal pathways. *Hippocampus* **2,** 269–278.

91. Renick, S. E., Kleven, D. T., Chan, J., Stenius, K., Milner, T. A., Pickel, V. M., and Fremeau, R. T. Jr. (1999) The mammalian brain high-affinity L-proline transporter is enriched preferentially in synaptic vesicles in a subpopulation of excitatory nerve terminals in rat forebrain. *J. Neurosci.* **19,** 21–33.

92. Cohen, S. M. and Nadler, J. V. (1997) Proline-induced potentiation of glutamate transmission. *Brain Res.* **761,** 271–282.

93. Choi, D. W. (1992) Bench to bedside: the glutamate connection. *Science* **258,** 241–243.

94. Borden, L. A., Smith, K. F., Gustafson, F. L., Branchek, T. A., and Weinshank, R. L. (1995) Cloning and expression of a betaine/GABA transporter from human brain. *J. Neurochem.* **64,** 977–984.

95. Rasola, A., Galietta, L. J. V., Barone, V., Romeo, G., and Bagnasco, S. (1995) Molecular cloning and functional characterization of a GABA/betaine transporter from human kidney. *FEBS Lett.* **373,** 229–233.

96. Matskevitch, I., Wagner, C. A., Stegen, C., Broer, S., Noll, B., Risler, T., et al. (1999) Functional characterization of the Betaine/gamma-aminobutyric

acid transporter BGT-1 expressed in Xenopus oocytes. *J. Biol. Chem.* **274,** 16,709–16,716.

97. Petty, C. N. and Lucero, M. T. (1999) Characterization of a Na+-dependent betaine transporter with Cl- channel properties in squid motor neurons. *J. Neurophys.* **81,** 1567–1574.

98. Takenaka, M., Preston, A. S., Kwon, H. M., and Handler, J. S. (1991) The tonicity-sensitive element that mediates increased transcription of the betaine transporter gene in response to hypertonic stress. *J. Biol. Chem.* **269,** 29,379–29,381.

99. Smith, K. E., Borden, L. A., Wang, C.-H. D., Hartig, P.R., Branchek, T. A., and Weinshank, R. L. (1992) Cloning and expression of a high affinity taurine transporter from rat brain. *Mol. Pharmacol.* **42,** 563–569.

100. Kulanthaivel, P., Cool, D. R., Ramamoorthy, S., Mahesh, V. B., Leibach, F. H., and Ganapathy, V. (1991) Transport of taurine and its regulation by protein kinase C in the JAR human placental choriocarcinoma cell line. *Biochem. J.* **277,** 53–58.

101. Brandsch, M., Miyamota, Y., Ganapathy V., and Leibach, F. H. (1993) Regulation of taurine transport in human colon carcinoma cell lines (HT-29 and Caco-2) by protein kinase C. *Am. J. Physiol.* **264,** G939–G946.

102. Han, X., Budreau, A. M., and Chesney, R. W. (1999) Ser-322 is a critical site for PKC regulation of the MDCK cell taurine transporter (pNCT). *J. Am. Soc. Neph.* **10,** 1874–1879.

103. Ramamoorthy, S., Leibach, F. H., Mahesh, V. B., Han, H., Yang-Feng, T., Blakely, R. D., and Ganapathy, V. (1994) Functional characterization and chromosomal localization of a cloned taurine transporter from human placenta. *Biochem. J.* **300,** 893–900.

104. Norman, J. A., Ansell, J., Stone, G. A., Wennogle, L. P., and Wasley, J. W. F. (1987) CGS 9343 B, a novel, potent, and selective inhibitor of calmodulin activity. *Mol. Pharmacol.* **31,** 535–540.

105. Satsu, H., Watanabe, H., Arai, S., and Shimizu, M. (1997) Characterization and regulation of taurine transport in Caco-2, human intestinal cells. *J. Biochem.* **121,** 1082–1087.

106. Takeuchi, K., Toyohara, H., and Sakaguchi, M. (2000) A hyperosmotic stress-induced mRNA of carp cell encodes Na(+)- and Cl(-)-dependent high affinity taurine transporter. *Biochim. Biophys. Acta.* **1464,** 219–230.

107. Sturman, J. A. (1990) Taurine deficiency. *Prog. Clin. Biol. Res.* **351,** 385–395.

108. Salceda, R. (1999) Insulin-stimulated taurine uptake in rat retina and retinal pigment epithelium. *Neurochem. Int.* **35,** 301–306.

109. Saransaari, P. and Oja, S. S. (1999) Taurine release modified by nitric oxide-generating compounds in the developing and adult mouse hippocampus. *Neuroscience* **89,** 1103–1111.

110. Guerrero-Ontiveros, M. L., and Wallimann, T. (1998) Creatine supplementation in health and disease. Effects of chronic creatine ingestion in vivo: down-regulation of the expression of creatine transporter isoforms in skeletal muscle. *Mol. Cell. Biochem.* **184,** 427–437.

111. Neubauer, S., Remkes, H., Spindler, M., Horn, M., Wiesmann, F., Prestle, J., et

al. (1999) Downregulation of the Na(+)-creatine cotransporter in failing human myocardium and in experimental heart failure. *Circulation* **100,** 1847–1850.

112. Guimbal, C. and Kilimann, M. W. (1993) A Na+-dependent creatine transporter in rabbit brain, muscle, heart, and kidney: cDNA cloning and functional expression. *J. Biol. Chem.* **268,** 8418–8421.

113. Dai, W. Vinnakota, S. Qian, X. Kunze, D. L., and Sarkar H. K. (1999) Molecular characterization of the human CRT-1 creatine transporter expressed in Xenopus oocytes. *Arch Biochem. Biophys.* **361,** 75–84.

114. Happe, H. K. and Murrin, L. C. (1995) In situ hybridization analysis of CHOT1, a creatine transporter, in the rat central nervous system. *J. Comp. Neurol.* **351,** 94–103.

115. Nash, S. R., Giros, B., Kingsmore, S. F., Rochelle, J. M., Suter, S. T., Gregor, P., et al. (1994) Cloning, pharmacological characterization, and genomic localization of the hurnan creatine transporter. *Recept. Chan.* **2,** 165–174.

116. Guimbal, C. and Kilimann, M. W. (1994) A creatine transporter cDNA from Torpedo illustrates structure/function relationships in the GABA / noradrenaline transporter family. *J. Mol. Biol.* **241,** 317–324.

117. Liu, Q.-R., Mandiyan, S., Lopez-Corcuera, B., Nelson, H., and Nelson, N. (1993) A rat brain cDNA encoding the neurotransmitter transporter with an unusual structure. *FEBS Lett.* **315,** 114–118.

118. Smith, K. E., Fried, S. G., Durkin, M. M., Gustafson, E. L., Borden, L. A., Branchek, T. A., and Weinshank, R. L. (1995) Molecular cloning of an orphan transporter. A new member of the neurotransmitter transporter family. *FEBS Lett.* **357,** 86–92.

119. Wasserman, J. C., Delpire, E., Tonidandel, W., Kojima, R., and Gullans, S. R. (1994) Molecular characterization of ROSIT, a renal osmotic stress-induced Na(+)-Cl(-)-organic solute cotransporter. *Am. J. Phys.* **267,** F688–F694.

120. Uhl, G. R., Kitayama, S., Gregor, P., Nanthakumar, E., Persico, A., and Shimada, S. (1992) Neurotransmitter transporter family cDNAs in a rat midbrain library: 'orphan transporters' suggest sizable structural variations. *Brain Res.* **16,** 353–359.

121. Luque, J. M., Jursky, F., Nelson, N., and Richards J. G. (1996) Distribution and sites of synthesis of NTT4, an orphan member of the Na+/Cl(-)-dependent neurotransmitter transporter family, in the rat CNS. *Eur. J. Neurosci.* **8,** 127–137.

122. Fischer, J., Bancila, V., Mailly, P., Masson, J., Hamon, M., El Mestikawy, S., and Conrath, M. (1999) Immunocytochemical evidence of vesicular localization of the orphan transporter RXT1 in the rat spinal cord. *Neuroscience* **92,** 729–743.

123. Masson, J., Riad, M., Chaudhry, F., Darmon, M., Aidouni, Z., Conrath, M., et al. (1999) Unexpected localization of the Na+/Cl—dependent-like orphan transporter, Rxt1, on synaptic vesicles in the rat central nervous system. *Eur. J. Neurosci.* **11,** 1349–1361.

7

Sodium-Coupled GABA and Glutamate Transporters

Structure and Function

Baruch I. Kanner

1. INTRODUCTION

Sodium-coupled neurotransmitter transporters, located in the plasma membrane of nerve terminals and glial processes, serve to keep the extracellular transmitter levels below those which are neurotoxic. They also help, in conjunction with diffusion, to terminate its action in synaptic transmission. Such a termination mechanism operates with most transmitters, including γ-aminobutyric acid (GABA), L-glutamate, glycine, dopamine (DA), serotonin, and norepinephrine (NE). Another termination mechanism is observed with cholinergic transmission. After dissociation from its receptor, acetylcholine is hydrolyzed into choline and acetate. The choline moiety is then recovered by sodium-dependent transport as described here. As the concentration of the transmitters in the nerve terminals is much higher than in the cleft—typically by four orders of magnitude—energy input is required. The transporters that are located in the plasma membranes of nerve endings and glial cells obtain this energy by coupling the flow of neurotransmitters to that of sodium. The $(Na^+ + K^+)$-ATPase generates an inwardly directed electrochemical sodium gradient which is utilized by the transporters to drive "uphill" transport of the neurotransmitters (reviewed in *1–3*). Neurotransmitter-uptake systems have been investigated in detail by using plasma membranes obtained upon osmotic shock of synaptosomes. It appears that these transporters are coupled not only to sodium, but also to additional ions such as potassium or chloride (Table 1).

Sodium-coupled neurotransmitter transporters are of considerable medical interest. Since they function to regulate neurotransmitter activity by removing it from the synaptic cleft, specific transporter inhibitors can be potentially used as novel drugs for neurological disease. For instance,

From: *Contemporary Neuroscience:*
Neurotransmitter Transporters: Structure, Function, and Regulation, 2nd Edition
Edited by: M. E. A. Reith © Humana Press Inc., Totowa, NJ

Table 1
Comparison of GABA and Glutamate Transporters

	GABA tp	Glutamate tp
Cosubstrates	Na$^+$, Cl$^-$	Na$^+$, K$^+$, H$^+$
Electrogenicity	+	+
Localization	Neuronal, glial	Neuronal, glial
Sociology	Belong to large family of transporters for several amino acids and all neurotransmitters excluding glutamate	Belong to separate small family of glutamate transporters
Relationship to bacterial transporters	Bacterial homologs identified but their function is not yet known	glt-P, glutamate transporter dct-A, dicarboxylic-acid transporter
Predicted topology	12 TMs; amino and carboxyl termini are cytoplasmic	8 TMs with reentrant loops and linker regions; amino and carboxyl termini are cytoplasmatic
Glycosylation	+	+
Possible regulation	Protein kinase C Syntaxin 1A	Protein kinase C Arachidonic acid

attenuation of GABA removal will prolong the effect of this inhibitory transporter, thereby potentiating its action. Thus, inhibitors of GABA transport could represent a novel class of anti-epileptic drugs. Well-known inhibitors which interfere with the functioning of biogenic amine transporters include antidepressant drugs and stimulants such as amphetamines and cocaine. The neurotransmitter glutamate—at excessive local concentrations—causes cell death by activating N-methyl-D-aspartic acid (NMDA) receptors and subsequent calcium entry. The transmitter has been implicated in neuronal destruction during ischaemia, epilepsy, stroke, amyotropic lateral sclerosis, and Huntington's disease. Neuronal and glial glutamate transporters may play a critical role in preventing glutamate from acting as an exitotoxin (4,5).

In the last decade, major advances in the cloning of these neurotransmitter transporters have been made. After the GABA transporter was purified (6), the ensuing protein sequence information was used to clone it (7). Subsequently the expression cloning of a norepinephrine transporter (NET) (8) provided evidence that these two proteins are the first members of a novel superfamily of neurotransmitter transporters. This result led—using polymerase chain reaction (PCR) and other technologies relying on sequence conservation—to the isolation of a growing list of neurotransmitter transporters (reviewed in *9–11*). This list includes various subtypes of GABA transporters as well as those for all the above-mentioned neurotransmitters, except glutamate. Most of the members of this superfamily are dependent on sodium and chloride, and by analogy with the GABA transporter *(12)* are likely to cotransport their transmitter with both sodium and chloride. Interestingly, sodium-dependent glutamate transport is not chloride dependent, but rather sodium and glutamate are countertransported with potassium *(13,14;* Table 1). In 1992, three distinct but closely related glutamate transporters were cloned *(15–17)*. Subsequently, two other subtypes were cloned, exhibiting a large chloride conductance *(18,19)*. These transporters represent a distinct family of transporters. The following describes the current status on two prototypes of these distinct families: the GABA and glutamate transporters, with an emphasis on the structural basis of function. Since in this edition the regulation of GABA transporters is covered in Chapter 6, this topic is not discussed here. For regulation of glutamate transport, *see* also ref. *20*.

2. MECHANISM

The GABA transporter GAT-1 cotransports the neurotransmitter with sodium and chloride in an electrogenic fashion *(1,12;* Table 1; Fig. 1, top). The available measurements include tracer fluxes *(12)* and electrophysiological approaches *(21,22)*. The latter approach reveals that at very negative (inside) potentials, the chloride dependency is not absolute *(22,23)*. At the present time, it is not clear what the mechanistic interpretation of this result is. One possibility is that under these conditions, another anion—such as hydroxyl—may take over the role of chloride. However, the outward GAT-1 current ("reverse transport mode") requires the presence of all three substrates on the cytoplasmic side *(23)*. In addition, a transient current is observed in the absence of GABA. This transient can be blocked by bulky GABA analogs, which can bind to the transporter, but are not translocated by it *(22)*. It probably reflects a conformational change of the transporter occurring after the sodium has bound. The transient can be blocked by internal chloride, and it appears that cytoplasmic chloride and extracellular sodium bind to the tranporter in a mutually exclusive fashion *(24)*, providing strong support for an alternating access mode *(25)*. A major role for external chloride is to increase the affinity for external sodium *(26)*. Comparison of the transient and steady-state currents facilitates a determination of the turnover number of the transporter. These estimates of a few cycles per second *(22)* are in agreement with biochemical ones *(6)*. Notably, that although both GABA and 5-HT transporters belong to the same transporter

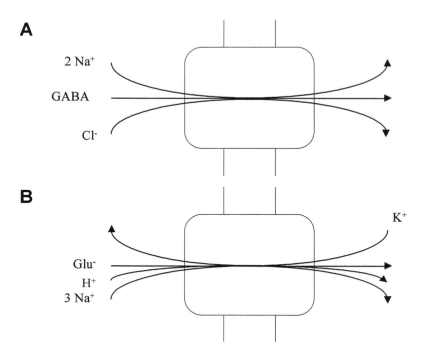

Fig. 1. Schematic representation of GABA and L-glutamate transport. The diagrams illustrate the coupled fluxes of GABA **(top)** and L-glutamate **(bottom)** with their cosubstrates.

superfamily (Table 1), the latter one appears to exhibit highly distinct properties *(27)*. It appears that the 5-HT transporter is electroneutral, but that a transporter-associated current can be detected. This is probably related to the observation that under some conditions this transporter may act as a channel. This transporter-channel appears to be less sodium-selective than the transporter mode, which is carrying 5-HT in a coupled fashion *(27)*.

The mechanism of sodium-dependent L-glutamate transport has been studied initially using tracer-flux studies employing radioactive glutamate. These studies indicated that the process is electrogenic, with positive charge moving in the direction of the glutamate *(13)*. This observation suggested that it is possible to monitor L-glutamate transport electrically using the whole-cell patch-clamp technique *(28)*. In addition to L-glutamate, D- and L-aspartate are transportable substrates with affinities in the lower micromolar range. The system is stereospecific with regard to glutamate, and the D-isomer is a poor substrate. Glutamate uptake is driven by an inwardly directed sodium-ion gradient, and potassium is moving outward at the same time. The potassium movement is not a passive movement in response to the charge carried by the transporter. Rather, it is an integral part of the translocation cycle catalyzed by the transporter as further described here.

Electrogenic ($Na^+ + K^+$)-coupled glutamate transporters, located in the plasma membranes of nerve terminals and glial cells, fulfill an important role in the process of excitatory transmission. They keep the extracellular level of the transmitter below neurotoxic levels *(2,29–31)*. Moreover, at some synapses glutamate transporters appear to be important in limiting the duration of synaptic excitation *(32–34)*. They achieve this through an electrogenic process *(13,28,35)* in which the transmitter is cotransported with three sodium ions and one proton *(36)*, followed by countertransport of a potassium ion *(14,36–38)*. It appears that the three sodium ions are not equivalent. Two of them are probably not entirely specific and can be replaced by lithium, but one is completely specific *(39)*.

3. RECONSTITUTION AND PURIFICATION

Using methodology which enables one to reconstitute many samples simultaneously and rapidly, one of the subtypes of the GABA *(6)* as well as of the L-glutamate transporter *(40)* have been purified to an apparent homogeneity. Both are glycoproteins, and both have an apparent molecular mass of 70–80 kDa. The two transporters retain all the properties observed in membrane vesicles. They are distinct, partly because of their different functional properties. Antibodies generated against the GABA transporter *(6)* react (as detected by immunoblotting) only with fractions containing GABA-transport activity and not with those containing L-glutamate-transport activity *(40)*. The opposite is true for antibodies generated against the glutamate transporter *(41)*. A glycine transporter (GLYT) has also been purified and reconstituted. Interestingly, it appears to be a larger protein than the GABA and glutamate transporters—approx 100 kDa in size *(42)* and appears to correspond to the cloned GLYT2 transporter *(43)*. The serotonin transporter (SERT) has also been purified, but these preparations, which contain a band approx 70 kDa, have been shown to be active only in the binding of (^3H)-imipramine but not in serotonin transport *(44,45)*.

4. THE LARGE SUPERFAMILY OF NA⁺-DEPENDENT NEUROTRANSMITTER AND AMINO-ACID TRANSPORTERS

Partial sequencing of the purified $GABA_A$ transporter allowed the cloning of GAT-1, the first member of the new family of Na-dependent neurotransmitter transporters *(7)*. After expression and cloning of the noradrenaline transporter *(8)*, it became clear that it had significant homology with GAT-1. The use of functional c-DNA expression assays and amplification of related sequences by polymerase chain reaction (PCR) resulted in the cloning of additional transporters which belong to this family, such as the dopamine *(46–48)* and serotonin *(49,50)* transporters, additional GABA transporters *(51–54)*, transporters of glycine *(43,55–57)*, proline *(58)*, taurine *(59,60)*, betaine *(61)*, and "orphan" transporters, whose substrates are still unknown. In addition, another family member which was originally believed to be a choline transporter *(62)*, probably is in fact a creatine transporter *(63)*. Recently, a novel human Na^+- and Cl^--dependent neutral and cationic amino-acid transporter B^{0+} has been cloned and expressed which belongs to the superfamily, and has the highest sequence similarity to glycine and proline transporters *(64)*. Furthermore, two other interesting family members, KAAT1 *(65)* and CAATCH1 *(66)*, amino-acid transporters from caterpillar absorptive epithelium, have been cloned and expressed. They can couple amino-acid transport to sodium as well as potassium *(65,66)*.

The deduced amino-acid sequences of these proteins reveal 30–65% identity between different members of the family. Based on these differences in homology. the family can be divided into four subgroups:

1. Transporters of biogenic amines (noradrenaline, dopamine. and serotonin);
2. Various GABA transporters as well as transporters of taurine and creatine;
3. Transporters of proline, glycine, and the B^{0+} transporter;
4. Amino-acid transporters coupled to either sodium or potassium;
5. Orphan transporters.

GAT-1, as well as the other members of the family, is predicted to have 12 transmembrane domains (TMDs) linked by hydrophilic loops, with the amino- and carboxyl-termini residing inside the cell *(7)*. Many aspects of this model were verified experimentally, although the topology of the amino-terminal third of the transporter was controversial *(67–70)*. Recent evidence suggests that the theoretical topology model *(7;* Fig. 2) is correct *(71)*.

Alignment of the deduced amino-acid sequences of the different members of this superfamily, whose substrates are known, revealed that some segments within these proteins share a higher degree of homology than others. The most highly conserved regions (>50% homology) are helix 1 together with the extracellular loop (EL) connecting it with helix 2, and helix 5 together with a short intracellular loop connecting it with helix 4 and a larger EL connecting it with helix 6. These domains may be involved in stabilizing a tertiary structure, which is essential for the function of all these transporters. Alternatively, they may be related to a common function of these transporters, such as the translocation of sodium ions. The region

extracellular

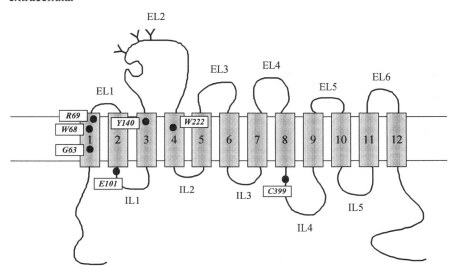

intracellular

Fig. 2. Proposed topology for the GAT-1 transporter. Putative transmembrane α-helices are indicated as cylinders and individual amino acids as small circles. Also indicated are consensus sites for asparagine-linked glycosylation (**Y**). Amino-acid residues critical for activity are indicated using the one letter code (*see* previous page).

stretching from helix 9 on is far less conserved than the segment containing the first 8 helices. Possibly, this domain contains some residues that are involved in translocating the different substrates. The least conserved segments are the amino- and carboxy-terminii. The "orphan" transporters differ from all other members of the family in three regions. They contain much larger ELs between helices 7–8 and helices 11–12 and have a shorter EL connecting helices 3–4.

5. MOLECULAR CLONING AND PREDICTED STRUCTURE OF GLUTAMATE TRANSPORTERS

Transporters for many neurotransmitters were cloned on the assumption that they are related to the GABA *(7)* and norepinephrine *(8)* transporters *(9–11)*. This approach was unsuccessful for the glutamate transporter. Three different glutamate transporters were cloned using different approaches: GLAST *(15)*, GLT-1 *(16)*, and EAAC 1 *(17)*. The former two appear to be of glial *(15,16,41,72)*, the latter of neuronal origin *(17,73)*, and the same is true for the later cloned EAAT-4 *(18)*. Indeed, these transporters are not related to the above superfamily *(15–17)*. On the other hand, they are very similar to each other, displaying ~50% identity and ~60% similarity. They also appear to be related to the proton-coupled glutamate transporter from *E. coli* and other bacteria (**glt-P** *[74]*) and the dicarboxylate transporter (**dct-A** *[75]*) of *Rhizobium meliloti*. In these cases, the identities are approx 25–30%. Thus, they form a distinct family. They contain between 500–600 amino acids. It has been shown that this family also encodes sodium-dependent transporters which do not use dicarboxylic acids as substrates, but rather neutral amino acids ASCT-1 *(76,77)* and ASCT-2 *(78,79)*. The three human homologues of the rat and rabbit glutamate transporters have been cloned *(80)*, as well as two novel subtypes that are characterized by a large substrate-induced chloride current *(18,19)*. A similar but smaller current, which is not thermodynamically coupled to glutamate transport, has been observed in several of the other subtypes as well *(81)* as well as in ASCT-1 *(82)*.

GLT-1, which encodes the glutamate transporter which was purified *(16,40,41)*, has 573 amino acids and a relative molecular mass of 64 kDa, in good agreement with the value of 65 kDa of the purified and deglycosylated transporter *(41)*. Hydrophaty plots are relatively straightforward at the amino-terminal side of the protein and the three different groups, who originally cloned GLAST-1, GLT-1, and EAAC-1, have predicted six transmembrane α-helices at very similar positions *(15–17)*. However, there is much more ambiguity at the carboxyl side where zero *(15)*, two *(16)*, or four *(17)* α-helices have been predicted. Recently, attempts have been made to determine the topology of the glutamate transporters experimentally. Studies of the highly conserved carboxyl terminal half of the glutamate transporters indicate a nonconventional topology (Fig. 3) containing two reentrant loops, two TMDs—7 and 8—long enough to span the membrane as α-helices, as well as an outward-facing hydrophobic linker *(83–85)*. A very recent study arrives at a somewhat different model, including the assignment of TMD 7 as a reentrant loop *(86)*. There is a basic disagreement related solely to the accessibility of only one of the engineered cysteines to impermeant sulfhydryl reagents *(85,87)*, and the reason for this is unclear at present.

6. STRUCTURE–FUNCTION RELATIONSHIPS IN THE SUPERFAMILY OF NEUROTRANSMITTER TRANSPORTERS

It has been shown that parts of amino- and carboxyl-terminii of the GABA transporter GAT-1 are not required for function *(88)*. In order to define these domains, a series of deletion mutants was studied in the GABA transporter *(89)*. Transporters truncated at either end, until just a distance of a few amino acids from the beginning of helix 1 and the end of helix 12, retained their

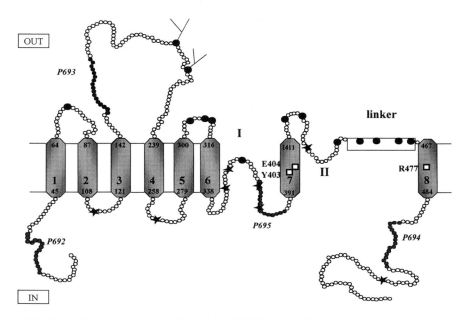

Fig. 3. Adjusted topological model of GLT-1. Residues that react with Biotin maleimide from the outside (**closed circles**) or from the inside (**stars**) are indicated by their position number. P692, P695 and P694 are stretches of amino acids that were shown to be internal by trypsinization studies and P693 external. N205 and N217 were shown to be used as native glycosylation sites (Y-shaped symbols). The transmembrane domains, that are long enough to cross the membrane as α-helices, are indicated by Arabic numerals and the reentrant loops by Roman numerals. The presumed boundaries of TM 1–8 are indicated as well. Some amino-acid residues that are critical for activity are indicated by the one-letter code (*see* previous page).

ability to catalyze sodium and chloride-dependent GABA transport. These deleted segments did not contain any residues conserved among the different members of the superfamily. Once the truncated segment included part of these conserved residues, such as arginine-44 and tryptophan-47, the transporter's activity was severely reduced. However, the functional damage was not caused by impaired turnover or impaired targeting of the truncated proteins *(89)*. Tryptophan-47 can be replaced only by aromatic residues without loss of activity. In the case of arginine-44, only when it is replaced by lysine, low activity levels—approx 15% of those of the wild-type—are observed. Using a reconstitution assay, it has been shown that mutants in which this residue is replaced by lysine or histidine exhibit sodium- and chloride-dependent GABA exchange similar to the wild-type. This finding indicates that these mutants are selectively impaired in the reorientation of the unloaded transporter, a step in the translocation cycle by which net flux and exchange differ. These observations suggest that the consensus sequence RXXW may control this step in the related transporters as well *(90)*.

Fragments of the (Na$^+$ + Cl$^-$)-coupled GABA$_A$ transporter have been produced by proteolysis of membrane vesicles and reconstituted preparations from the rat brain *(91)*. The former were digested with pronase, and the latter with trypsin. Fragments with different apparent molecular masses were recognized by sequence-directed antibodies raised against this transporter. When GABA was present in the digestion medium, the generation of these fragments was almost entirely blocked *(91)*. At the same time, the neurotransmitter largely prevented the loss of activity caused by the protease. The effect was specific for GABA; protection was not afforded by other neurotransmitters. It was only observed when the two cosubstrates, sodium and chloride, were present on the same side of the membrane as GABA *(91)*. The results indicate that the transporter may exist in two conformations. In the absence of one or more of the substrates, multiple sites located throughout the transporter are accessible to the proteases. In the presence of all three substrates—conditions favoring the formation of the translocation complex—the conformation is changed such that these sites become inaccessible to protease action.

The author's group has investigated the role of the hydrophilic loops connecting the putative transmembrane α-helices connecting GAT-1. Deletions of randomly picked nonconserved single amino acids in the loops connecting helices 7 and 8, or 8 and 9, result in inactive transport upon expression in Hela cells. However, transporters that exist where these amino acids are replaced with glycine retain significant activity. The expression levels of the inactive mutant transporters was similar to that of the wild-type, but one of these, ΔVal-348, appears to be defectively targeted to the plasma membrane. Our data are compatible with the idea that a minimal length of the loops is required, presumably to enable the TMDs to interact optimally with each other *(92)*. Furthermore, it is not only the length of the loops which is important, but also the sequence itself. For instance, it has been shown that EL V—connecting TMDs 9 and 10—contains determinants for the sensitivity of GABA transporters toward the analog β-alanine *(93)*. Moreover, cysteine-399 of GAT-1, located on intracellular loop IV—connecting TMDs 8 and 9—is a major determinant of the sensitivity of the transporter to sulfhydryl modification. The accessibility of cysteine-399 is highly dependent on the conformation of GAT-1 *(94)*.

The substrate translocation performed by the various members of the superfamily is sodium-dependent, and is usually chloride-dependent. In addition, some of the substrates contain charged groups as well. Therefore, charged amino acids in the membrane domain of the transporters may be essential for their normal function. This theory was tested using the GABA transporter *(95)*. Of five charged amino acids within its membrane domain only one, arginine-69 in helix 1, is absolutely essential for activity (Fig. 2). It is not merely the positive charge which is important, as even its substitution to other positively charged amino acids does not restore activity. The functional damage is not the result of impaired turnover or impaired targeting of the mutated protein. The three other positively charged amino acids and the only negatively charged one are not critical *(95)*. It is possible that the arginine-69 residue may be involved in chloride binding.

The transporters of biogenic amines contain an additional negatively charged residue in helix 1: aspartate-79 (dopamine-transporter numbering). Replacement of aspartate-79 in the dopamine transporter (DAT) with alanine, glycine, or glutamate significantly reduced the transport of dopamine, MPP^+ (Parkinsonism-inducing neurotoxin), and the binding of CFT (cocaine analogue) without affecting B_{max}. Apparently, aspartate-79 in helix 1 interacts with dopamine's amine during the transport process. Serine-356 and serine-359 in helix 7 may also be involved in dopamine binding and translocation, perhaps by interacting with the hydroxyls on the catechol *(96)*.

Studies of other proteins indicate that in addition to charged amino acids, aromatic amino acids containing π-electrons are also involved in maintaining the structure and function of these proteins *(97)*. Therefore, tryptophan residues in the membrane domain of the GABA transporter were mutated into serine as well as leucine *(98)*. Mutations at the 68 and 222 position (in helix 1 and helix 4, respectively; Fig. 2) led to a decrease of over 90% of the GABA uptake. The replacement of tryptophan-68 by leucine results in an increased affinity of the transporter for sodium *(26)*. This strongly suggests the involvement of this residue in sodium binding. A glutamate residue, critical for GAT-1 function and possibly in sodium binding, has been identified *(99)*. TMD I of GAT-1 also contains a residue, glycine-63, which is completely conserved in all members of the superfamily, except for the biogenic amine transporters DAT, NET, and SERT. This finding led to the idea that the carboxyl group of the aspartate, which replaces glycine in these three transporters, is interacting with the amino group of the biogenic amine substrates *(100)*. Interestingly, substitution of this aspartate to glutamate in the serotonin transporter SERT, which extends the acidic acid side chain by one carbon, selectively increased the potency of two analogues with shorter alkylamine side-chains. The mutation also perturbed however other parameters such as ion dependence, as well as substrate-induced currents. This indicates again that TMD 1 plays a critical role in defining the permeation pathway of the transporters *(100)*. Nevertheless, the amino group is maintained in almost all substrates of the transporter family, suggesting that other residues may be involved in the recognition of this amino-group. A candidate residue is tyrosine-140 of GAT-1. It is completely conserved throughout the superfamily, and even substitution by the other aromatic amino acids, phenylalanine (Y140F) and tryptophan (Y140W), results in completely inactive transporters *(101)*. Electrophysiological characterization reveals that both mutant transporters exhibit the sodium-dependent transient currents associated with sodium binding, as well as the chloride-dependent lithium leak currents characteristic of GAT-1. However, in both mutants GABA cannot to induce a steady-state transport current or block their transient currents. The nontransportable analog SKF 100330A potently inhibits the sodium-dependent transient in the wild-type GAT-1, but not in the Y140W transporter. It also partly blocks the transient of Y140F. Thus, although sodium and chloride binding are unimpaired in the tyrosine mutants, they have a specific defect in the binding of GABA *(101)*. The total conservation of the residue throughout the family suggests that tyrosine 140 may be involved in the ligand binding of the amino group, the moiety common to all of the neurotransmitters. Independently it was shown in the related SERT that tyrosine-176, which occupies the same position as tyrosine-140 on GAT-1, as well as isoleucine-172, which would be one turn of the α-helix away, are in proximity to the binding site of serotonin and cocaine *(102)*.

7. STRUCTURE–FUNCTION RELATIONSHIPS
FOR THE GLUTAMATE TRANSPORTERS

Substrate-induced conformational changes in the GLT-1 transporter have been detected, as revealed by the altered accessibility of trypsin-sensitive sites to the protease *(39)*. These experiments indicate that lithium can occupy one of the sodium-binding sites, and also that there are at least two transporter-glutamate-bound states *(39)*.

Two adjacent amino-acid residues of GLT-1 located in TMD 7, tyrosine-403 and glutamate-404, appear to be involved in potassium binding and are close to one of the sodium-binding sites *(38,103; see* Fig. 3). Because of the sequential nature of the transport process *(14,37,38)*, mutations in these residues cause the transporter to be locked in an obligatory exchange mode *(38,103)*. Moreover, tyrosine-403 behaves as if it is alternately accessible to either side of the membrane *(104)*. Analysis of GLT-1 mutants where serine-440, located in one of the reentrant loops, has been modified indicates that at least part of this loop is crucial for the coupling of sodium and glutamate fluxes, and that it is close to the glutamate binding site *(105)*.

We have very recently identified a residue which controls the binding of the γ-carboxyl group of glutamate *(106)*. Conradt and Stoffel *(107)* had already noted that arginine-479 of GLAST-1 located in TMD 8 *(83,85)*, is conserved in all dicarboxylic acid transporters of the family, but not in the small neutral amino-acid transporter ASCT1 *(76,77)*, whose substrates have only a single carboxyl group. They found that mutation of arginine-479 to threonine, which occupies the same position in ASCT-1, abolished glutamate uptake, but they noted that there may be other reasons for such a defect *(107)*. Nevertheless the idea of a role of the arginine in glutamate binding remained viable, especially after the subsequent cloning and sequencing of the isotransporter ASCT-2 revealed that it contains a cysteine residue at the equivalent position *(78,79)*. We reasoned that mutation of this arginine in a glutamate transporter which also transports a non-dicarboxylic acid substrate, might leave the transport of this substrate intact. It was shown that EAAT-3—the human homolog of EAAC-1 *(80)*—also exhibits considerable transport of cysteine *(108)*. We have found that mutation of the equivalent arginine-447 of EAAC-1 to neutral or negative amino-acid residues completely abolishes transport of L-glutamate and D- and L-aspartate, without impairing cysteine transport. Surprisingly, this cysteine transport is electroneutral rather than electrogenic. This appears to be caused by a defective interaction with potassium. We propose that arginine-447, by sequentially participating in the binding of glutamate and potassium, enables the coupling of their fluxes *(106)*.

ACKNOWLEDGMENTS

I wish to thank Mrs. Beryl Levene for expert secretarial assistance. The work from the author's laboratory was supported by the Bernard Katz Minerva Center for Cell Biophysics, and by grants from the US–Israel Binational Science Foundation, the Basic Research Foundation administered by the Israel Academy of Sciences and Humanities, the NINDS, National Institutes of Health and the Federal Ministry of Education, Science, Research, and Technology of Germany, and its International Bureau at the Deutsches Zentrum für Luft und Raumfahrt.

REFERENCES

1. Kanner, B. I. (1983) Bioenergetics of neurotransmitter transport. *Biochim. Biophys. Acta* **726,** 293–316.
2. Kanner, B. I. and Schuldiner, S. (1987) Mechanism of transport and storage of neurotransmitters. *CRC Crit. Rev. Biochem.* **22,** 1–39.
3. Kanner, B. I. (1989) Ion-coupled neurotransmitter transport. *Curr. Opin. Cell Biol.* **1,** 735–738.

4. Johnston, G. A. R. (1981) Glutamate uptake and its possible role in neurotransmitter inactivation. in: *Glutamate: Transmitter in the Central Nervous System* (Roberts, P. J., Storm-Mathisen, J., and Johnston, G. A. R., eds.), John Wiley and Sons, Chichester/New York/Brisbane/Toronto, pp. 77–87.

5. McBean, G. J. and Roberts, P. J. (1985) Neurotoxicity of glutamate and DL-*threo*-hydroxy–aspartate in the rat striatum. *J. Neurochem.* **44**, 247–254.

6. Radian, R., Bendahan, A., and Kanner, B. I. (1986) Purification and identification of the functional sodium- and chloride-coupled gamma-aminobutyric acid transport glycoprotein from rat brain. *J. Biol. Chem.* **261**, 15,437–15,441.

7. Guastella, J., Nelson, N., Nelson, H., Czyzyk, L., Keynan, S., Miedel, M. C., Davidson, N., Lester, H., and Kanner, B. I. (1990) Cloning and expression of rat brain GABA transports. *Science* **249**, 1303–1306.

8. Pacholczyk, T., Blakely, R. D., and Amara, S. G. (1992) Expression cloning of a cocaine- and antidepressant-sensitive human noradrenaline transporter. *Nature* **350**, 350–353.

9. Uhl, G. R. (1992) Neurotransmitter transporters (plus): a promising new gene family, *Trends Neurosci.* **15**, 265–268.

10. Schloss, P., Mayser, W., and Betz, H. (1992) Neurotransmitter transporters. A novel family of integral plasma membrane proteins. *FEBS Lett.* **307**, 76–78.

11. Amara, S. G. and Kuhar, M. J. (1993) Neurotransmitter transporters: recent progress. *Ann. Rev. Neurosci.* **16**, 73–93.

12. Keynan, S. and Kanner, B. I. (1988) Gamma-aminobutyric acid transport in reconstituted preparations from rat brain: coupled sodium and chloride fluxes, *Biochemistry* **27**, 12–17.

13. Kanner, B. I. and Sharon, I. (1978) Active transport of L-glutamate by membrane vesicles isolated from rat brain. *Biochemistry* **17**, 3949–3953.

14. Kanner, B. I. and Bendahan, A. (1982) Binding order of substrates to the sodium and potassium ion coupled L-glutamate transporter from rat brain. *Biochemistry* **21**, 6327–6330.

15. Storck, T., Schulte, S., Hofmann, K., and Stoffel, W. (1992) Structure, expression and functional analysis of a Na^+-dependent glutamate/aspartate transporter from rat brain. *Proc. Natl. Acad. Sci. USA* **89**, 10955–10959.

16. Pines, G., Danbolt, N. C., Bjoras, M., Zhang, Y., Bendahan, A., Eide, L., et al. (1992) Cloning and expression of a rat brain L-glutamate transporter. *Nature* **360**, 464–467.

17. Kanai, Y. and Hediger, M. A. (1992) Primary structure and functional characterization of a high affinity glutamate transporter. *Nature* **360**, 467–471.

18. Fairman, W. A., Vandenberg, R. J., Arriza, J. L., Kavanaugh, M. P., and Amara, S. G. (1995) An excitatory amino-acid transporter with properties of a ligand-gated chloride channel. *Nature* **375**, 599–603.

19. Arriza, J. L., Eliasof, S., Kavanaugh, M. P., and Amara, S. G. (1997) Excitatory amino acid transporter 5, a retinal glutamate transporter coupled to a chloride conductance. *Proc. Natl. Acad. Sci USA* **94**, 4155–4160.

20. Sims, K. D. and Robinson, M. B. (1999) Expression patterns and regulation of glutamate transporters in the developing and adult nervous system. *Crit. Rev. in Neurobiol.* **13**, 169–197.

21. Kavanaugh, M. P., Arriza, J. L., North, R. A., and Amara, S. G. (1992) Electrogenic uptake of γ-aminobutyric acid by a cloned transporter expressed in oocytes. *J. Biol. Chem.* **267**, 22,007–22,009.

22. Mager, S. J., Naeve, J., Quick, M., Guastella, J., Davidson, N., and Lester, H. A. (1993) Steady states, charge movements and rates for a cloned GABA transporter expressed in *Xenopus* oocytes. *Neuron* **10**, 177–188.

23. Lu, C. C. and Hilgemann, D. W. (1999) GAT-1 (GABA:Na$^+$:Cl$^-$) cotransport function. Steady state studies in giant *Xenopous* oocyte membrane patches. *J. Gen. Physiol.* **114**, 429–444.

24. Lu, C. C. and Hilgemann, D. W. (1999) GAT-1 (GABA: Na$^+$:Cl$^-$) cotransport function. Kinetic studies in giant *Xenopus* oocyte membrane patches. *J. Gen. Physiol.* **114**, 445–457.

25. Hilgemann, D. W. and Lu, C. C. (1999) GAT-1 (GABA: Na$^+$:Cl$^-$) cotransport function. Database reconstruction with an alternating access model. *J. Gen. Physiol.* **114**, 459–475.

26. Mager, S., Kleinberger-Doron, N., Keshet, G. I., Dvidson, N., Kanner, B. I., and Lester, H. A. (1996) Ion binding and permeation at the GABA transporter GAT-1. *J. Neurosci.* **16**, 5405–5414.

27. Mager, S., Min, C., Henry, D. J., Chavkin, L., Hoffman, B. J., Davidson, N., and Lester, H. A. (1994) Conducting states of a mammalian serotonin transporter. *Neuron* **12**, 845–859.

28. Brew, H. and Atwell, D. (1987) Electrogenic glutamate uptake is a major current carrier in the membrane of axolotl retinal glial cells. *Nature* **327**, 707–709.

29. Nicholls, D. and Attwell, D. (1990) The release and uptake of excitatory amino acids. *Trends. Pharmacol. Sci.* **11**, 462–468.

30. Rothstein, J. D., Dykes Hoberg, M., Pardo, C. A., Bristol, L. A., Jin, L., Kuncl, R. W., et al. (1996) Knockout of glutamate transporters reveals a major role for astroglial transport in excitotoxicity and clearance of glutamate. *Neuron* **16**, 675–686.

31. Tanaka, K., Watase, K., Manabe, T., Yamada, K., Watanabe, M., Takahashi, K., et al. (1997) Epilepsy and exacerbation of brain injury in mice lacking the glutamate transporter GLT-1. *Science* **276**, 1699–1702.

32. Mennerick, S. and Zorumski, C. F. (1994) Glial contributions to excitatory neurotransmission in cultured *hippocampel* cells. *Nature* **368**, 59–62.

33 Otis, T. S., Wu, Y. C., and Trussell, L. O. (1996) Delayed clearance of transmitter and the role of glutamate transporters at synapses with multiple release sites. *J. Neurosci* **16**, 1634–1644.

34. Diamond, J. S. and Jahr, C. E. (1997) Transporters buffer synaptically released glutamate in a submillisecond time scale. *J. Neurosci* **17**, 4672–4687.

35. Wadiche, J. I., Arriza, J. L., Amara, S. G., and Kavanaugh, M. P. (1995) Kinetics of a human glutamate transporter. *Neuron* **14**, 1019–1027.

36. Zerangue, N. and Kavanaugh, M. P. (1996) Flux coupling in a neuronal glutamate transporter. *Nature* **383**, 634–637.

37. Pines, G. and Kanner, B. I. (1990) Counterflow of L-glutamate in plasma membrane vesicles and reconstituted preparations from rat brain. *Biochemistry* **29**, 11,209–11,214.

38. Kavanaugh, M. P., Bendahan, A., Zerangue, N., Zhang, Y., and Kanner, B. I. (1997) Mutation of an amino acid residue influencing potassium coupling in the glutamate transporter GLT-1 induces obligate exchange. *J. Biol. Chem.* **272,** 1703–1708.

39. Grunewald, M. and Kanner, B. (1995) Conformational changes monitored on the glutamate transporter GLT-1 indicate the existence of two neurotransmitter-bound sites. *J. Biol Chem.* **270,** 17,017–17,024.

40. Danbolt, N. C., Pines, G., and Kanner, B. I. (1990) Purification and reconstitution of the sodium- and potassium-coupled glutamate transport glycoprotein from rat brain. *Biochemistry* **29,** 6734–6740.

41. Danbolt, N. C., Storm-Mathisen, J., and Kanner, B. I. (1992) An (Na$^+$ + K$^+$) coupled L–transporter purified from rat brain is located in glial cell processes. *Neuroscience* **51,** 295–310.

42. Lopez-Corcuera, B., Vazquez, J., and Aragon, C. (1991) Purification of the sodium- and chloride-coupled glycine transporter from central nervous system. *J. Biol. Chem.* **266,** 24,809–24,814.

43. Liu, Q. R., Lopez-Corcuera, B., Mandiyan, S., Nelson, H., and Nelson, N. (1993) A rat brain cDNA encoding the neurotransmitter transporter with an unusual structure. *J. Biol. Chem.* **268,** 22,802–22,808.

44. Launay, J. M., Geoffroy, C., Mutel, V., Buckle, M., Cesura, A., Alouf, J. E., and Da-Prada, M. (1992) One-step purification of the serotonin transporter located at the human platelet plasma membrane. *J. Biol. Chem.* **267,** 11,344–11,351.

45. Graham, D., Esnaud, H., and Langer, S. Z. (1992) Partial purification and characterization of the sodium-ion-coupled 5–hydroxytryptamine transporter of rat cerebral cortex. *Biochem. J.* **286,** 801–805.

46. Shimada, S., Kitayama, S., Lin, C. L., Patel, A., Nanthakumar, E., Gregor, P., et al. (1991) Cloning and expression of a cocaine-sensitive dopamine transporter complementary DNA. *Science* **254,** 576–578.

47. Kilty, J. E., Lorang, D., and Amara, S. G. (1991) Cloning and expression of a cocaine-sensitive rat dopamine transporter. *Science* **254,** 578–579.

48. Usdin, T. B., Mezey, E., Chen, C., Brownstein, M. J., and Hoffman, B. J. (1991) Cloning of the cocaine-sensitive bovine dopamine transporter, *Proc. Natl. Acad. Sci. USA* **88,** 11,168–11,171.

49. Hoffman, B. J., Mezey, E., and Brownstein, M. J. (1991) Cloning of a serotonin transporter affected by antidepressants. *Science* **254,** 579,580.

50. Blakely, R. D., Benson, H. E., Fremeau, R. T. Jr., Caron, M. G., Peek, M. M., Prince, H. K., and Bradley, C. C. (1991) Cloning and expression of a functional serotonin transporter from rat brain. *Nature* **353,** 66–70.

51. Clark, J. A., Deutch, A. Y., Gallipoli, P. Z., and Amara, S. G. (1992) Functional expression and CNS distribution of a b–alanine-sensitive neuronal GABA transporter. *Neuron* **9,** 337–348.

52. Borden, L. A., Smith, K. E., Hartig, P. R., Branchek, T. A., and Weinshank, R. L. (1992) Molecular heterogeneity of the GABA transport system. *J. Biol. Chem.* **267,** 21,098–21,104.

53. Lopez-Corcuera, B., Liu, Q. R., Mandiyan, S., Nelson, H., and Nelson, N. (1992) Expression of a mouse brain cDNA encoding novel γ-amino-butyric acid transporter. *J. Biol. Chem.* **267,** 17,491–17,493.

54. Liu, Q. R., Lopez-Corcuera, B., Mandiyan, S., Nelson, H., and Nelson, N. (1993) Molecular characterization of four pharmacology distinct γ-aminobutyric acid transporters in mouse brain. *J. Biol. Chem.* **268,** 2104–2112.

55. Smith, K. E., Borden, L. A., Hartig, P. A., Branchek, T., and Weinshank, R. L. (1992) Cloning and expression of a glycine transporter reveal colocalization with NMDA receptors, *Neuron* **8,** 927–935.

56. Liu, Q. R., Nelson, H., Mandiyan, S., Lopez-Corcuera, B., and Nelson, N. (1992) Cloning and expression of a glycine transporter from mouse brain. *FEBS Lett.* **305,** 110–114.

57. Guastella, J., Brecha, N., Weigmann, C., and Lester, H. A. (1992) Cloning, expression and localization of a rat brain high affinity glycine transporter. *Proc. Natl. Acad. Sci. USA* **89,** 7189–7193.

58. Fremeau, R. T. Jr., Caron, M. G., and Blakely, R. D. (1992) Molecular cloning and expression of a high affinity *l*-proline transporter expressed in putative glutamatergic pathways of rat brain. *Neuron* **8,** 915–926.

59. Uchida, S., Kwon, H. M., Yamauchi, A., Preston, A. S., Marumo, F., and Handler, J. S. (1992) Molecular cloning of the cDNA for an MDCK cell Na^+- and Cl^--dependent taurine transporter that is regulated by hypertonicity. *Proc. Natl. Acad. Sci. USA* **89,** 8230–8234.

60. Liu, Q. R., Lopez-Corcuera, B., Nelson, H., Mandiyan, S., and Nelson, N. (1992) Cloning and expression of a cDNA encoding the transporter of taurine and b-alanine in mouse brain. *Proc. Natl. Acad. Sci. USA* **89,** 12,145–12,149.

61. Yamauchi, A., Uchida, S., Kwon, H. M., Preston, A. S., Robey, R. B., Garcia-Perez, A., Burg, M. B., and Handler, J. S. (1992) Cloning of a Na^+- and Cl^--dependent betaine transporter that is regulated by hypertonicity. *J. Biol. Chem.* **267,** 649–652.

62. Mayser, W., Schloss, P., and Betz, H. (1992) Primary structure and functional expression of a choline transporter expressed in the rat nervous system. *FEBS Lett.* **305,** 31–36.

63. Guimbal, C. and Kilimann, M. W. (1993) A Na^+-dependent creatine transporter in rabbit brain, muscle, heart and kidney. cDNA cloning and functional expression. *J. Biol. Chem.* **268,** 8418–8421.

64. Sloan, J. L. and Mager, S. (1999) Cloning and functional expression of a human Na^+ and Cl^- dependent neutral and cationic amino acid transporter B^{0+}. *J. Biol. Chem.* **274,** 23,740–23,745.

65. Castagna, M., Shayakul, C., Trotti, D., Sacchi, V. F., Harvey, W. R., and Hediger, M. A. (1998) Cloning and characterisation of a potassium-coupled amino acid transporter. *Proc. Natl. Acad. Sci. USA* **95,** 5395–5400.

66. Feldman, D. H., Harvey, W. R., and Stevens, B. R. (2000) A novel electrogenic amino acid transporter is activated by K^+ or Na^+, is alkaline dependent and is Cl^--independent. *J. Biol. Chem.* **275,** 24,518–24,526.

67. Bennett, E. R. and Kanner, B. I. (1997) The membrane topology of GAT-1, a (Na^+ + Cl^-)-coupled g-amino butyric acid transporter from rat brain. *J. Biol. Chem.* **272,** 1203–1210.

68. Olivares, L., Aragon, C., Gimenez, C., and Zafra, F. (1997) Analysis of the transmembrane topology of the glycine transporter GlyT1. *J. Biol. Chem.* **272,** 1211–1217.

69. Clark, J. A. (1997) Analysis of the transmembrane topology and membrane assembly of the GAT-1 *gamma*-aminobutyric acid transporter. *J. Biol. Chem.* **272,** 14,695–14,704.
70. Yu, N., Cao, Y., Mager, S., and Lester, H. A. (1998) *FEBS Lett.* Topological organization of cysteine 74 in the GABA transporter, GAT-1, and its importance in ion-binding and permeation. **426,** 174–178.
71. Chen, J. G., Liu-Chen, S., and Rudnick, G. (1998) Determination of external loop topology in the serotonin transporter by site-directed chemical labeling. *J. Biol. Chem.* **273,** 12,675–12,681.
72. Lehre, K. P., Levy, L. M., Ottersen, O. P., Storm-Mathisen, J., and Danbolt, N. C. (1995) Differential expression of two glial glutamate transporters in the rat brain: quantitative and immunocytochemical observations. *J. Neurosci.* **15,** 1835–1853.
73. Rothstein, J. D., Martin, L., Levey, A. I., Dykes–Hoberg, M., Jun, L., Wu, D., et al. (1994) Localization of neuronal and glial glutamate transporters. *Neuron* **13,** 713–725.
74. Tolner, B., Poolman, B., Wallace, B., and Konings, W. N. (1992) Revised nucleotide sequence of the gltP gene, which encodes the proton-glutamate-aspartate transport protein of *Escherichia coli* K–12. *J. Bacteriol.* **174,** 2391–2393.
75. Jiang, J., Gu, B., Albright, L. M., and Nixon, B. T. (1989) Conservation between coding and regulatory elements of *Rhizobium meliloti* and *Rhizobium leguminosarum* dct genes, *J. Bacteriol.* **171,** 5244–5253.
76. Shafqat, S., Tamarappoo, B. K., Kilberg, M. S., Puranam, R. S., McNamara, J. O., GuadaÒo-Ferraz, A., and Fremeau, R. T. (1993) Cloning and expression of a novel Na+–dependent neutral amino acid transporter structurally related to mammalian Na/glutamate cotransporters. *J. Biol. Chem.* **268,** 15,351–15,355.
77. Arriza, J. L., Kavanaugh, M. P., Fairman, W. A., Wu, Y.-N., Murdoch, G. H., North, R. A., and Amara, S. G. (1993) Cloning and expression of a human neutral amino acid transporter with structural similarity to the glutamate transporter gene family. *J. Biol. Chem.* **268,** 15,329–15,332.
78. Utsunomiya-Tate, N., Endo, H., and Kanai, Y. (1996) Cloning and functional characterization of a system ASC-like Na+ dependent neutral amino acid transporter. *J. Biol. Chem.* **271,** 14883–14890.
79. Kekuda, R., Prasad, R. D., Fei, Y. -J., Torres-Zamorano, V., Sinha, S., Yang-Feng, T. L., Leibach, F. H., and Ganapathy, V. (1996) Cloning of the sodium-dependent, broad scope, neutral amino acid transporter B^0 from a human placental carcinoma cell line. *J. Biol. Chem.* **271,** 18,657–18,661.
80. Arriza, J. L., Fairman, W. A., Wadiche, J. I., Murdoch, G. H., Kavanaugh, M. P., and Amara, S. G. (1994) Functional comparisons of three glutamate transporter subtypes cloned from human motor cortex. *J. Neurosci.* **14,** 5559–5569.
81. Wadiche, J. I., Amara, S. G., and Kavanaugh, M. P. (1995) Ion fluxes associated with excitatory amino acid transport. *Neuron* **15,** 721–728.
82. Zerangue, N. and Kavanaugh, M. P. (1996) ASCT-1 is a neutral amino acid exchanger with chloride channel activity. *J. Biol. Chem.* **271,** 27,991–27,994.

83. Grunewald, M., Bendahan, A., and Kanner, B. I. (1998) Biotinylation of single cysteine mutants of the glutamate transporter GLT-1 from rat brain reveals its unusual topology. *Neuron* **21,** 623–632.

84. Slotboom, D. J., Sobczak, I., Konings, W. N., and Lolkema, J. S. (1999) A conserved serine-rich stretch in the glutamate transporter family forms a substrate-sensitive reentrant loop. *Proc. Natl. Acad. Sci. USA* **96,** 14,282–14,287.

85. Grunewald, M. and Kanner, B. I. (2000) The accessibility of a novel reentrant loop of the glutamate transporter GLT-1 is restricted by its substrate. *J. Biol. Chem.* **275,** 9684–9689.

86. Seal, R. P., Leighton, B. H., and Amara, S. G. (2000) A model for the topology of excitatory amino acid transporters determined by the extracellular accessibility of substituted cysteines. *Neuron* **25,** 695–706.

87. Seal, R. P. and Amara, S. G. (1998) A reentrnat lop domain in the glutamate carrier EAAT-1 participates in subsrate binding and translocation. *Neuron* **21,** 1487–1498.

88. Mabjeesh, N. J. and Kanner, B. I. (1992) Neither amino nor carboxyl termini are required for function of the sodium- and chloride-coupled gamma-aminobutyric acid transporter from rat brain. *J. Biol. Chem.* **267,** 2563–2568.

89. Bendahan, A. and Kanner, B. I. (1993) Identification of domains of a cloned rat brain GABA transporter which are not required for its functional expression. *FEBS Lett.* **318,** 41–44.

90. Bennett, E. R., Su, H., and Kanner, B. I. (2000) Mutation of arginine-44 of the (Na$^+$ + Cl$^-$)-coupled GABA transporter GAT-1 impairs net flux, but not exchange. *J. Biol. Chem.,* **275,** 34,106–34,113.

91. Mabjeesh, N. J. and Kanner, B. I. (1993) The substrates of a sodium- and chloride-coupled g-aminobutyric acid transporter protect multiple sites throughout the protein against proteolytic cleavage. *Biochemistry* **32,** 8540–8546.

92. Kanner, B. I., Bendahan, A., Pantanowitz, S., and Su, H. (1994) The number of amino acid residues in hydrophillic loops connecting transmembrane domains of the GABA transporter GAT-1 is critical for its function. *FEBS Lett.* **356,** 192–194.

93. Tamura, S., Nelson, H., Tamura, A., and Nelson, N. (1995) Short external loops as potential substrate binding site of γ-aminobutyric acid transporters. *J. Biol. Chem.* **270,** 28,712–28,715.

94. Golevanevsky, V. and Kanner, B. I. (1999) The reactivity of the γ-aminobutyric acid transporter GAT-1 toward sulfhydryl reagents is conformationally sensitive. *J. Biol. Chem.* **274,** 23,020–23,026.

95. Pantanowitz, S., Bendahan, A., and Kanner, B. I. (1993) Only one of the charged amino acids located in the transmembrane a-helices of the γ-aminobutyric acid transporter (subtype A) is essential for its activity. *J. Biol. Chem.* **268,** 3222–3225.

96. Kitayama, S., Shimada, S., Xu, H., Markham, L., Donovan, D. M., and Uhl, G. R. (1992) Dopamine transporter site-directed mutations differentially alter substrate transport and cocaine binding. *Proc. Natl. Acad. Sci. USA* **89,** 7782–7785.

97. Sussman, J. L. and Silman, I. (1992) Acetylcholinesterase: structure and use as a model for specific cation-protein interactions. *Curr. Opin. Struc. Biol.* **2,** 721–729.

98. Kleinberger-Doron, N. and Kanner, B. I. (1994) Identification of tryptophan residues critical for the function and targeting of the γ-aminobutyric acid transporter (subtype A). *J. Biol. Chem.* **269,** 3063–3067.

99. Keshet, G. I., Bendahan, A., Su, H., Mager, S., Lester, H. A., and Kanner, B. I. (1995) Glutamate 101 is critical for the function of the sodium and chloride-coupled GABA transporter GAT-1. *FEBS Lett.* **371,** 39–42.

100. Barker, E. L., Moore, K. R., Rakhshan, F., and Blakely, R. D. (1999) Transmembrane domain 1 contributes to the permeation pathway for serotonin and ions in the serotonin transporter. *J. Neurosci.* **19,** 4705–4717.

101. Bismuth, Y., Kavanaugh, M. P., and Kanner, B. I. (1997) Tyrosine-140 of the γ-aminobutyric acid transporter GAT-1 plays a critical role in neurotransmitter recognition. *J. Biol. Chem.* **272,** 16,096–16,102.

102. Chen, J. G., Sachpatzides, A., and Rudnick, G. (1997) The third transmembrane domain of the serotonin transporter contains residues associated with substrate and cocaine binding. *J. Biol. Chem.* **272,** 28,231–28,237.

103. Zhang, Y., Bendahan, A., Zarbiv, R., Kavanaugh, M. P., and Kanner, B. I. (1998) Molecular determinant of ion selectivity of a $(Na^+ + K^+)$-coupled rat brain glutamate transporter. *Proc. Natl. Acad. Sci. USA* **95,** 751–755.

104. Zarbiv, R., Grunewald, M., Kavanaugh, M. P., and Kanner, B. I. (1998) Cysteine scanning of the surroundings of an alkali-ion binding site of the glutamate transporter GLT-1 reveals a conformationally sensitive residue. *J. Biol. Chem.* **273,** 14,231–14,237.

105. Zhang, Y. and Kanner, B. I. (1999) Two serine residues of the glutamate transporter GLT-1 are crucial for coupling the fluxes of sodium and the neurotransmitter. *Proc. Natl. Acad. Sci. USA* **96,** 1710–1715.

106. Bendahan, A., Armon, A., Madani, N., Kavanaugh, M. P., and Kanner, B. I. (2000) Arginine-447 plays a pivotal role in subsrate interactions in a neuronal glutamate transporter. *J. Biol. Chem.,* **275,** 37,436–37,442.

107. Conradt, M. and Stoffel, W. (1995) Functional analysis of the high affinity, Na^+-dependent glutamate transporter GLAST-1 by site-directed mutagenesis. *J. Biol. Chem.* **270,** 25,207–25,212.

108. Zerangue, N. and Kavanaugh, M. P. (1996) Interaction of L-cysteine with a human excitatory amino acid transporter. *J. Physiol.* **493,** 419–423.

8

The High-Affinity Glutamate and Neutral Amino-Acid Transporter Family

Structure, Function, and Physiological Relevance

Yoshikatsu Kanai, Davide Trotti, Urs V. Berger, and Matthias A. Hediger

1. INTRODUCTION

The classical studies in transport physiology revealed that neurons, glial cells, and epithelial cells possess unique Na^+- and K^+-dependent glutamate transporters with either a high affinity (Km =1–50 μM) or a low affinity (Km >100 μM) for glutamate *(1–8)*. The high-affinity transporters have been identified through various approaches. They belong to a single gene superfamily, called SLC1, which also includes genes encoding the neutral amino-acid transporters (Table 1). The genes encoding low-affinity glutamate transporters *(7,8)* have not yet been identified. The high-affinity transporters are distinct from the recently identified vesicular glutamate transporters which belong to the SLC17 family of transporters, originally believed to be type I Na^+/phosphate transporters *(9–11)*. Briefly, the vesicular glutamate transporters mediate glutamate uptake into synaptic vesicles, driven by the vesicular-membrane potential, which is maintained by the vacuolar H^+-ATPase. There are two vesicular glutamate transporter isoforms, called VGluT1 and VGluT2 (*see* Chapter 9). Both of these transporters are exclusively expressed in glutamatergic synapses, with complementary tissue distribution within the central nervous system (CNS). The main purpose of this chapter is to review recent insights into the structure, in vivo functional roles, and pathological implications of the mammalian high-affinity glutamate transporters (SLC1 family). The chapter also explores recent progress in characterizing related neutral amino-acid transporters.

From: *Contemporary Neuroscience:*
Neurotransmitter Transporters: Structure, Function, and Regulation, 2nd Edition
Edited by: M. E. A. Reith © Humana Press Inc., Totowa, NJ

Table 1
The SLC1 Superfamily of Glutamate and Neutral Amino-Acid Transporters

Gene nomenclature		Gene product	Predominant substrates	Transport type[a]/ coupling ions	Gene locus	Tissue or cellular distribution	Link to disease[b]	Refs.
SLC1	A1	EAAC1/EAAT3	L-Glu, D,L-Asp	C/Na+, H+, and K+	9p24	Brain (neurons), intestine, kidney, liver, heart	Dicarboxylic amino-aciduria?[G]	(1–5)
	A2	GLT-1/EAAT2	L-Glu, D,L-Asp	C/Na+, H+, and K+	11p13-p12	Brain (astrocytes), liver	Amyotrophic lateral sclerosis[A]	(6–10)
	A3	GLAST1/EAAT1	L-Glu, D,L-Asp	C/Na+, H+, and K+	5p13 5p11-p12	Brain (astrocytes), heart, Sk. muscle, placenta		(11–14)
	A4	ASCT1/SATT	L-Ala, L-Ser, L-Cys	C/Na+, E/amino acids	2p15-p13	Widespread		(15–18)
	A5	AAAT/ASCT2	L-Ala, L-Ser, L-Thr, L-Cys, and L-Gln	C/Na+, E/amino acids	19q13.3	Lung, sk. muscle, large intestine, kidney, testis, adipose tissue		(19–21)
	A6	EAAT4	L-Glu, D,L-Asp	C/Na+, H+, and K+	19	Cerebellum (Purkinje cells)		(22,23)
	A7	EAAT5	L-Glu, D,L-Asp	C/Na+, H+, and K+		Retina		(24)

Nomenclature based on the HUGO Human Gene Nomenclature Committee database (see http://www/gene.ucl.ac.uk/nomenclature/).

[a]C, Cotransporter; E, exchanger.

[b]A, Acquired defect; G, genetic defect.

1. Kanai, Y. and Hediger, M. A. (1992) Primary structure and functional characterization of a high-affinity glutamate transporter. *Nature* **360,** 467–471.
2. Kanai, Y., Smith, C. P., and Hediger, M. A. (1993) The elusive transporters with a high affinity for glutamate. *Trends Neurosci.* **16,** 365–730

molecular cloning, characterization and expression in human brain. *Brain Res.* **662**, 245–250.

4. Smith, C. P., Weremowicz, S., Kanai, Y., Stelzner, M., Morton, C. C., and Hediger, M. A. (1994) Assignment of the gene coding for the human high-affinity glutamate transporter EAAC1 to 9p24: potential role in dicarboxylic aminoaciduria and neurodegenerative disorders. *Genomics* **20**, 335,336.

5. Peghini, P., Janzen, J., and Stoffel, W. (1997) Glutamate transporter *EAAC-1*-deficient mice develop dicarboxylic aminoaciduria and behavioral abnormalities but no neurodegeneration. *EMBO J.* **16**, 3822–3832.

6. Pines, G., Danbolt, N. C., Bjoras, M., Zhang, Y., Bendahan, A., Eide, L., et al. (1992) Cloning and expression of a rat brain l-glutamate transporter. *Nature* **360**, 464–467.

7. Krishnan, S. N., Desai, T., Wyman, R. J., and Haddad, G. G. (1993) Cloning of a glutamate transporter from human brain. *Soc. Neurosci.* **19**, 219.

8. Li, X. and Francke, U. (1995) Assignment of the gene SLC1A2 coding for the human glutamate transporter EAAT2 to human chromosome 11 bands P13-P12. cytogenet. *Cell Genet.* **71(3)**, 212,213.

9. Lin, C. L., Bristol, L. A., Jin, L., Dykes-Hoberg, M., Crawford, T., Clawson, L., and Rothstein, J. D. (1998) Aberrant RNA processing in a neurodegenerative disease: the cause for absent EAAT2, a glutamate transporter, in amyotrophic lateral sclerosis. *Neuron* **20(3)**, 589–602.

10. Tanaka, K., Watase, K., Manabe, T., Yamada, K., Watanabe, M., Takahashi, K.,et al. (1997) Epilepsy and exacerbation of brain injury in mice lacking the glutamate transporter GLT-1. *Science* **276**, 1699–1702.

11. Storck, T., Schulte, S., Hofmann, K., and Stoffel, W. (1992) Structure, expression, and functional analysis of a Na(+)- dependent glutamate/aspartate transporter from rat brain. *Proc. Natl. Acad.Sci. USA* **89**, 10,955–10,959.

12. Arriza, J. L., Fairman, W. A., Wadiche, J. I., Murdoch, G. H., Kavanaugh, M. P., and Amara, S. G. (1994) Functional comparisons of three glutamate transporter subtypes cloned from human motor cortex. *J. Neurosci.* **14**, 5559–5569.

13. Stoffel, W., Sasse, J., Duker, M., Muller, R., Hofmann, K., Fink, T., and Lichter, P. (1996) Human high affinity, Na$^+$-dependent L-glutamate/L-aspartate transporter GLAST-1 (EAAT-1): gene structure and localization to chromosome 5p11-P12. *FEBS Lett.* **386**, 189–193.

14. Takai, S., Yamada, K., Kawakami, H., Tanaka, K., and Nakamura, S. (1995) Localization of the gene (SLC1A3) encoding human glutamate transporter (GluT-1) to 5p13 by fluorescence *in situ* hybridization. cytogenet. *Cell Genet.* **69**, 209,210.

15. Arriza, J. L., Kavanaugh, M. P., Fairman, W. A., Wu, Y. N., Murdoch, G. H., North, R. A., and Amara, S. G. (1993) Cloning and expression of a human neutral amino acid transporter with structural similarity to the glutamate transporter gene family. *J. Biol. Chem.* **268**, 15,329–15,332.

16. Shafqat, S., Tamarappoo, B. K., Kilberg, M. S., Puranam, R. S., McNamara, J. O., Guadano-Ferraz, A., and Fremeau, R. T., Jr. (1993) Cloning and expression of a novel Na(+)-dependent neutral amino acid transporter structurally related to mammalian Na$^+$/glutamate cotransporters. *J. Biol. Chem.* **268**, 15,351–15,355.

Table 1 (continued)

17. Hofmann, K., Duker, M., Fink, T., Lichter, P., and Stoffel, W. (1994) Human neutral amino acid transporter ASCT1: structure of the gene (SLC1A4) and localization to chromosome 2p13-P15. *Genomics* **24**, 20–26.

18. Zerangue, N. and Kavanaugh, M. P. (1996) ASCT-1 is a neutral amino acid exchanger with chloride channel activity. *J. Biol. Chem.* **271**, 27,991–27,994.

19. Liao, K. and Lane, M. D. (1995) Expression of a novel insulin-activated amino acid transporter gene during differentiation of 3T3-L1 preadipocytes into adipocytes. *Biochem. Biophys. Res. Commun.* **208**, 1008–1015.

20. Utsunomiya-Tate, N., Endou, H., and Kanai, Y. (1996) Cloning and functional characterization of a system ASC-like Na⁺-dependent neutral amino acid transporter. *J. Biol. Chem.* **271**, 14,883–14,890.

21. Jones, E. M., Menzel, S., Espinosa, R. 3rd, Le Beau, M. M., Bell, G. I., and Takeda, J. (1994) Localization of the gene encoding a neutral amino acid transporter-like protein to human chromosome band 19q13.3 and characterization of a simple sequence repeat DNA polymorphism. *Genomics* **23**, 490,491.

22. Fairman, W. A., Vandenberg, R. J., Arriza, J. L., Kavanaugh, M. P., and Amara, S. G. (1995) An excitatory amino-acid transporter with properties of a ligand-gated chloride channel. *Nature* **375**, 599–603.

23. Yamada, K., Watanabe, M., Shibata, T., Tanaka, K., Wada, K., and Inoue, Y. (1996) EAAT4 is a post-synaptic glutamate transporter at Purkinje cell synapses. *Neuroreporter* **7**, 2013–2017.

24. Arriza, J. L., Eliasof, S., Kavanaugh, M. P., and Amara, S. G. (1997) Excitatory amino acid transporter 5, a retinal glutamate transporter coupled to a chloride conductance. *Proc. Natl. Acad. Sci. USA* **94**, 4155–4160.

2. MOLECULAR PROPERTIES OF NA⁺ AND K⁺-DEPENDENT GLUTAMATE AND NEUTRAL AMINO-ACID TRANSPORTERS

The identification of the genes encoding high-affinity glutamate transporters began in 1992, when three isoforms (EAAC1, GLT1, and GLAST) were identified using different approaches. Expression cloning with *Xenopus* oocyutes was used in our laboratory to isolate a cDNA encoding the neuronal and epithelial high-affinity glutamate transporter, known as EAAC1 *(12)*. Kanner and colleagues *(13)* purified a 73-kDa glycoprotein (GLT1) from crude synaptosome fraction P_2, which was shown to exhibit high-affinity glutamate transport when reconstituted into liposome. An antibody was then raised against the purified protein and used to isolate a clone from a rat-brain cDNA library that encodes the glial glutamate transporter, known as GLT1 *(14)*. Stoffel and colleagues *(15)* copurified a 66-kDa hydrophobic glycoprotein, the glial glutamate transporter GLAST, during the isolation of UDPgalactose:ceramide galactosyltransferase. These three high-affinity glutamate transporters share 50–55% amino-acid sequence identity with each other.

Several additional members of this novel mammalian transporter family were subsequently identified. Using a cloning approach which was based on the PCR-amplification with degenerate oligonucleotides, Amara and colleagues isolated a cDNA from a human motor cortex which encodes the glutamate-transporter subtype EAAT4 *(16)*. EAAT4 exhibits 58%, 39%, and 51% amino-acid sequence identity to the human glutamate transporters GLAST, GLT1, and EAAC1, respectively. A fifth isoform of the glutamate-transporter family (EAAT5) was cloned from the human retina cDNA library with a probe obtained by polymerase chain reaction (PCR)-amplification of salamander retina cDNA, using degenerate oligonucleotide primers *(17)*. EAAT5 exhibits 51%, 44%, 48%, and 49% amino-acid sequence identity to the human glutamate transporters GLAST, GLT1, EAAC1, and EAAT4, respectively.

The HUGO gene names for the human high-affinity glutamate transporter are: SLC1A1 for EAAC1 (also known as EAAT3), SLC1A2 for GLT1 (also known as EAAT2), SLC1A3 for GLAST (also known as EAAT1), SLC1A4 for EAAT4, and SLC1A5 for EAAT5 (*see* http://www.gene.ucl.ac.uk/nomenclature/) (also *see* Table 1). SLC1 refers to "solute carrier family" number 1, and A1 to family-member number 1. There are currently more than 35 different human transporter families.

The neutral amino-acid transporter ASCT1 was isolated by two different laboratories *(18,19)*. ASCT1 has the properties of the previously described neutral amino-acid transport system ASC (Ala, Ser, Cys) *(20)*. A second neutral amino-acid transporter, ASCT2, was isolated from mouse testis. ASCT2 also has the properties characteristic of system ASC, yet its functional properties and tissue distribution are distinct from those of ASCT1 (for review, *see* ref. *21*). ASCT2 exhibits 79% identity to another member of this family, known as hATB^0, which was isolated from a human placental choriocarcinoma cell line *(22)*. Although hATB^0 is proposed to correspond to the Na$^+$-dependent amino-acid transport system B^0 *(23)*, it is reasonable to assume that hATB^0 is the ASCT2 ortholog in humans. The neutral amino-acid transporters ASCT1 and ASCT2 are 57% identical to each other, and show 40–44% sequence identity to the cloned glutamate transporters.

All these mammalian glutamate and ASC transporters also have significant homology to the H$^+$-coupled GLTP glutamate transporters of *Escherichia coli*, *B. stearothermophilus*, and *B. caldotenax*, and to the DCTA dicarboxylate transporter of *Rhizobium meliloti*, with sequence identities ranging from 27 to 32% (*see* ref. *24*). Interestingly, there is no significant homology with other Na$^+$-coupled transporters such as members of the Na$^+$- and Cl$^-$-dependent GABA/neurotransmitter-transporter family (*see* Chapters 3, 6, and 7) or members of the Na$^+$/glucose cotransporter family *(25,26)*.

All eukaryotic and prokaryotic high-affinity glutamate-transporter family members appear to have a similar topology in the membrane, with approximately 10 transmembrane domains (TMDs) (Fig. 1). The presence of a large hydrophobic stretch near the C-terminus (residues 356-438) (Fig. 1) is unique and suggests non-α-helical TMDs.

3. FUNCTIONAL AND ELECTROGENIC PROPERTIES OF THE SLC1 FAMILY MEMBERS

High-affinity glutamate transporters are coupled to the inwardly directed electrochemical potential gradients of Na$^+$ and H$^+$, and to the outwardly directed gradient of K$^+$. This unique coupling stoichiometry allows efficient removal of glutamate from extracellular fluids such as cerebrospinal fluid (CSF), the intestinal lumen, and the lumen of renal proximal tubules. Knowledge of the precise coupling stoichiometry is important, because it determines the concentrating capacity of these transporters and also has pathological implications. During ischemia—for example, after a stroke—extracellular K$^+$ rises, extracellular Na$^+$ decreases, and the membrane depolarizes. As a result, the electrochemical gradients for the coupling ions of these glutamate transporters fail to mop up extracellular glutamate from the synaptic cleft, resulting in the accumulation of extracellular glutamate to toxic levels.

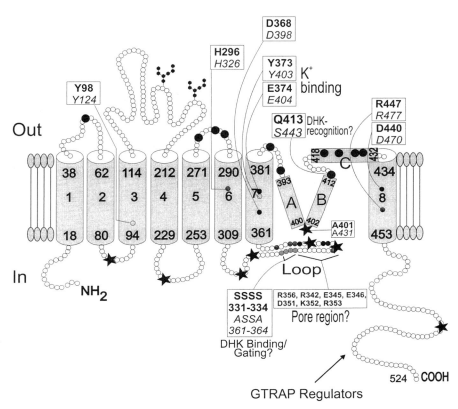

Fig. 1. Membrane topology model of glutamate transporters. The topology shown is based on experiments involving biotinylation of single-cysteine mutants of the glutamate transporter GLT1 from the rat brain *(61,203)*. Rat GLT1 numbering is shown in italics. The numbering in bold refers to the human EAAC1 glutamate transporter. Residues that were mutated to cysteine residues in GLT1 are indicated by their position number. The results of labeling with a biotinylation reagent are shown *(see* Chapter 7). Full circles refer to labeling with the biotinylation agent in the absence of permeabilization. Stars refer to labeling after permeabilization. The glycosylation sites N205 and N217 were shown to be used as native N-glycosylation sites. The TMDs which are long enough to span the membrane as α-helices are indicated by arabic numbers. The legs of the reentrant loop are indicated by **"A"** and **"B"**. **"C"** corresponds to a "loop" which is predicted to extend partially into the "translocation pore" between TMDs 7 and 8. The residues highlighted by boxes correspond to conserved regions and/or residues involved in the transport cycle.

3.1. Determination of the Coupling Stoichiometry of Glutamate Transporters

The coupling stoichiometry of high-affinity glutamate transporters was originally estimated using a variety of preparations, including a neuronal cell line from mouse cerebellum *(27)*, salamander retinal glial cells *(28–30)*, rabbit glial cells *(31)*, kidney brush-border-membrane vesicles *(2,32,33)*, proteoliposomes containing partially purified renal brush-border-membrane glutamate transporters *(34)*, and eel intestinal brush-border-membrane vesicles *(35)*. The Na$^+$, K$^+$, and glutamate stoichiometries were investigated based on Hill analyses of the ion-dependencies of glutamate uptake or the glutamate-evoked currents, and direct comparison of the ion and glutamate fluxes. Studies of glutamate transport in glial cells from salamander retina *(28–30)* and rabbit brain *(31)* suggested a Na$^+$-to-glutamate coupling ratio of 2:1. Glutamate uptake in salamander retinal glial cells was not chloride-dependent *(36)*.

To study whether K$^+$ is involved in glutamate transport, Attwell and colleagues *(28,30)* used whole-cell voltage-clamp analysis of salamander retinal glial cells *(36)*. By changing the intracellular K$^+$ concentration, these investigators showed that a strong dependence of glutamate evoked currents on intracellular K$^+$, indicating that glutamate uptake is coupled to the countertransport of K$^+$.

A similar approach was used to study the pH dependence of retinal glial glutamate uptake. The studies suggested that the transport of glutamate is coupled to the countertransport OH$^-$ ions rather than to the cotransport of H$^+$. This conclusion was based on the observation that anion efflux accounted for intracellular acidification. The data showed that the OH$^-$ anion on the intracellular membrane surface can be replaced by other small anions such as NO$_3^-$, HCO$_3^-$, and ClO$_4^-$ *(37,38)*, and that glutamate uptake resulted in the exit of these ions. Recent studies, however, indicated that high-affinity glutamate transporters—particularly GLAST and EAAT4—exhibit glutamate-gated anion conductances. Subsequently, it was concluded that the observed countertransport of NO$_3^-$, HCO$_3^-$, and ClO$_4^-$ in salamander retinal glia was related to this anion conductance.

The individual studies described here did not provide clear overall stoichiometries. Nevertheless, their combined information provided evidence that high-affinity glutamate transporters link uphill glutamate transport to the cotransport of two or three Na$^+$ ions *(29)*, the countertransport of one K$^+$ ion *(28,30)*, and the cotransport of one H$^+$ ion or the countertransport of one OH$^-$ *(38)*.

The coupling stoichiometry of the cloned glutamate transporter (EAAC1) was analyzed in our laboratory using the *Xenopus* oocyte expression system *(39)*. The stoichiometry was studied by comparing the charge flux, the H$^+$ flux, and the initial rates of the ^{22}Na$^+$ and ^{14}C– glutamate uptakes. Two electrode voltage-clamp analysis of glutamate-evoked currents gave a first-order dependence of the current on extracellular glutamate concentration, indicating that one glutamate molecule is translocated during each transport cycle. Hill analysis of the Na^+-dependence of the glutamate- evoked currents in oocytes yielded Hill coefficients for Na^+ that were strongly dependent on an extracellular glutamate concentration. The Hill coefficient was ~1.2 at 1 mM glutamate, 2 at 200 μM glutamate, and larger than 2 at glutamate below 40 μM. Since the Hill coefficient is generally considered to be an indicator of the coupling stoichiometry, this finding suggests that the Na^+-to-glutamate coupling ratio depends on extracellular glutamate concentration. However, flux measurements using $^{22}Na^+$ and ^{14}C-glutamate yielded a constant Na^+ to glutamate coupling ratio of 2:1, independent of extracellular glutamate concentration. Based on this finding, the initial proposal was that EAAC1-mediated glutamate uptake is coupled to the cotransport of two Na^+-ions. The unexpected result from Hill plot analysis was assumed to be the result of complex cooperativity between the Na^+ and glutamate-binding sites *(40)*.

To test whether transport mediated by EAAC1 is coupled to the cotransport of H^+ (or the countertransport of OH^-), the intracellular pH in oocytes was measured using pH-sensitive microelectrodes impaled into oocytes. The studies revealed that EAAC1-mediated transport is associated with an intracellular acidification. The H^+-to-charge coupling ratio was   estimated from the rate of pH decrease and depolarization, and gave approximately a 1:1 coupling ratio. The K^+-countertransport of glutamate transport was confirmed in oocytes expressing EAAC1 using capillary-zone electrophoresis to measure the concentration of intracellular cations *(40)*. Together with studies involving measurement of ^{22}Na and ^{14}C-glutamate flux studies, the initially proposed overall stoichiometry for EAAC1-mediated glutamate transport was one glutamate to (approximately) two Na^+ to 1H^+ (or OH^-) to one K^+.

Using voltage clamping with a pH-sensitive fluorescent dye to monitor electrical currents and pH changes associated with flux of glutamate mediated by the human neuronal glutamate transporter EAAC1, Zerangue and Kavanaugh demonstrated that, unlike L-glutamate and L-cysteate, transport of an equivalent amount of L-cysteine, a neutral amino-acid substrate of EAAC1 *(41)*, did not result in the marked intracellular acidification *(42)*. If OH^- is countertransported, it would be expected that the transport of L-cysteine, as well as L-glutamate and L-cysteate, acidifies the cells. Therefore, it was proposed that H^+ is cotransported with amino acids as a thiolate, sulphate, or carboxylate ion pair, leading to an intracellular pH change dependent on the pK of the amino acids *(42)*. After intracellular release, cysteine (pK = 8.3) remains predominantly protonated, whereas glutamic and cycteic acids (pK < 5) release the proton, because the intracellular pH is approx 7.3.

These observations support the concept that glutamate uptake is coupled to the cotransport with H^+ rather than the countertransport with OH^-. Alternatively, Attwell and colleagues recently proposed that translocation of an H^+ occurs within the K^+-transporting portion of the transporter cycle, when glutamate is not bound to the transporter *(43)*. Thus, although the H^+ "cotransport" model is now generally accepted, further studies are required to establish the exact role of protons during the glutamate transport cycle.

Zerangue and Kavanaugh re-examined the stoichiometry of human EAAC1 (which they named EAAT3). They determined the reversal potential as a function of the concentration of coupling ions. The studies revealed that three rather than two Na^+ ions are cotransported with each glutamate molecule (43). This finding implies that two positive charges instead of one are translocated with each glutamate molecule.

As indicated previously, direct comparison in our laboratory of ^{14}C-glutamate influx and accompanying $^{22}Na^+$ influx in *Xenopus* oocytes expressing rabbit EAAC1 suggested that two Na^+ ions are cotransported with each glutamate molecule *(40)*. Also, the charge-to-glutamate ratio in that study was 1:1. It is of interest to understand the discrepancy between our data and those of Zerangue et al., considering the importance of Na^+ coupling in driving glutamate uptake. Recent re-evaluation in our laboratory of the charge-to-glutamate ratio in EAAC1-injected oocytes indicated that the three Na^+ cotransport model must be correct (Xing-Zhen Chen and Matthias Hediger, unpublished data). Experiments were performed under voltage-clamp conditions (-50 mV) in the absence of chloride to avoid any contribution of the glutamate-evoked current by the glutamate-gated anion conductance. In these experiments, extracellular chloride was replaced by gluconate for 20 hr, and oocytes were dialyzed in this chloride-free solution to reduce intracellular chloride. Currents evoked by 100 μM glutamate were recorded between pH 5.5 and 9.0. A charge-to-glutamate ratio close to 2.0 was obtained at any given pH.

Taken together, these data have led to the current conclusion that EAAC1-mediated transport involves transport of 1 glutamate molecule, presumably paired with one H^+, cotransport with three Na^+ ions, and countertransport with one K^+ ion (*see* Fig. 3., bottom). Attwell and colleagues recently reported the identical stoichiometry for GLT1-expressed Chinese Hamster ovary (CHO) cells *(44)*. This stoichiometry is probably true for all other glutamate-transporter isoforms of the SLC1 family. Based on stoichiometry, it was calculated that glutamate transporters can concentrate glutamate 5×10^6-fold inside cells under physiological conditions *(42)*. Thus, assuming a concentration of 10 mM inside glutamatergic neurons, the extracellular concentration which can be achieved at equilibrium is ~2 nM.

3.2. Steady-State and Pre-Steady-State Kinetics and Transport Cycle

The detailed analysis of the steady-state and pre-steady-state currents displayed by glutamate transporters in response to step changes of the membrane potential has yielded important information on their dynamics (i.e., the kinetics of the transitions between the empty transporters and the transporter-substrate complexes) and on their structures (i.e., whether the substrate-binding sites are within or outside the membrane electric field) *(45–47)*. A hypothetical kinetic model of glutamate transporters is presented in Figure 2. Such approaches were particularly successful for the determination of the kinetic properties of individual transport steps and to estimate the turnover rates of the Na^+/glucose cotransporter and the GABA transporter *(45,46)*. In response to sudden changes in membrane potential, two components of electric currents were usually observed in oocytes expressing the transporters—the transient "pre-steady-state" currents and the "steady-state" currents. With respect to glutamate transporters such as EAAC1, the accurate determination of pre-steady-state currents has been difficult because they are much smaller than those of the Na^+/glucose cotransporter and the GABA transporter, and useful inhibitors that "freeze" the transporter-inhibitor complex in a nontransporting form were mostly unavailable until recently *(40,45–48)*. Wadiche et al. used kainate, a nontransported inhibitor of GLT1, to isolate pre-steady-state currents for human GLT1 (EAAT2) *(47)*. The pre-steady-state currents had the properties of a nonlinear capacitive current, and the total charge transfer was obtained as the time integral of the current fit to the Boltzmann distribution. The charge transfer is therefore analogous in its properties to those obtained for the Na^+/glucose cotransporter and the GABA transporter. Based on the Na^+-dependence of the pre-steady-state currents, the investigators concluded that the currents reflect the voltage-dependent binding and unbinding of Na^+ near the extracellular surface of the transporter. In contrast, the pre-steady-state transient currents of the Na^+/glucose cotransporter were believed to reflect mainly the voltage-driven rapid conformational change of the negatively charged empty carrier within the membrane electric field, after unbinding of Na^+ *(45)*. The pre-steady-state currents of GLT1 did not appear to reflect conformational changes of charged residues of the transporter molecule in the membrane electrical field *(47)*. This finding is in agreement with the previous speculation that the empty glutamate transporters carrier is "electroneutral" *(32)*.

Analysis of the steady-state currents of the glutamate transporters EAAC1, GLT1, and GLAST revealed a strong voltage-dependence of the glutamate-evoked currents *(16,39,47–49)*. A specific voltage-dependent

Glutamate Transporters

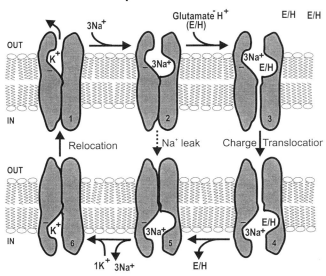

ASC Transporters

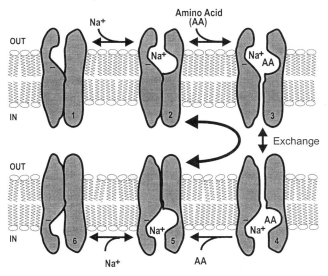

Fig. 2. Kinetic models for glutamate (top) and ASC transporters (bottom) (hypothetical). Under normal conditions, glutamate transport involves loading the empty glutamate carrier with glutamate⁻/H⁺ and three Na⁺, following translocation of the fully loaded carrier across the plasma membrane (charge translocation step) and

step, the so-called "charge translocation step," was proposed to be rate-limiting during the transport cycle *(39,48; see* Fig. 2). The countertransport of K^+ appears to speed up the relocation of the empty carrier so that the relocation step is faster than the charge translocation step.

3.3. Anion Conductance

The functional characterization of EAAT4 has led to the identification of an additional feature of high-affinity glutamate transporters, a substrate-gated anion conductance. This feature is also displayed by the other SLC1 glutamate-transporter family members, and the anion permeabilities decrease in the order EAAT4/5>GLAST>EAAC1>>GLT1. The substrate-evoked currents of EAAT4 expressed in *Xenopus* oocytes reverse at approx –20 mV, and they consist of two components: an electrogenic-transporter current reflecting translocation of substrate across the membrane, and a Cl^- current. The latter has the characteristics of a substrate-gated anion channel with a selectivity order $NO_3^- > I^- > Br^- > Cl^- > F^-$. Cl^- is mainly translocated in the presence of glutamate or related substrates. Cl^- movement is not thermodynamically coupled to the substrate transport. Cl^- is therefore unnecessary for substrate translocation. Although the possibility that EAAT4 couples to an endogenous oocyte Cl^--channel cannot be completely ruled out, the results from Amara and colleagues strongly indicate that EAAT4 itself functions as a Cl^- channel *(16)*. By using rapid applications of glutamate to outside-out patches excised from transfected human embryonic kidney 293 cells, Otis and Kavanaugh further demonstrated that both anion and stoichiometric currents display similar kinetics, suggesting that anion-channel gating and stoichiometric charge movement are linked to early transitions in the transport cycle *(50)*. One possible explanation of the anion-conductance phenomenon is that the conformation of EAAT4, when transporting glutamate or aspartate, may have a "loose" structure in which a path for Cl^- is created. Consistent with this hypothesis is the observation that the Cl^- conductance is larger when aspartate is transported instead of glutamate, because aspartate is smaller than glutamate *(16,23)*. Amara and colleagues also observed that the other glutamate-transporter isoforms, par-

---

Fig. 2. *(continued)* release of the substrates at the intracellular face. Thereafter, K^+ binds to the carrier inside and promotes the relocation of the empty carrier. For net uptake of glutamate, glutamate transporters must complete this cycle. If it is not completed because the empty carrier cannot enter the relocation step, the empty carrier binds Na^+ and glutamate again at the inside of the cells and translocates back in the reverse direction. In this case, the transporter behaves like an exchanger. Because the GLT1 mutant (Glu404Asp) and the ASC transporters lack the K^+-coupling step, they cannot enter the relocation step and they will only work in the exchange mode.

ticularly and GLAST (EAAT1), exhibit current reversals to various degrees when the membrane potential is shifted to more positive potentials *(51)*. More recently, it was shown that EAAT5 exhibits prominent chloride conductance compared with its amino-acid fluxes *(52)*. Thus, glutamate transporters appear to possess structures that can function as chloride channels.

3.4. Exchange Mode

The glutamate transporter EAAC1 can also facilitate substrate exchange, in addition to electrogenic glutamate uptake. However, under normal conditions, the exchange component is small compared to the uptake component *(41)*. In contrast, the neutral amino-acid transporters ASCT1 and ASCT2 mediate exclusively Na^+-dependent exchange of substrate amino acids *(53,54)*. Site-directed mutagenesis studies have provided the answer to the question of why glutamate transporters mediate two transport modes, whereas ASC transporters mediate only the exchange mode: Mutation of Glu 404 or Tyr residue 403 resulted in loss of K^+-coupling in rat GLT1, (*see* Fig. 1) *(55,56)*. Surprisingly, those GLT1 mutants displayed only the exchange mode without regular uptake transport, similar to ASC transporters. Interestingly, the Glu and Tyr residues are conserved in all glutamate transporters (Fig. 1) but not in ASCT1 and ASCT2. ASC transporters were reported not to be coupled to the K^+ *(53)*. In conclusion, these Glu and Tyr residues are crucial for the K^+-coupling and the mutagenesis studies support the concept that K^+ drives the relocation step of glutamate transporters. Because K^+-coupling was disrupted in those Glu and Tyr mutants, the mutant GLT1 was not able to facilitate the relocation step of the empty carrier after releasing glutamate at the intracellular side, and the only relocation was the reversal of the Na^+-coupled uptake step, resulting in glutamate exchange (*see* Fig. 2).

3.5. Properties of the Neutral Amino-Acid Transporters

The ASC neutral amino-acid transporters of the SLC1 family exhibit the properties of the classical Na^+-dependent ASC transporters *(18,19,57)*. Glutamate transporters generally transport L-glutamate and D- and L-aspartate, and do not accept neutral amino acids. On the other hand, ASC transporters have a high affinity for alanine, serine, threonine, and cysteine, and the preferred substrates are amino acids without highly branched or bulky side chains *(58)*. Interestingly, the two ASC transporters ASCT1 and ASCT2 exhibit distinct substrate selectivities *(18,19,57)*. In addition to the common substrates of ASC transporters, ASCT2 also accepts glutamine and asparagine as high affinity, and methionine, leucine, and glycine as low-affinity substrates, whereas ASCT1 does not accept these substrates.

Despite the distinctive substrate selectivities displayed among glutamate and ASC transporters, they still exhibit common properties in substrate recognition, which reflect their structural similarity. For example, glutamate transporters, particularly EAAC1, transport the neutral amino acid cysteine *(41)* and, conversely, the neutral amino acid transporter ASCT2 transports glutamate although with low affinity *(57)*. Glutamate transport via ASCT2 is enhanced at low pH *(41)*, and ASCT1 becomes inhibited by acidic amino acids such as glutamate, aspartate, cysteate, and cysteinesulfinate by lowering pH *(59)*. These findings, together with the high conservation of amino-acid sequences between glutamate transporters and ASC transporters, suggest that both transporter types have structurally similar substrate-binding sites. The difference in substrate selectivity may be explained by the existence of amino-acid residues in the binding sites of glutamate transporters, which specifically interact with a negative charge in the side chains of acidic amino acids *(57)*.

Since substrate transport via transporters is associated with a Cl^--conductance, ASC transporters also exhibit anion conductance *(53,54)*. In both glutamate and ASC transporters, chloride currents are detected in the presence of substrate amino acids, and the transporters behave as ligand-gated chloride channels *(16,17,51,53,54)*. As already noted, in contrast to glutamate transporters, ASC transporters do not couple to K^+-countertransport *(53,54)*. As for H^+-coupling, ASC transporters—in contrast to glutamate transporters *(53,54)*—do not appear to couple with H^+-transport either *(53,54)*. ASC transporters are therefore proposed to function exclusively as Na^+-dependent amino-acid exchangers. By contrast, glutamate transporters possess both the uptake and the exchange modes, although the contribution of the exchange mode seems negligible under normal physiological conditions.

4. STRUCTURE-FUNCTION RELATIONSHIP OF GLUTAMATE TRANSPORTERS

The SLC1 family members have a unique, highly conserved and long hydrophobic stretch near the C-terminus (*see* Fig. 1), and we previously proposed that this region may be responsible for binding and translocation of the transport substrates *(60)*. Although the structure-function relationship of glutamate transporters has not yet been studied in great detail, some information is now available from site-directed mutagenesis and pharmacological studies. Grunewald and Kanner preformed cysteine-scanning mutagenesis to examine the accessibility of amino-acid residues in the hydrophobic stretch using sulfhydyl-reactive reagents *(61)*. Based on these studies, the membrane model shown in Fig. 1 was constructed (*see* also

Chapter 7). A similar topology was also deduced from studies of Amarea and colleagues, based on extracellular accessibility of substituted cysteines in GLT1 (EAAT1) *(62)*. The model in Fig. 1 features eight α-helical TMSs (1–8), a large extracellular glycosylated loop between TMDs 3 and 4, a "re-entrant loop" (A/B) between TMDs 7 and 8 and a "loop" which is predicted to extend partially into the "translocation pore" between TMDs 7 and 8. In ion channels, the re-entrant loop is considered to be the ion-permeating pore, but such a structure has not yet been identified in any other SLC-transporter family. It will be interesting to determine whether this structure is associated with the permeation paths for the transport substrates and cotransported ions.

In the seventh TMD, there are the two amino-acid residues, Tyr 403 and Glu 404 (rat GLT1 numbering), which are important for the coupling with K^+ (*see* **Subheading 3.4.**).

With respect to the coupling to inorganic cations, Kanner and colleagues reported interesting observations based on site-directed mutagenesis studies. In the previously mentioned Tyr 403 mutation, the mutant GLT1 functions not only in the presence of Na^+, but also in the presence of Li^+ or Cs^+, in contrast to wild-type GLT1, which cannot use Li^+ or Cs^+ as a coupling ion *(56)*. Based on this observation, it was proposed that Tyr403 is not only important for the K^+-coupling, but that it is also associated with the Na^+-binding site or the structure responsible for the selectivity of inorganic cations. Another important finding regarding the coupling to inorganic cations is that the mutation of Ser 440, which resides at the extracellular face of the re-entrant loop (Fig. 1B) alters the selectivity to inorganic cations. Threrefore, Ser 440 is also important for the Na^+-coupling mechanism. In addition, the Ser 440 mutation alters sensitivity to the glutamate-transport inhibitors *(63)*. Thus, it is suggested that this residue is tightly associated with the substrate-binding site of glutamate transporters.

The unique sequence located in this region which consists of several consecutive serine residues (SSSS in EAAC1 and GLAST and ASSA in GLT1) (*see* Fig. 1; 331–334 in EAAC1) is reminiscent of three consecutive serine residues in the mGluR1 metabotropic glutamate receptor, which have been suggested to be involved in hydrogen bonding to glutamate *(64)*. Because this motif is common to all prokaryotic and eukaryotic glutamate-transporter family members, it is likely that it is involved in substrate binding and translocation in glutamate transporters. Further studies will be needed to address this hypothesis.

4.1. Pentameric Structure of Glutamate Transporters

The three-dimensional structure of glutamate transporters is not available presently. However, freeze-fracture electron microscopy studies of *Xenopus* oocytes overexpressing human EAAC1 (EAAT3) revealed distinct 10-nm freeze-fracture particles, which appeared in the protoplasmic face only after the EAAC1 expression *(65)*. The total number of the 10-nm particles was linearly correlated with maximum carrier-mediated charge of EAAC1, suggesting that the particles represent functional EAAC1 in the plasma membrane. The cross-sectional area of the human EAAC1 in the plasma membrane (48 ± 5 nm^2) predicted 35 ± 3 transmembrane α-helices in the transporter complex. This information, along with secondary-structure models (6–10 transmembrane α-helices) suggested that human EAAC1 particles were pentagonal, in which five domains could be identified. Based on the fivefold symmetry and the projections from all five domains to the extracellular space, the transporter complex was predicted to be in the appearance of a penton-based pyramid. It is speculated that, although the EAAC1 monomer can perform secondary active transport, the chloride-channel mode observed in glutamate transporters is related to the oligomeric assembly, in analogy to multimeric ion channels whose surrounding subunits contribute to the lining of the ion-permeation pathway.

5. DISTRIBUTION OF EXPRESSION OF GLUTAMATE AND ASC TRANSPORTERS

Initial studies addressing the distribution of EAAC1, GLT1, and GLAST in the CNS indicated that EAAC1 is expressed in neurons, whereas GLT1 and GLAST are expressed in astrocytic glial cells (*see* Figs. 3 and 4). Over the past several years, more detailed analyses have revealed that this distinction is not completely correct. In addition, studies using cultured neuronal cells identified a number of factors that can influence the expression of these transporter genes. The following summary focuses on the results obtained from normal animals, using primarily *in situ* hybridization, immunocytochemistry, and immuno-electron microscopy. We would also like to refer the reader to a recent review on the regulation of expression of glutamate transporters in cultured cells *(66)*.

5.1. Tissue Distribution of EAAC1 (EAAT3)

Initial studies of the expression of EAAC1 in CNS demonstrated strong mRNA signals in neuronal populations in the hippocampus and cerebral cortex *(12)* and in spinal-cord motorneurons *(67)*. The neuronal expression of EAAC1 was confirmed subsequently in several immunocytochemical *(68–70)* and *in situ* hybridization studies *(71–77*; Fig. 4). As is the case for GLAST and GLT1, EAAC1 is expressed at higher levels in the forebrain than in the brain stem. Areas with very high levels of EAAC1 expression include hippocampus, cerebral cortex, olfactory bulb, striatum, superior colliculus, and, at least at the mRNA level, the thalamus *(68,77)*. In the

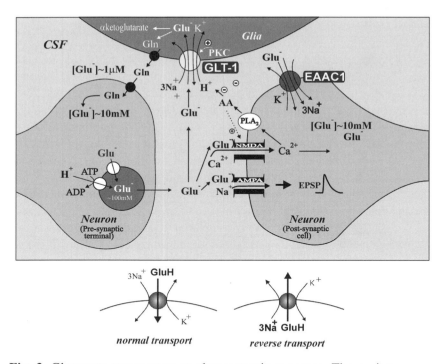

Fig. 3. Glutamate transporters at glutamatergic synapses. The excitatory neurotransmitter L-glutamate is stored in synaptic vesicles at presynaptic terminals. Glutamate is transported into these vesicles by the VGluT vesicular glutamate transporters. Glutamate is released into the synaptic cleft to act on glutamate receptors. The AMPA receptors mediate rapid excitatory postsynaptic potentials (EPSP,) whereas the NMDA receptors possess a cation channel which is permeable to Ca^{2+}. High-affinity glutamate transporters play essential roles in removing released glutamate from the synaptic cleft. These transporters are also crucial for maintaining the extracellular glutamate concentration of the cerebrospinal fluid (CSF) below neurotoxic levels. The high accumulative power of glutamate transporters is provided by the coupling transport to the co- or countertransport of the ions Na^+, H^+, and K^+. The figure shows the glial glutamate transporter GLT1 and the neuronal glutamate transporter EAAC1.

Glutamate taken up into glial cells via GLT1 or GLAST is metabolized to glutamine by glutamine synthase (with hydrolysis of ATP) and to α-ketoglutarate by glutamate dehydrogenase or glutamate oxaloacetate transaminase. Glial cells in turn supply the nerve terminal with glutamine and α-ketoglutarate, which serve as precursors of glutamate synthesis. Glutamine exits neurons via the system N transporter system and then enters the neuron via the system A transporter system *(204,205)*.

Arachidonic acid (AA) and ROS are both liberated in response to glutamate-receptor activation. Certain forms of familial amyotrophic lateral sclerosis (ALS) are caused by genetic defects of the cytosolic superoxide dismutase (SOD1), and the pathogenesis of motor-neuron degeneration is believed to involve chronic

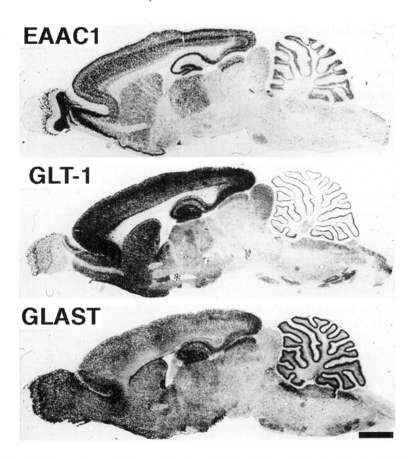

Fig. 4. Comparison of the distribution of mRNA for EAAC1, GLT1, and GLAST in sagittal sections of the rat brain. The development time for the GLT1 labeling was 4 h, and the one for EAAC1 and GLAST labeling was 18 h. Note the relatively higher labeling intensities with all probes in the cerebral cortex, hippocampus and ventral forebrain than in the midbrain and brain stem. Also, GLT1 labeling is relatively low in the olfactory bulb, and that of GLAST is of the same intensity in olfactory bulb than cortex. *, tear in section. Bar = 2.5 mm.

Fig. 3. *(continued)* overstimulating of non-NMDA receptors and activation of voltage-dependent Ca^{2+} channels, leading to the accumulation of oxygen free radicals through activation of xanthine oxidase and decreased SOD1 oxygen free-radical scavenging capacity, and to neuronal death. Arachidonic acid and ROS were shown to inhibit or modulate the function of high-affinity glutamate transporters, which can lead to the elevation of extracellular glutamate to neurotoxic levels. PLA, phospholipase A.

cerebellum, EAAC1 protein is prominent in the dendrites, and axons of the Purkinje cells. Analysis of EAAC1 expression at the mRNA level suggests that a gradient in EAAC1 expression exists between lightly labeled dorsal Purkinje cells and more strongly labeled ventral Purkinje cells *(77)*. Ultrastructural analysis of EAAC1 immunostaining has revealed that EAAC1 is primarily expressed in postsynaptic elements such as dendritic shafts, spines, and axons, whereas presynaptic glutamatergic elements such as cortico-striate terminals in the caudate putamen appeared to be devoid of EAAC1 *(68)*. These findings were initially taken as evidence that EAAC1 is a postsynaptic transporter that provides glutamate for metabolic purposes (Fig. 3). Recently, significant EAAC1 expression has also been observed over the presynaptic membrane in the synaptic complex *(78)*, suggesting that EAAC1 may contribute to presynaptic glutamate reuptake, and thus to termination of glutamatergic neurotransmission.

High-sensitivity *in situ* hybridization has revealed that EAAC1 mRNA is present in the majority of neurons in the brain *(77*; Fig. 4). In addition, it led to the identification of a unique population of strongly labeled cells that is scattered throughout grey and white matter areas, and that may represent oligodendroglial precursor cells *(77)*. Subsequent immunocytochemical studies by Matute and colleagues *(79,80)* have shown, through double-staining, that EAAC1 is indeed also expressed in immature oligodendrocytes. Furthermore, Kugler and Schmitt have reported that EAAC1 mRNA and protein can also be identified in regular oligodendrocytes *(82)*. EAAC1 expression in astrocytes has generally not been observed; in particular, double-*in situ* hybridization experiments could not provide evidence for EAAC1 expression in white-matter astrocytes *(77*; Fig. 5). However, one study has reported that a set of polyclonal antibodies could detect rare EAAC1 protein expression in white- and grey-matter astrocytes at the light and electron-microscopic level *(81)*. Moreover, although these investigators could not find evidence for EAAC1 expression in astrocytes within the CNS, Kugler and Schmitt did observe EAAC1 protein expression by satellite cells in dorsal-root ganglia, which are the corollary for astrocytes in the peripheral nervous system *(82)*. Interestingly, EAAC1 expression has also been found in the rat glial cell line C6 and in human astrocytic tumor cells *(83)*, suggesting that under certain conditions EAAC1 is also expressed in astrocytic glial cells. EAAC1 expression in neurons in the retina has also been reported *(72,84,85)*.

EAAC1 is also an epithelial transporter which, after having been originally isolated from rabbit small intestine, was also found to be expressed in a wide variety of other peripheral tissues, including the kidney, liver, heart, muscle, lung, and placenta *(12,86–88)*.

5.2. Tissue Distribution of GLT1 (EAAT2)

In the initial characterization of GLT1 expression in the brain, GLT1 immunoreactivity was attributed at the light- and electron-microscopic levels to astrocytes *(13)* (*see* Fig. 3). Subsequent *in situ* hybridization *(77,89–92)* and immunocytochemical *(68,91,93,94)* studies confirmed the primarily astrocytic origin of GLT1 (Fig. 4). Like GLAST, GLT1 expression in brain shows unique regional variations. In contrast to GLAST, GLT1 expression is about 10-fold higher in the cerebral cortex and hippocampus than in the olfactory bulb or cerebellum *(94)*. Other subcortical structures show more intermediate expression levels for GLT1. In spinal cord, GLT1 immunoreactivity is more widely found in gray matter than that of GLAST *(68,94)*. Electron-microscopic studies also indicated that all astrocytes in the brain express GLT1, and, as is the case for GLAST, astrocytic membranes facing nerve terminals, axons, and spines express higher GLT1 levels *(95)*. A detailed quantitative immunoblotting analysis revealed that the number of astroglial transporters GLT1 and GLAST per μm^3 tissue are 12,000 and 3200 in the stratum radiatum of adult rat hippocampus (CA1), and 2800 and 18,000 in the cerebellar molecular layer, respectively *(96)*.

We have shown that astrocytes in the circumventricular organs, with the exception of the pineal gland, express very little GLT1 mRNA or GLT1 protein compared to GLAST (Fig. 6), suggesting that outside the blood-brain barrier GLAST is the primary astrocytic glutamate transporter, whereas inside the blood-brain barrier the opposite is true *(97)*. In the pineal gland, GLT1 is expressed both by astrocytes *(98)* and by pinealocytes *(97,99)*.

Until recently, GLT1 immunostaining has been found in the CNS only in astrocytes. In contrast, *in situ* hybridization studies revealed that GLT1 mRNA is also present in some neuronal subgroups. Noticeable GLT1 expression was first reported in CA3 pyramidal cells *(89)*. Subsequently, neuronal GLT1 mRNA expression was also noticed in layer V and VI neurons in the cerebral cortex *(90)*, as well as in neurons of the endopiriform nucleus, and in neurons in the medioventral thalamus *(91)*. Expression of GLT1 mRNA by layer II cortical neurons was also reported *(92)*. In 1998, using a novel method of differential double-*in situ* hybridization, we demonstrated that the distribution of GLT1 mRNA in neurons is even more widespread *(77)*. We found GLT1 mRNA expression in the majority of neurons in the cerebral cortex, in neurons in the external plexiform layer in the olfactory bulb, in neurons in dorsal and ventral parts of the anterior olfactory nucleus, in the majority of neurons in the anteromedial thalamic nuclei, the CA3 pyramidal neurons in the hippocampus, and neurons in the inferior olive *(77)*. The discrepancy between the presence of transporter mRNA and absence of transporter protein

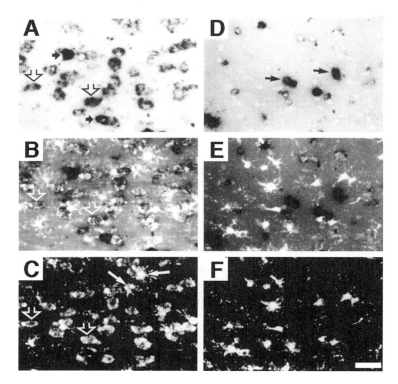

Fig. 5. Co-expression of EAAC1 and GLT1 mRNA in the cortex but not in the corpus callosum. The section was double-hybridized to a digoxigenin-labeled EAAC1 and a FITC-labeled GLT1 cRNA probes. The EAAC1 probe was visualized at the bright-field level with alkaline phosphatase histochemistry, and the GLT1 probe was visualized at the fluorescent level with the help of tyramine signal amplification. **(A)** In the deep part of layer VI of cortex, EAAC1 is expressed moderately in the majority of neurons; a few neurons are strongly labeled (small arrows). **(B)** Simultaneous EAAC1 and GLT1 labeling in neurons; open arrows depict example of neurons that are double-labeled for EAAC1 and GLT1. **(C)** GLT1 labeling is present in neurons and astrocytes (filled arrows show two examples of astrocytes). **(D)** Strongly labeled EAAC1-positive cells in corpus callosum; these cells are probably oligodendrocyte precursor cells. **(E)** Simultaneous EAAC1 and GLT1 labeling, showing that EAAC1-positive cells and GLT1-positive astrocytes are separate. **(F)** GLT1 labeling of astrocytes. Bar = 40 μm.

has initially been explained as being possibly caused by posttranscriptional regulation. However, recent studies by Rosenberg and colleagues *(101)* suggest that this may be caused by the existence of GLT1 isoforms. They have found that antibodies against a GLT1 isoform called GLT1b are able to stain

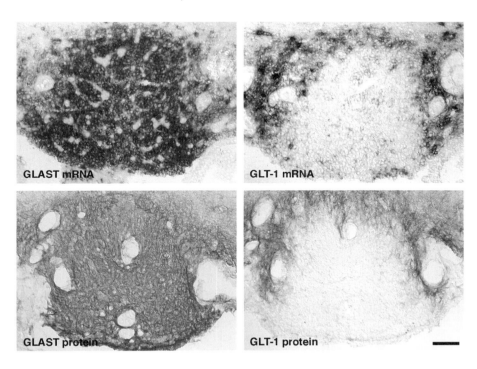

Fig. 6. Distribution of GLAST and GLT1 mRNA and protein in the subfornical organ, as detected by non-isotopic *in situ* hybridization and immunocytochemistry. GLAST mRNA and protein are strongly expressed in astrocytes in the subfornical organ, whereas GLT1 mRNA and protein are absent. This finding suggests that GLAST is the primary glutamate transporter in astrocytes outside the blood-brain barrier. Magnification bar = 100 μm.

neurons in the brain *(101)*. Future studies will be required to demonstrate whether the neuronal subgroups shown to express GLT1 mRNA can be correlated with the neurons exhibiting positive GLT1b immunostaining. Interestingly, in the retina, GLT1 is localized in cone photoreceptors and in a subclass of amacrine and bipolar neurons, but not in glial cells *(85,102,103)*.

GLT1 mRNA and protein expression has recently been demonstrated in a unique subgroup of tanycytes in the ependymal lining of the third ventricle *(104*; Fig. 7).

Outside the brain, GLT1 has been found in liver and pancreas *(88,105,106)*, placenta *(107)*, bone *(108)*, lymph nodes, and spleen (Berger and Hediger, unpublished observations).

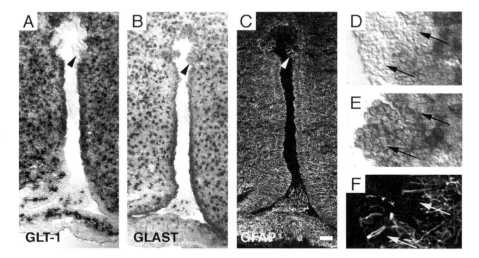

Fig. 7. Labeling for GLT1 mRNA, GLAST mRNA, and GFAP protein of the third-ventricle ependyma at the level of the infundibular recess in three adjacent 16-μm sections. (**A**) GLT1 mRNA is expressed in dorsal walls but not in the cuboidal cells (arrows in **D**) of the roof. (**B**) GLAST mRNA is expressed throughout the walls, lateral parts of the floor, and cuboidal cells (arrows in **E**). (**C**) GFAP staining, which is a marker for tanycytes, is present in the walls and floor but not in cuboidal cells of the roof (arrows in **F**). (**D–F**) Higher magnification of the cuboidal ependyma indicated by arrowheads in **A**, **B**, and **C**, respectively. In **F**, strongly stained subependymal astrocytes extend processes through the cuboidal cell layer to the lumen. Magnification bar (**C**) = 100 μm.

5.3. Tissue Distribution of GLAST (EAAT1)

The distribution of GLAST mRNA in the brain was initially reported by Stoffel and colleagues *(15)*. A strong labeling of the Purkinje cell/Bergmann glia-cell layer in cerebellum was noted, as well as a more diffuse labeling in forebrain, which suggested the expression of GLAST by glial cells. The expression by astrocytic glial cells was then confirmed by subsequent in situ hybridization *(77,89,90,109;* Fig. 4) and immunocytochemical studies *(68,70,94,95,109–111)*. In their initial immunocytochemical studies, *(68)* Rothstein et al. reported that GLAST immunoreactivity is also found in neurons, but this observation was retracted later in ref. *112*. Lehre et al. *(94)* performed a quantitative immunoblot analysis of the distribution of GLAST and GLT1 protein in different regions of the rat brain. They demonstrated that the expression level of GLAST is approx 5× higher in the cerebellum than in the olfactory bulb, hippocampus, and cerebral cortex, with even lower levels in subcortical and brain-stem regions. A similar decreasing gradient from the cerebellar Bergmann glia, to the cerebral cortex and olfactory bulb astrocytes, to the subcortical and brain-stem astrocytes, can be observed with *in situ* hybridization analysis *(77,89*; Fig. 4) or with immunocytochemistry *(94,95,109)*. In the spinal cord, GLAST protein is strongly expressed in the substantia gelatinosa of the dorsal horn and around the central canal, but comparatively weakly in other gray-and-white matter areas *(68,94,109)*. Differences in regional expression levels of GLAST protein or mRNA were also noted in the thalamus and other subcortical areas. Thus, for example, astrocytes within the paraventricular nucleus of the hypothalamus express higher levels of mRNA than astrocytes in the areas immediately surrounding it *(77)*. Immunocytochemical studies at the electron microscopic level showed that all astrocytes in the brain express GLAST to some extent, and that astrocytic membranes facing capillaries, pia, or stem dendrites have lower amounts of GLAST protein than those facing nerve terminals, axons, and spines *(95)*. In a recent study of the glutamate-transporter expression in the circumventricular organs *(97)*, we demonstrated that astrocytes in the subfornical organ, the organum vasculosum of the lamina terminalis, the area postrema and the pineal gland, and pituicytes in the posterior pituitary all express relatively high amounts of GLAST mRNA and protein, whereas little GLT1 was found in these regions *(see* Fig. 6). The expression of GLAST by pineal astrocytes was also reported by Redecker and Pabst *(98)*, and GLAST has also been identified as the glutamate transporter of the Mueller cells of the retina *(85,113,114)*.

Schmitt et al. *(109)* first noted that GLAST expression is not restricted to astrocytes. They found GLAST mRNA as well as protein in ependymal cells lining the cerebral ventricles. We confirmed this ependymal expression of GLAST *(77)*, and also demonstrated that GLAST mRNA and protein are expressed by meningeal cells *(97)* and tanycytes *(104*; Fig. 7). In addition, Domercq and colleagues reported that oligodendrocytes from either rat optic nerve or bovine corpus callosum stain also positively with several C- or N-terminal GLAST antibodies *(79,80)*. This oligodendroglial expression of GLAST has thus far not been observed for GLAST mRNA when using *in situ* hybridization *(77,89,109)*. However, GLAST-like transcripts have been identified in the rat optic nerve using RT-PCR *(80)*. Outside the CNS, GLAST mRNA and protein expression have been found in rat cochlear neurons, where GLAST is important for hearing, and in bone-marrow cells *(108,115,116)*, respectively.

Unpublished observations from our laboratory indicate the expression of GLAST by immunological cells in thymus and spleen, by epithelial cells in kidney and bladder, and by Leydig cells in the testis.

5.4. Tissue Distribution of EAAT4

In the initial cloning paper, Northern blot analysis revealed EAAT4 expression selectively in the cerebellum *(16)*. Immunocytochemical and *in situ* hybridization studies subsequently identified the cerebellar Purkinje cells as the main cell type expressing EAAT4 *(117–122)*. Subcellular analysis revealed that EAAT4 is primarily expressed on postsynaptic dendritic spines of Purkinje cells *(119)*, and that cell spines and thin dendrites are more strongly labeled than large-diameter dendrites and cell bodies *(122)*. EAAT4 is present at low concentrations in the synaptic membrane, but is highly enriched in the parts of the dendritic and spine membranes facing astrocytes *(120,122)*. The density of EAAT4 molecules in spines has been calculated as 1800 molecules/μm^2 membrane *(122)*. In contrast to EAAC1, EAAT4 is not expressed in Purkinje-cell axons *(70)*. Furthermore, unlike EAAC1, EAAT4 immunostaining forms banding patterns in the molecular layer that are correlated with the banding patterns for zebrin II *(119,122)*.

EAAT4 is also expressed at lower levels in the forebrain, particularly in the cerebral cortex, hippocampus, and striatum *(70,119,121,122)*. In these forebrain regions, EAAT4 immunostaining was primarily seen in the neuropil. At the subcellular level in the hippocampus and cortex, EAAT4 protein localized to small dendrites, and very rarely EAAT4 immunoreactivity was also found in astrocytic-cell processes *(118)*.

Outside the CNS, EAAT4 has only been found in the placenta so far *(16)*.

5.5. Tissue Distribution of EAAT5

The initial cloning study showed the expression of EAAT5 to be restricted to the retina within the CNS. Expression at a much lower level was also detected in the liver *(17)*. Whereas *in situ* hybridization analysis of EAAT5 has not been reported, immunocytochemical studies in the rat retina have shown that EAAT5 is associated with rod photoreceptors and with some bipolar cells *(123,124)*. In contrast, immunostaining for EAAT5 in salamander retina has been observed in both glial and neuronal cells *(125)*. In the rat, GLAST1, GLT1, EAAC1, and EAAT5 are all expressed in the retina, where they exhibit unique localization and functional properties in glutamatergic neurotransmission *(126)*.

5.6. Tissue Distribution of ASCT1 and ASCT2

Only two recent studies have explored the distribution patterns of ASCT1 and ASCT2 in brain thus far, with differing results. Dykes-Hoberg et al. *(127)* have found that ASCT1 immunoreactivity appears to be specific to astrocytes, whereas Weiss et al. *(100)* have found ASCT1 immunoreactivity to be associated with neurons and astrocytes. Immunoreactivity for ASCT2 has also been described in neurons and glial cells *(127)*.

Both transporters also exist in peripheral organs. ASCT1 appears to be ubiquitous, whereas ASCT2 has been found to be expressed in the lung, kidney, skeletal muscle, and testis.

6. PHYSIOLOGICAL SIGNIFICANCE OF GLUTAMATE TRANSPORTERS

6.1. The Role of Glutamate Transporters in Glutamatergic Transmission

Glutamate transporters are believed to play important roles in the termination of synaptic transmission (*128*; Fig. 3). This idea stems from the general concept that synapses which do not have mechanisms for enzymatically degrading released neurotransmitters possess presynaptic reuptake mechanisms for transmitter removal and for terminating their actions on the postsynaptic neurotransmitter receptors. At GABAergic and serotonergic synapses, inhibitors of presynaptic neurotransmitter uptake were shown to prolong the postsynaptic currents, indicating that presynaptic reuptake carriers largely contribute to the termination of synaptic transmission (*129,130*). In glutamatergic synapses, however, the inhibition of glutamate transporters did not prolong the decay time-courses of rapid excitatory postsynaptic currents (EPSC) in hippocampal CA1 pyramidal cells and cerebellar granule cells when using brain slices or culture preparations (*131–134*). Based on studies using single neuron micro-islands prepared from hippocampus, Mennerick and Zorumski explained why glutamatergic synapses are different from other synapses and why glutamate uptake inhibitors did not affect the shape of the excitatory postsynaptic potential (*132*). In their preparation, they inhibited glutamate uptake by glial depolarization rather than with uptake inhibitors, which are known also to attenuate NMDA receptor-mediated responses. They found that inhibition of glutamate uptake prolongs autoptically induced NMDA responses which are known to exhibit slow or little desensitization, whereas the inhibition did not affect the time-course of the EPSC mediated by non-NMDA receptors, which exhibit rapid desensitization. In contrast, if desensitization of non-NMDA receptor-mediated responses was slowed down by cyclothiazide, the investigators were able to show that glutamate-transporter inhibition significantly prolonged the time-course of non-NMDA receptor-mediated autoptic EPSC. Also consistent with these findings, the non-NMDA receptor-mediated EPSC recorded in cerebellar Purkinje cells after parallel-fiber stimulation was prolonged by glutamate-uptake inhibition because the EPSC of Purkinje cells normally exhibits a slower decay time-course (*135*). Taken together, these data demonstrate that, because of the rapid desensitization of non-NMDA receptor-mediated responses, glutamate transporters usually do not significantly affect the decay time-course of the rapid EPSC. However, glutamate transporters play important roles in removing released glutamate from the synaptic cleft.

Clements and colleagues empirically estimated the decay time-course of the glutamate concentration in the synaptic cleft, and concluded that it is slower than that derived from theoretical calculation based on simple diffusion out of the synaptic cleft *(136)*. Consequently, these investigators proposed that the diffusion of glutamate from the synaptic cleft is restricted by neighboring structures. This indicated that altering the location, density, and/ or the function of glutamate transporters could influence the time-course of the transmitter clearance in the synaptic cleft. Kavanaugh and colleagues recently determined the turnover rate of the human glutamate-transporter GLT1 based on the analysis of steady-state and pre-steady-state currents in response to voltage jumps *(47)*. They found that the time constant for a complete cycle of transport at –80 mV and 22°C is approx 70 ms—significantly slower than the estimated glutamate-decay time constant in hippocampal synapses which is 1–2 ms *(136)*. This difference was predicted to be true also at physiological temperatures *(47)*. Therefore, there is some controversy as to whether glutamate transporters really constitute a major mechanism for removing released glutamate *(47)*. Binding of glutamate to glutamate transporters, however, exhibits rapid kinetics. Tong and Jahr demonstrated that glutamate binding to glutamate transporters significantly contributes to the glutamate clearance in the synaptic cleft *(134)*. Recently, by applying the laser-pulse photolysis technique of caged glutamate with a time resolution of 100 µs, steady-state and pre-steady-state kinetics of EAAC1 were analyzed *(137)*. It was demonstrated that EAAC1-mediated pre-steady-state currents are composed of two components: a transport current generated by substrate-coupled charge translocation across the membrane and an anion current that is not thermodynamically coupled to glutamate transport. It was concluded that glutamate translocation occurs within a few milliseconds after binding. The transition to an anion-conducting state, however, is delayed with respect to the onset of glutamate transport.

Thus, although the turnover rate is low, as indicated by the studies of Kavanaugh and colleagues *(47)*, the glutamate translocation step itself is believed to be rapid enough to participate in the removal of the released glutamate from the synaptic cleft. Consistent with this theory, Auger and Attwell demonstrated—based on the electric current recorded on the whole-cell-clamped Purkinje cells in cerebellar slices—that when glutamate binds to postsynaptic transporters at the cerebellar climbing-fiber synapse, it evokes a conformational change and inward current that reflect glutamate removal from the synaptic cleft within a few milliseconds, a time-scale much faster than the overall cycle time *(52)*.

By examining the effects of the inhibition of glial glutamate transporters on the synaptic currents in a hippocampal microculture, Menneric and colleagues provided further evidence to support a major role of substrate translocation in determining the time-course of the glutamate concentration transient at excitatory synapses *(138)*.

As Clements and colleagues predicted, the geometry of the synapse is also an important determinant of the decay time-courses of the glutamate concentration at the synaptic cleft. For examaple, in cerebellar Purkinje cells, the non-NMDA receptor-mediated EPSC exhibits a slower decay time-course, and this is believed to be the result of impaired transmitter diffusion, and thereby the continued presence of glutamate in the synaptic cleft *(135)*.

Thus, based on our current knowledge, both the diffusion of glutamate out of the synaptic cleft and binding of glutamate to glutamate transporters are the major factors affecting the decay time-course of the glutamate concentration in the synaptic cleft.

6.2. Role of Glutamate Transporters in Maintaining Extracellular Glutamate Concentration

Based on the stoichiometry of glutamate transporters (cotransport of one glutamate-H^+ with three Na^+ ions and the countertransport of one K^+ ion) and the prevailing ionic environment, it can be calculated that glutamate transporters concentrate glutamate more than 10,000-fold across cell membranes *(44)*. For example, the glutamate concentration in the CSF is maintained at ~1 μM, whereas the intracellular glutamate concentration in neurons is as high as 10 mM. Because of this high concentrating capacity of high-affinity glutamate transporters, these proteins are believed to play a major role in maintaining the extracellular glutamate concentration at low levels, and to protect neurons from the excitotoxic action of glutamate.

The importance of glutamate transporters in protecting neurons from glutamate excitotoxicity was experimentally demonstrated by Rothstein and colleagues who treated organotypic rat spinal-cord culture with the glutamate uptake inhibitors THA or PDC and observed a slight increase in glutamate concentration in the culture medium *(139)*. Under this condition, motor-neuron specific degeneration was observed, characterized by a slow onset and gradually progressing.

In order to determine which glutamate transporters are most important to protect neurons from glutamate excitotoxicity, Rothstein and colleagues used organotypic culture preparation and incubated the spinal-cord explant with antisense oligonucleotides corresponding to the N-terminal part of cloned glutamate transporters *(140)*. They found that antisense oligonucleotides corresponding to the glial glutamate transporters GLT1 and GLAST induced similar motor neuron-specific degeneration, as was observed when they used glutamate uptake inhibitors, whereas antisense oligonucleotides corresponding to the neuronal glutamate transporter EAAC1 did not induce neurodegeneration. The investigators also applied antisense oligonucleotides to the cerebroventricle of alert rats. Cerebroventricular administration of antisense GLT1 or GLAST oligonucleotides also resulted in the degeneration of neurons, whereas EAAC1 antisense oligonucleotides did not induce neurodegeneration *(140)*. Thus, primarily the glial but not the neuronal glutamate transporters protect neurons from glutamate excitotoxicity. The importance of glial glutamate transporters in protecting neurons from the glutamate was also demonstrated in the glutamate transporter knockout transgenic mice. Consistent with the results from antisense oligonucleotide-knockout experiments, the GLT1- and GLAST-knockout mice exhibited increased susceptibility to glutamate-mediated brain injury, whereas EAAC1- knockout mice did not show neurodegeneration.

Antisense oligonucleotide inhibition was also used to determine the contribution of each glutamate-transporter isoform to the total glutamate uptake in synaptosomes isolated from rat striatum. The contribution was 60% for GLT1, 20% for GLAST, and 20% for EAAC1 *(140)*. An important question arises in regard to why only knockouts of glial but not neuronal glutamate transporters resulted in neurodegeneration. The answer to this question is related to the different functional roles of glial and neuronal glutamate transporters in the CNS. In neurons, the intracellular glutamate concentration is up to ~ 10 mM. Considering the low extracellular glutamate concentration of ~ 1 µM, this results in a steep glutamate concentration gradient across neuronal-cell membranes, which is consistent with the concentration capacity of high-affinity glutamate transporters. Consequently, under normal conditions, neuronal glutamate transporters are almost at equilibrium, and have little capacity to take up glutamate. It is therefore reasonable that the knockout of neuronal EAAC1 did not induce neurodegeneration. In glial cells, glutamate is taken up continuously, and is then rapidly converted to glutamine by glutamine synthetase, an enzyme which is present in glial cells but not in neurons. The intracellular glutamate concentration in glia is as low as ~ 50 µM. Therefore, glial glutamate transporters are not at equilibrium and keep pumping glutamate into glial cells, which generates a continuous flow of glutamate from glutamatergic synapse ("source") to glial cells ("sink") *(see* also ref. *140a)*.

Storm-Mathisen and colleagues observed that extracellularly applied D-aspartate was not taken up into glutamatergic terminals, whereas large amounts of D-aspartate were taken up in glial cells *(141)*. This observation is consistent with the model that diffusion of glutamate out of the synaptic cleft is an important determinant of the decay time-courses of glutamate in the cleft. This result further raised the question of whether presynaptic glutamate transporters really exist. The existence of presynaptic glutamate uptake systems was originally postulated based on glutamate-uptake studies of the synaptosomal fraction P2. However, this fraction was subsequently found to contain large amounts of glial components *(13,142)*. It is currently still unclear whether there really exists a presynaptic glutamate transporter. The possible candidate proteins are EAAC1, which was recently observed in presynaptic membrane *(78)*, GLT1 *(77)* or its splice variants *(101)* or even possibly the VGlut vesicular glutamate transporters.

What are the major in vivo roles of the neuronal glutamate transporter EAAC1? This transporter most likely functions to keep the intracellular neuronal glutamate concentration at high levels for use as a neurotransmitter and/or as a precursor for various metabolic reactions. We also propose that, in GABAergic neurons such as cerebellar Purkinje cells, EAAC1 provides glutamate as a precursor for GABA synthesis *(72)*. Rothstein and colleagues found that rats which were repeatedly administered with EAAC1 antisense oligonucleotide into the cerebroventricle exhibited epileptic seizures within 1 wk *(140)*. Although the mechanism for induction of these seizures is not clear, it is reasonable to hypothesize that depletion of EAAC1 at GABAergic terminals results in a decrease in GABA synthesis because of the decreased supply of glutamate to GABAergic terminals as the precursor of GABA synthesis.

7. GLUTAMATE TRANSPORTERS AND ISCHEMIA

Glutamate neurotoxicity in brain areas of severe ischemia is mainly caused by reversal of glutamate transporters, most likely neuronal glutamate transporters *(143)*. Because glutamate transport is driven by free energy stored in the form of electrochemical-potential gradients across plasma membranes, the disturbance of the ionic gradients caused by insufficient energy supply which occurs during brain ischemia results in a decrease in the concentrating capacity of glutamate transporters and favors the reversed operation of glutamate transporter driven by the intracellular glutamate concentration. Because neurons have a much higher content of glutamate than glia, neuronal glutamate transporters are more likely to run in reverse in ischemia and to contribute to the extracellular rise in glutamate to excitotoxic levels. Selective inhibitors of neuronal glutamate transporters may therefore be of therapeutic interest to prevent reversed glutamate transport without affecting the capability of glial glutamate transporters to keep the extracellular glutamate concentration at low levels. The availability of selective chemical and subtype-selective inhibitors would be required to test this hypothesis.

Because glial high-affinity glutamate transporters play a critical role in keeping the extracellular glutamate concentration below neurotoxic levels, alteration of their function or expression levels may affect the postischemic vulnerability of neurons. Trop et al. showed that the GLT1 message and protein content were decreased in the hippocampal CA1 region in rats following transient forebrain ischemia, but the expression of the other glial isoform GLAST and the neuronal isoform EAAC1 were unaltered *(144)*. Raghavendra et al. observed that death of hippocampal neurons and corresponding loss of EAAC1 in gerbils following transient global cerebral ischemia is preceded by severe downregulation of GLT1 *(145)*. It is therefore likely that the GLT1 glutamate-transporter decrease is one of several factors that contribute to the high sensitivity of neurons to the postischemic damage. Aberrant functioning of glutamate transporters in ischemic conditions may also be directly mediated by polyunsaturated fatty acids, particularly by arachidonic acid, and reactive oxygen species (ROS), which can be produced and liberated in response to excessive glutamate-receptor activation (*see* **Subheading 12**).

8. GLUTAMATE-TRANSPORTER-KNOCKOUT MICE

Following the initial studies on glutamate-transporter knockout by the application of antisense oligonucleotides in vivo and in vitro, glutamate-transporter-knockout mice were generated for GLT1, GLAST, and EAAC1. The analysis of the phenotypes of these mice provided important insight into the physiological and pathological implications of glutamate transporters.

8.1. GLT1-Knockout Mice

Homozygous mice deficient in GLT1 showed dramatic lethal spontaneous epileptic seizures with a behavioral pattern similar to those of NMDA-induced seizures. The mice also exhibited increased susceptibility to acute cortical injury. Histological examination of GLT1-knockout mice revealed selective neuronal degeneration in the hippocampal CA1 region. These observations confirmed the major roles of the glial glutamate transporter GLT1 in neuroprotection *(120)*.

Electrophysiological analysis of the CA1 pyramidal neurons of these knockout mice revealed that the decay time-course of the synaptically evoked non-NMDA receptor-mediated EPSC are not altered compared with that of wild-type mice, consistent with the concept that the time-course of EPSC is determined by the properties of glutamate receptors and that glutamate transporters do not contribute to shape of the EPSC. In contrast, the inhibition of the NMDA-receptor-mediated EPSC in the presence of the rapidly dissociating NMDA-receptor antagonist was significantly less in the knockout mice than that in the wild-type mice, indicating that GLT1 significantly contributes to the removal of glutamate from the synaptic cleft. This confirmed the observation of Tong and Jahr *(134)*, based on experiments using glutamate-transporter inhibitors in hippocampal slices.

In the retina, GLT1 is found only in cones and various types of bipolar cells. GLT1-deficient mice show almost normal electroretinograms and mildly increased damage after ischemia, in contrast to the prominent changes of electroretinograms and severe retinal damage after ischemia in mice with normal levels of GLT1.

8.2. GLAST-Knockout Mice

Although GLAST-deficient mice developed normally and could manage simple coordinated tasks, they exhibited motor discoordination with more difficult tasks. This observation is consistent with the abnormality found in the cerebellum. Electrophysiologically, it was revealed that cerebellar Purkinje cells in the knockout mice remained to be innervated by climbing fibers, even at the adult stage. Furthermore, the knockout mice exhibited increased susceptibility to cerebellar injury. In the knockout mice, edema-related volume increases occurred after cerebellar injury. These observations indicate that GLAST plays critical roles in the cerebellar climbing-fiber synapse formation and in preventing excitotoxic cerebellar damage after acute brain injury *(146)*.

In the retina, GLAST is expressed in Muller cells. In GLAST-deficient mice, the electroretinogram beta-wave and oscillatory potentials are reduced, indicating that GLAST is required for normal signal transmission between photoreceptors and bipolar cells. In addition, retinal damage after ischemia is exacerbated in the GLAST-knockout mice, indicating that GLAST also plays a neuroprotective role during ischemia in the retina *(147)*.

GLAST-knockout mice were further examined for noise-induced damage to the auditory system, because GLAST is a glutamate transporter highly expressed in the cochlea. In the peripheral auditory system, where glutamate is the most likely neurotransmitter for afferent synapses, it was proposed that glutamate excitotoxicity may be involved in noise trauma caused by acoustic overstimulation. GLAST-deficient mice showed an increased accumulation of glutamate in perilymphs after acoustic overstimulation, resulting in greater hearing loss. Therefore, it is suggested that GLAST plays an important role in maintaining the concentration of glutamate in the perilymph at a nontoxic level during acoustic overstimulation *(148)*.

The effect of GLAST knockout was also investigated on epileptogenesis. In GLAST-knockout mice, the generalized seizure duration of amygdala-kindled seizures was significantly prolonged compared with that of wild-type mice. Furthermore, GLAST-knockout mice showed more severe stages of pentylenetetrazole-induced seizures than wild-type mice, and the latency to the onset of seizures was significantly shorter for the mutant mice, indicating that GLAST is one of the determinants of seizure susceptibility *(149)*.

8.3. EAAC1-Knockout Mice

In contrast to GLT1- and GLAST-knockout mice, EAAC1-knockout mice did not develop remarkable neurological symptoms or neurodegeneration during a period of more than 12 mo, except that homozygous mutants display a significantly reduced spontaneous locomotor activity. EAAC1-knockout mice developed dicarboxylic aminoaciduria, confirming the role of EAAC1 in the reabsorption of glutamate from the renal proximal tubules *(150)*.

9. REGULATION OF GLUTAMATE TRANSPORTERS

Several studies indicate that neuronal and non-neuronal factors are required for the expression and maintenance of functionally active glutamate transporters *(151–154)*. A search for the compounds that mimic the action of neuron-conditioned media has revealed several soluble (poly-) peptides, including hormones and growth and trophic factors, which are capable of regulating the expression of glutamate transporters. For example, GLT1 can be induced in astroglial cultures by the pituitary adenylate cyclase-activating peptide (PACAP), a neuron-derived peptide *(155)*, by brain-derived neurotrophic factor (BDNF), a neurotrophin that is predominantly expressed in neurons *(66)*, or by epidermal growth factor (EGF) receptor activation *(156)*. In C6 glioma, the Wnt-1 gene product, an autocrine- and paracrine-soluble factor, induces GLT1 expression *(157)*. Growth hormone stimulates GLT1 expression in the mouse placenta, whereas insulin-like growth factor II (IGF-II) downregulates EAAT4. Physiological concentrations of IGF-II ensure maintenance of GLT1, GLAST, and EAAC1 at normal levels *(158)*. Several growth factors that are neuroprotective also increase transport activity. For example, platelet-derived growth factor (PDGF) increases cell-surface expression of EAAC1 in C6 glioma cells, but has no effect on transporter expression *(159)*. Activity-dependent neurotrophic factor (ADNF) enhances basal glutamate transport in neocortical synaptosomes and attenuates oxidative stress-induced impairment of glutamate uptake, as does basic fibroblast growth factor (bFGF) and nerve-growth factor (NGF) *(160)*. Excitotoxic neuronal death triggered by inhibition of glutamate transporters can be attenuated by IGF-1, GDNF, and NT4/5 *(161)* and some of the protective effects of these factors may be caused by induction of the synthesis of new transporter molecules.

9.1. Protein Kinases in Glutamate-Transporter Regulation: Direct and Indirect Effects on Expression and Activity

Most of the agents discussed here activate a variety of intracellular signaling pathways, and several studies have examined the effects of direct activation of these pathways. Most of these studies have focused on the effects of protein kinase C (PKC). For example, activators of PKC have been shown to increase, decrease, or have no effect on GLT1 expression in various systems. Although it is presently unclear why there are such different effects, the simplest explanation is that there are differences in the cellular signal-transduction machinery in different systems, which could include variations in the expression of subtypes of PKC and/or accessory proteins required for protein trafficking. With the exception of GLAST, there is limited evidence that regulation by any of the protein kinases involves direct phosphorylation of the transporters.

9.2. Regulated Cell-Surface Expression of Glutamate Transporters

The functional activity of many membrane proteins, including various transporters and receptors, is rapidly regulated by changing their expression at the plasma membrane. This regulated trafficking of proteins is generally independent of protein synthesis and is not directly related to degradation, but is instead related to redistribution of receptors, transporters, or other proteins to or from the plasma membrane. For example, the signaling through many different types of receptors is regulated by changing the cell-surface expression. Agonist binding and receptor activation often causes internalization of the receptor and serves to attenuate receptor signaling. In recent years, several studies suggest that the cell-surface expression of most of the subtypes of glutamate transporters can also be regulated. In earlier studies, several groups provided evidence that PKC may regulate the activity of glutamate transporters. In C6 glioma cells which endogenously express only EAAC1, phorbol 12-myristate 13-acetate (PMA) doubles glutamate transport activity within minutes *(162)*. Using a membrane-impermeant biotinylation reagent that labels the cell-surface proteins, it was found that slightly less than one-half of the EAAC1 immunoreactivity is biotinylated (or presumably is on the cell surface) *(163)*. Within 30 min, PMA causes an increase in the amount of biotinylated (cell-surface) transporter and a decrease in the non-biotinylated (intracellular) fraction. This redistribution of transporter is also apparent in optical cross-sections of the cells. Therefore, the PMA-induced increase in activity is correlated with an increase in cell-surface expression.

As indicated previously, this regulated trafficking of EAAC1 is not unique to EAAC1. For example, the cell-surface expression of both GLAST and EAAT4 is increased in response to substrates *(66,164)*. The cell-surface expression of GLT1 and GLAST is decreased in response to activation of PKC *(164a)*. Together, these data suggest that there are several mechanisms to rapidly regulate individual subtypes of glutamate transporters.

9.3. Regulation of EAAT4 and EAAC1 by Interacting Proteins

In addition to growth factors or protein phosphorylation, glutamate transporters are regulated by their associated proteins. By means of yeast two-hybrid screening, Rothstein and colleagues identified distinct proteins that interact with EAAT4 and EAAC1. Two proteins, called GTRAP41 and GTRAP48 (for glutamate-transporter EAAT4-associated proteins) were found to specifically interact with the intracellular carboxy-terminal domain of EAAT4 and to modulate its glutamate-transport activity *(165)*. GTRAP41 is a 2,388 amino-acid-residue protein which exhibits 87% identity with β-spectrin III. GTRAP48 is a 1,527 amino-acid-residue protein with significant homology to p115 RhoGEF. It has a PDZ domain and some distinctive domains characteristic of guanine-nucleotide-exchange factors of the Rho family of G-proteins. The expression of either GTRAP41 or GTRAP48 resulted in an increase in V_{max} of glutamate transport without altering the Km value. It is speculated that EAAT4 couples to the actin cytoskeleton and to a Rho GTPase signaling via GTRAP41 and GTRAP48. Interestingly, EAAC1 interacts with a completely different protein *(166)*. Rothstein and colleagues identified a protein called GTRAP3-18 which interacts with the intracellular carboxy-terminal domain of EAAC1. GTRAP3-18 is a predicted 188 amino-acid hydrophobic protein with four possible TMDs. GTRAP3-18 possesses 95% sequence identity to "E18" encoded by a vitamin-A-responsive gene. In contrast to GTRAP41 and GTRAP48 which activate EAAT4-mediated transport, GTRAP3-18 reduces the EAAC1-mediated glutamate transport by decreasing the affinity of the transporter for glutamate. It was also observed that retinoic acid upregulates GTRAP3-18 expression and consequently inhibits EAAC1-mediated transport. The effect of retinoic acid is specific to EAAC1, because retinoic acid has no effect on the GLT1 or EAAT4-mediated transport.

10. ROLE OF GLUTAMATE TRANSPORTERS DURING DEVELOPMENT

Each of the five glutamate transporters shows a unique developmental expression pattern, suggesting that they play important roles in brain development and that each transporter subtype plays a distinct role in this process. GLAST mRNA and immunoreactivity is prominently expressed in the ventricular zone of the fetal brain as early as embryonic day E11 *(167,168)*. GLAST immunoreactivity is also associated with radial glial fibers between E11 and E18 *(168)*. At E15, cells expressing GLAST mRNA first appear in the mantle zone. From E18 to postnatal day 7, GLAST mRNA or its immunoreactivity gradually decrease from the ventricular zone and disappear from radial processes *(90,167)*, whereas they become stronger in astrocytes in early postnatal weeks. Quantitative immunoblot analysis of postnatal GLAST expression showed marked increases of GLAST protein in cerebellum and the cerebral cortex, only a small increase in the striatum and no increase in the spinal cord *(70)*.

GLT1 mRNA and immunoreactivity is also present in the ventricular zone in later embryonal stages *(70,90,167,169)*. In contrast to GLAST, GLT1 immunoreactivity has not been described in radial glial cells. Instead, GLT1 immunoreactivity and GLT1 mRNA has been observed in the embryonal brain along axonal pathways and in neuronal cell bodies *(70,169,170)*. This neuronal expression of GLT1 disappears after birth. Also, in contrast to GLAST, GLT1 expression in the proliferating cells of the subventricular zone persists into adulthood *(90)*. Immunoblot analysis demonstrated the strongest increase in postnatal GLT1 protein in striatum, followed by the cerebral cortex and spinal cord, and only a modest increase in the cerebellum *(70)*. A marked increase in GLT1 immunoreactivity in the postnatal brain was also noted *(110)*.

No significant expression of EAAC1 mRNA in the ventricular zone and only low levels were found in the mantle zone in the embryonal brain in an *in situ* hybridization study *(167)*, possibly as a result of low detection sensitivity. In contrast, immunocytochemical analysis demonstrated abundant expression of EAAC1 in the embryonal forebrain, including the marginal zone and cortical plate, neocortex, hippocampus, and striatum *(70)*. In the postnatal brain, EAAC1 immunoreactivity increased throughout the forebrain until P10, but then decreased in intensity by adulthood *(70)*. Immunoblot analysis also revealed a decrease in protein abundance in the cortex, striatum, cerebellum, and spinal cord between the first postnatal week and adulthood *(70)*.

EAAT4 mRNA was first detected at E13 in the Purkinje cell layer of the caudal cerebellum *(171)*. During late fetal and neonatal periods, EAAT4 mRNA is expressed only in the Purkinje-cell layer in the caudal cerebellum. By P7, EAAT4 mRNA is found in the Purkinje cells of the entire cerebellum. EAAT4 immunoreactivity was first observed at E18 in Purkinje cells perikarya of the caudal cerebellum, and this pattern of immunostaining was maintained at P1 and P7. At P14 and thereafter, the molecular layer in the entire cerebellum became immunopositive for EAAT4. EAAT4 immunoreactivity was also observed at low levels in the forebrain and thalamus by P10, and by P24 was uniformly enriched in the cerebellar cortex, but was expressed at low levels in the neocortex, striatum, hippocampus, and thalamus *(118)*. Immunoblot analysis also revealed a marked increase in EAAT4 levels in the second postnatal week in the cerebellum, and a minute increase in neocortical EAAT4 levels around P10 and P16 compared to neonatal or adult levels *(70)*.

EAAT5 immunostaining in photoreceptors is first found on postnatal day P7 in the central areas of the retina *(124)*. This time-point coincides with the arrival of photoreceptor synapses. By P10, the staining appeared strong in both the central and peripheral retina. By P21, EAAT5 staining also appeared in the bipolar cells of the inner nuclear layer, and became prominent in these cells by adulthood. Antigen retrieval also localized EAAT5 staining to bipolar-cell processes and terminals. GLAST, which is expressed in Mueller glial cells, is also switched on at P7 *(172)*. No developmental studies have been performed in the retina for the other glutamate transporters.

11. ROLE OF GLUTAMATE TRANSPORTERS IN AMYOTROPHIC LATERAL SCLEROSIS

ALS is a progressive neurological disorder characterized by degeneration of upper and lower motor neurons. Approximately 10% of the cases are inherited as an autosomal dominant trait. About 25% of these arise because of mutations in the gene for Cu/Zn superoxide dismutase (SOD1) *(173–175)*. Although the primary pathogenic trigger is still unknown, mounting evidence implicates a role for glutamate-mediated excitotoxicity in the disorder. An abnormality of the glutamate transport system has been reported in ALS patients. A decrease in the glutamate transporter activity was observed in synaptic preparation from the motor and sensory cortex of sporadic ALS patients *(176)* that was subsequently ascribed to reduced levels of the glutamate transporter isoform EAAT2, as assessed by immunodetection techniques *(112)*. Recently, aberrant splicing of the EAAT2 transcript was suggested to be the cause for a reduced expression of EAAT2 in ALS *(177)*. One of the novel transcripts contains an intronic sequence at the 3' end of the exon 7 sequence, leading to a truncated protein. A second transcript was characterized by skipping of the protein coding exon 8 of the EAAT2 gene. In vitro expression studies suggested that proteins translated from these transcripts were rapidly degraded, and show a dominant-negative effect on normal EAAT2 protein expression. However, the inolvement of aberrant splicing processes for the EAAT2 gene in ALS is still unclear. The exon-skipping transcript and the three other splice products were identified and cloned from normal human brain RNA by others *(178–182)*. Furthermore, it was shown that the alternative spliced forms are expressed in the spinal cord of ALS patients and controls *(183)*. Alternative splicing of the EAAT2 transcript was also found for untranslated exons in humans and in other species. The splicing process may be part of the EAAT2 gene regulation, leading to altered protein activity, protein localization, and RNA stability. In conclusion, the EAAT2 gene regulation and its relevance in ALS await further investigations.

The case of ALS illustrates the involvement of glutamate transporters in a neurodegenerative pathology that is also characterized by an oxidative stress condition. Studies by Rosen and colleagues *(175,184,185)* linked ALS to reactive oxygen species toxicity, since they showed that 25% of familial ALS patients carry missense mutations in the gene encoding copper-zinc superoxide dismutase (SOD1). Mice made transgenic for mutant SOD1 develop selective motor-neuron pathology which strongly resembles ALS *(186)*. Signs of oxidative stress have been detected in tissues from sporadic patients, suggesting that the etiology of the sporadic form may involve mechanisms similar to the SOD1-dependent defects in familial ALS. Some of the mutations in SOD1 not only reduce the capacity to detoxify superoxide, but also catalyze the formation of more deleterious oxidants, such as hydroxyl radicals.

Several lines of evidence suggest a link between oxidative stress mediated by the SOD1 mutants and glutamate-induced cell death. SOD1 mutant-induced damage may produce excessive glutamatergic stimulation, because of insufficient clearance of glutamate from the extracellular space by the glial glutamate transporter EAAT2. A loss of EAAT2 has been observed in post-mortem tissue *(112)* and in the spinal cord of transgenic SOD1-G85R mice *(187)*. Evidence has suggested that the loss of EAAT2 was caused by oxidative damage induced by the SOD1 mutation: EAAT2 is vulnerable to oxidative stress, and immunoprecipitation studies have revealed that EAAT2 is one of the proteins that is oxidatively damaged by 4-hydroxynonenal *(188)*. More direct evidence came from studies reporting that the activity of mutant SOD1 proteins directly impairs glutamate transport by EAAT2 in the presence of hydrogen peroxide *(189)*. This sensitivity to oxidation is peculiar to EAAT2, since the neuronal glutamate transporter EAAT3 was not inactivated. The oxidant-vulnerable site of EAAT2 resides within its intracellular carboxyl-terminal domain. Whether the oxidation involves intra- or intermolecular reactions or both is not known. The precise site(s) of oxidant attach within that domain have not yet been defined, but it seems unlikely that oxidation of a single residue is responsible for the loss of activity. Oxidation did not alter mechanistic properties of EAAT2 such as the affinity for glutamate and coupling coefficient, suggesting that the oxidized molecules are either nonfunctional or form a more rigid structure that impairs the transporter dynamics. Proteins exposed to oxidative damage have altered structure, and are likely to undergo spontaneous internalization and increased proteolitic fragmentation. This process may explain the loss of EAAT2 immunoreactivity detected in a transgenic model for ALS as well as in human patients.

11.1. Glutamate Transporter Mutations and Amyotrophic Lateral Sclerosis

Using single-strand conformation polymorphism (SSCP) analysis of genomic DNA, Brown and colleagues reported a mutation in the GLT1 gene associated with sporadic ALS. This mutation substitutes an asparagine for a serine at position 206 (N206S) *(190)*. The mutation suppressed the glycosylation of the glutamate transporter GLT1 at position 206 and affected its functional properties *(190a)*. Using a combined approach involving biophysical and cell biology techniques, we were able to show that the GLT1-N206S mutant has reduced glutamate transport activity and a dominant-negative impact on wild-type GLT1 activity. The decreased rate of transport was mainly caused by a reduction of the GLT1-N206S mutant in the plasma membrane. Moreover, the GLT1-N206S exhibited an increased reverse transport capacity. These combined effects are predicted to impair the ability of GLT1 to mop up glutamate in vivo, at the synaptic cleft of neurons.

The evidence that a missense mutation of GLT1 is present in a patient with sporadic ALS, and that such a mutation affects the capacity of a cell to regulate the glutamate concentration at synapses, supports the concept that this mutation contributes to excitotoxicity that occurs in ALS.

12. GLUTAMATE TRANSPORTERS, OXIDATIVE STRESS, AND ARACHIDONIC ACID

An event common to several neurodegenerative pathologies and pathologic conditions such as ischemia after a stroke is activation of phospholipases with release of free fatty acids, particularly of polyunsaturated fatty acids such as arachidonic acid, from membrane phospholipids *(191)*. In most of these situations, fatty acids tend to accumulate in brain tissue, because reacylation in phospholipids—which is an ATP-dependent process—is compromised by the energy failure. Moreover, ROS, including superoxide anion (O_2^-), hydrogen peroxide (H_2O_2), and hydroxyl radical (OH), may escape the impaired antioxidant defenses and target vital cell components *(192)*. Several lines of evidence suggest their participation in the mechanisms of neuronal damage *(193)*. However, dysregulation of glutamatergic transmission is known to be one of the major components leading to neurotoxicity, and elevated extracellular levels of glutamate, which were observed for both in vitro and in vivo models of ischemia, trauma, and seizures, were directly correlated with damage *(194)*. Thus, excess stimulation of different types of glutamate receptors results in several forms of neuronal damage by both Ca^{2+}-dependent and Ca^{2+}-independent mechanisms (broadly termed "excitotoxicity"), and often leads to irreversible degeneration of nerve cells *(195)*. Arachidonic acid release, ROS formation, and extracellular glutamate accumulation are interdependent phenomena, and not just parallel events. Arachidonic acid and ROS are both liberated in response to glutamate-receptor activation *(128)*. Arachidonic acid and ROS oppose the removal of glutamate from the extracellular space by inhibiting certain high-affinity glutamate transporter isoforms. Thus, their feedback inhibition on glutamate uptake is predicted to trigger a vicious cycle which likely contributes to the elevation of extracellular glutamate to neurotoxic levels.

12.1. Inhibition of Glutamate Uptake by Arachidonic Acid

By using synaptosomes and cultured astrocytes from rat cerebral cortex, Volterra and colleagues reported that arachidonic acid leads to a rapid (within 30 s) and largely reversible reduction in glutamate uptake activity *(196)*. Inhibition of glutamate uptake could also be achieved through the modulation of the extracellular levels of endogenous arachidonic acid by using compounds that interfere with its deacylation-reacylation cycle into the membrane phospholipids such as melittin (a phospholipase A_2 activator) and thimerosal (an inhibitor of the reacylation pathway).

Inhibition of glutamate uptake into salamander retinal glial cells by arachidonic acid was shown by monitoring the inward current induced by the electrogenic glutamate transport *(197)*. Moreover, arachidonic acid does not act through mechanisms mediated by eicosanoid formation or PKC activation *(196,197)*.

The inhibitory effect of arachidonic acid on glutamate uptake was also demonstrated in a simple system, consisting of purified glutamate transporter protein reconstituted in liposomes *(198)*. Therefore, it was possible to exclude indirect actions of arachidonic acid such as changes in the ion gradients (inhibition of Na^+/K^+-ATPase and modification of ion channels), protein kinase C activation, formation of downstream mediators (enzymatic derivatives of arachidonic acid, briefly eicosanoids) or changes in membrane lipid composition caused by active esterification of arachidonic acid in phospholipids by acetyltransferase enzymes. Using this model, it was demonstrated that inhibition of glutamate uptake by arachidonic acid is caused by reversible binding of free aqueous arachidonic acid to either the transporter protein or the protein-lipid boundary. The binding, which requires both a hydrophobic *cis*-polyunsaturated carbon chain and a free carboxyl group, is probably of low affinity, since it easily reversed by simple dilution. This procedure does not remove arachidonic acid bound to liposomal lipids (approx 95% of the added arachidonic acid), suggesting that only the arachidonic acid partitioning to the aqueous phase affects transport by binding to the transporter protein, yet the binding to the phospholipid membranes does not seem to affect the function of the glutamate transporter. Arachidonic acid exerts an opposite effect on the uptake by 2 out of 4 human glutamate transporters *(199)*. Micromolar amounts of arachidonic acid inhibited glutamate uptake mediated by human GLAST (EAAT1) by reducing the maximal transport rate. In contrast, arachidonic acid increased transport mediated by human GLT1 (EAAT2) by causing an increase in the apparent affinity for glutamate more than twofold. Human EAAC1 (EAAT3) activity was not affected, yet AA increased selectively the proton conductance associated to the transport operation of EAAT4 *(200)*. The effect of AA on EAAT5 has not yet been reported.

12.2. Inhibition of Glutamate Uptake by Oxygen Free Radicals

In pathological conditions such as ischemia after a stroke, oxygen free radical production overcome the endogenous protection system consisting of scavenger enzymes and antioxidant molecules. As a consequence, reactive species, including superoxide anion and, particularly, hydroxyl radicals, can attack cellular components. The high-affinity glutamate transporters are important targets of ROS. In cortical glial cultures, Volterra and colleagues have demonstrated that glutamate uptake is persistently inhibited by brief exposures to reactive oxygen species such as superoxide anion and H_2O_2 *(201)*. This effect is antagonized and partly reversed (>50%) by disulfide-reducing agents (e.g., dithiothreitol [DTT]), suggesting that cysteine residues of glutamate transporters are the target for inhibition. H_2O_2 attenuates the electrogenic glutamate uptake current in voltage-clamped astrocytes with a minor or no effect on resting membrane conductance. Therefore, H_2O_2 probably directly reacts with glutamate transporters. Although both arachidonic acid and ROS appear to directly modify glutamate transporters, their mechanism of action must be distinct and independent because their effects were fully additive and different pharmacological agents selectively blocked their action *(202)*.

REFERENCES

1. Fonnum, F. (1984) Glutamate: a neurotransmitter in mammalian brain. *J. Neurochem.* **42(1),** 1–11.
2. Fukuhara, Y. and Turner, R. J. (1985) Cation dependence of renal outer

cortical brush border membrane 1- glutamate transport. *Am. J. Physiol.* **248(6 Pt 2)**, F869–F875.

3. Lerner, J. (1987) Acidic amino acid transport in animal cells and tissues. comp. biochem. Physiol B 87(3), 443–457.

4. Nicholls, D. and Attwell, D. (1990) The release and uptake of excitatory amino acids. *Trends Pharmacol. Sci.* **11(11)**, 462–468.

5. Schousboe, A. (1981) Transport and metabolism of glutamate and GABA in neurons are glial cells. *Int. Rev. Neurobiol.* **22**, 1–45.

6. Somohano, F. and Lopez-Colome, A. M. (1991) Characteristics of excitatory amino acid uptake in cultures from neurons and glia from the retina. *J. Neurosci. Res.* **28(4)**, 556–562.

7. Wingrove, T. G. and Kimmich, G. A. (1988) Low-affinity intestinal L-aspartate transport with 2:1 coupling stoichiometry for Na+/Asp. *Am. J. Physiol* **255(6 Pt 1)**, C737–C744.

8. Levi, G. and Raiteri, M. (1973) Detectability of high and low affinity uptake systems for GABA and glutamate in rat brain slices and synaptosomes. *Life Sci. I.* **12(2)**, 81–88.

9. Otis, T. S. (2001) Vesicular glutamate transporters in cognito. *Neuron* **29(1)**, 11–14.

10. Bellocchio, E. E., Reimer, R. J., Fremeau, R. T., Jr., and Edwards, R. H. (2000) Uptake of glutamate into synaptic vesicles by an inorganic phosphate transporter. *Science* **289(5481)**, 957–960.

11. Takamori, S., Rhee, J. S., Rosenmund, C., and Jahn, R. (2000) Identification of a vesicular glutamate transporter that defines a glutamatergic phenotype in neurons. *Nature* **407(6801)**, 189–194.

12. Kanai, Y. and Hediger, M. A. (1992) Primary structure and functional characterization of a high-affinity glutamate transporter. *Nature* **360**, 467–471.

13. Danbolt, N. C., Storm-Mathisen, J., and Kanner, B. I. (1992) An [Na$^+$ + K$^+$]coupled L-glutamate transporter purified from rat brain is located in glial cell processes. *Neuroscience* **51(2)**, 295–310.

14. Pines, G., Danbolt, N. C., Bjoras, M., Zhang, Y., Bendahan, A., Eide, L., et al. (1992) Cloning and expression of a rat brain L-glutamate transporter. *Nature* **360(6403)**, 464–467.

15. Storck, T., Schulte, S., Hofmann, K., and Stoffel, W. (1992) Structure, expression, and functional analysis of a Na(+)- dependent glutamate/aspartate transporter from rat brain. *Proc. Natl. Acad. Sci. USA* **89(22)**, 10,955–10,959.

16. Fairman, W. A., Vandenberg, R. J., Arriza, J. L., Kavanaugh, M. P., and Amara, S. G. (1995) an excitatory amino-acid transporter with properties of a ligand- gated chloride channel. *Nature* **375(6532)**, 599–603.

17. Arriza, J. L., Eliasof, S., Kavanaugh, M. P., and Amara, S. G. (1997) Excitatory amino acid transporter 5, a retinal glutamate transporter coupled to a chloride conductance. *Proc. Natl. Acad. Sci. USA* **94(8)**, 4155–4160.

18. Shafqat, S., Tamarappoo, B. K., Kilberg, M. S., Puranam, R. S., McNamara, J. O., Guadano-Ferraz, A., et al. (1993) Cloning and expression of a novel Na($^+$)-dependent neutral amino acid transporter structurally related to mam-

malian Na⁺/glutamate cotransporters. *J. Biol. Chem.* **268(21)**, 15,351–15,355.

19. Arriza, J. L., Kavanaugh, M. P., Fairman, W. A., Wu, Y. N., Murdoch, G. H., North, R. A., and Amara, S. G. (1993) Cloning and expression of a human neutral amino acid transporter with structural similarity to the glutamate transporter gene family. *J. Biol. Chem.* **268(21)**, 15,329–15,332.

20. Vadgama, J. V. and Christensen, H. N. (1984) Wide distribution of pH-dependent service of transport system ASC for both anionic and zwitterionic amino acids. *J. Biol. Chem.* **259(6)**, 3648–3652.

21. Kanai, Y. (1997) Family of neutral and acidic amino acid transporters: molecular biology, physiology and medical implications. *Curr. Opin. Cell Biol.* **9(4)**, 565–572.

22. Kekuda, R., Prasad, P. D., Fei, Y. J., Torres-Zamorano, V., Sinha, S., Yang-Feng, T. L., et al. (1996) Cloning of the sodium-dependent, broad-scope, neutral amino acid transporter bo from a human placental choriocarcinoma cell line. *J. Biol. Chem.* **271(31)**, 18,657–18,661.

23. Kilberg, M. S., Stevens, B. R., and Novak, D. A. (1993) Recent advances in mammalian amino acid transport. *Annu. Rev. Nutr.* **13**, 137–165.

24. Kanai, Y., Smith, C. P., and Hediger, M. A. (1993) The elusive transporters with a high affinity for glutamate. *Trends Neurosci.* **16**, 365–730.

25. Diez-Sampedro, A., Eskandari, S., Wright, E. M., and Hirayama, B. A. (2001) Na+-to-sugar stoichiometry of sglt3. *Am. J. Physiol. Renal Physiol.* **280(2)**, F278–F282.

26. Hediger, M. A. and Rhoads, D. B. (1994) Molecular physiology of sodium-glucose cotransporters. *Physiol. Rev.* **74(4)**, 993–1026.

27. Stallcup, W. B., Bulloch, K., and Baetge, E. E. (1979) Coupled transport of glutamate and sodium in a cerebellar nerve cell line. *J. Neurochem.* **32(1)**, 57–65.

28. Barbour, B., Brew, H., and Attwell, D. (1988) Electrogenic glutamate uptake in glial cells is activated by intracellular potassium. *Nature* **335(6189)**, 433–435.

29. Brew, H. and Attwell, D. (1987) Electrogenic glutamate uptake is a major current carrier in the membrane of axolotl retinal glial cells [Published Erratum in *Nature* (1987) **328(6132)**, 742]. *Nature* **327(6124)**, 707–709.

30. Sarantis, M. and Attwell, D. (1990) Glutamate uptake in mammalian retinal glia is vol. *Brain Res.* **516(2)**, 322–325.

31. Kimelberg, H. K., Pang, S., and Treble, D. H. (1989) Excitatory amino acid-stimulated uptake of 22Na+ in primary astrocyte cultures. *J. Neurosci.* **9(4)**, 1141–1149.

32. Heinz, E., Sommerfeld, D. L., and Kinne, R. K. (1988) Electrogenicity of sodium/L-glutamate cotransport in rabbit renal brush- border membranes: a reevaluation. *Biochim. Biophys. Acta* **937(2)**, 300–308.

33. Nelson, P. J., Dean, G. E., Aronson, P. S., and Rudnick, G. (1983) Hydrogen ion cotransport by the renal brush border glutamate transporter. *Biochemistry* **22(23)**, 5459–5463.

34. Koepsell, H., Korn, K., Ferguson, D., Menuhr, H., Ollig, D., and Haase, W. (1984) Reconstitution and partial purification of several Na+ cotransport

systems from renal brush-border membranes. properties of the L-glutamate transporter in proteoliposomes. *J. Biol. Chem.* **259(10)**, 6548–6558.

35. Romano, P. M., Ahearn, G. A., and Storelli, C. (1989) Na-dependent L-glutamate transport by EEL intestinal BBMV: role of K+ and Cl-. *Am. J. Physiol.* **257(1 Pt 2)**, R180–R188.

36. Barbour, B., Brew, H., and Attwell, D. (1991) Electrogenic uptake of glutamate and aspartate into glial cells isolated from the salamander (Ambystoma) retina. *J. Physiol.* **436**, 169–193.

37. Billups, B. and Attwell, D. (1996) Modulation of non-vesicular glutamate release by pH. *Nature* **379(6561)**, 171–174.

38. Bouvier, M., Szatkowski, M., Amato, A., and Attwell, D. (1992) The glial cell glutamate uptake carrier countertransports pH-changing anions. *Nature* **360(6403)**, 471–474.

39. Kanai, Y., Nussberger, S., Romero, M. F., Boron, W. F., Hebert, S. C., and Hediger, M. A. (1995) Electrogenic properties of the epithelial and neuronal high affinity glutamate transporter. *J. Biol. Chem.* **270(28)**, 16,561–16,568.

40. Nussberger, S., Foret, F., Hebert, S. C., Karger, B. L., and Hediger, M. A. (1996) Nonradioactive monitoring of organic and inorganic solute transport into single Xenopus oocytes by capillary zone electrophoresis. *Biophys. J.* **70(2)**, 998–1005.

41. Zerangue, N. and Kavanaugh, M. P. (1996) Interaction of L-cysteine with a human excitatory amino acid transporter. *J. Physiol.* **493(Pt 2)**, 419–423.

42. Zerangue, N. and Kavanaugh, M. P. (1996) Flux coupling in a neuronal glutamate transporter. *Nature* **383(6601)**, 634–637.

43. Auger, C. and Attwell, D. (2002) Fast removal of synaptic glutamate by postsynaptic transporters. *Neuron* **28**, 547–558.

44. Levy, L. M., Warr, O., and Attwell, D. (1998) Stoichiometry of the glial glutamate transporter GLT-1 expressed inducibly in a chinese hamster ovary cell line selected for low endogenous Na+-dependent glutamate uptake. *J. Neurosci.* **18(23)**, 9620–9628.

45. Loo, D. D., Hazama, A., Supplisson, S., Turk, E., and Wright, E. M. (1993) Relaxation kinetics of the Na+/glucose cotransporter. *Proc. Natl. Acad. Sci. USA* **90(12)**, 5767–5771.

46. Mager, S., Naeve, J., Quick, M., Labarca, C., Davidson, N., and Lester, H. A. (1993) Steady states, charge movements, and rates for a cloned GABA transporter expressed in Xenopus oocytes. *Neuron* **10(2)**, 177–188.

47. Wadiche, J. I., Arriza, J. L., Amara, S. G., and Kavanaugh, M. P. (1995) Kinetics of a human glutamate transporter. *Neuron* **14(5)**, 1019–1027.

48. Kanai, Y., Stelzner, M., Nussberger, S., Khawaja, S., Hebert, S. C., Smith, C. P., and Hediger, M. A. (1994) The Neuronal and epithelial human high affinity glutamate transporter. Insights into structure and mechanism of transport. *J. Biol. Chem.* **269(32)**, 20,599–20,606.

49. Klockner, U., Storck, T., Conradt, M., and Stoffel, W. (1993) Electrogenic L-glutamate uptake in Xenopus laevis oocytes expressing a cloned rat brain L-glutamate/L-aspartate transporter (GLAST-1). *J. Biol. Chem.* **268(20)**, 14,594–14,596.

50. Otis, T. S. and Kavanaugh, M. P. (2000) Isolation of current components and partial reaction cycles in the glial glutamate transporter EAAT2. *J. Neurosci.* **20(8),** 2749–2757.
51. Wadiche, J. I., Amara, S. G., and Kavanaugh, M. P. (1995) Ion fluxes associated with excitatory amino acid transport. *Neuron* **15(3),** 721–728.
52. Auger, C. and Attwell, D. (2000) Fast removal of synaptic glutamate by postsynaptic transporters. *Neuron* **28(2),** 547–558.
53. Zerangue, N. and Kavanaugh, M. P. (1996) ASCT-1 is a neutral amino acid exchanger with chloride channel activity. *J. Biol. Chem.* **271(45),** 27,991–27,994.
54. Broer, A., Wagner, C., Lang, F., and Broer, S. (2000) Neutral amino acid transporter SCT2 displays substrate-induced Na+ exchange and a substrate-gated anion conductance. *Biochem. J.* **346(Pt 3),** 705–710.
55. Kavanaugh, M. P., Bendahan, A., Zerangue, N., Zhang, Y., and Kanner, B. I. (1997) Mutation of an amino acid residue influencing potassium coupling in the glutamate transporter GLT-1 induces obligate exchange. *J. Biol. Chem.* **272(3),** 1703–1708.
56. Zhang, Y., Bendahan, A., Zarbiv, R., Kavanaugh, M. P., and Kanner, B. I. (1998) Molecular determinant of ion selectivity of a (Na+ + K+)-coupled rat brain glutamate transporter. *Proc. Natl. Acad. Sci. USA* **95(2),** 751–755.
57. Utsunomiya-Tate, N., Endou, H., and Kanai, Y. (1996) Cloning and functional characterization of a system ASC-like Na+- dependent neutral amino acid transporter. *J. Biol. Chem.* **271(25),** 14,883–14,890.
58. Christensen, H. N. (1990) Role of amino acid transport and countertransport in nutrition and metabolism. *Physiol. Rev.* **70(1),** 43–77.
59. Tamarappoo, B. K., McDonald, K. K., and Kilberg, M. S. (1996) Expressed human hippocampal ASCT1 amino acid transporter exhibits a pH-dependent change in substrate specificity. *Biochim. Biophys. Acta* **1279(2),** 131–136.
60. Kanai, Y., Smith, C. P., and Hediger, M. A. (1993) A new family of neurotransmitter transporters: the high-affinity glutamate transporters. *FASEB J.* **7(15),** 1450–1459.
61. Grunewald, M. and Kanner, B. I. (2000) The accessibility of a novel reentrant loop of the glutamate transporter GLT-1 is restricted by its substrate. *J. Biol. Chem.* **275(13),** 9684–9689.
62. Seal, R. P., Leighton, B. H., and Amara, S. G. (2000) A model for the topology of excitatory amino acid transporters determined by the extracellular accessibility of substituted cysteines. *Neuron* **25(3),** 695–706.
63. Zhang, Y. and Kanner, B. I. (1999) Two serine residues of the glutamate transporter GLT-1 are crucial for coupling the fluxes of sodium and the neurotransmitter. *Proc. Natl. Acad. Sci. USA* **96(4),** 1710–1715.
64. O'Hara, P. J., Sheppard, P. O., Thogersen, H., Venezia, D., Haldeman, B. A., McGrane, V., et al. (1993) The ligand-binding domain in metabotropic glutamate receptors is related to bacterial periplasmic binding proteins. *Neuron* **11(1),** 41–52.
65. Eskandari, S., Kreman, M., Kavanaugh, M. P., Wright, E. M., and Zampighi, G. A. (2000) Pentameric assembly of a neuronal glutamate transporter. *Proc. Natl. Acad. Sci. USA* **97(15),** 8641–8646.

66. Gegelashvili, G., Dehnes, Y., Danbolt, N. C., and Schousboe, A. (2000) The high-affinity glutamate transporters GLT1, GLAST, and EAAT4 are regulated via different signalling mechanisms. *Neurochem. Int.* **37(2-3)**, 163–170.

67. Meister, B., Arvidsson, U., Zhang, X., Jacobsson, G., Villar, M. J., and Hokfelt, T. (1993) Glutamate transporter mRNA and glutamate-like immunoreactivity in spinal motoneurones. *Neuroreport* **5(3)**, 337–340.

68. Rothstein, J. D., Martin, L., Levey, A. I., Dykes-Hoberg, M., Jin, L., Wu, D., Nash, N., and Kuncl, R. W. (1994) Localization of neuronal and glial glutamate transporters. *Neuron* **13(3)**, 713–725.

69. Shashidharan, P., Huntley, G. W., Murray, J. M., Buku, A., Moran, T., Walsh, M. J., Morrison, J. H., and Plaitakis, A. (1997) Immunohistochemical localization of the neuron-specific glutamate transporter EAAC1 (EAAT3) in rat brain and spinal cord revealed by a novel monoclonal antibody. *Brain Res.* **773(1-2)**, 139–148.

70. Furuta, A., Rothstein, J. D., and Martin, L. J. (1997) Glutamate transporter protein subtypes are expressed differentially during rat CNS development. *J. Neurosci.* **17(21)**, 8363–8375.

71. Shashidharan, P., Huntley, G. W., Meyer, T., Morrison, J. H., and Plaitakis, A. (1994) Neuron-specific human glutamate transporter: molecular cloning, characterization and expression in human brain. *Brain Res.* **662**, 245–250.

72. Kanai, Y., Bhide, P. G., DiFiglia, M., and Hediger, M. A. (1995) Neuronal high-affinity glutamate transport in the rat central nervous system. *Neuroreport* **6(17)**, 2357–2362.

73. Kiryu, S., Yao, G. L., Morita, N., Kato, H., and Kiyama, H. (1995) Nerve injury enhances rat neuronal glutamate transporter expression: identification by differential display PCR. *J. Neurosci.* **15(12)**, 7872–7878.

74. Bjoras, M., Gjesdal, O., Erickson, J. D., Torp, R., Levy, L. M., Ottersen, O. P., et al. Cloning and expression of a neuronal rat brain glutamate transporter. *Brain Res. Mol. Brain Res.* **36(1)**, 163–168.

75. Velaz-Faircloth, M., McGraw, T. S., alandro, M. S., Fremeau, R. T., Jr., Kilberg, M. S., and Anderson, K. J. (1996) Characterization and distribution of the neuronal glutamate transporter EAAC1 in rat brain. *Am. J. Physiol.* **270(1 Pt 1)**, C67–C75.

76. Torp, R., Hoover, F., Danbolt, N. C., Storm-Mathisen, J., and Ottersen, O. P. (1997) Differential distribution of the glutamate transporters GLT1 and REAAC1 in rat cerebral cortex and thalamus: an in situ hybridization analysis. *Anat. Embryol. (Berl)* **195(4)**, 317–326.

77. Berger, U. V. and Hediger, M. A. (1998) Comparative analysis of glutamate transporter expression in rat brain using differential double in situ hybridization. *Anat. Embryol. (Berl.)* **198(1)**, 13–30.

78. He, Y., Janssen, W. G., Rothstein, J. D., and Morrison, J. H. (2000) Differential synaptic localization of the glutamate transporter EAAC1 and glutamate receptor subunit GluR2 in the rat hippocampus. *J. Comp Neurol.* **418(3)**, 255–269.

79. Domercq, M. and Matute, C. (1999) Expression of glutamate transporters in the adult bovine corpus callosum. *Brain Res. Mol. Brain Res.* **67(2)**, 296–302.

80. Domercq, M., Sanchez-Gomez, M. V., Areso, P., and Matute, C. (1999) Expression of glutamate transporters in rat optic nerve oligodendrocytes. *Eur. J. Neurosci.* **11(7),** 2226–2236.
81. Conti, F., DeBiasi, S., Minelli, A., Rothstein, J. D., and Melone, M. (1998) EAAC1, a high-affinity glutamate tranporter, is localized to astrocytes and gabaergic neurons besides pyramidal cells in the rat cerebral cortex. *Cereb. Cortex* **8(2),** 108–116.
82. Kugler, P. and Schmitt, A. (1999) Glutamate transporter EAAC1 is expressed in neurons and glial cells in the rat nervous system. *Glia* **27(2),** 129–142.
83. Palos, T. P., Ramachandran, B., Boado, R., and Howard, B. D. (1996) Rat C6 and human astrocytic tumor cells express a neuronal type of glutamate transporter. *Brain Res. Mol. Brain Res.* **37(1-2),** 297–303.
84. Schultz, K. and Stell, W. K. (1996) Immunocytochemical localization of the high-affinity glutamate transporter, EAAC1, in the retina of representative vertebrate species. *Neurosci. Lett.* **211(3),** 191–194.
85. Rauen, T., Rothstein, J. D., and Wassle, H. (1996) Differential expression of three glutamate transporter subtypes in the rat retina. *Cell Tissue Res.* **286(3),** 325–336.
86. Arriza, J. L., Fairman, W. A., Wadiche, J. I., Murdoch, G. H., Kavanaugh, M. P., and Amara, S. G. (1994) Functional comparisons of three glutamate transporter subtypes cloned from human motor cortex. *J. Neurosci.* **14(9),** 5559–5569.
87. Mukainaka, Y., Tanaka, K., Hagiwara, T., and Wada, K. (1995) Molecular cloning of two glutamate transporter subtypes from mouse brain. *Biochim. Biophys. Acta* **1244(1),** 233–237.
88. Nakayama, T., Kawakami, H., Tanaka, K., and Nakamura, S. (1996) Expression of three glutamate transporter subtype mRNAs in human brain regions and peripheral tissues. *Brain Res. Mol. Brain Res.* **36(1),** 189–192.
89. Torp, R., Danbolt, N. C., Babaie, E., Bjoras, M., Seeberg, E., Storm-Mathisen, J., and Ottersen, O. P. (1994) Differential expression of two glial glutamate transporters in the rat brain: an in situ hybridization study. *Eur. J. Neurosci.* **6(6),** 936–942.
90. Sutherland, M. L., Delaney, T. A., and Noebels, J. L. (1996) Glutamate transporter mRNA expression in proliferative zones of the developing and adult murine CNS. *J. Neurosci.* **16(7),** 2191–2207.
91. Schmitt, A., Asan, E., Puschel, B., Jons, T., and Kugler, P. (1996) Expression of the glutamate transporter GLT1 in neural cells of the rat central nervous system: non-radioactive in situ hybridization and comparative immunocytochemistry. *Neuroscience* **71(4),** 989–1004.
92. Torp, R., Hoover, F., Danbolt, N. C., Storm-Mathisen, J., and Ottersen, O. P. (1997) Differentiated distribution of the glutamate transporters GLT1 and rEAAC1 in rat cerebral cortex and thalamus: an in site hybridization analysis. *Anat. Embryol.* **195(4),** 317–326.
93. Levy, L. M., Lehre, K. P., Rolstad, B., and Danbolt, N. C. (1993) A Monoclonal Antibody raised against an [Na(+)+K+]coupled L-glutamate trans-

porter purified from rat brain confirms glial cell localization. *FEBS Lett.* **317(1-2)**, 79–84.

94. Lehre, K. P., Levy, L. M., Ottersen, O. P., Storm-Mathisen, J., and Danbolt, N. C. (1995) differential expression of two glial glutamate transporters in the rat brain: quantitative and immunocytochemical Observations. *J. Neurosci.* **15(3 Pt 1)**, 1835–1853.

95. Chaudhry, F. A., Lehre, K. P., van Lookeren, Campagne M., Ottersen, O. P., Danbolt, N. C., and Storm-Mathisen, J. (1995) Glutamate transporters in glial plasma membranes: highly differentiated localizations revealed by quantitative ultrastructural immunocytochemistry. *Neuron* **15(3)**, 711–720.

96. Lehre, K. P. and Danbolt, N. C. (1998) The number of glutamate transporter subtype molecules at glutamatergic synapses: chemical and stereological quantification in young adult rat brain. *J. Neurosci.* **18(21)**, 8751–8757.

97. Berger, U. V. and Hediger, M. A. (2000) Distribution of the glutamate transporters GLAST and GLT-1 in rat circumventricular organs, meninges, and dorsal root ganglia. *J. Comp. Neurol.* **421(3)**, 385–399.

98. Redecker, P. and Pabst, H. (2000) Immunohistochemical study of the glutamate transporter proteins GLT-1 and GLAST in rat and gerbil pineal gland. *J. Pineal Res.* **28(3)**, 179–184.

99. Yamada, H., Yatsushiro, S., Yamamoto, A., Hayashi, M., Nishi, T., Futai, M., et al. (1997) Functional expression of a GLT-1 type Na+-dependent glutamate transporter in rat pinealocytes. *J. Neurochem.* **69(4)**, 1491–1498.

100. Weiss, M. S., Derazi, S., Kilberg, M., and Anderson, K., J. (2001) Ontogeny and cellular localization of the neutral amino acid transporter ASCT1 in the rat brain. *Brain. Res. Dev. Brain. Res.* **130(2)**, 183–190.

101. Chen, W., Aoki, C., Gruber, C., Hadley, R., Wang, G., Blitzblau, R., et al. (2000) Molecular cloning, functional characterization, and neuronal localization of a variant form of the glutamate transporter GLT1. *Soc. Neurosci. Abstr.* **26.**

102. Grunert, U., Martin, P. R., and Wassle, H. (1994) Immunocytochemical analysis of bipolar cells in the Macaque monkey retina. *J. Comp. Neurol.* **348(4)**, 607–627.

103. Rauen, T. and Kanner, B. I. (1994) Localization of the glutamate transporter GLT-1 in rat and macaque monkey retinae. *Neurosci. Lett.* **169(1-2)**, 137–140.

104. Berger, U. V. and Hediger, M. A. (2001) Differential distribution of the glutamate transporters GLT-1 and GLAST in tanycytes of the third ventricle. *J. Comp Neurol.* **433(1)**, 101–114.

105. Kirschner, M. A., Copeland, N. G., Gilbert, D. J., Jenkins, N. A., and Amara, S. G. (1994) Mouse excitatory amino acid transporter EAAT2: isolation, characterization, and proximity to neuroexcitability loci on mouse chromosome 2. *Genomics* **24(2)**, 218–224.

106. Manfras, B. J., Rudert, W. A., Trucco, M., and Boehm, B. O. (1994) Cloning and characterization of a glutamate transporter cDNA from human brain and pancreas. *Biochim. Biophys. Acta* **1195(1)**, 185–188.

107. Matthews, J. C., Beveridge, M. J., Malandro, M. S., Rothstein, J. D., Campbell-Thompson, M., Verlander, J. W., Kilberg, M. S., and Novak, D.

A. (1998) Activity and protein localization of multiple glutamate transporters in gestation day 14 vs. day 20 rat placenta. *Am. J. Physiol* **274(3 Pt 1),** C603–C614.

108. Mason, D. J., Suva, L. J., Genever, P. G., Patton, A. J., Steuckle, S., Hillam, R. A., and Skerry, T. M. (1997) Mechanically regulated expression of a neural glutamate transporter in bone: a role for excitatory amino acids as osteotropic agents? *Bone* **20(3),** 199–205.

109. Schmitt, A., Asan, E., Puschel, B., and Kugler, P. (1997) Cellular and regional distribution of the glutamate transporter GLAST in the CNS of rats: nonradioactive in situ hybridization and comparative immunocytochemistry. *J. Neurosci.* **17(1),** 1–10.

110. Ullensvang, K., Lehre, K. P., Storm-Mathisen, J., and Danbolt, N. C. (1997) Differential developmental expression of the two rat brain glutamate transporter proteins GLAST and GLT. *Eur. J. Neurosci.* **9(8),** 1646–1655.

111. Bar-Peled, O., Ben Hur, H., Biegon, A., Groner, Y., Dewhurst, S., Furuta, A., and Rothstein, J. D. (1997) Distribution of glutamate transporter subtypes during human brain development. *J. Neurochem.* **69(6),** 2571–2580.

112. Rothstein, J. D., Van Kammen, M., Levey, A. I., Martin, L. J., and Kuncl, R. W. (1995) Selective loss of glial glutamate transporter GLT-1 in amyotrophic lateral sclerosis. *Ann. Neurol.* **38(1),** 73–84.

113. Lehre, K. P., Davanger, S., and Danbolt, N. C. (1997) Localization of the glutamate transporter protein GLAST in rat retina. *Brain Res.* **744(1),** 129–137.

114. Rauen, T., Taylor, W. R., Kuhlbrodt, K., and Wiessner, M. (1998) High-affinity glutamate transporters in the rat retina: a major role of the glial glutamate transporter GLAST-1 in transmitter clearance. *Cell Tissue Res.* **291(1),** 19–31.

115. Li, H. S., Niedzielski, A. S., Beisel, K. W., Hiel, H., Wenthold, R. J., and Morley, B. J. (1994) Identification of a glutamate/aspartate transporter in the rat cochlea. *Hear. Res.* **78(2),** 235–242.

116. Furness, D. N. and Lehre, K. P. (1997) Immunocytochemical localization of a high-affinity glutamate-aspartate transporter, GLAST, in the rat and guinea-pig cochlea. *Eur. J. Neurosci.* **9(9),** 1961–1969.

117. Yamada, K., Watanabe, M., Shibata, T., Tanaka, K., Wada, K., and Inoue, Y. (1996) EAAT4 is a post-synaptic glutamate transporter at Purkinje cell synapses. *Neuroreporter* **7(12),** 2013–2017.

118. Furuta, A., Martin, L. J., Lin, C. L., Dykes-Hoberg, M., and Rothstein, J. D. (1997) Cellular and synaptic localization of the neuronal glutamate transporters excitatory amino acid transporter 3 and 4. *Neuroscience* **81(4),** 1031–1042.

119. Nagao, S., Kwak, S., and Kanazawa, I. (1997) EAAT4, a Glutamate transporter with properties of a chloride channel, is predominantly localized in Purkinje cell dendrites, and forms parasagittal compartments in rat cerebellum. *Neuroscience* **78(4),** 929–933.

120. Tanaka, K., Watase, K., Manabe, T., Yamada, K., Watanabe, M., Takahashi, K., Iet al. (1997) Epilepsy and exacerbation of brain injury in mice lacking the glutamate transporter GLT-1. *Science* **276(5319),** 1699–1702.

121. Lin, C. L., Tzingounis, A. V., Jin, L., Furuta, A., Kavanaugh, M. P., and

Rothstein, J. D. (1998) Molecular cloning and expression of the rat EAAT4 glutamate transporter subtype. *Brain Res. Mol. Brain Res.* **63(1)**, 174–179.

122. Dehnes, Y., Chaudhry, F. A., Ullensvang, K., Lehre, K. P., Storm-Mathisen, J., and Danbolt, N. C. (1998) The glutamate transporter EAAT4 in rat cerebellar purkinje cells: a glutamate-gated chloride channel concentrated near the synapse in parts of the dendritic membrane facing astroglia. *J. Neurosci.* **18(10)**, 3606–3619.

123. Pow, D. V., Barnett, N. L., and Penfold, P. (2000) Are neuronal transporters relevant in retinal glutamate homeostasis? *Neurochem. Int.* **37(2-3)**, 191–198.

124. Pow, D. V. and Barnett, N. L. (2000) Developmental expression of excitatory amino acid transporter 5: a photoreceptor and bipolar cell glutamate transporter in rat retina. *Neurosci. Lett.* **280(1)**, 21–24.

125. Eliasof, S., Arriza, J. L., Leighton, B. H., Amara, S. G., and Kavanaugh, M. P. (1998) Localization and function of five glutamate transporters cloned from the salamander retina. *Vision Res.* **38(10)**, 1443–1454.

126. Rauen, T. (2000) Diversity of glutamate transporter expression and function in the mammalian retina. *Amino. Acids* **19(1)**, 53–62.

127. Dykes-Hoberg, M., Ganel, R., and Rothstein, J. D. (2000) Localization of the neutral amino acid transporters ASCT1 and ASCT2 in the rat CNS. *Soc. Neurosci. Abstr.* **26.**

128. Dumuis, A., Sebben, M., Haynes, L., Pin, J. P., and Bockaert, J. (1988) NMDA receptors activate the arachidonic acid cascade system in striatal neurons. *Nature* **336(6194)**, 68–70.

129. Bruns, D., Engert, F., and Lux, H. D. (1993) A fast activating presynaptic reuptake current during serotonergic transmission in identified neurons of hirudo. *Neuron* **10(4)**, 559–572.

130. Thompson, S. M. and Gahwiler, B. H. (1992) Effects of the GABA uptake inhibitor tiagabine on inhibitory synaptic potentials in rat hippocampal slice cultures. *J. Neurophysiol.* **67(6)**, 1698–1701.

131. Isaacson, J. S. and Nicoll, R. A. (1993) The uptake inhibitor L-Trans-PDC enhances responses to glutamate but fails to alter the kinetics of excitatory synaptic currents in the hippocampus. *J. Neurophysiol.* **70(5)**, 2187–2191.

132. Mennerick, S. and Zorumski, C. F. (1994) Glial contributions to excitatory neurotransmission in cultured hippocampal cells. *Nature* **368(6466)**, 59–62.

133. Sarantis, M., Ballerini, L., Miller, B., Silver, R. A., Edwards, M., and Attwell, D. (1993) Glutamate uptake from the synaptic cleft does not shape the decay of the non-NMDA component of the synaptic current. *Neuron* **11(3)**, 541–549.

134. Tong, G. and Jahr, C. E. (1994) Block of glutamate transporters potentiates postsynaptic excitation. *Neuron* **13(5)**, 1195–1203.

135. Barbour, B., Keller, B. U., Llano, I., and Marty, A. (1994) Prolonged presence of glutamate during excitatory synaptic transmission to cerebellar Purkinje cells. *Neuron* **12(6)**, 1331–1343.

136. Clements, J. D., Lester, R. A., Tong, G., Jahr, C. E., and Westbrook, G. L. (1992) The time course of glutamate in the synaptic cleft. *Science* **258(5087)**, 1498–1501.

137. Grewer, C., Watzke, N., Wiessner, M., and Rauen, T. (2000) Glutamate translocation of the neuronal glutamate transporter EAAC1 occurs within milliseconds. *Proc. Natl. Acad. Sci. USA* **97(17),** 9706–9711.

138. Mennerick, S., Shen, W., Xu, W., Benz, A., Tanaka, K., Shimamoto, K., Isenberg, K. E., Krause, J. E., and Zorumski, C. F. (1999) Substrate turnover by transporters curtails synaptic glutamate transients. *J. Neurosci.* **19(21),** 9242–9251.

139. Rothstein, J. D., Jin, L., Dykes-Hoberg, M., and Kuncl, R. W. (1993) Chronic inhibition of glutamate uptake produces a model of slow neurotoxicity. *Proc. Natl. Acad. Sci. USA* **90(14),** 6591–6595.

140. Rothstein, J. D., Dykes-Hoberg, M., Pardo, C. A., Bristol, L. A., Jin, L., Kuncl, R. W., et al. (1996) Knockout of glutamate transporters reveals a major role for astroglial transport in excitotoxicity and clearance of glutamate. *Neuron* **16(3),** 675–686.

140a. Kanai, Y. and Hediger, M. A. (1995) High-affinity glutamate transporters: physiological and pathophysiological relevance in the central nervous system, in: *Excitatory Amino Acids: Their Role in Neuroendocrine Function* (Brann, D. W. and Mahesh, V. B., eds.), CRC Press, Boca Raton, FL, pp. 103–131.

141. Gundersen, V., Shupliakov, O., Brodin, L., Ottersen, O. P., and Storm-Mathisen, J. (1995) Quantification of excitatory amino acid uptake at intact glutamatergic synapses by immunocytochemistry of exogenous D-aspartate. *J. Neurosci.* **15(6),** 4417–4428.

142. Kanner, B. I. and Schuldiner, S. (1987) Mechanism of transport and storage of neurotransmitters. *CRC Crit. Rev. Biochem.* **22(1),** 1–38.

143. Rossi, D. J., Oshima, T., and Attwell, D. (2000) Glutamate release in severe brain ischaemia is mainly by reversed uptake. *Nature* **403(6767),** 316–321.

144. Torp, R., Lekieffre, D., Levy, L. M., Haug, F. M., Danbolt, N. C., Meldrum, B. S., and Ottersen, O. P. (1995) Reduced postischemic expression of a glial glutamate transporter, GLT1, in the rat hippocampus. *Exp. Brain Res.* **103(1),** 51–58.

145. Raghavendra, Rao, V, Rao, A. M., Dogan, A., Bowen, K. K., Hatcher, J., Rothstein, J. D., and Dempsey, R. J. (2000) Glial glutamate transporter GLT-1 down-regulation precedes delayed neuronal death in gerbil hippocampus following transient global cerebral ischemia. *Neurochem. Int.* **36(6),** 531–537.

146. Watase, K., Hashimoto, K., Kano, M., Yamada, K., Watanabe, M., Inoue, Y., et al. (1998) Motor discoordination and increased susceptibility to cerebellar injury in GLAST mutant mice. *Eur. J. Neurosci.* **10(3),** 976–988.

147. Harada, T., Harada, C., Watanabe, M., Inoue, Y., Sakagawa, T., Nakayama, N., et al. (1998) Functions of the two glutamate transporters GLAST and GLT-1 in the retina. *Proc. Natl. Acad. Sci. USA* **95(8),** 4663–4666.

148. Hakuba, N., Koga, K., Gyo, K., Usami, S. I., and Tanaka, K. (2000) Exacerbation of noise-induced hearing loss in mice lacking the glutamate transporter GLAST. *J. Neurosci.* **20(23),** 8750–8753.

149. Watanabe, T., Morimoto, K., Hirao, T., Suwaki, H., Watase, K., and Tanaka, K. (1999) Amygdala-kindled and pentylenetetrazole-induced seizures in glutamate transporter GLAST-deficient mice. *Brain Res.* **845(1),** 92–96.

150. Peghini, P., Janzen, J., and Stoffel, W. (1997) Glutamate transporter EAAC-1-deficient mice develop dicarboxylic aminoaciduria and behavioral abnormalities but no neurodegeneration. *EMBO J.* **16(13)**, 3822–3832.

151. Drejer, J., Meier, E., and Schousboe, A. (1983) Novel neuron-related regulatory mechanisms for astrocytic glutamate and GABA high affinity uptake. *Neurosci. Lett.* **37(3)**, 301–306.

152. Swanson, R. A., Liu, J., Miller, J. W., Rothstein, J. D., Farrell, K., Stein, B. A., and Longuemare, M. C. (1997) Neuronal regulation of glutamate transporter subtype expression in astrocytes. *J. Neurosci.* **17(3)**, 932–940.

153. Gegelashvili, G. and Schousboe, A. (1997) High affinity glutamate transporters: regulation of expression and activity. *Mol. Pharmacol.* **52(1)**, 6–15.

154. Schlag, B. D., Vondrasek, J. R., Munir, M., Kalandadze, A., Zelenaia, O. A., Rothstein, J. D., and Robinson, M. B. (1998) Regulation of the glial Na⁺-dependent glutamate transporters by cyclic amp analogs and neurons. *Mol. Pharmacol.* **53(3)**, 355–369.

155. Figiel, M. and Engele, J. (2000) Pituitary adenylate cyclase-activating polypeptide (PACAP), a neuron-derived peptide regulating glial glutamate transport and metabolism. *J. Neurosci.* **20(10)**, 3596–3605.

156. Zelenaia, O., Schlag, B. D., Gochenauer, G. E., Ganel, R., Song, W., Beesley, J. S., et al. (2000) Epidermal growth factor receptor agonists increase expression of glutamate transporter GLT-1 in astrocytes through pathways dependent on phosphatidylinositol 3-kinase and transcription factor NF-kappaB. *Mol. Pharmacol.* **57(4)**, 667–678.

157. Palos, T. P., Zheng, S., and Howard, B. D. (1999) Wnt signaling induces GLT-1 expression in rat C6 glioma cells. *J. Neurochem.* **73(3)**, 1012–1023.

158. Matthews, J. C., Beveridge, M. J., Dialynas, E., Bartke, A., Kilberg, M. S., and Novak, D. A. (1999) Placental anionic and cationic amino acid transporter expression in growth hormone overexpressing and null IGF-II or null IGF-I receptor mice. *Placenta* **20(8)**, 639–650.

159. Sims, K. D., Straff, D. J., and Robinson, M. B. (2000) Platelet-derived growth factor rapidly increases activity and cell surface expression of the EAAC1 subtype of glutamate transporter through activation of phosphatidylinositol 3-kinase. *J. Biol. Chem.* **275(7)**, 5228–5237.

160. Guo, Z. H. and Mattson, M. P. (2000) Neurotrophic factors protect cortical synaptic terminals against amyloid and oxidative stress-induced impairment of glucose transport, glutamate transport and mitochondrial function. *Cereb. Cortex* **10(1)**, 50–57.

161. Corse, A. M., Bilak, M. M., Bilak, S. R., Lehar, M., Rothstein, J. D., and Kuncl, R. W. (1999) Preclinical testing of neuroprotective neurotrophic factors in a model of chronic motor neuron degeneration. *Neurobiol. Dis.* **6(5)**, 335–346.

162. Dowd, L. A. and Robinson, M. B. (1996) Rapid stimulation of EAAC1-mediated Na+-dependent L-glutamate transport activity in C6 glioma cells by phorbol ester. *J. Neurochem.* **67(2)**, 508–516.

163. Davis, K. E., Straff, D. J., Weinstein, E. A., Bannerman, P. G., Correale, D. M., Rothstein, J. D., and Robinson, M. B. (1998) Multiple signaling path-

ways regulate cell surface expression and activity of the excitatory amino acid carrier 1 subtype of Glu transporter in C6 glioma. *J. Neurosci.* **18(7)**, 2475–2485.

164. Duan, S., Anderson, C. M., Stein, B. A., and Swanson, R. A. (1999) Glutamate induces rapid upregulation of astrocyte glutamate transport and cell-surface expression of GLAST. *J. Neurosci.* **19(23)**, 10,193–10,200.

164a. *Soc. Neurosci. Abstr.* **25**, 427.

165. Jackson, M., Song, W., Liu, M. Y., Jin, L., Dykes-Hoberg, M., Lin, C. I., Bowers, W. J., Federoff, H. J., Sternweis, P. C., and Rothstein, J. D. (2001) Modulation of the neuronal glutamate transporter EAAT4 by two interacting proteins. *Nature* **410(6824)**, 89–93.

166. Lin, C. I., Orlov, I., Ruggiero, A. M., Dykes-Hoberg, M., Lee, A., Jackson, M., and Rothstein, J. D. (2001) Modulation of the neuronal glutamate transporter EAAC1 by theinteracting protein GTRAP3-18. *Nature* **410(6824)**, 84–88.

167. Shibata, T., Watanabe, M., Tanaka, K., Wada, K., and Inoue, Y. (1996) Dynamic changes in expression of glutamate transporter mRNAs in developing brain. *Neuroreport* **7(3)**, 705–709.

168. Shibata, T., Yamada, K., Watanabe, M., Ikenaka, K., Wada, K., Tanaka, K., and Inoue, Y. (1997) Glutamate transporter GLAST is expressed in the radial glia-astrocyte lineage of developing mouse spinal cord. *J. Neurosci.* **17(23)**, 9212–9219.

169. Yamada, K., Watanabe, M., Shibata, T., Nagashima, M., Tanaka, K., and Inoue, Y. (1998) Glutamate transporter GLT-1 is transiently localized on growing axons of the mouse spinal cord before establishing astrocytic expression. *J. Neurosci.* **18(15)**, 5706–5713.

170. Northington, F. J., Traystman, R. J., Koehler, R. C., and Martin, L. J. (1999) GLT1, glial glutamate transporter, is transiently expressed in neurons and develops astrocyte specificity only after midgestation in the ovine fetal brain. *J. Neurobiol.* **39(4)**, 515–526.

171. Yamada, K., Wada, S., Watanabe, M., Tanaka, K., Wada, K., and Inoue, Y. (1997) Changes in expression and distribution of the glutamate transporter EAAT4 in developing mouse Purkinje cells. *Neurosci. Res.* **27(3)**, 191–198.

172. Pow, D. V. and Barnett, N. L. (1999) Changing patterns of spatial buffering of glutamate in developing rat retinae are mediated by the muller cell glutamate transporter GLAST. *Cell Tissue Res.* **297(1)**, 57–66.

173. Brown, R. H., Jr. (1995) Amyotrophic lateral sclerosis: recent insights from genetics and transgenic mice. *Cell* **80(5)**, 687–692.

174. Robberecht, W., Sapp, P., Viaene, M. K., Rosen, D., McKenna-Yasek, D., Haines, J., et al. (1994) Cu/Zn superoxide dismutase activity in familial and sporadic amyotrophic lateral sclerosis. *J. Neurochem.* **62(1)**, 384–387.

175. Rosen, D. R., Bowling, A. C., Patterson, D., Usdin, T. B., Sapp, P., Mezey, E., et al. (1994) A frequent Ala 4 to Val superoxide dismutase-1 mutation is associated with a rapidly progressive familial amyotrophic lateral sclerosis. *Hum. Mol. Genet.* **3(6)**, 981–987.

176. Rothstein, J. D., Martin, L. J., and Kuncl, R. W. (1992) Decreased glutamate

transport by the brain and spinal cord in amyotrophic lateral sclerosis. *N. Engl. J. Med.* **326(22),** 1464–1468.

177. Lin, C. L., Bristol, L. A., Jin, L., Dykes-Hoberg, M., Crawford, T., Clawson, L., and Rothstein, J. D. (1998) Aberrant RNA processing in a neurodegenerative disease: the cause for absent EAAT2, a glutamate transporter, in amyotrophic lateral sclerosis. *Neuron* **20(3),** 589–602.

178. Meyer, T., Fromm, A., Munch, C., Schwalenstocker, B., Fray, A. E., Ince, P. G., et al. (1999) The RNA of the Glutamate Transporter EAAT2 is variably spliced in amyotrophic lateral sclerosis and normal individuals. *J. Neurol. Sci.* **170(1),** 45–50.

179. Meyer, T., Lenk, U., Kuther, G., Weindl, A., Speer, A., and Ludolph, A. C. (1995) Studies of the coding region of the neuronal glutamate transporter gene in amyotrophic lateral sclerosis. *Ann. Neurol.* **37(6),** 817–819.

180. Meyer, T., Munch, C., Knappenberger, B., Liebau, S., Volkel, H., and Ludolph, A. C. (1998) Alternative splicing of the glutamate transporter EAAT2 (GLT-1). *Neurosci. Lett.* **241(1),** 68–70.

181. Meyer, T., Munch, C., Liebau, S., Fromm, A., Schwalenstocker, B., Volkel, H., and Ludolph, A. C. (1998) Splicing of the Glutamate Transporter EAAT2: a candidate gene of amyotrophic lateral sclerosis. *J. Neurol. Neurosurg. Psych.* **65(6),** 954.

182. Meyer, T., Speer, A., Meyer, B., Sitte, W., Kuther, G., and Ludolph, A. C. (1996) The glial glutamate transporter complementary DNA in patients with amyotrophic lateral sclerosis. *Ann. Neurol.* **40(3),** 456–459.

183. Shaw, P. J. and Eggett, C. J. (2000) Molecular factors underlying selective vulnerability of motor neurons to neurodegeneration in amyotrophic lateral sclerosis. *J. Neurol.* **247 Suppl,** I17–I27.

184. Rosen, D. R., Sapp, P., O'Regan, J., McKenna-Yasek, D., Schlumpf, K. S., Haines, J. L., Gusella, J. F., Horvitz, H. R., and Brown, R. H., Jr. (1994) Genetic linkage analysis of familial amyotrophic lateral sclerosis using human chromosome 21 microsatellite DNA Markers. *Am. J. Med. Genet.* **51(1),** 61–69.

185. Rosen, D. R., Siddique, T., Patterson, D., Figlewicz, D. A., Sapp, P., Hentati, A., et al. (1993) Mutations in Cu/Zn superoxide dismutase gene are associated with familial amyotrophic lateral sclerosis. *Nature* **362(6415),** 59–62.

186. Gurney, M. E., Pu, H., Chiu, A. Y., Dal Canto, M. C., Polchow, C. Y., Alexander, D. D., et al. (1994) Motor neuron degeneration in mice that express a human Cu,Zn superoxide dismutase mutation [see Comments] [Published erratum appears in *Science* (1995) **269(5221),** 149]. *Science* **264(5166),** 1772–1775.

187. Bruijn, L. I., Becher, M. W., Lee, M. K., Anderson, K. L., Jenkins, N. A., Copeland, N. G., et al. (1997) ALS-linked SOD1 mutant G85R mediates damage to astrocytes and promotes rapidly progressive disease with SOD1-containing inclusions. *Neuron* **18(2),** 327–338.

188. Pedersen, W. A., Fu, W., Keller, J. N., Markesbery, W. R., Appel, S., Smith, R. G., Kasarskis, E., and Mattson, M. P. (1998) Protein modification by the lipid peroxidation product 4-hydroxynonenal in the spinal cords of amyotrophic lateral sclerosis patients. *Ann. Neurol.* **44(5),** 819–824.

189. Trotti, D., Rolfs, A., Danbolt, N. C., Brown, R. H., Jr., and Hediger, M. A. (1999) SOD1 mutants linked to amyotrophic lateral sclerosis selectively inactivate a glial glutamate transporter. *Nat. Neurosci.* **2(9)**, 848.

190. Aoki, M., Lin, C. L., Rothstein, J. D., Geller, B. A., Hosler, B. A., Munsat, T. L., et al. (1998) Mutations in the glutamate transporter EAAT2 Gene do not cause abnormal EAAT2 transcripts in amyotrophic lateral sclerosis. *Ann. Neurol.* **43(5)**, 645–653.

190a. Jin, L., Dykes-Hoberg, M., Kuncl, M., and Rothstein, J. D. (1994) Selective loss of glutamate transporter subtypes in amyotrophic lateral sclerosis. *Soc. Neurosci. Abstr.* **20**, 927.

191. Farooqui, A. A. and Horrocks, L. A. (1991) Excitatory amino acid receptors, neural membrane phospholipid metabolism and neurological disorders. *Brain Res. Brain Res. Rev.* **16(2)**, 171–191.

192. Siesjo, B. K., Agardh, C. D., and Bengtsson, F. (1989) Free radicals and brain damage. *Cerebrovasc. Brain Metab Rev.* **1(3)**, 165–211.

193. Hall, E. D. and Braughler, J. M. (1989) Central nervous system trauma and stroke. II. Physiological and pharmacological evidence for involvement of oxygen radicals and lipid peroxidation. *Free Radic. Biol. Med.* **6(3)**, 303–313.

194. Choi, D. W. (1988) Glutamate neurotoxicity and diseases of the nervous system. *Neuron* **1(8)**, 623–634.

195. Choi, D. W. (1992) Excitotoxic cell death. *J. Neurobiol.* **23(9)**, 1261–1276.

196. Volterra, A., Trotti, D., Cassutti, P., Tromba, C., Salvaggio, A., Melcangi, R. C., and Racagni, G. (1992) High sensitivity of glutamate uptake to extracellular free arachidonic acid levels in rat cortical synaptosomes and astrocytes. *J. Neurochem.* **59(2)**, 600–606.

197. Barbour, B., Szatkowski, M., Ingledew, N., and Attwell, D. (1989) Arachidonic acid induces a prolonged inhibition of glutamate uptake into glial cells. *Nature* **342(6252)**, 918–920.

198. Trotti, D., Volterra, A., Lehre, K. P., Rossi, D., Gjesdal, O., Racagni, G., and Danbolt, N. C. (1995) Arachidonic acid inhibits a purified and reconstituted glutamate transporter directly from the water phase and not via the phospholipid membrane. *J. Biol. Chem.* **270(17)**, 9890–9895.

199. Zerangue, N., Arriza, J. L., Amara, S. G., and Kavanaugh, M. P. (1995) Differential modulation of human glutamate transporter subtypes by arachidonic acid. *J. Biol. Chem.* **270(12)**, 6433–6435.

200. Fairman, W. A., Sonders, M. S., Murdoch, G. H., and Amara, S. G. (1998) Arachidonic acid elicits a substrate-gated proton current associated with the glutamate transporter EAAT4. *Nat. Neurosci.* **1(2)**, 105–113.

201. Volterra, A., Trotti, D., Tromba, C., Floridi, S., and Racagni, G. (1994) Glutamate uptake inhibition by oxygen free radicals in rat cortical astrocytes. *J. Neurosci.* **14(5 Pt 1)**, 2924–2932.

202. Volterra, A., Trotti, D., and Racagni, G. (1994) Glutamate uptake is inhibited by arachidonic acid and oxygen radicals via two distinct and additive mechanisms. *Mol. Pharmacol.* **46(5)**, 986–992.

203. Grunewald, M., Bendahan, A., and Kanner, B. I. (1998) Biotinylation of

single cysteine mutants of the glutamate transporter GLT-1 from rat brain reveals its unusual topology [In Process Citation]. *Neuron* **21(3),** 623–632.

204. Yao, D., Mackenzie, B., Ming, H., Varoqui, H., Zhu, H., Hediger, M. A., and Erickson, J. D. (2000) A novel system a isoform mediating Na+/neutral amino acid cotransport. *J. Biol. Chem.* **275(30),** 22,790–20,797.

205. Reimer, R. J., Chaudhry, F. A., Gray, A. T., and Edwards, R. H. (2000) Amino acid transport system a resembles system N in sequence but differs in mechanism. *Proc. Natl. Acad. Sci. USA* **97(14),** 7715–7720.

9

Vesicular Neurotransmitter Transporters*
Pharmacology, Biochemistry, and Molecular Analysis

Rodrigo Yelin and Shimon Schuldiner

1. TRANSPORTERS AND NEUROTRANSMISSION

Synaptic transmission involves the regulated release of transmitter molecules to the synaptic cleft, where they interact with postsynaptic receptors which subsequently transduce the information. Removal of the transmitter from the cleft enables termination of the signal, which usually occurs through reuptake back to the presynaptic terminal or into glial elements in a sodium-dependent process. This process assures constant and high levels of neurotransmitters in the neuron and low concentrations in the cleft.

The storage of neurotransmitters in subcellular organelles ensures their regulated release, and is crucial for protecting the accumulated molecules from leakage or intraneuronal metabolism and the neuron from possible toxic effects of the transmitters. In addition, the removal of intraneuronal molecules into the storage system effectively lowers the concentration gradient across the neuronal membrane and thus acts as an amplification stage for the overall process of uptake. Drugs that interact with either transport system have profound pharmacological effects, as they modify the levels of neurotransmitter in the cleft.

Vesicular transport has been observed for several classical transmitters, including acetylcholine (ACh), the monoamines, glutamate, g-aminobutyric acid (GABA), and glycine (reviewed in ref. 1). All the transporters are driven by the proton electrochemical gradient ($\Delta\tilde{\mu}_{H^+}$) generated by the vacuolar type ATPase (V-ATPase). This ubiquituous enzyme utilizes the energy of cytoplasmic ATP to translocate H^+ ions into the vesicle, and generates a $\Delta\tilde{\mu}_{H^+}$ (acid and positive inside). The vesicular neurotransmitter transpor–ters (VNTs) utilize this energy by exchanging one or more protons with a neurotransmitter molecule (for reviews on the bioenergetics of the transport

---

* Chapter authored in August 2000.

From: *Contemporary Neuroscience:*
Neurotransmitter Transporters: Structure, Function, and Regulation, 2nd Edition
Edited by: M. E. A. Reith © Humana Press Inc., Totowa, NJ

system, see refs. 1–4). The vesicular monoamine (VMAT) and ACh (VAChT) transporters have been the most intensively studied, and are the ones for which most molecular information has been obtained. In this chapter, we will review the most salient features of both. In both cases, the key for this knowledge resides in the availability of excellent experimental paradigms for their study and potent and specific inhibitors. For other reviews on the topic, also see refs. 1,5–8.

Molecular information on vesicular glutamate transport is begining to emerge only recently. Thus, glutamate transport activity was shown *(9,10)* for a brain-specific Na^+-dependent inorganic phosphate transporter (BNPI) *(11)*. Loss-of-function mutations in eat-4, the *C. elegans* orthologue of BNPI, cause defective glutamatergic chemical transmission *(12)*. In addition, BNPI was found to be associated with small synaptic vesicles and to localize to excitatory terminals *(13)*. BNPI exhibits a conductance for chloride that is blocked by glutamate *(9,10)*.

2. PHARMACOLOGY OF VNTs

The best-characterized inhibitors of VNTs are reserpine and tetrabenazine, the two principal agents that inhibit vesicular monoamine transport *(14,15)*. Vesamicol, an inhibitor of VAChT, has been introduced in the last decade and studied extensively *(16–18)*.

2.1. Molecular Mechanism of Reserpine Action

Reserpine is believed to bind at the site of amine recognition, as shown by the fact that it inhibits transport in a seemingly competitive way, with $K$is in the subnanomolar range *(19,20)*, it binds to the transporter with a Kd similar to its Ki *(19,20)* and transport substrates prevent its association in a concentration range similar to the range of their apparent Km values *(20)*. Its effect in vitro is practically irreversible *(21)*, in line with the in vivo effect of the drug, which is extremely long-lasting and is only relieved when new vesicles replace the ones that were hit *(22)*. As a result of this action, it depletes monoamine reserves, providing considerable information on the physiological role of biogenic amines in the nervous system *(23)*. Reserpine has been in clinical use because it potently reduces blood pressure; however, it frequently produces a disabling effect of lethargy that resembles depression and has limited its clinical utility *(24)*. This observation has given rise to the amine hypothesis of affective disorders, which in modified form still produces a useful framework for considering this group of major psychiatric disorders.

The time-course of reserpine binding is relatively slow. This low rate of association is consistent with a similar time-course for inhibition of monoamine transport *(21)*. Reserpine binding is accelerated by $\Delta\tilde{\mu}_{H^+}$, whether generated by the H^+-ATPase *(20,25)* or artificially imposed *(21)*. This acceleration is also observed in proteoliposomes reconstituted with the purified protein *(26)*.

In every case, in the presence or absence of $\Delta\tilde{\mu}_{H^+}$, and in the native as well in the purified protein, two distinct populations of sites have been detected *(20,26)*. A high affinity site, Kd = 0.5 nM, B_{max} = 7–10 pmol/mg protein in the native chromaffin-granule membrane vesicle preparation (0.3 and 310, respectively for the purified protein), and a low-affinity site, Kd = 20 nM, B_{max} = 60 pmol/mg protein in the native system (30 and 4200, respectively for the purified preparation). Surprisingly the apparent Kd does not change with an imposition of a $\Delta\tilde{\mu}_{H^+}$, although the on-rate increases by several fold *(20)*. It has to be assumed that the off-rate changes also accordingly, although it is so slow that it has not been possible to measure. A Kd of 30 pM has been measured under conditions in which the concentration of ligand-binding sites does not exceed the value of the dissociation constant. This is about approx 10× higher affinity than previously estimated *(19)*.

The reserpine-binding rate is less sensitive than transport to changes in the ΔpH, and is stimulated equally efficiently by ΔpH and $\Delta\psi$ *(21)*. These findings suggest that fewer protons are translocated in the step which generates the high-affinity binding site than in the overall transport cycle. Changes in binding rate probably reflect changes in the availability of reserpine-binding sites, and translocation of a single H^+ generates the binding form of the transporter. The high-affinity form of the transporter is apparently achieved by either protonation of VMAT or by H^+ translocation. The energy invested in the transporter may be released by ligand binding and converted into vectorial movement of a substrate molecule across the membrane, or directly into binding energy in the case of reserpine. In the case of a substrate, a second conformational change results in the ligand-binding site being exposed to the vesicle interior, where the substrate can dissociate. The second H^+ in the cycle may be required to facilitate the conformational change or to allow for release of the positively charged substrate from the protein. In the model, this second H^+ binding and release is arbitrarily located, because there is no information about the order of the reactions. Intriguing findings indicate that the apparent affinity of the transporter for substrates drops when the pH decreases *(27,28)*. This process could reflect a mechanism for releasing the substrate in the acidic lumenal milieu. Substrate and H^+ release regenerates the transporter, which can now start a new cycle. However, in the case of reserpine, its structure (bulk of its side chain?) restricts the conformational change so that instead of releasing the ligand on the interior, the complex becomes trapped in a state from which reserpine cannot readily dissociate, and which cannot translocate another H^+ to regenerate the high affinity form. It is not known whether the slow binding of reserpine requires also protonation of VMAT. If this is the case, protonation in the absence of $\Delta\tilde{\mu}_{H^+}$ would be the rate-limiting step.

2.2. A Second Site for Inhibitor Action on VMAT

Tetrabenazine (TBZ) is another potent inhibitor of the transporter. Radiolabeled dihydrotetrabenazine (TBZOH) has been used in binding studies to characterize the protein (20,29), to study the regulation of its synthesis (30–32) and its distribution in various tissues. Binding of TBZOH to the transporter is not modified by the imposition of a $\Delta\tilde{\mu}_{H^+}$, as shown for reserpine. In addition, binding is not inhibited by reserpine at concentrations which fully inhibit transport. Moreover, transport substrates block binding only at concentrations 100-fold higher than their apparent Km values. These findings have led Henry and colleagues to suggest that TBZOH binds to a site on the protein which is different from the reserpine and substrate-binding site (19,29,33). It has been suggested that both sites are mutually exclusive—that VMAT exists in two different conformations, each conformation binding only one type of ligand, TBZ or reserpine. According to this interpretation, the addition of TBZ would pull the conformational equilibrium toward the TBZ-binding conformation, which is unable to bind reserpine. Indeed, under proper conditions (low protein concentration and short incubation times) 50 nM TBZOH inhibits reserpine binding by 70%. ATP, through the generation of $\Delta\tilde{\mu}_{H^+}$, would pull the conformational equilibrium toward the reserpine-binding site (19). Although elegant and attractive, this model has yet to explain the lack of effect of $\Delta\tilde{\mu}_{H^+}$ on TBZ binding, which should be inhibited, if the two forms are mutually exclusive. Also, the concentrations of reserpine required to inhibit TBZ binding are higher than those required for site occupancy (binding and inhibition of transport).

2.3. Acetylcholine Transporter

An important development in the study of VAChT was the discovery of a specific inhibitor, trans-2-(4-phenylpiperidino)cyclohexanol—code-named AH5183 and now called vesamicol—which blocks neuromuscular transmission and exhibits unusual characteristics of action. Marshall (16) hypothesized that AH5183 blocks storage by synaptic vesicles, and indeed it inhibits ACh storage by purified Torpedo synaptic vesicles with an IC_{50} of 40 nM. The drug was the most potent inhibitor found among at least 80 compounds initially screened (34,35) (see ref. 36 for review). [^{3}H]Vesamicol binding showed an apparent Kd of 34 nM (36–41). ACh inhibits vesamicol binding only at very high concentrations (20–50 mM) and other high-affinity analogs were shown to competitively inhibit binding. However, in all cases in which the analogs are transported, the inhibition constant is approx 20-fold higher than the apparent constants for transport. Non-transported analogs show the same efficiency for inhibition of both ACh transport and [³H]vesamicol binding *(42)*. A kinetic model has been suggested in which it is assumed that vesamicol binds to an allosteric site to form a dead-end complex when ACh is not bound. As described previously, the existence of two sites with similar properties has been observed also for VMAT. In the latter, TBZ is a potent inhibitor of monoamine accumulation, but its binding is inhibited by monoamines only at very high concentrations. In the case of VMAT, both binding sites are present in one protein, since the purified and the recombinant proteins show high sensitivity to TBZ. In the case of VAChT, the existence of a "receptor" has been postulated, which could lie on the same protein or in a separate one. The vesamicol "receptor" has been extensively studied by Parsons and colleagues, it has been purified *(43)* and photoaffinity-labeled *(36,44,45)*. The "receptor" solubilized in cholate and stabilized with glycerol and a phospholipid mixture was purified to yield a specific binding of 4400 pmol/mg protein, a purification factor of approx 15. Unfortunately, the purified receptor exhibits highly heterogeneous electrophoretic mobility in SDS PAGE with a very diffuse stain at about 240 kDa. This is the typical behavior of membrane glycoproteins that are not fully monodispersed because of the detergent used and boiled—treatments known to induce aggregation of membrane proteins.

Recently, glycoproteins from various species which bind vesamicol with high affinity have been expressed in CV1-fibroblasts *(46–48)*. In addition, the rat VAChT expressed in CV1-fibroblasts catalyzes vesamicol-sensitive ACh accumulation *(46)*. As will be seen here, the evidence available now clearly demonstrates that vesamicol binds to the vesicular ACh transporter itself.

3. IDENTIFICATION OF FUNCTIONAL TRANSPORTERS

Native bovine VMAT was purified in a functional form. The high stability of the complex [³H]reserpine-transporter has been used to label the transporter and follow its separation through a variety of procedures. In these experiments, a small amount of Triton X-100-extracts from prelabeled membranes was mixed with a four- to fivefold higher amount of extract from unlabeled membranes. Purification of the material labeled in this way has revealed the presence of two proteins that differ in p*I*, a very acidic one (p*I* 3.5) and a moderately acidic one (p*I* 5.0) *(26)*. Reconstitution in proteoliposomes has shown that both proteins catalyze monoamine transport with the expected properties. The more acidic isoform is a glycoprotein of 80 kDa, which has been purified and reconstituted in proteoliposomes. It catalyzes transport of serotonin with an apparent Km of 2 mM and a V_{max} of 140 nmol/ mg/min, about 200-fold higher than the one determined in the native system. Transport is inhibited by reserpine and tetrabenazine, ligands which bind to two distinct sites on the transporter. In addition, the reconstituted purified transporter binds reserpine with a biphasic kinetic behavior, which is typical of the native system. These results demonstrate that a single polypeptide is required for all the activities displayed by the transporter, i.e., reserpine- and tetrabenazine-sensitive, ΔpH-driven serotonin accumulation, and binding of reserpine in an energy-dependent and independent way. Based on these and additional findings, it was estimated that the transporter represents approx 0.2–0.5% of the chromaffin-granule-membrane vesicle, and has a turnover of about 30 min^{-1}.

The assignment of the activity to the 80-kDa polypeptide and the localization of the tetrabenazine and reserpine-binding sites in the same polypeptide are confirmed by several independent approaches, including direct sequencing of the purified protein *(49,50)* and cloning and analysis of the recombinant protein *(51–53)*. Vincent and Near purified a TBZOH-binding glycoprotein from bovine adrenal medulla, using a protocol identical to the one used to purify the functional transporter. The TBZOH-binding protein displays an apparent Mr of 85 kDa *(54)*. In addition, Henry and colleagues, labeled bovine VMAT with 7-azido-8-[^{125}I] iodoketanserin, a photoactive derivative of ketanserin, which is believed to interact with the TBZ-binding site. The labeled polypeptide displayed an apparent Mr of 70 kDa and a pI ranging from 3.8 to 4.6 *(55)*. In each case, broad diffuse bands which are characteristic of membrane proteins were detected so that the differences in the Mrs reported (70-, 80-, and 85-kDa) are probably caused by a different analysis of the results and not by innate variations. Sequencing of the azido-Iodoketanserine-labeled protein confirmed that it is identical with the functional transporter *(49)*.

The basis of the difference between the two isoforms has not yet been studied and could be the result of either covalent modification (i.e., phosphorylation or different glycosylation levels) of the same polypeptide backbone, to limited proteolysis during preparation, or to a different polypeptide backbone. Since we now know that there are two types of VMATs— VMAT2, which is sensitive to TBZ, and VMAT1, which is less sensitive— it should be determined whether the activity of the high pI form is less sensitive to TBZ. The sequence of 26 N-terminal amino acids of the purified protein (low pI, high TBZ sensitivity) is practically identical to the predicted sequence of the bovine adrenal VMAT2 *(49,50,56)*. Antibodies raised against a synthetic peptide based on the described sequences specifically recognize the pure protein on Western blots, and immunoprecipitate reserpine-binding activity under conditions in which the 80-kDa protein alone is precipitated *(50)*.

4. CLONING AND FUNCTIONAL EXPRESSION OF VESICULAR NEUROTRANSMITTER TRANSPORTERS (VNTS)

4.1. Vesicular Monoamine Transporter (VMAT)

Several sequences of VMAT and VAChT from various species are now available. Rat VMAT was cloned by Edwards et al. *(52)* and Erickson et al. *(51)* practically at the same time, using different strategies reviewed elsewhere *(1,5)*. Erickson and colleagues *(51)* used expression cloning in CV-1 cells transfected with cDNA prepared from rat basophilic leukemia (RBL) cell mRNA. Edwards et al. utilized the ability of VMAT1 to render Chinese Hamster ovary (CHO) cells resistant to MPP$^+$ by means of its ability to transport the neurotoxin into intracellular acidic compartments, thereby lowering its effective concentration in the cytoplasm (*52*; Fig. 1).

Sequence analysis of the cDNA conferring MPP$^+$ resistance (VMAT1) shows a single large open reading frame, which predicts a 521-amino acid protein. Analysis by the method of hydrophobic moments predicts 12 putative transmembrane (TM) segments. A large hydrophilic loop occurs between transmembrane domains (TMDs) 1 and 2 and contains three potential sites for N-linked glycosylation. According to the 12-TM model *(52)* this loop faces the lumen of the vesicle, and both termini face the cytoplasm.

Biochemical and quantitative evidence for the identity of the cloned cDNA was provided by developing a cell-free system in which membranes were assayed for dopamine transport and reserpine binding *(52,53)*.

VMAT1 expressed in MPP$^+$-resistant CHO cells accounts for approx 0.1% of the total cell-membrane protein *(53)*, and the bovine vesicular transporter accounts for 0.2–0.5% of the chromaffin-granule membrane *(26)*.

A transporter distinct from VMAT1 has been identified in the rat brain (VMAT2) *(52)* and in RBL cells *(51)*. The predicted protein shows an overall identity of 62% and a similarity of 78% to VMAT1. The major sequence divergences occur at the large lumenal loop located between the first and the second TMDs and at the N- and C-termini.

4.1.1. The Two VMAT Subtypes Differ Also in Some Functional Properties

A comprehensive comparison between the functions of rat VMAT1 and VMAT2 has been performed by Edwards and colleagues in membrane vesicles prepared from CHO stable transformed cell lines in which the respective proteins are expressed *(57)*. According to these studies, VMAT2 has a consistently higher affinity for all the monoamine substrates tested. In the case of serotonin, dopamine, and epinephrine, the apparent Km of

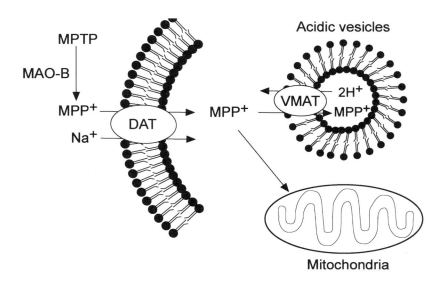

Fig. 1. VMAT renders cells resistant to the neurotoxin MPP⁺ because it removes the neurotoxin away from its presumed target. The ability of the neurotoxin MPTP to induce Parkinsonism has been explained by its ability to cross the blood-brain barrier and its oxidation in astrocytes to MPP⁺ by monoamine oxidase B (MAO-B). The higher sensitivity of dopaminergic cells to MPP⁺ is probably caused by its accumulation into these cells by DAT, the Na⁺-coupled plasma-membrane dopamine transporter. VMAT renders CHO cells resistant to MPP⁺ by means of its ability to transport the neurotoxin into intracellular acidic compartments, thereby lowering its effective concentration in the cytoplasm and in its presumed target, the respiratory chain of the mitochondria *(5,52)*. These findings have suggested a novel type of drug resistance in eukaryotic cells: protection of the devices essential for life by compartmentalization *(156)*.

VMAT1 for ATP-dependent transport is 0.85, 1.56, and 1.86 μ*M* respectively, and the corresponding *K*m values measured for VMAT2 are four- to fivefold lower: 0.19, 0.32, and 0.47 μ*M* respectively. Although the affinities are slightly different, the rank order for the various monoamines is similar. Also other substrates such as MPP⁺ (1.6 vs 2.8 μ*M*, respectively) and methamphetamine (2.7 vs 5.5 μ*M*, respectively, as estimated from measurements of the ability to inhibit reserpine binding) display a similar pattern. One striking difference is detected for histamine: 3 μ*M* for VMAT2 and 436 μ*M* for VMAT1. Also, VMAT1 is significantly less sensitive to tetrabenazine— IC$_{50}$ = 3–4 μ*M* *(52,57)*—than either VMAT2 (IC$_{50}$ = 0.3–0.6 μ*M*) *(51,57)* or the native *(15,19)* and the purified transporter (IC$_{50}$ = 25 n*M*) *(26)* from bovine adrenal medulla. The apparent affinities determined in mammalian expression systems or in proteoliposomes reconstituted with the purified transporter are higher than those determined in chromaffin-granules membrane vesicles (*see* ref. 2). The turnover number (TO) of the recombinant protein has been calculated based on the V_{max} for serotonin transport and the number of reserpine-binding sites. The TO for VMAT1 is 10 min^{-1}, and that of VMAT2 is 40 min^{-1} (57). A similar analysis of the purified bovine transporter (VMAT2 type) showed a TO of 30 min^{-1} (26). These values are lower than the 135 min^{-1} estimated for intact bovine chromaffin granules (58), but coincide well with the values obtained from brain regions (10–35 min^{-1}) (59) and other estimates (15 and 35) in chromaffin granules (33).

4.1.2. Tissue Distribution and Subcellular Targeting of VMAT

Tissue distribution of rat VMAT subtypes has been studied very intensively using a variety of techniques: Northern analysis (51,52), *in situ* hybridization (52) and immunohistochemistry (60–63). From these studies, it is concluded that the expression of VMAT1 and VMAT2 is mutually exclusive: VMAT1 is restricted to non-neuronal cells, and VMAT2 to neuronal cells. VMAT1 is expressed in endocrine/paracrine cells: in the adrenal medulla chromaffin cells, in the intestine and gastric mucose in serotonin and histamine-containing endocrine and paracrine cells, and in dopamine-containing SIF cells of sympathetic ganglia. VMAT2 is expressed in neuronal cells throughout, including in the intestine and stomach. There are two exceptions for this restriction: a subpopulation of VMAT2-expressing chromaffin cells in the adrenal medulla and a population of VMAT2, chromogranin A-positive endocrine cells of the oxyntic mucose of the rat stomach (63).

Although the studies with rats are very definitive, the situation is very different for other species. In human pheochromocytoma, mRNA for both subtypes is found (63). The bovine adrenal medulla expresses a VMAT2-type transporter, and its message has also been detected in the brain (49,56). VMAT2 is the main adrenal medulla transporter, as judged from direct protein purification studies (26,49,50,56). The purified transporter from bovine adrenal is VMAT2, and accounts for at least 60% of the activity in the gland.

Storage of monoamines differs from that of other classical neurotransmitters. Thus, although the latter are stored in small synaptic vesicles, monoamines in the adrenal medulla are stored with neural peptides in large, dense-core chromaffin granules. In the CNS, neurons store monoamines in small vesicles that may contain a dense core. The difference in the storage of the monoamines as compared to that of classical transmitters may reflect differential sorting of the VNTs. Sorting of VMAT was studied with immunohistochemical and biochemical tools *(64)*. In CHO cells, rVMAT1 is targeted to a population of recycling vesicles, and colocalizes with the transferrin receptor. Thus, localization in CHO cells is similar to that of other neuronal-vesicle proteins such as synaptophysin and SV2 *(65)*. In PC12 cells, endogenous VMAT1, occurs principally in large, dense-core granules (LDCV). Only small amounts are found in synaptic-like microvesicles (SLMV) and in endosomes *(65)*. In the rat adrenal medulla, immunoreactivity for VMAT1 occurs at several sites in the secretory pathway, but most prominently in the chromaffin granules, supporting the results found with PC12 cells *(65)*. In central neurons, the localization of rVMAT2s was studied in the nuclei of the solitary tract, a region known to contain a dense and heterogenous population of monoaminergic neurons *(64)*. VMAT2 localizes primarily to LDCV in axon terminals. It is also detected in less prominent amounts in small synaptic vesicles, the trans-Golgi network, and other sites of vesicle transport and recycling. In serotonergic raphe neurons, rVMAT2 localizes mainly to small synaptic vesicles (SSV) rather than to LDCV *(66)*. Thus, VMAT2 can be sorted to LDCV and SSV depending on the cell type, and VMAT1 is primarily sorted to LDCV.

4.1.3. Regulation of Expression of VMAT

Evidence for the regulation of expression of VMAT was first obtained in insulin-shocked rats in which an increase in the number of [^{3}H]TBZOH-binding sites in the adrenal medulla was detected. The increase was maximal after 4–6 d *(32)*. A similar increase was observed in vitro in bovine chromaffin cells in culture in the presence of carbamylcholine or depolarizing concentrations of potassium ions *(30)*. The response was mimicked by forskolin and by phorbol esters and was blocked by actinomycin and cycloheximide, suggesting the involvement of transcriptional activation.

The link to transcriptional activation was supported by the detection of an increase in message for VMAT2 after 6 h depolarization *(49)*. After 5 d, the cells contained less secretory granules, and remaining cells had a higher density, suggesting that they were newly synthesized and immature. Although the catecholamine, chromogranin A, and cytochrome b561 content decreased, [^{3}H]TBZOH-binding sites increased about 1.5-fold. The physiological significance of these findings is unclear. This phenomenon may reflect the fact that vesicular uptake is rate-limiting, and thereby an increase in VMAT would be needed to accelerate refilling. However, Mahata et al. have suggested that there is no increase in other membrane proteins in the rat granule *(67,68)*. They have reported no changes in the level of VMAT1 message (the main subtype in the rat adrenal) under conditions at which mRNA for the matrix peptide NPY increased *(67,68)*. Since the

[³H]TBZOH binding sites increased in the same system, the latter findings may suggest a novel mode of regulation of activity of pre-existing protein.

4.1.4. Heterologous Expression of VMAT

High-level production of a protein is a prerequisite to the performance of biophysical and structural studies. Vesicular transporters were expressed in a functional form in several mammalian cell lines, including CHO, CV-1, PC12, COS, and Hela. Yet, although the expression renders active transporter, the amount of protein produced is fairly limited. Higher yields of purified homogenous protein are usually obtained using heterologous expression systems. The VMAT was heterologously expressed in insect and yeast cells *(69,70)*. Expression of rVMAT1 in the yeast *S. cerevisiae* was largely dependent on the codons used for the N-terminus of the protein. Temperature also heavily influenced protein expression and processing. As temperature of growth decreased, expression levels significantly increased and greater fractions of the transporter underwent core glycosylation *(70)*. The bovine VMAT2 (bVMAT2) was also expressed in yeast, but its expression was less dependent on the temperature. The expressed bVMAT2 showed high-affinity tetrabenazine binding, with characteristics similar to those found in the native protein from bovine chromaffin granules. The binding activity could also be assayed on solubilized bVMAT2, immobilized on Ni^{2+}-NTA beads through a C-terminal polyhistidine tag. Despite the tetrabenazine binding activity, transport activity and reserpine binding could not be detected for the yeast-expressed rVMAT1 or bVMAT2 at any conditions tested *(70)*.

Sievert and colleagues reported the heterologous expression of rVMAT2 in Spodoptera frugiperda (Sf9) cells infected with recombinant baculovirus. Rat VMAT2 tagged with a polyhistidine epitope was purified using immobilized Ni^{2+}-affinity chromatography followed by a lectin column (Concanavalin A) (135). The purified-polypeptide bound [³H]TBZOH and was covalently modified and labeled by [¹⁴C]-DCCD. In addition, the Sf9-expressed rVMAT2 was photoaffinity labelled by [¹²⁵I]-iodoketanserin ([¹²⁵I]AZIK) *(69,71)* and by a photoactive tetrabenazine derivative ([¹²⁵I]-TBZ-AIPP) *(71)*. [¹²⁵I]AZIK bound to Lys-20 in the amino terminus just prior to TM 1; essentially similar results were achieved by Sagne and colleagues using bovine chromaffin granules *(72)*. [¹²⁵I]-TBZ-AIPP derived a segment between Gly-408 and Cys-431 in TM 10 and 11, respectively. Based on these results, the authors proposed a model in which the TM 1 is juxtaposed to TM 10/11 *(71)*. Despite high-affinity tetrabenazine and ketanserin binding, no reserpine binding or transport activity were reported in these studies. In our experience, the expression of rVMAT1 using the

Sf9/baculovirus system renders high levels of recombinant protein, which does not catalyze either serotonin transport or reserpine binding (RY and SS, unpublished observations). The results illustrate the difficulties in expressing fully functional VMATs and presumably other neurotransmitter transporters as well. This may be caused by the complex folding of these proteins or by the lack of essential factors limiting or absent in non-mammalian heterologous systems. The native binding site for tetrabenazine and ketanserin is accurately reproduced using various heterologous expression systems. The results suggest that the form required for high-affinity binding of TBZ is less dependent on the folding machinery. In contrast, the reserpine/substrate-binding site, which is directly affected by the H^+-electrochemical gradient, requires a more delicate arrangement not yet achieved in most heterologous systems.

4.2. Vesicular Acetylcholine Transporter (VAChT)

The powerful genetics of the nematode *Caenorhabditis elegans* has provided important information regarding VAChT. The elegant approach used by Rand and colleagues *(73)* was based on the analysis of one of the mutants described by Brenner 25 yr ago *(74)*. Mutations in the *unc-17* gene of the nematode result in impaired neuromuscular function, which suggests that cholinergic processes may be defective in the mutant. In addition, *unc-17* mutants were resistant to cholinesterase inhibitors *(74)*, a resistance which may result from decreased synthesis or release of the transmitter. Moreover, *unc-17* was found to be closely linked to *cha-1* gene, which encodes choline acetyltransferase *(75)*. The genomic region of *unc-17* was cloned by walking from the *cha-1* gene, and thereafter cDNA was isolated from a library *(73)*. Injection of a cosmid containing the complete coding sequence of the isolated cDNA rescues the mutant phenotypes of *unc-17* animals. A protein with 532 amino acids is predicted from the isolated DNA sequences. This protein (UNC-17 = VAChT) is 37% identical to VMAT1 and 39% identical to VMAT2. These findings strongly suggest that UNC-17 is a VAChT. This theory is supported by the fact that antibodies against specific peptides stain most regions of the nervous system. Within individual cells, staining has been punctate and concentrated near synaptic regions. Double-labeling with anti-synaptotagmin has showed colocalization of the two antigens. In addition, in *unc-104* mutants, a mutation in a kinesin-related protein required for the axonal transport of synaptic vesicles, synaptic vesicles accumulate in cell bodies. In these animals, anti-UNC-17 staining was restricted to neuronal-cell bodies. More than 20 alleles, viable as homozygotes, have now been identified. Their phenotypes vary from mild to severe. In two of these mutants, staining was dramatically decreased throughout the nervous system. Two other alleles were isolated which are lethal as homozygotes, and they seem to represent the null *unc-17* phenotype. This is the first demonstration that the function encoded by a VNT is essential for survival *(73)*.

Homology screening with a probe from *unc17* allowed for the isolation of DNA clones from *Torpedo marmorata* and *Torpedo ocellata (48)*. The *Torpedo* proteins display approx 50% identity to UNC17 and 43% identity to VMAT1 and VMAT2. Message is specifically expressed in the brain and the electric lobe. The *Torpedo* protein, expressed in CV-1 fibroblast cells, binds vesamicol with high affinity ($Kd = 6$ nM). Interestingly, the UNC17 protein expressed in the same cells also binds vesamicol, although with a lower affinity (124 nM) *(48)*. Mammalian VAChTs (human and rat) have been identified *(46,47)*. The predicted sequences of both proteins are highly similar to those of the *Torpedo* and *C. elegans* counterparts. The rat VAChT has been shown to bind vesamicol with high affinity ($Kd = 6$ nM). It also catalyzes proton-dependent, vesamicol-sensitive, ACh accumulation in transfected CV1 cells *(46)*. The distribution of rat VAChT mRNA coincides with that reported for choline acetyltransferase (ChAT), the enzyme required for ACh biosynthesis, in the peripheral and central cholinergic nervous systems. The human VAChT gene localizes to chromosome 10q11.2, which is also the location of the ChAT gene. The entire sequence of the human VAChT coding area is contained uninterrupted within the first intron of the ChAT gene locus *(46)*. Transcription of both genes from the same or contiguous promoters provides a novel mechanism for coordinate regulation of two proteins whose expression is required to establish a phenotype.

VAChT-positive cell bodies were visualized in the cerebral cortex, basal forebrain, medial habenula, striatum, brain stem, and spinal cord by using a polyclonal anti-VAChT antiserum. VAChT-immunoreactive fibers and terminals were also visualized in these regions and in the hippocampus, at neuromuscular junctions within skeletal muscle, and in sympathetic and parasympathetic autonomic ganglia and target tissues (*see* refs. *76* and *77*). In nerve terminals, VAChT is concentrated within neuronal arborizations in which secretory vesicles are clustered. Subcellular localization studies indicate that VAChT is targeted to small synaptic vesicles *(76,77)*.

4.3. Vesicular GABA/Glycine Transporter (VGAT, VIAAT)

The *C. elegans* mutant *unc-47* is selectively impaired in GABAergic transmission. The defect was defined as presynaptic and included all the GABA-mediated behaviors *(78,79)*. GABA was normally synthesized but not released, suggesting that the loading of synaptic vesicles was deficient *(78)*. Taking advantage of the *unc-47* mutant, two groups independently cloned a transporter that shows glycine-sensitive GABA uptake activity

(80,81). McIntire and collaborators screened for cosmids spanning the *unc-47* mapped region, able to rescue the mutant behavior. The rescuing sequence was further localized to a genomic fragment containing a single open reading frame of 486 amino acids *(80)*. Sagne et al. identified the same open reading frame using an elegant bioinformatic approach. They searched the *unc-47* mapped genomic zone for open reading frames coding for polypeptides containing various TMDs (hydropathy profile criterion), discriminating between the highly hydrophobic proteins by their expected size and homology to other proteins *(81)*. Using the predicted peptide sequence, a mouse EST was identified and served to clone the mouse *(81)* and rat orthologs *(80)*. The amino-acid sequence anticipates 10 TMDs, with amino and the carboxy termini facing the cystosol *(80,81)* and one uncertain N-glycosylation site near TM 6 *(81)*. An unusually large N-terminal hydrophilic domain containing several consensus phosphorylation sites is also predicted. The rat and mouse proteins (525 and 521 amino acids long, respectively) are almost identical, and differ only at their short carboxy-terminal. The amino-acid sequence shows no similarity to VMAT or VAChT, but is similar to a group of sequences from *S. cerevisiae*, *Arabidopsis thaliana*, and *C. elegans* apparently involved in transport of amino acids. PC12 cells expressing the cloned proteins show enhanced [^{3}H]GABA accumulation, inhibitable by GABA analogs *(80)*, nipecotic acid, and glycine, but not by glutamate *(80,81)*. Previous studies have shown that the vesicular GABA transport relies equally on the electrical component ($\Delta\psi$) and chemical component (ΔpH) of the $\Delta\bar{\mu}_{H^+}$ *(82–84)* whereas the monoamine transport is more dependent on the ΔpH *(1)*. Indeed, GABA uptake was less susceptible to ΔpH collapse than serotonin uptake by VMAT *(80)*. The rat ortholog was termed vesicular GABA transporter (VGAT) by McIntire et al. for its ability to transport GABA *(80)*. Howbeit, Sagne and colleagues proposed that the protein transports in vivo either GABA and/or glycine and therefore coined it vesicular inhibitory amino acid transporter (VIAAT) *(81)*. This group based its hypothesis on the following observations: glycine inhibited GABA uptake (IC$_{50}$ = 27.5 m*M*) *(80)*; a small but reproducible [^{3}H]glycine uptake could be measured in mVIAAT-expressing cells; and results from *in situ* hybridization studies showed that not only GABAergic but also glycinergic terminals were labeled by the mVIAAT mRNA *(81)*. Such a hypothesis is supported by a myriad of biochemical *(82,85–89)* and electron-microscopy studies which show the colocalization of GABA and glycine in some terminals *(90–93)*. Recently, specific antibodies against VGAT/VIAAT were used in immunohistochemical and

immunoelectron-microscopy analysis *(94,95)*. VGAT/VIAAT immunore-activity was restricted to presynaptic boutons exhibiting classical inhibitory features, and within the boutons, concentrated over synaptic-vesicle clusters *(94,95)*. Multiple detection of the transporter, GABA, and glycine on serial sections of the spinal cord or cerebellar cortex indicated that VGAT/VIAAT was present in glycine-, GABA-, or GABA- and glycine-containing nerve terminals *(95)*. These results support the contention that a single vesicular transporter is responsible for the storage of these two inhibitory transmitters. The regulation of which neurotransmitter is to be released may depend on several factors. As in the case of monoamines, the nature of the transmitter released relies on the expression of specific biosynthetic enzymes and plasma-membrane transporters. Since GABA synthesis depends upon the expression of glutamic-acid decarboxylase (GAD), the release of glycine can be regarded as default. Still, the affinity of VGAT/VIAAT for glycine is rather low. An interesting possibility is that regulatory events occurring directly at the level of the transporter alter its affinity toward either substrate or allowing their co-storage in the case of mixed GABA- and-glycine synapses. Changes in affinity for substrates were already implicated in the regulation of VMAT by the G-protein Go2 *(96)*. In addition, the presence of multiple phosphorylation sites at the large amino terminus provides several potential targets for detailed regulation. Different sites may govern the level of activity, the affinity to substrates, and even the subcellular targeting, and thus the mode of release.

Nevertheless, the existence of a specific transporter for glycine cannot be completely overruled *(80,94)*. In addition to the biochemical data *(84)*; Chaudry et al. found that GABAergic nerve endings contain higher levels of VGAT/VIAAT than glycinergic terminals. Besides, certain glutamatergic regions known to contain GABA such as hippocampal mossy-fiber boutons and subpopulations of nerve endings rich in GABA or glycine lack VGAT/VIAAT immunoreactivity *(94)*. Therefore, additional vesicular transporters or alternative modes of release may contribute to the inhibitory neurotransmission mediated by these two amino acids on the above terminals *(94)*.

VGAT/VIAAT define a novel gene family with a predicted topology that differs from that of VMAT and VAChT (also *see* Fig. 2). The structural divergence might be associated to the different bionergetic dependence of the transport process on the chemical and electrical components of the $\Delta^-\mu H^+$. In this respect, the yet-uncloned vesicular glutamate transporter is expected to more closely resemble the VGAT/VIAAT family.

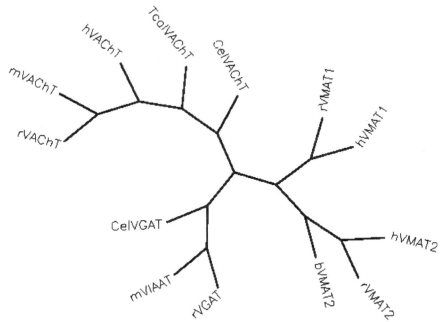

Fig. 2. Unrooted phylogenetic tree (PHYLIP) *(160)*. The VGAT/VIAAT family is evolutionary distant, and the VAChT and VMAT families are closely related. Note also that the two VMAT isoforms cluster together. Abreviations used are: m, *Mus musculus*; r, *Rattus norvegicus*; h, *Homo sapiens*; b, *Bos taurus*; Cel, *Caenorhabditis elegans*; Tcal, *Torpedo californica*. The tree was drawn using CLUSTAL W calculated similarities *(161)* using Workbench 3.2 software (http://workbench.sdsc.edu/).

5. STRUCTURE-FUNCTION STUDIES: IDENTIFICATION OF RESIDUES/DOMAINS WITH PUTATIVE ROLES IN STRUCTURE AND FUNCTION

Hydropathic analysis of the protein sequence of the VMAT and VAChT (Fig. 3) predicts 12 putative transmembrane segments and a large hydrophilic loop between TMDs 1 and 2. The loop contains potential sites for N-linked glycosylation. Previous studies have demonstrated that all the glycan moieties in glycoproteins of chromaffin granules and ACh storage vesicles face the lumen. Therefore, according to the latter finding and the model, the loop faces the lumen, and both termini face the cytoplasm. These assumptions have been confirmed in several studies.

The identification of functional residues in VNTs is based on studies using site-directed mutagenesis, chimeric molecules, and relatively specific

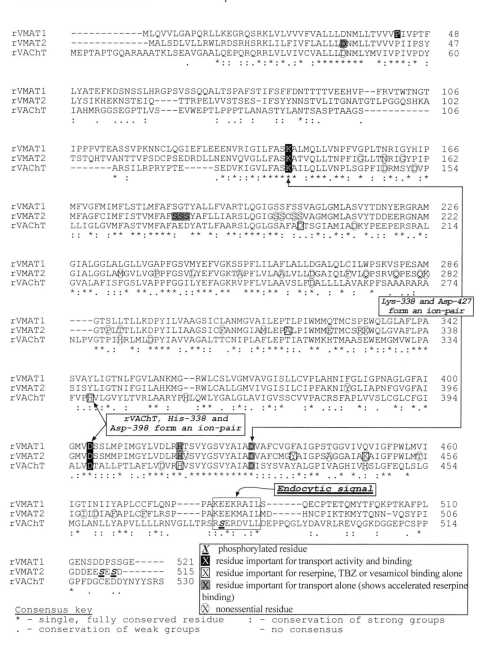

Fig. 3. Summary of mutational experiments shown in an alignment of rVMAT1, rVMAT2, and rVAChT. More details of the effect of the mutations are given in the text. The endocytic signal is also shown with black shading.

chemical modifiers. These studies are facilitated by the availability of sequences from different species and subtypes. It is usually assumed that residues which play central roles in catalysis are conserved throughout species. The degree of conservation in the VMAT-VAChT group is rather high. The members which are farther away in the group (human VMAT2 and the C. elegans VAChT) are still 38% identical and 63% similar (*1*; Fig. 2). The highest divergence is detected in the N-and C-termini and in the glycosylation loop between putative TM 1 and 2. The highest identity is observed in TM 1 ("LxxVxxAxLLDNMLxxVxVPIxP"), TM 2 ("GxLFASKAxxQxxxNP"), TM 4, ("ARxLQGxGS") and TM 11, where at least 11 amino acids are fully conserved. In TM 11 the conservation is particularly striking because practically all the amino acids conserved are in a contiguous stretch "SVYGSVYAIAD." Substitutions of some of these residues have been made, as discussed in this present chapter.

5.1. Molecular Basis of the Differential Targeting of VMAT and VAChT

Despite the similarities in structure, function, and sequence homology, VMATs and VAChT localize to distinct secretory organelles. In PC12 cells, although VMATs are preferentially targeted to LDCV, VAChTs predominantly occur in synaptic-like microvesicles (SLMV) *(65,97)*. LDCV and SLMV differ in their biogenesis, fate, protein cargo, and mode of release *(98)*. Therefore, the differential trafficking of VMATs and VAChTs poses an interesting paradigm to identify the signals implicated in targeting. To address this question, Varoqui and Erickson prepared chimeras of hVMAT2 and hVAChT in which their carboxy terminus were exchanged *(99)*. The targeting of the chimeras was examined in transformants of PC12 cells. The hVMAT2 chimera which contained the cytoplasmic tail of hVAChT was preferentially targeted to synaptic vesicles. The reciprocal hVAChT chimera carrying the cytoplasmic tail of hVMAT2 revealed a marked shift in its distribution toward that observed for chromogranin B (CgB, an LDCV marker) or the wild-type hVMAT2 *(99)*. In parallel, using a similar approach followed by deletions and site-directed mutagenesis, Tan and colleagues identified a short sequence within the carboxy-terminal domain critical for the internalization of the transporters from the plasma membrane. The rVMAT2 sequence—"KEEKMAIL" and the rVAChT counterpart "RSERDVLL"— was found to be sufficient and essential for efficient endocytosis of the transporters *(100)*. Furthermore, addition of these sequences to interleukin 2 receptor α-subunit (interleukin-2 or Tac), a plasma-membrane protein *(101,102)*, induced its efficient internalization. Although single replacements were ineffectual, replacement of the isoleucine-leucine pair of rVMAT2's

KEEKMAIL or leucine-leucine at rVAChT's RSERDVLL by alanine-alanine reduced transporter endocytosis by 50%. Similarly, in the chimeric proteins Tac-KEEKMAIL and Tac-RSERDVLL, double substitution of the Ile-Leu and Leu-Leu pair for Ala-Ala restrained protein internalization *(100)*. Leucine-based motifs are known to interact with the clathrin-adaptor protein AP-2 *(103)*, suggesting that VMATs and VAChTs may internalize via clathrin-coated pits. For some proteins, the acidic residues (**KEEK**MAIL) at positions -4 and -5 from the dileucine motif contribute to the endocytic signal *(101,103)*. Yet, for rVMAT2 simultaneous replacement of both glutamates for alanine did not influence internalization *(100)*. The acidic residues may instead have a role in binding to other adaptor proteins involved in trafficking to different secretory organelles. Indeed, differences in the sequence preceding the dileucine motif were recently addressed *(104,105)*. VAChT is phosphorylated in PC12 and COS cells at a serine residue *(104,105)*. Five serine residues exist at the C-terminal cytoplasmic tail of rVAChT. Two of these, Ser-478 and Ser-480, are conserved in vertebrates and exist in the vicinity of the endocytic signal. Serines 478 and 480, at positions −7 and −5 relative to the dileucine motif (**SRS**ERDVLL) represent consensus sites for casein kinase II, and serine 480 is also a consensus site for protein kinase C (PKC). Replacement of Ser-480—but not Ser-478—with Ala dramatically reduced the extent of phosphorylation in both COS and PC12 cells. Consistent with a previous study *(106)*, both groups maintain that VAChT is mainly phosphorylated by protein kinase C, and that this residue plays a role in transporter targeting *(104,105)*. Although the fraction of phosphorylated transporters was not estimated, the lack of phosphorylation at Ser480Ala had no effect on protein level, transport activity *(104,105)*, or vesamicol binding *(104)* indicating that phosphorylation may have a role distinct from regulation of activity. For example, in two proteins unrelated to VAChT, phosphorylation of a serine residue at position −5 from the dileucine motif dramatically influences their internalization through clathrin-coated pits *(107,108)*. In addition to alanine, serine-480 was also mutated to glutamate *(105)* and to aspartate *(104)*, resembling the acidic residue found in VMAT endocytic signal. Differences between the fractionation patterns shown for the Ser480Ala mutant and the wild-type VAChT are minor and subtle. Alhough both groups reach conflicting conclusions regarding the single replacements, the role of Ser-480 in targeting was established. Additional analysis using neuronal cell lines for expression or refined sedimentation procedures should be made to further clarify this point.

The acidic residues near the dileucine motif in the endocytic signal of VMATs contribute to the sorting to LDCV. Simultaneous replacement of Glu-478 and Glu-379 for Ala partially redistributes rVMAT2 to lighter membranes and away from LDCV *(105)*. In view of these results, the glutamates of the endocytic signal are important to direct VMATs to the precise secretory organelle. Negatively charged phosphoserine may mimic the effects of the acidic residues present in the VMAT signal. This raises the possibility that VAChT phosphorylation at Ser-480 relocates the transporter from synaptic vesicles (or SLMV) to other secretory organelles. Such a mechanism will change the mode of transmitter release and the quantal size, increasing the potential of presynaptic plasticity *(105)*.

5.2. Post-Translational Modifications and Regulation of Activity

The purification process of the VMAT from chromaffin granules included lectin chromatography, indicating that the protein is glycosylated in vivo *(26)*. In VMATs as well as in VAChTs, the highly divergent hydrophilic loop between membrane domains 1 and 2 contains potential sites for N-linked glycosylation. The number of sites is not uniform, and varies between two and four in different species and isoforms. The importance of N-linked glycosylation for rVMAT1 was addressed using site-directed mutagenesis to modify Asn to Gln in the consensus sequences (Asn-X-Ser/Thr) *(109)*. The results showed that rVMAT1 is glycosylated at all three putative sites. Glycosylation was not essential for activity or substrate recognition, as a mutant transporter devoid of all glycosylation sites ("Triple" mutant) retained the ability to transport serotonin, to bind reserpine, and to protect the cells against MPP+ *(109)*. In contrast to the findings for plasma-membrane neurotransmitter transporters for which the unglycosylated form is not efficiently targeted, the "Triple" mutant was found to localize with a pattern similar to that of the wild-type protein. Because of the low levels of expression of the "Triple" mutant, it was suggested that glycosylation may confer stability to the mature protein. The extremely conserved stretch in TM 1 seems to be critical for glycosylation and folding of the transporter. Replacement of a conserved proline to leucine in this region (VFXALLLDNMLXT VVVP$_{43}$IXP, consensus for all mammalian VMATs) renders an inactive protein that is not glycosylated despite the presence of all glycosylation sites. In addition, consistent with the fate of misfolded proteins, the Pro43Leu mutant is not efficiently targeted, and is probably retained at the endoplasmic reticulum (ER) *(109)*. The results indicate an important structural role for Pro-43, and probably also for part of the conserved stretch at TM1. Since the transporters share extensive homology in sequence and topology, it is reasonable to presume that the roles of glycosylation and the residue Pro-43 are relevant to all the proteins in the family.

Phosphorylation is a reversible ubiquitous mode of regulating protein activity. Plasma-membrane neurotransmitter transporters are often regulated by phosphorylation (*see* Chapters 1 and 10). In the case of vesicular transporters, emerging evidence now suggests that the regulation of transporters is a component of synaptic plasticity. Addition of the PKC activator phorbol myristate acetate (PMA) to hippocampal slices decreased the inhibitory effect of vesamicol on acetylcholine release, but did not alter the release in the absence of the inhibitor *(106)*. Besides, in the presence of PMA, VAChT phosphorylation was increased concomitant with a decrease in binding of 4-aminobenzovesamicol *(106,110)*. Interestingly, the effects were detected only when activation of protein kinase C occurred prior to exposure to the vesamicol analog *(110)*. As detailed here, it was shown that phosphorylation of VAChT by PKC occurs at Ser-480, and its effect concerns only regulation of transporter targeting *(104,105)*. The fraction of phosphorylated transporter was not determined in either case; thus, the relationship between vesamicol binding and PKC-induced phosphorylation remains elusive.

Addition of dibutyryl cyclic AMP inhibited the vesicular transport of monoamines in PC12 cells. This effect was antagonized by the protein kinase inhibitor K252a, and enhanced by the phosphatase inhibitor okadaic acid *(111)*. Direct evidence of phosphorylation of VMAT was provided by Krantz et al. *(112)*. Rat VMAT2, but not rVMAT1, is constitutively phosphorylated at the carboxyl-terminus by the acidotropic kinases casein kinase I (CKI) and casein kinase II (CKII). The main site of phosphorylation is Ser-512 and to a lesser extent Ser-514. Simultaneous replacement of Ser-512 and Ser-514 for Ala did not significantly influence the transporter activity or its affinity for substrates *(112)*. Phosphorylation in the double mutant was greatly reduced but not completely abolished, indicating the presence of other potential sites for phosphorylation. Differential phosphorylation of the two VMAT isoforms may influence its subcellular localization (as found for VAChT) or facilitate other regulatory events *(112)*.

Regulation of VMAT activity by heterotrimeric G proteins was recently described *(66,96)*. In streptolysin O-permeabilized PC12 cells, non-hydrolyzable GTP-analogs and AlF_4- inhibited noradrenaline uptake. Inhibition was prevented by the addition of antibodies specific to $G\alpha o$ and by pertussis toxin and induced only by activated $G_{\alpha o 2}$. Reserpine binding was also inhibited by GTPγS and AlF_4- activated $G_{\alpha o 2}$. G-protein mediated downregulation appears to act on the transporter and not on the acidification process essential for building the energy-gradient that sustains the uptake activity. Downregulation appears to decrease the affinity of the transporter toward substrates, because increasing concentrations of noradrenaline partially compensate the inhibition *(96)*. In line with a role of G-proteins in regulation of vesicular content, $G_{\alpha o 2}$ was found to be associated with the membrane of chromaffin granules *(113)*, with LDCV from (BON) cells and small synaptic vesicles (SSV) from rat raphe neurons *(66)*. Taking advantage of the different affinity toward histamine and tetrabenazine it was found that rVMAT2 is more sensitive to G-protein regulation than rVMAT1 *(66)*. Rat VMAT2 inhibition by $G_{\alpha o2}$ was measured in a synaptic-vesicle preparation from prefrontal cortex, and in permeabilized BON cells and raphe neurons. Therefore, VMATs can be modulated by Go2 in neurons and neuroendocrine cells if the transporter is localized to LDCV or SSV *(66)*. Still unknown are the factors involved in the upstream regulation of this process and whether the G-protein interacts directly with the transporter or through an unidentified component.

5.3. Acidic Residues

Particularly striking are four transmembrane-conserved Asp residues in VMATs and VAChTs: D34, D267, D404, and D431 (numbers of rVMAT1) in the middle of TM 1, 6, 10, and 11 (Fig. 3). In addition to their being conserved charged residues in the membrane, available biochemical evidence indicated that N,N'-dicyclohexylcarbodiimide (DCC) inhibits VMAT mediated transport *(114–116)*. DCC reacts with a carboxyl residue whose availability is influenced by the occupancy of the tetrabenazine-binding site *(116)*. Reaction with this carboxyl residue inhibits overall transport activity as well as TBZ and reserpine binding, suggesting that the residue plays a role in one of the first steps of the transport cycle. As with all chemical modifiers, indirect effects such as steric hindrance by the DCC moiety or an indirect effect on the structure of the protein cannot be ruled out at present. Therefore, mutagenesis studies of the roles of these four Asp residues on VMAT should be instructive.

Conservative replacement of Asp33 with Glu in rVMAT2 reduced activity, but did not abolish it *(117)*. Replacement with Asn abolished transport, indicating the crucial role of a negative charge at this position. However, VMAT could still bind [^{3}H]reserpine, and binding was accelerated by $\Delta \tilde{\mu}_{H^+}$ *(117)*. In addition, Asp-46, the corresponding residue in rVAChT can be substituted with Asn, and the resulting protein retains both acetylcholine transport and vesamicol binding *(118)*. Inhibition of reserpine binding in D33N by serotonin differs dramatically from the wild-type, suggesting interference with substrate recognition. Similarly, replacements in positions 404, 431 located in TM 10 and 11 had dramatic effects on the activity of rVMAT1, rVMAT2,and rVAChT *(118–120)*.

In rVMAT1, replacement of the corresponding Asp 404 with Glu generated a very interesting protein; the pH optimum of transport was shifted by approx 1 pH unit to the acid side. In addition, the affinity of the mutant protein to TBZ dramatically increased to levels comparable to those observed in the VMAT2 subtype. These results indicate that the environment around the carboxyl moiety at position 404 is critical for the recognition of TBZ, and influences the pK of one or more groups in the protein involved in the last steps of the catalytic cycle. Replacement of Asp-404 with either Cys or Ser yielded proteins that displayed no transport or reserpine binding *(120)*. In line with the results for rVMAT1, conservative replacement of Asp-398 with Glu in rVAChT retains almost full transport activity but none vesamicol binding, D398N does not shows neither transport or binding activity *(118)*. Replacement of the corresponding Asp-400 with Asn totally abolished transport *(119)*.

In the case of 431 (425 in rVAChT), even a conservative replacement with Glu led to transport inhibition of both rVMAT1 and rVAChT *(118,120)*. The Asp-431 replacements (rVMAT1), D431E, and D431S—but not D431C, which was not expressed at detectable levels—bound [^{3}H]reserpine normally, and binding was accelerated by $\Delta\tilde{\mu}_{H^+}$. The corresponding rVAChT replacements D425N and D425E retained normal vesamicol binding *(118)*. In rVMAT2, the substitution of Asp-427 with Asn eliminated transport activity. Yet simultaneous replacement of Asp-427 with Asn and Lys-139 with Ala (K139A/D427N) restores serotonin transport *(119)*. The charges of these residues are suggested to form a salt-bridge or ion pair *(119)*.

Replacement of Asp-263 in rVMAT2 with Asn had no effect on serotonin transport *(119)*. Similar substitution of Asp-255 in rVAChT also had no effect on transport or vesamicol binding *(118)*. However, Asp-193, a membranal residue conserved only in VAChTs, drastically reduces vesamicol binding without effect on transport when substituted for Asn *(118)*. It was previously reported that the Asp193Asn mutant is inactive, yet in this study transport activity was measured indirectly *(121)*.

Five additional conserved aspartate residues located in cytoplasmic loops were studied in rVAChT (Asp-147, 152, 202, 288, and 410). Asp-288 and Asp-410 are also fully conserved in VMATs. Individual replacement of those residues with Asn (D147N, D152N, D202N, D288N, or D410N) did not diminish either acetylcholine transport activity or vesamicol binding *(118)*.

An important conclusion from these studies is that some replacements of negative charges in the middle of putative transmembrane segments may have dramatic effects on transport, but do not necessarily induce major changes in the protein structure, since some of the mutants still bind a high-affinity ligand and respond to $\Delta\tilde{\mu}_{H^+}$. The ability to measure partial reactions in VMAT allows for a more sophisticated analysis of mutagenized proteins than previously possible. Thus, although it does not transport, it was shown that the rVMAT2 D33N mutant protein has a lower affinity to serotonin than the wild-type. An aspartic residue at the first TM in plasma-membrane monoamine transporters was proposed to serve in substrate recognition *(122)*. In line with this theory, although Asp-33 was found essential for VMAT activity, its replacement in VAChT had no effect. Also, the rVMAT1 mutant D404E shows an acid shift on the overall cycle but not on partial reactions, suggesting an effect on a pK_a important for the final steps of transport. It is tempting to speculate that rVMAT1 D404 plays a direct role in substrate recognition and/or the translocation of the second H^+ needed for the overall cycle. Direct proof for this theory requires further experimentation.

5.4. Basic Residues

His residues have also been suggested to play a role in H^+ translocation and sensing in other H^+-coupled transporters *(123,124)*. In VMATs, there is only one His conserved (His419) in loop 10, between TM 9 and 10. This His is immediately behind an Arg residue conserved throughout the whole VNT family, and very close to the longest conserved sequence stretch. Although also present in the rat, human and *C. elegans* VAChT, it is replaced by a Phe in the Torpedo VAChT. Biochemical evidence also suggests a role for His in VMAT. Phenylglyoxal (PG) and diethylpyrocarbonate (DEPC) are reagents relatively specific for Arg and His residues, respectively. They both inhibit serotonin accumulation in chromaffin-granule-membrane vesicles in a dose dependent manner (IC_{50} of 8 and 1 mM, respectively) *(125,126)*. The inhibition by DEPC was specific for His groups, since transport could be restored by hydroxylamine *(126)*. PG and DEPC failed to inhibit binding of either reserpine or tetrabenazine, indicating that the inhibition of transport is not caused by a direct interaction with either of the known binding sites. Interestingly, however, the acceleration of reserpine binding by a transmembrane H^+ gradient was inhibited by both reagents *(126)*. These results suggest that either proton transport or a conformational change induced by proton transport is inhibited by both types of reagents. In the case of VAChT, DEPC causes time-dependent inhibition of [^{3}H]vesamicol binding. Acetylcholine is unable to protect from the effect of DEPC, probably because it binds to a different site *(127)*. Interestingly, the authors proposed that DEPC binds to a fully conserved histidine (His-338). Indeed, when this residue was substituted for Ala or Lys, the transporter retained full transport activity but no vesamicol binding *(128)*. This finding indicates that His-338 is not involved in proton translocation, and that substitutions for other residues results in conformational changes that affect vesamicol binding.

A more direct analysis of the role of histidines in VMAT has been carried out by site-directed mutagenesis of rVMAT1 *(129)* and rVAChT *(128)*. Replacement of His-419 with either Cys (H419C) or Arg (H419R) completely abolishes transport, as measured in permeabilized CV1 cells transiently transformed with the mutants DNA. In the absence of $\Delta \tilde{\mu}_{H^+}$, reserpine binding to the mutant proteins is at levels comparable to those detected in the wild-type. However, acceleration of binding in the presence of $\Delta \tilde{\mu}_{H^+}$ is not observed in either H419C or H419R. These results suggest that His-419 plays a role in a step other than binding and may be associated directly with H^+ translocation or in conformational changes that occurr after substrate binding. His-413 is the residue in rVAChT which corresponds to His-419 of rVMAT1. This residue is functionally different in VAChT activity, since it can be substituted for Ala, Arg, or Cys, rendering an active protein displaying full transport and half of the normal vesamicol binding *(128)*. In addition, the His at this position is naturally substituted for Tyr in Drosophila and Torpedo VAChTs. His-444 of rVAChT, is located in the boundary of TM 12 and is conserved in all vertebrate transporters. Its replacement with Ala, Arg, or Cys had no influence on acetylcholine transport or vesamicol binding. Furthermore, double mutation of VAChT membranous histidines (H338, H444), or mutation of all three luminal-loop histidines (H283, H354, and H413) showed no change in acetylcholine transport *(128)*.

Lys-139 of rVMAT2, located in the core of TM 2, is conserved in all VMATs and VAChTs. Substitution of Lys-139 with Ala completely eliminated transport of serotonin *(119)*. Replacement of Lys-131 of rVAChT with Arg or His renders a functional protein, whereas substitution with either Ala or Leu abolishes transport activity while retaining full vesamicol binding *(128)*. These findings indicate that at this position, a positive charge is important for transport activity *(128)*. This lysine residue is located in the middle of a transmembrane region. In several cases, the existence of ion pairs between transmembrane-positive and negatively charged residues has been proposed as a means of avoiding an energetically unfavorable environment. Replacement of either charged residue creates an unpaired charge which can lead to functional defects. Simultaneous neutralization of both charges or reversal of polarity may restore the function lost in the single mutants *(130–132)*. In rVMAT2, charge neutralization but not polarity reversal restored transport activity. The mutant carrying a double substitution K139A/D427N shows significant serotonin transport, yet a fivefold decrease in the affinity toward serotonin and reserpine *(119)*. The charge-reversal K139D/D427K mutant did not show transport activity, yet retained accelerated reserpine binding. These results suggest that Lys-139 and Asp-427 form an ion pair in the wild-type protein, and that a single unpaired charge prevents transport *(119)*.

In the case of rVAChT, simultaneous neutralization of Lys-131 and Asp-398 (in K131A/D398N) or charge reversal in K131D/D398K and K131D/D425K did not restore vesamicol binding *(128)*. Interestingly, although the individual replacements retain vesamicol binding *(118,128)* the double-neutralization mutant K131A/D425N does not. The results of rVAChT K131A/D425N suggest that such an ion-pair may also exist in VAChT. In addition, charge reversal in the double-mutant H338D/D398H, but not neutralization of charges, restores the vesamicol-binding activity lost with the individual substitutions, indicating that in VAChT, His-338, and Asp-398 may form an ion pair *(118)*.

5.5. Serine Residues

Ser-480 of rVACHT was shown to undergo phosphorylation, and its effect seems to influence protein targeting *(104,105)*. In the β-adrenergic receptor and in the dopamine plasma-membrane transporter *(133)*, it has been suggested that serines play a role in ligand recognition. Two groups of serine residues occur in VMAT2: in TM 3 and TM 4. Simultaneous replacement of four serines in TM 4 (S197, 198, 200, and 201) with alanine does not affect transport activity. However, mutant VMAT2 in which serines 180, 181, and 182 (in TM 3) were replaced with alanine, showed no transport activity. Moreover, binding of [^{3}H]reserpine was at normal levels, and it was accelerated by $\Delta\bar{\mu}_{H^+}$. However, in contrast to wild-type, and similar to D33N, binding was not inhibited by serotonin, even at concentrations of 500 μM. These findings suggest a possible role of Ser 180–182 in substrate recognition *(117)*. In the same study, mutations at four residues conserved in VMAT (T154A, N155Q, G151L, and G158L; numbers of rVMAT2) had no effect on transport activity *(117)*.

5.6. Chimeric Transporters

The use of chimeras between homologous proteins that differ in their pharmacological profile is a powerful tool to identify the domains responsible for various affinities. Peter and colleagues prepared chimeras between rVMAT1 and rVMAT2 and analyzed the recombinant proteins for their susceptibility to tetrabenazine and histamine inhibition and for their affinity to serotonin *(134)*. Two different domains were identified, the first spanning TM 5 to TM 8, and the second from TM 9 to TM 12. Both domains were found to provide part of the interactions, and seemed to cooperate to confer the high-affinity interactions of VMAT2 toward tetrabenazine and histamine. A strong correlation between tetrabenazine and histamine inhibition was found, supporting the contention that both compounds bind to the same site *(134)*. Erickson created a hVMAT2/1/2 sandwich chimera composed of half of TM 6 up to TM 10 from hVMAT1, flanked by hVMAT2 sequences. This protein shows enhanced tetrabenazine inhibition, but the affinity of rVMAT1 to histamine *(135)*. Taken together, the results support the presence of important residues in TM 11–12 for histamine recognition, and probably also for TBZ inhibition. A role for residues of TM 1 in substrate recognition was suggested by the finding that an rVMAT1-based chimera ("S37C") that contains only the N-terminal and TM 1 of rVMAT2 shows an increased affinity toward TBZ and histamine *(134)*. In line with this hypothesis, a recombinant hVMAT2 transporter containing only the TM 1, the glycosylation loop, and part of TM 2 of hVMAT1 is not inhibited by TBZ *(135)*. In addition, a chimera between hVMAT2 and hVAChT containing the hVAChT sequences except the N-terminal up to TM 2 of hVMAT2 shows a sevenfold decrease in the affinity for acetylcholine, but retains vesamicol binding *(136)*.

Since the contribution of each domain is mediated by only a few scattered residues, attempts to distinguish the critical amino acids have been made *(137,138)*. Residues of VMAT2 were replaced for the corresponding rVMAT1 residues. The majority of the replacements made were ineffectual *(137,138)*. Yet some substitutions provoked a twofold increase in the IC_{50} of tetrabenazine: Y434F, K446Q, D461N *(137)*, and A315T *(138)*. Y434F also displays a threefold decrease in the affinity toward histamine. The combination of the influencing substitutions in Y434F/D461N, K446Q/D461N had an additive effect on the affinity to TBZ. The triple combination (Y434F/K446Q/D461N) shows half of the IC_{50} of VMAT1 *(137)* and the quadruple combination (A315T/Y434F/K446Q/D461N) displays the same half inhibitory concentration (IC_{50}) as the wild-type rVMAT1 *(138)*. Yet the same effect conferred by A315T to the quadruple mutant A315T/Y434F/K446Q/D461N is achieved by the addition of K328E in K328E/Y434F/K446Q/D461N, although K328E alone has no effect either in the affinity for tetrabenazine or histamine *(138)*. The model involving those residues alone in TBZ and histamine recognition raises several questions. First, steric distant effects caused by certain replacements may only slightly distort the conformation of the interacting residues at the catalytic site, and therefore cause a small decrease in the affinity toward substrates. Since those structural effects are presumably additive, a corresponding cumulative decrease in the affinity is expected to occur. Second, considering that the interactions with the substrate at the binding site are usually accomplished by very few residues, even conservative replacements of interacting amino acids are expected to induce drastic changes in affinity. The observation that the rVMAT2 mutant Y434A displays a 25-fold decrease in affinity toward TBZ, a sixfold decrease for histamine, and a fourfold increase in the affinity for serotonin supports a role for Tyr-434 in substrate recognition *(137)*. On the other hand, since D461N is functional and displays less than two fold decrease in the affinity for TBZ *(137)*, it is difficult to conceive that this residue has a role distinct from structural. The recently identified *C. elegans* VMAT shows a pharmacological profile similar to that of VMAT2 *(139)*. Interestingly, this transporter shows a 20-fold higher affinity for dopamine than vertebrate VMATs, and is inhibited by octopamine. Apart from Lys-446, which is also conserved in *C. elegans* VMAT and hVMAT1 *(60)*, Ala-315, Tyr-434, and Asp-461 of rVMAT2 are replaced in the *C. elegans* VMAT by Ser, Phe, and Ser, respectively. Yet *C. elegans* VMAT is inhibited by tetrabenazine, and shows the apparent affinity of VMAT2 toward histamine.

If such residues are directly involved in conferring to VMAT2 the structural or catalytic interactions required for histamine and tetrabenazine recognition, the introduction of those amino acids to VMAT1 should increase its affinity toward both substrates. Yet the reciprocal mutants of rVMAT1 (F434Y/Q446K/N461D and A237P/F434Y/Q446K/N461D) show an even lesser affinity for tetrabenazine and histamine *(138)*.

Finally, rVMAT1 shows a fivefold higher affinity to tryptamine than rVMAT2 *(137)*. The rVMAT1 mutant F434Y/Q446K/N461D shows the affinity of rVMAT2 for tryptamine. In this case, the reciprocal rVMAT2 mutant Y434F/K446Q/D461N shows the affinity of rVMAT1 for tryptamine. However, the same increase in affinity is displayed by the single replacements Y434F, P237A, and A315T; the mutant Y434A shows increased affinity for serotonin, dopamine, and tryptamine *(137,138)*.

It is clear that the interacting residues at the catalytic site are imperative for substrate recognition, as well as neighboring and distant amino acids, which affect the disposition of the interacting residues at the binding sites. Therefore, even if the truly interacting residues are inserted in VMAT1, tetrabenazine binding and histamine recognition are not guaranteed.

6. VMAT2-KNOCKOUT MICE

To assess the role of the VMAT in gestation and development, the groups of Uhl *(140)*, Edwards *(141)*, and Caron *(142)* created transgenic VMAT2-knockout mice. Homozygous mice lacking the *VMAT2* gene $^{(-/-)}$ were born at the expected Mendelian ratio, indicating that VMAT2 activity is not essential for gestation. Yet newborns were poorly viable, slightly smaller and hypoactive. They fed poorly; more than one-half died by the first postnatal day, and all died by postnatal day 14. Examinations of homozygous mice revealed good subcutaneous brown fat and normal organs. Histopathological studies showed no remarkable differences in the general morphology and density of cell populations between the brains of homozygous animals and those of wild-type littermates *(140–142)*. However, the monoamine content in brains from $VMAT2^{(-/-)}$ animals was drastically reduced. The measured levels of dopamine, serotonin, and norepinephrine were less than 5% of the wild-type levels *(141,142)*. In contrast, the levels of monoamine metabolites were very similar in both animal populations.

The results indicate that despite increased synthesis, transmitter that is not stored in vesicles undergoes rapid metabolism, implying a central role for the vesicular transport in the protection of newly synthesized transmitter. Using monoamine oxidase (MAO) inhibitors, it was shown that MAO-A and MAO-B account for the low levels of monoamines in *VMAT2*-knockout mice *(141)*. No release of dopamine was measured after K^+-evoked depolarization using primary midbrain cultures prepared from $VMAT2^{-/-}$ mice. Yet, a 30-fold increase in released dopamine was measured in the same cultures upon exposure to D-amphetamine *(141,142)*. The non-exocytotic mechanism by which amphetamine releases the cytosolic dopamine is flux reversal through plasma-membrane transporters *(143)*. Systemic administration of amphetamine to newborn $VMAT2^{-/-}$ mice prolonged survival, increased directed locomotor behavior, and promoted feeding and walking by the third week until they eventually died *(141)*, indicating that monoaminergic exocytotic release is essential for the adult animal to develop and thrive.

As stated earlier, homozygous knockout embryos develop normally despite the lack of VMAT2. Furthermore, monoaminergic regions form the same projections found in wild-type brains. VMAT1, the non-neuronal isoform, was shown to be expressed in neurons during embryonic development *(144)*. Therefore its expression, could compensate the lack of VMAT2 and allow brain formation. Another possibility is that flux reversal through plasma-membrane transporters releases enough transmitter to sustain development of neural connections. Conversely, monoaminergic projections can occur independently of signaling by monoamines and respond to other intrinsic developmental mechanisms *(141)*.

Heterozygous mice for *VMAT2* gene $^{(+/-)}$ are viable into adult life, showing locomotor activities, habituation, fertility, and weight gain similar to wild-type mice *(140–142)*. $VMAT2^{(+/-)}$ animals show moderately elevated blood pressure and heart rate *(140)*, and are more prone to sudden death in midlife. Uhl and colleagues found that electrocardiograms from heterozygous mice displayed prolonged QT intervals, which may lead to lethal arrythmias *(145)*. Given these observations, severe cardiovascular failure can be a plausible cause of death of homozygous mice soon after birth. Heterozygous mice display VMAT2 protein levels one-half that of wild-type values and lower monoamine content. In the brains of $VMAT2^{+/-}$ mice, levels of dopamine are 58% of those in wild-type animals, and the figures for norepinephrine and serotonin are 77 and 66%, respectively *(140–142,146)*. Under conditions of limited time for vesicular filling the level of transporter expression would largely influence the amount of transmitter released. Surprisingly, despite prolonged culture incubation (4 wk), primary midbrain cultures from heterozygous mice released only one-half of the dopamine released by cultures prepared from wild-type animals *(141)*. A possible explanation for the findings is that because of the low VMAT2 expression in heterozygotes, not all the vesicles carry the transporter, causing a general decrease in the ability of the cells to store and release transmitters. The neurotoxic potency effects of the MPP$^+$ precursor N-methyl-4-phenyl-1,2,3,6-tetrahydropyridine (MPTP) was nearly doubled in *VMAT2$^{+/-}$* mice causing more than twice the dopamine cell losses found in wild-type mice *(140,147)*. Also the metamphetamine-induced dopaminergic neurotoxicity was increased in the striatum of VMAT2 heterozygous mice compared with wild-type mice *(148)*. The hypothesis that transporter expression regulates quantal size *(141,149,150)* is also supported by the fact that elevated expression of VMAT2 reduces the toxic effects of the neurotoxin MPP$^+$, and overexpression of VAChT in *Xenopus laevis* embryos causes a significant increase in postsynaptic quantal events in cocultured myocytes, reflecting a 10-fold increase in the vesicular packaging of acetylcholine *(121)*.

The results with VMAT2-knockout mice have confirmed several of the previously described roles attributed to vesicular transporters. Vesicular transport is obligatory for the exocytotic release of transmitters. In knockout mice, VMAT2 activity is essential for mouse locomotor activity, feeding, survival, and general development. The vesicular transport activity protects the synthesized transmitters from cytoplasmic oxidation. It also protects the cell from the possible cytotoxic effects of transmitters and from exogenous chemicals (i.e., MPP$^+$). In addition, the knockout results confirmed the relationship between the amount of transporter expressed and the quantal size. Regulation of the expression levels, together with the ability to maneuver the activity of the transporters by endogenous proteins *(66,96)*, define a new component which contributes to the synaptic plasticity.

7. TEXANS

The VMAT-VAChT subfamily is part of a family which includes at least 40 proteins from prokaryotes and eukaryotes, and has been surveyed in several monographs *(1,7,151,152)*. Based on an analysis of the evolutionary relationships, four subgroups have been identified. All the proteins present in microorganisms are presumed to be exporters located in the cytoplasmic membrane of these cells, which confer resistance to a large list of compounds because of their ability to actively remove them from the cell. The VNTs, however, are located in intracellular vesicles. The bacterial transporters extrude the toxic compounds to the medium, and those presently known in mammals remove neurotransmitters from the cytoplasm into intracellular storage compartments. In both cases, as a result of their functioning, the concentration of the substrates in the cytoplasm is reduced. When the substrate of the VNTs is cytotoxic, such as the case of MPP$^+$, a substrate of

VMAT, the removal of the toxic compound from the cytoplasm, away from its presumed target will ameliorate the toxicity of the compound. Indeed, CHO cells expressing VMAT are more resistant to MPP$^+$ *(52)*. These findings suggest a novel type of drug resistance in eukaryotic cells: protection of the devices essential to life by compartmentalization. Similar strategies have previously been suggested to explain tolerance to high salt and toxic compounds in plants and yeast *(11,153,154)*. In addition, heterologous expression of a bacterial multidrug transporter in *S. cerevisiae* confers resistance to toxic compounds by active removal into the cell vacuole *(155)*.

Most transporters of the family have a very broad specificity for substrates. All of the substrates are aromatic compounds, usually bearing an ionizable or permanently charged nitrogen moiety. In some substrates, however, carboxylic groups are also present (i.e., norfloxacin); in others, a phosphonium moiety is present (i.e., TPP$^+$), and yet in others, no positive charge is present at all (i.e., actinorhodin, uncouplers). Many of the substrates of the multidrug transporters also interact with VMATs *(156)*. This large overlap in substrate recognition may help produce clues to common solutions to the problem of recognition of multiple substrates with high affinity.

All the transporters of the toxin extruding antiporter (TEXAN) family are located in membranes, across which H$^+$ electrochemical gradients exist. The gradients are generated by primary pumps, such as the bacterial respiratory chain or the H$^+$-translocating ATPases of both bacteria and intracellular storage organelles. The gradient is utilized by the protein through the exchange of a substrate molecule with one or more hydrogen ions. All the neurotransmitter storage vesicles studied thus far in the brain, platelets, mast cells, and the adrenal medulla contain a vacuolar type H$^+$ pumping ATPase, similar in composition to the ATPase of lysosomes, endosomes, Golgi membranes, and clathrin-coated vesicles *(157–159)*. In all these organelles, the activity of this proton pump generates an H$^+$ electrochemical gradient ($\Delta\bar{\mu}_{H^+}$ acid and positive inside). In synaptic vesicles and neurotransmitter storage organelles, the proton electrochemical gradient is utilized by the VNT's, which couple efflux of H$^+$ ions to the uptake of a neurotransmitter molecule (for review, *see* ref. *1*). In the case of *Escherichia coli* Tet proteins, it has been suggested that the exchange is between a Metal-Tetracycline complex and one proton in an electroneutral process. Also *B. subtilis* multidrug transporter (BMR) mediated drug efflux is apparently driven by a transmembrane pH gradient. Although very little is known about the other transporters, the fact that they all display sequence similarities and none of them show any ATP-binding domains—in addition to the fact that they are all found in with H$^+$ gradients—suggests that they all are antiporters which exchange one or more H$^+$ ions with a substrate molecule.

ACKNOWLEDGMENTS

Work in the authors' laboratory was supported by grants from the National Institute of Health (NS16708) and the National Institute of Psychobiology in Israel.

REFERENCES

1. Schuldiner, S., Shirvan, A., and Linial, M. (1995) Vesicular neurotransmitter transporters: from bacteria to human. *Phys. Rev.* **75**, 369–392.
2. Johnson, R. (1988) Accumulation of Biological Amines in Chromaffin Granules: A Model for Hormone and Neurotransmitter Transport. *Physiol. Rev.* **68**, 232–307.
3. Njus, D., Kelley, P. M., and Harnadek, G. J. (1986) Bioenergetics of secretory vesicles. *Biochim. Biophys. Acta* **853**, 237–265.
4. Njus, D., Knoth, J., and Zallakian, M. (1981) Proton-linked transport in chromaffin granules. *Curr. Top. Bioenerg.* **11**, 107–147.
   Ohya, Y., Umemoto, N., Tanida, I., Ohta, A., Iida, H., and Anraku, Y. (1991) Calcium sensitive *cls* mutants of *Saccharomyces cerevisiae* showing a Pet-phenotypeare ascribable to defects of vacuolar membrane H+-ATPase activity. *J. Biol. Chem.* **266**, 13,971–13,977.
5. Edwards, R. (1992) The transport of neurotransmitters into synaptic vesicles. *Curr. Opin. Neurobiol.* **2**, 586–594.
6. Liu, Y. and Edwards, R. H. (1997) The role of vesicular transport proteins in synaptic transmission and neural degeneration. *Annu. Rev. Neurosci.* **20**, 125–156.
7. Schuldiner, S. (1994) A molecular glimpse of vesicular monoamine transporters. *J. Neurochem.* **62**, 2067–2078.
8. Usdin, T. B., Eiden, L. E., Bonner, T. I., and Erickson, J. D. (1995) Molecular biology of the vesicular ACh transporter. *Trends. Neurosci.* **18**, 218–224.
9. Bellocchio, E. E., Reimer, R. J., Fremeau, R. T., Jr., and Edwards, R. H. (2000) Uptake of glutamate into synaptic vesicles by an inorganic phosphate transporter. *Science* **289**, 957–960.
10. Takamori, S., Rhee, J. S., Rosenmund, C., and Jahn, R. (2000) Identification of a vesicular transporter that defines a glutamatergic phenotype in neurons. *Nature* **407**, 189–194.
11. Ni, B., Rosteck, P. R., Jr., Nadi, N. S., and Paul, S. M. (1994) Cloning and expression of a cDNA encoding a brain-specific Na(+)-dependent inorganic phosphate cotransporter. *Proc. Natl. Acad. Sci. USA* **91**, 5607–5611.
12. Lee, R. Y., Sawin, E. R., Chalfie, M., Horvitz, H. R., and Avery, L. (1999) EAT-4, a homolog of a mammalian sodium-dependent inorganic phosphate cotransporter, is necessary for glutamatergic neurotransmission in caenorhabditis elegans. *J. Neurosci.* **19**, 159–167.
13. Bellocchio, E. E., Hu, H., Pohorille, A., Chan, J., Pickel, V. M., and Edwards, R. H. (1998) The localization of the brain-specific inorganic phosphate transporter suggests a specific presynaptic role in glutamatergic transmission. *J. Neurosci.* **18**, 8648–8659.
14. Kirshner, N. (1962) Uptake of catecholamines by a particulate fraction of the adrenal medulla. *J. Biol. Chem.* **237**, 2311–2317.

15. Pletscher, A. (1977) Effect of neuroleptics and other drugs on monoamine uptake by membrane of adrenal chromaffin granules. *Br. J. Pharmacol.* **59,** 419–424.

16. Marshall, I. (1970) Studies on the blocking action of 2-(4-phenyl piperidino) cyclohexanol (AH5183). *Br. J. Pharmacol.* **38,** 503–516.

17. Marshall, I. and Parsons, S. (1987) The vesicular acetylcholine transport system. *Trends. Neurosci.* **10,** 174–177.

18. Parsons, S., Bahr, B., Rogers, G., Clarkson, E., Noremberg, K., and Hicks, B. (1993) Acetylcholine transporter-vesamicol receptor pharmacology and structure. *Prog. Brain Res.* **98,** 175–181.

19. Darchen, F., Scherman, D., and Henry, J. P. (1989) Reserpine binding to chromaffin granules suggests the existence of two conformations of the monoamine transporter. *Biochemistry* **28,** 1692–1697.

20. Scherman, D. and Henry, J. P. (1984) Reserpine binding to bovine chromaffin granule membranes. Characterization and comparison with dihydrotetrabenazine binding. *Mol. Pharmacol.* **25,** 113–122.

21. Rudnick, G., Steiner-Mordoch, S. S., Fishkes, H., Stern-Bach, Y., and Schuldiner, S. (1990) Energetics of reserpine binding and occlusion by the chromaffin granule biogenic amine transporter. *Biochemistry* **29,** 603–608.

22. Stitzel, R. E. (1977) The Biological Fate of Reserpine. *Pharm. Rev.* **28,** 179–205.

23. Carlsson, A. (1965) Drugs which block the storage of 5-hydroxytryptamine and related amines. *Hand. Exp. Pharmacol.* **19,** 529–592.

24. Frize, E. (1954) Mental depression in hypertensive patients treated for long periods with high doses of reserpine. *N. Engl. J. Med.* **251,** 1006–1008.

25. Weaver, J. A.,and Deupree, J. D. (1982) Conditions required for reserpine binding to the catecholamine transporter on chromaffin granule ghosts. *Eur. J. Pharm.* **80,** 437,438.

26. Stern-Bach, Y., Greenberg-Ofrath, N., Flechner, I., and Schuldiner, S. (1990) Identification and purification of a functional amine transporter from bovine chromaffin granules. *J. Biol. Chem.* **265,** 3961–3966.

27. Darchen, F., Scherman, D., Desnos, C., and Henry, J. P. (1988) Characteristics of the transport of the quaternary ammonium 1-methyl-4-phenyl-pyridinium by chromaffin granules. *Biochem. Pharmacol.* **37,** 4381–4387.

28. Scherman, D. and Henry, J. P. (1981) pH-dependence of the ATP-driven uptake of noradrenaline by bovine chromaffin-granule ghosts. *Eur. J. Biochem.* **116,** 535–539.

29. Henry, J. P. and Scherman, D. (1989) Radioligands of the vesicular monoamine transporter and their use as markers of monoamine storage vesicles. *Biochem. Pharmacol.* **38,** 2395–2404.

30. Desnos, C., Laran, M. P., and Scherman, D. (1992) Regulation of the chromaffin granule catecholamine transporter in cultured bovine adrenal-medullary cells - stimulus biosynthesis coupling. *J. Neurochem.* **59,** 2105–2112.

31. Desnos, C., Raynaud, B., Vidal, S., Weber, M. J., and Scherman, D. (1990) Induction of the vesicular monoamine transporter by elevated potassium concentration in cultures of rat sympathetic neurons. *Dev. Brain. Res.* **52,** 161–166.

32. Stietzen, M., Schober, M., Fischer-Colbrie, R., Scherman, D., Sperk, G., and Winkler, H. (1987) Rat adrenal medulla: levels of chromogranins, enkepha-

lins, dopamine b-hydroxylase and of the amine transporter are changed by nervous activity and by hypophysectomy. *Neuroscience* **22,** 131–139.

33. Scherman, D., Jaudon, P., and Henry, J. P. (1983) Characterization of the monoamine carrier of chromaffin granule membrane by binding of [2-3H]dihydrotetrabenazine. *Proc. Natl. Acad. Sci. USA* **80,** 584–588.

34. Rogers, G. A. and Parsons, S. M. (1989) Inhibition of acetylcholine storage by acetylcholine analogs in-vitro. *Mol. Pharmacol.* **36,** 333–341.

35. Rogers, G. A., Parsons, S. M., Anderson, D. C., Nilsson, L. M., Bahr, B. A., Kornreich, W. D., Kaufman, R., Jacobs, R. S., and Kirtman, B. (1989) Synthesis, invitro acetylcholine-storage-blocking activities, and biological properties of derivatives and analogs of trans- 2-(4-phenylpiperidino)cyclohexanol (vesamicol). *J. Med. Chem.* **32,** 1217–1230.

36. Parsons, S. M., Bahr, B. A., Rogers, G. A., Clarkson, E. D., Noremberg, K., and Hicks, B. W. (1993) Acetylcholine Transporter Vesamicol Receptor Pharmacology and Structure. *Prog. Brain Res.* **98,** 175–181.

37. Bahr, B., and Parsons, S. (1986) Acetylcholine transport and drug inhibition kinetics in Torpedo synaptic vesicles. *J. Neurochem.* **46,** 1214–1218.

38. Bahr, B. and Parsons, S. (1986) Demonstration of a receptor in Torpedo synaptic vesicles for the acetylcholine storage blocker L-trans-2-(4-phenyl[3,4-3H]piperidino) cyclohexanol. *Proc. Natl. Acad. Sci. USA* **83,** 2267–2270.

39. Bahr, B. A., Clarkson, E. D., Rogers, G. A., Noremberg, K., and Parsons, S. M. (1992) A kinetic and allosteric model for the acetylcholine transporter-vesamicol receptor in synaptic vesicles. *Biochemistry* **31,** 5752–5762.

40. Bahr, B. A., Noremberg, K., Rogers, G. A., Hicks, B. W., and Parsons, S. M. (1992) Linkage of the acetylcholine transporter vesamicol receptor to proteoglycan in synaptic vesicles. *Biochemistry* **31,** 5778–5784.

41. Kaufman, R., Rogers, G. A., Fehlmann, C., and Parsons, S. M. (1989) Fractional vesamicol receptor occupancy and acetylcholine active-transport inhibition in synaptic vesicles. *Mol. Pharmacol.* **36,** 452–458.

42. Clarkson, E. D., Rogers, G. A., and Parsons, S. M. (1992) Binding and active-transport of large analogs of acetylcholine by cholinergic synaptic vesicles invitro. *J. Neurochem.* **59,** 695–700.

43. Bahr, B. A. and Parsons, S. M. (1992) Purification of the vesamicol receptor. *Biochemistry* **31,** 5763–5769.

44. Rogers, G. A. and Parsons, S. M. (1992) Photoaffinity-labeling of the acetylcholine transporter. *Biochemistry* **31,** 5770–5777.

45. Rogers, G. A. and Parsons, S. M. (1993) Photoaffinity-labeling of the vesamicol receptor of cholinergic synaptic vesicles. *Biochemistry* **32,** 8596–8601.

46. Erickson, J. D., Varoqui, H., Schafer, M. K., Modi, W., Diebler, M. F., Weihe, E., Rand, J., Eiden, L. E., Bonner, T. I., and Usdin, T. B. (1994) Functional identification of a vesicular acetylcholine transporter and its expression from a "cholinergic" gene locus. *J. Biol. Chem.* **269,** 21,929–21,932.

47. Roghani, A., Feldman, J., Kohan, S. A., Shirzadi, A., Gundersen, C. B., Brecha, N., and Edwards, R. H. (1994) Molecular cloning of a putative vesicular transporter for acetylcholine. *Proc. Natl. Acad. Sci. USA* **91,** 10,620–10,624.

48. Varoqui, H., Diebler, M. F., Meunier, F. M., Rand, J. B., Usdin, T. B., Bonner, T. I., Eiden, L. E., and Erickson, J. D. (1994) Cloning and expression of the vesamicol binding protein from the marine ray Torpedo. Homology with the putative vesicular acetylcholine transporter UNC-17 from Caenorhabditis elegans. *FEBS Lett.* **342,** 97–102.

49. Krejci, E., Gasnier, B., Botton, D., Isambert, M. F., Sagne, C., Gagnon, J., Massoulie, J., and Henry, J. P. (1993) Expression and regulation of the bovine vesicular monoamine transporter gene. *FEBS Lett.* **335,** 27–32.

50. Stern-Bach, Y., Keen, J. N., Bejerano, M., Steiner-Mordoch, S., Wallach, M., Findlay, J. B., and Schuldiner, S. (1992) Homology of a vesicular amine transporter to a gene conferring resistance to 1-methyl-4-phenylpyridinium. *Proc. Natl. Acad. Sci. USA* **89,** 9730–9733.

51. Erickson, J. D., Eiden, L. E., and Hoffman, B. J. (1992) Expression cloning of a reserpine-sensitive vesicular monoamine transporter. *Proc. Natl. Acad. Sci. USA* **89,** 10,993–10,997.

52. Liu, Y., Peter, D., Roghani, A., Schuldiner, S., Prive, G., Eisenberg, D., Brecha, N., and Edwards, R. (1992) A cDNA that suppresses MPP+ toxicity encodes a vesicular amine transporter. *Cell* **70,** 539–551.

53. Schuldiner, S., Liu, Y., and Edwards, R. H. (1993) Reserpine binding to a vesicular amine transporter expressed in Chinese hamster ovary fibroblasts. *J. Biol. Chem.* **268,** 29-34.

54. Vincent, M. and Near, J. (1991) Purification of a [H-3]dihydrotetrabenazine-binding protein from bovine adrenal medulla. *Mol. Pharmacol.* **40,** 889–894.

55. Isambert, M. F., Gasnier, B., Botton, D., and Henry, J. P. (1992) Characterization and purification of the monoamine transporter of bovine chromaffin granules. *Biochemistry* **31,** 1980–1986.

56. Howell, M., Shirvan, A., Stern-Bach, Y., Steiner-Mordoch, S., Strasser, J. E., Dean, G. E., and Schuldiner, S. (1994) Cloning and functional expression of a tetrabenazine sensitive vesicular monoamine transporter from bovine chromaffin granules. *FEBS Lett.* **338,** 16–22.

57. Peter, D., Jimenez, J., Liu, Y., Kim, J., and Edwards, R. H. (1994) The chromaffin granule and synaptic vesicle amine transporters differ in substrate recognition and sensitivity to inhibitors. *J. Biol. Chem.* **269,** 7231–7237.

58. Scherman, D. and Boschi, G. (1988) Time required for transmitter accumulation inside monoaminergic storage vesicles differs in peripheral and in central systems. *Neuroscience* **27,** 1029–1035.

59. Scherman, D. (1986) Dihydrotetrabenazine binding and monoamine uptake in mouse brain regions. *J. Neurochem.* **47,** 331–339.

60. Erickson, J. D., Schafer, M. K., Bonner, T. I., Eiden, L. E., and Weihe, E. (1996) Distinct pharmacological properties and distribution in neurons and endocrine cells of two isoforms of the human vesicular monoamine transporter. *Proc. Natl. Acad. Sci. USA* **93,** 5166–5171.

61. Nirenberg, M. J., Chan, J., Liu, Y., Edwards, R. H., and Pickel, V. M. (1997) Vesicular monoamine transporter-2: immunogold localization in striatal axons and terminals. *Synapse* **26,** 194–198.

62. Peter, D., Liu, Y., Sternini, C., de Giorgio, R., Brecha, N., and Edwards, R.

H. (1995) Differential expression of two vesicular monoamine transporters. *J. Neurosci.* **15,** 6179–6188.

63. Weihe, E., Schafer, M. K., Erickson, J. D., and Eiden, L. E. (1994) Localization of vesicular monoamine transporter isoforms (VMAT1 and VMAT2) to endocrine cells and neurons in rat. *J. Mol. Neurosci.* **5,** 149–164.

64. Nirenberg, M. J., Liu, Y., Peter, D., Edwards, R. H., and Pickel, V. M. (1995) The vesicular monoamine transporter 2 is present in small synaptic vesicles and preferentially localizes to large dense core vesicles in rat solitary tract nuclei. *Proc. Natl. Acad. Sci. USA* **92,** 8773–8777.

65. Liu, Y., Schweitzer, E. S., Nirenberg, M. J., Pickel, V. M., Evans, C. J., and Edwards, R. H. (1994) Preferential localization of a vesicular monoamine transporter to dense core vesicles in PC12 cells. *J. Cell. Biol.* **127,** 1419–1433.

66. Holtje, M., von Jagow, B., Pahner, I., Lautenschlager, M., Hortnagl, H., Nurnberg, B., Jahn, R., and Ahnert-Hilger, G. (2000) The neuronal monoamine transporter VMAT2 is regulated by the trimeric GTPase Go(2). *J. Neurosci.* **20,** 2131–2141.

67. Mahata, S. K., Mahata, M., Fischercolbrie, R., and Winkler, H. (1993) Reserpine causes differential changes in the messenger RNA levels of chromogranin-B, secretogranin-II, carboxypeptidase-H, alpha-amidating monooxygenase, the vesicular amine transporter and of synaptin/synaptophysin in rat brain. *Mol. Brain. Res.* **19,** 83–92.

68. Mahata, S. K., Mahata, M., Fischercolbrie, R., and Winkler, H. (1993) Vesicle monoamine transporter-1 and transporter-2-differential distribution and regulation of their messenger RNAs in chromaffin and ganglion cells of rat adrenal medulla. *Neurosci. Lett.* **156,** 70–72.

69. Sievert, M. K., Thiriot, D. S., Edwards, R. H., and Ruoho, A. E. (1998) High-efficiency expression and characterization of the synaptic-vesicle monoamine transporter from baculovirus-infected insect cells. *Biochem. J.* **330,** 959–966.

70. Yelin, R. and Schuldiner, S. (2000) Vesicular monoamine transporters heterologously expressed in the yeast *Saccharomyces cerevisiae* display high affinity tetrabenazine binding. *Biochem. Biophys. Acta.* **1510,** 426–441.

71. Sievert, M. K. and Ruoho, A. E. (1997) Peptide mapping of the [125I]Iodoazidoketanserin and [125I]2-N-[(3'- iodo-4'-azidophenyl)-propionyl]tetrabenazine binding sites for the synaptic vesicle monoamine transporter. *J. Biol. Chem.* **272,** 26,049–26,055.

72. Sagne, C., Isambert, M. F., Vandekerckhove, J., Henry, J. P., and Gasnier, B. (1997) The photoactivatable inhibitor 7-azido-8-iodoketanserin labels the N terminus of the vesicular monoamine transporter from bovine chromaffin granules. *Biochemistry* **36,** 3345–3352.

73. Alfonso, A., Grundahl, K., Duerr, J. S., Han, H. P., and Rand, J. B. (1993) The Caenorhabditis elegans Unc-17 Gene - A Putative Vesicular Acetylcholine Transporter. *Science* **261,** 617–619.

74. Brenner, S. (1974) The genetics of *Caenorhabditis elegans. Genetics* **77,** 71–94.

75. Rand, J. (1989) Genetic analysis of the *cha1-unc17* gene complex in *Caenorhabditis. Genetics* **122,** 73–80.

76. Weihe, E., Tao-Cheng, J. H., Schafer, M. K., Erickson, J. D., and Eiden, L. E. (1996) Visualization of the vesicular acetylcholine transporter in cholinergic nerve terminals and its targeting to a specific population of small synaptic vesicles. *Proc. Natl. Acad. Sci. USA* **93**, 3547–3552.

77. Gilmor, M. L., Nash, N. R., Roghani, A., Edwards, R. H., Yi, H., Hersch, S. M., and Levey, A. I. (1996) Expression of the putative vesicular acetylcholine transporter in rat brain and localization in cholinergic synaptic vesicles. *J. Neurosci.* **16**, 2179–2190.

78. McIntire, S. L., Jorgensen, E., and Horvitz, H. R. (1993) Genes required for GABA function in *Caenorhabditis elegans. Nature* **364**, 334–337.

79. McIntire, S. L., Jorgensen, E., Kaplan, J., and Horvitz, H. R. (1993) The GABAergic nervous system of *Caenorhabditis elegans. Nature* **364**, 337–341.

80. McIntire, S. L., Reimer, R. J., Schuske, K., Edwards, R. H., and Jorgensen, E. M. (1997) Identification and characterization of the vesicular GABA transporter. *Nature* **389**, 870–876.

81. Sagne, C., El Mestikawy, S., Isambert, M. F., Hamon, M., Henry, J. P., Giros, B., and Gasnier, B. (1997) Cloning of a functional vesicular GABA and glycine transporter by screening of genome databases. *FEBS Lett.* **417**, 177–183.

82. Burger, P., Hell, J., Mehl, E., Krasel, C., Lottspeich, F., and Jahn, R. (1991) Gaba and glycine in synaptic vesicles - storage and transport characteristics. *Neuron* **7**, 287–293.

83. Hell, J. W., Maycox, P. R., and Jahn, R. (1990) Energy dependence and functional reconstitution of the gamma-aminobutyric acid carrier from synaptic vesicles. *J. Biol. Chem.* **265**, 2111–2117.

84. Kish, P. E., Fischer-Bovenkerk, C., and Ueda, T. (1989) Active transport of gamma-aminobutyric acid and glycine into synaptic vesicles. *Proc. Natl. Acad. Sci. USA* **86**, 3877–3881.

85. Christensen, H. and Fonnum, F. (1991) Uptake of glycine, GABA and glutamate by synaptic vesicles isolated from different regions of rat CNS. *Neurosci. Lett.* **129**, 217–220.

86. Christensen, H., Fykse, E. M., and Fonnum, F. (1990) Uptake of glycine into synaptic vesicles isolated from rat spinal cord. *J. Neurochem.* **54**, 1142–1147.

87. Christensen, H., Fykse, E. M., and Fonnum, F. (1991) Inhibition of gamma-aminobutyrate and glycine uptake into synaptic vesicles. *Eur. J. Pharmacol.* **207**, 73–79.

88. Fykse, E. M. and Fonnum, F. (1988) Uptake of gamma-aminobutyric acid by a synaptic vesicle fraction isolated from rat brain. *J. Neurochem.* **50**, 1237–1242.

89. Hell, J. W., Maycox, P. R., Stadler, H., and Jahn, R. (1988) Uptake of GABA by rat brain synaptic vesicles isolated by a new procedure. *EMBO J.* **7**, 3023–3029.

90. Kolston, J., Osen, K. K., Hackney, C. M., Ottersen, O. P., and Storm-Mathisen, J. (1992) An atlas of glycine- and GABA-like immunoreactivity and colocalization in the cochlear nuclear complex of the guinea pig. *Anat. Embryol. (Berl)* **186**, 443–465.

91. Ornung, G., Shupliakov, O., Linda, H., Ottersen, O. P., Storm-Mathisen, J., Ulfhake, B., and Cullheim, S. (1996) Qualitative and quantitative analysis of

glycine- and GABA-immunoreactive nerve terminals on motoneuron cell bodies in the cat spinal cord: a postembedding electron microscopic study. *J. Comp. Neurol.* **365,** 413–426.

92. Ornung, G., Shupliakov, O., Ottersen, O. P., Storm-Mathisen, J., and Cullheim, S. (1994) Immunohistochemical evidence for coexistence of glycine and GABA in nerve terminals on cat spinal motoneurones: an ultrastructural study. *Neuroreport* **5,** 889–892.

93. Ottersen, O. P., Storm-Mathisen, J., and Somogyi, P. (1988) Colocalization of glycine-like and GABA-like immunoreactivities in Golgi cell terminals in the rat cerebellum: a postembedding light and electron microscopic study. *Brain. Res.* **450,** 342–353.

94. Chaudhry, F. A., Reimer, R. J., Bellocchio, E. E., Danbolt, N. C., Osen, K. K., Edwards, R. H., and Storm-Mathisen, J. (1998) The vesicular GABA transporter, VGAT, localizes to synaptic vesicles in sets of glycinergic as well as GABAergic neurons. *J. Neurosci.* **18,** 9733–9750.

95. Dumoulin, A., Rostaing, P., Bedet, C., Levi, S., Isambert, M. F., Henry, J. P., Triller, A., and Gasnier, B. (1999) Presence of the vesicular inhibitory amino acid transporter in GABAergic and glycinergic synaptic terminal boutons. *J. Cell. Sci.* **112,** 811–823.

96. Ahnert-Hilger, G., Nurnberg, B., Exner, T., Schafer, T., and Jahn, R. (1998) The heterotrimeric G protein Go2 regulates catecholamine uptake by secretory vesicles. *EMBO J.* **17,** 406–413.

97. Liu, Y. and Edwards, R. H. (1997) Differential localization of vesicular acetylcholine and monoamine transporters in PC12 cells but not CHO cells. *J. Cell. Biol.* **139,** 907–916.

98. Kelly, R. B. (1993) Storage and release of neurotransmitters. *Cell* **72 Suppl,** 43–53.

99. Varoqui, H. and Erickson, J. D. (1998) The cytoplasmic tail of the vesicular acetylcholine transporter contains a synaptic vesicle targeting signal. *J. Biol. Chem.* **273,** 9094–9098.

100. Tan, P. K., Waites, C., Liu, Y., Krantz, D. E., and Edwards, R. H. (1998) A leucine-based motif mediates the endocyosis of vesicular monoamine and acetylcholine transporters. *J. Biol. Chem.* **273,** 17,351–17,360.

101. Letourneur, F. and Klausner, R. D. (1992) A novel di-leucine motif and a tyrosine-based motif independently mediate lysosomal targeting and endocytosis of CD3 chains. *Cell* **69,** 1143–1157.

102. Rubin, L. A., Kurman, C. C., Biddison, W. E., Goldman, N. D., and Nelson, D. L. (1985) A monoclonal antibody 7G7/B6, binds to an epitope on the human interleukin-2 (IL-2) receptor that is distinct from that recognized by IL-2 or anti-Tac. *Hybridoma* **4,** 91–102.

103. Dietrich, J., Kastrup, J., Nielsen, B. L., Odum, N., and Geisler, C. (1997) Regulation and function of the CD3gamma DxxxLL motif: a binding site for adaptor protein-1 and adaptor protein-2 in vitro. *J. Cell. Biol.* **138,** 271–281.

104. Cho, G. W., Kim, M. H., Chai, Y. G., Gilmor, M. L., Levey, A. I., and Hersh, L. B. (2000) Phosphorylation of the rat vesicular acetylcholine transporter. *J. Biol. Chem.* **275,** 19,942–19,948.

105. Krantz, D. E., Waites, C., Oorschot, V., Liu, Y., Wilson, R. I., Tan, P. K., Klumperman, J., and Edwards, R. H. (2000) A phosphorylation site regulates sorting of the vesicular acetylcholine transporter to dense core vesicles. *J. Cell. Biol.* **149,** 379–396.

106. Barbosa, J., Jr., Clarizia, A. D., Gomez, M. V., Romano-Silva, M. A., Prado, V. F., and Prado, M. A. (1997) Effect of protein kinase C activation on the release of [3H]acetylcholine in the presence of vesamicol. *J. Neurochem.* **69,** 2608–2611.

107. Dietrich, J., Hou, X., Wegener, A. M., and Geisler, C. (1994) CD3 gamma contains a phosphoserine-dependent di-leucine motif involved in down-regulation of the T cell receptor. *EMBO J.* **13,** 2156–2166.

108. Shin, J., Dunbrack, R. L., Jr., Lee, S., and Strominger, J. L. (1991) Phosphorylation-dependent down-modulation of CD4 requires a specific structure within the cytoplasmic domain of CD4. *J. Biol. Chem.* **266,** 10,658–10,665.

109. Yelin, R., Steiner-Mordoch, S., Aroeti, B., and Schuldiner, S. (1998) Glycosylation of a vesicular monoamine transporter: a mutation in a conserved proline residue affects the activity, glycosylation, and localization of the transporter. *J. Neurochem.* **71,** 2518–2527.

110. Clarizia, A. D., Gomez, M. V., Romano-Silva, M. A., Parsons, S. M., Prado, V. F., and Prado, M. A. M. (1999) Control of the binding of a vesamicol analog to the vesicular acetylcholine transporter. *Neuroreport* **10,** 2783–2787.

111. Nakanishi, N., Onozawa, S., Matsumoto, R., Kurihara, K., Ueha, T., Hasegawa, H., and Minami, N. (1995) Effects of protein kinase inhibitors and protein phosphatase inhibitors on cyclic AMP-dependent down-regulation of vesicular monoamine transport in pheochromocytoma PC12 cells. *FEBS Lett.* **368,** 411–414.

112. Krantz, D. E., Peter, D., Liu, Y., and Edwards, R. H. (1997) Phosphorylation of a vesicular monoamine transporter by casein kinase II. *J. Biol. Chem.* **272,** 6752–6759.

113. Ahnert-Hilger, G., Schafer, T., Spicher, K., Grund, C., Schultz, G., and Wiedenmann, B. (1994) Detection of G-protein heterotrimers on large dense core and small synaptic vesicles of neuroendocrine and neuronal cells. *Eur. J. Cell. Biol.* **65,** 26–38.

114. Gasnier, B., Scherman, D., and Henry, J. P. (1985) Dicyclohexylcarbodiimide inhibits the monoamine carrier of bovine chromaffin granule membrane. *Biochemistry* **24,** 1239–1244.

115. Schuldiner, S., Fishkes, H., and Kanner, B. I. (1978) Role of a transmembrane pH gradient in epinephrine transport by chromaffin granule membrane vesicles. *Proc. Natl. Acad. Sci. USA* **75,** 3713–3716.

116. Suchi, R., Stern-Bach, Y., Gabay, T., and Schuldiner, S. (1991) Covalent modification of the amine transporter with N,N'-dicyclohexylcarbodiimide. *Biochemistry* **30,** 6490–6494.

117. Merickel, A., Rosandich, P., Peter, D., and Edwards, R. H. (1995) Identification of residues involved in substrate recognition by a vesicular monoamine transporter. *J. Biol. Chem.* **270,** 25,798–25,804.

118. Kim, M. H., Lu, M., Lim, E. J., Chai, Y. G., and Hersh, L. B. (1999) Mutational analysis of aspartate residues in the transmembrane regions and cytoplasmic loops of rat vesicular acetylcholine transporter. *J. Biol. Chem.* **274,** 673–680.

119. Merickel, A., Kaback, H. R., and Edwards, R. H. (1997) Charged residues in transmembrane domains II and XI of a vesicular monoamine transporter form a charge pair that promotes high affinity substrate recognition. *J. Biol. Chem.* **272,** 5403–5408.

120. Steiner-Mordoch, S., Shirvan, A., and Schuldiner, S. (1996) Modification of the pH profile and tetrabenazine sensitivity of rat VMAT1 by replacement of aspartate 404 with glutamate. *J. Biol. Chem.* **271,** 13,048–13,054.

121. Song, H., Ming, G., Fon, E., Bellocchio, E., Edwards, R. H., and Poo, M. (1997) Expression of a putative vesicular acetylcholine transporter facilitates quantal transmitter packaging. *Neuron* **18,** 815–826.

122. Blakely, R. D. and Bauman, A. L. (2000) Biogenic amine transporters: regulation in flux. *Curr. Opin. Neurobiol.* **10,** 328–336.

123. Gerchman, Y., Olami, Y., Rimon, A., Taglicht, D., Schuldiner, S., and Padan, E. (1993) Histidine-226 is part of the pH sensor of NhaA, a Na+/H+ antiporter in Escherichia coli. *Proc. Natl. Acad. Sci. USA* **90,** 1212–1216.

124. Puttner, I. B., Sarkar, H. K., Padan, E., Lolkema, J. S., and Kaback, H. R. (1989) Characterization of site-directed mutants in the lac permease of Escherichia coli. 1. Replacement of histidine residues. *Biochemistry* **28,** 2525–2533.

125. Isambert, M. F. and Henry, J. P. (1981) Effect of diethylpyrocarbonate on pH-driven monoamine uptake by chromaffin granule ghosts. *FEBS Lett.* **136,** 13–18.

126. Suchi, R., Stern-Bach, Y., and Schuldiner, S. (1992) Modification of arginyl or histidyl groups affects the energy coupling of the amine transporter. *Biochemistry* **31,** 12,500–12,503.

127. Keller, J. E. and Parsons, S. M. (2000) A critical histidine in the vesicular acetylcholine transporter. *Neurochem. Int.* **36,** 113–117.

128. Kim, M. H., Lu, M., Kelly, M., and Hersh, L. B. (2000) Mutational analysis of basic residues in the rat vesicular acetylcholine transporter. Identification of a transmembrane ion pair and evidence that histidine is not involved in proton translocation. *J. Biol. Chem.* **275,** 6175–6180.

129. Shirvan, A., Laskar, O., Steiner-Mordoch, S., and Schuldiner, S. (1994) Histidine-419 plays a role in energy coupling in the vesicular monoamine transporter from rat. *FEBS Lett.* **356,** 145–150.

130. Dunten, R. L., Sahin-Toth, M., and Kaback, H. R. (1993) Role of the charge pair aspartic acid-237-lysine-358 in the lactose permease of Escherichia coli. *Biochemistry* **32,** 3139–3145.

131. King, S. C., Hansen, C. L., and Wilson, T. H. (1991) The interaction between aspartic acid 237 and lysine 358 in the lactose carrier of Escherichia coli. *Biochim. Biophys. Acta* **1062,** 177–186.

132. Sahin-Toth, M., Dunten, R. L., Gonzalez, A., and Kaback, H. R. (1992) Functional interactions between putative intramembrane charged residues in the lactose permease of Escherichia coli. *Proc. Natl. Acad. Sci. USA* **89,** 10,547–10,551.

133. Kitayama, S., Shimada, S., Xu, H., Markham, L., Donovan, D., and Uhl, G. (1992) Dopamine transporter site-directed mutations differentially alter substrate transport and cocaine binding. *Proc. Natl. Acad. Sci. USA* **89,** 7782–7785.

134. Peter, D., Vu, T., and Edwards, R. H. (1996) Chimeric vesicular monoamine

transporters identify structural domains that influence substrate affinity and sensitivity to tetrabenazine. *J. Biol. Chem.* **271**, 2979–2986.

135. Erickson, J. D. (1998) A chimeric vesicular monoamine transporter dissociates sensitivity to tetrabenazine and unsubstituted aromatic amines. *Adv. Pharmacol.* **42**, 227–232.

136. Varoqui, H. and Erickson, J. D. (1998) Dissociation of the vesicular acetylcholine transporter domains important for high-affinity transport recognition, binding of vesamicol and targeting to synaptic vesicles. *J. Physiol. Paris* **92**, 141–144.

137. Finn, J. P., 3rd and Edwards, R. H. (1997) Individual residues contribute to multiple differences in ligand recognition between vesicular monoamine transporters 1 and 2. *J. Biol. Chem.* **272**, 16,301–16,307.

138. Finn, J. P., 3rd and Edwards, R. H. (1998) Multiple residues contribute independently to differences in ligand recognition between vesicular monoamine transporters 1 and 2. *J. Biol. Chem.* **273**, 3943–3947.

139. Duerr, J. S., Frisby, D. L., Gaskin, J., Duke, A., Asermely, K., Huddleston, D., Eiden, L. E., and Rand, J. B. (1999) The cat-1 gene of Caenorhabditis elegans encodes a vesicular monoamine transporter required for specific monoamine-dependent behaviors. *J. Neurosci.* **19**, 72–84.

140. Takahashi, N., Miner, L. L., Sora, I., Ujike, H., Revay, R. S., Kostic, V., Jackson-Lewis, V., Przedborski, S., and Uhl, G. R. (1997) VMAT2 knockout mice: heterozygotes display reduced amphetamine-conditioned reward, enhanced amphetamine locomotion, and enhanced MPTP toxicity. *Proc. Natl. Acad. Sci. USA* **94**, 9938–9943.

141. Fon, E. A., Pothos, E. N., Sun, B. C., Killeen, N., Sulzer, D., and Edwards, R. H. (1997) Vesicular transport regulates monoamine storage and release but is not essential for amphetamine action. *Neuron* **19**, 1271–1283.

142. Wang, Y. M., Gainetdinov, R. R., Fumagalli, F., Xu, F., Jones, S. R., Bock, C. B., Miller, G. W., Wightman, R. M., and Caron, M. G. (1997) Knockout of the vesicular monoamine transporter 2 gene results in neonatal death and supersensitivity to cocaine and amphetamine. *Neuron* **19**, 1285–1296.

143. Rudnick, G. and Wall, S. C. (1992) *p*-Chloroamphetamine induces serotonin release through serotonin transporters. *Biochemistry* **31**, 6710–6718.

144. Hansson, S. R., Hoffman, B. J., and Mezey, E. (1998) Ontogeny of vesicular monoamine transporter mRNAs VMAT1 and VMAT2. I. The developing rat central nervous system. *Brain. Res. Dev. Brain. Res.* **110**, 135–158.

145. Itokawa, K., Sora, I., Schindler, C. W., Itokawa, M., Takahashi, N., and Uhl, G. R. (1999) Heterozygous VMAT2 knockout mice display prolonged QT intervals: possible contributions to sudden death. *Brain. Res. Mol. Brain. Res.* **71**, 354–357.

146. Travis, E. R., Wang, Y. M., Michael, D. J., Caron, M. G., and Wightman, R. M. (2000) Differential quantal release of histamine and 5-hydroxytryptamine from mast cells of vesicular monoamine transporter 2 knockout mice. *Proc. Natl. Acad. Sci. USA* **97**, 162–167.

147. Gainetdinov, R. R., Fumagalli, F., Wang, Y. M., Jones, S. R., Levey, A. I.,

Miller, G. W., and Caron, M. G. (1998) Increased MPTP neurotoxicity in vesicular monoamine transporter 2 heterozygote knockout mice. *J. Neurochem.* **70,** 1973–1978.

148. Fumagalli, F., Gainetdinov, R. R., Wang, Y. M., Valenzano, K. J., Miller, G. W., and Caron, M. G. (1999) Increased methamphetamine neurotoxicity in heterozygous vesicular monoamine transporter 2 knock-out mice. *J. Neurosci.* **19,** 2424–2431.

149. Henry, J. P., Sagne, C., Bedet, C., and Gasnier, B. (1998) The vesicular monoamine transporter: from chromaffin granule to brain. *Neurochem. Int.* **32,** 227–246.

150. Reimer, R. J., Fon, E. A., and Edwards, R. H. (1998) Vesicular neurotransmitter transport and the presynaptic regulation of quantal size. *Curr. Opin. Neurobiol.* **8,** 405–412.

151. Griffith, J. K., Baker, M. E., Rouch, D. A., Page, M. G. P., Skurray, R. A., Paulsen, I., Chater, K. F., Baldwin, S. A., and Henderson, P. J. F. (1992) Evolution of transmembrane transport: relationships between transport proteins for sugars, carboxylate compounds, antibiotics and antiseptics. *Curr. Opin. Cell. Biol.* **4,** 684–695.

152. Paulsen, I. T. and Skurray, R. A. (1994) Topology, structure and evolution of two families of proteins involved in antibiotic and antiseptic resistance in eukaryotes and prokaryotes- an anlysis. *Gene* **124,** 1–11.

153. Flowers, T. J., Troke, P. F., and Yeo, A. R. (1977) The mechanism of salt tolerance in halophytes. *Ann. Rev. Plant. Physiol.* **28,** 89–121.

154. Wink, M. (1993) The plant vacuole: A multifunctional Compartment. *J. Exp. Botany* **44,** 231–246.

155. Yelin, R., Rotem, D., and Schuldiner, S. (1999) EmrE, a small Escherichia coli multidrug transporter, protects Saccharomyces cerevisiae from toxins by sequestration in the vacuole. *J. Bacteriol.* **181,** 949–956.

156. Yelin, R. and Schuldiner, S. (1995) The pharmacological profile of the vesicular monoamine transporter resembles that of multidrug transporters. *FEBS Lett.* **377,** 201–207.

157. Gluck, S. L. (1992) The structure and biochemistry of the vacuolar H+ ATPase in proximal and distal urinary acidifcation. *J. Bioenerg. Biomemb.* **24,** 351–359.

158. Gogarten, J. P., Kibak, H., Dittrich, P., Taiz, L., Bowman, E. J., Bowman, B. J., et al. (1989) Evolution of the vacuolar H+-ATPase: Implications for the origin of eukaryotes. *Proc. Natl. Acad. Sci. USA* **86,** 6661–6665.

159. Nelson, N. (1992) Structural Conservation and functional diversity of V-ATPases. *J. Bioenerg. Biomemb.* **24,** 407–414.

160. Felsenstein, J. (1989) PHYLIP — phylogeny inference package. *Cladistics* **5.**

161. Thompson, J. D., Higgins, D. G., and Gibson, T. J. (1994) CLUSTAL W: improving the sensitivity of progressive multiple sequence alignment through sequence weighting, position-specific gap penalties and weight matrix choice. *Nucleic Acids. Res.* **22,** 4673–4680.

10
Neurotransmitter-Transporter Proteins
Post-Translational Modifications

Amrat P. Patel and Maarten E.A. Reith

1. INTRODUCTION

The neurotransmitter-transporter protein or reuptake carrier is the most important component in the termination of the synaptic activity. A number of these neurotransmitter transporters have been cloned *(1)*. The cloned transporters contain consensus sequences for multiple N-linked glycosylation sites in the large extracellular loop (EL) between transmembrane regions 3 and 4. Consensus sites for phosphorylation by several protein kinases are located in the putative cytosolic domains.

It is widely accepted that proteins undergo several forms of post-translational modifications—for example, glycosylation and phosphorylation. This chapter addresses the role of post-translational modifications of neurotransmitter-transporter proteins, especially the effects of glycosylation on transporter function and stability, when expressed in foreign host-cell systems compared to that in the native state. An overview of other post-translational processes is also described.

2. GLYCOSYLATION

Glycosylation is a major cotranslational and post-translational modification experienced by a nascent polypeptide. The two major forms of glycosylations are N-linked and O-linked, and a third form is an attachment of a glycolipid anchor to the C- terminus of the protein (Fig. 1). The N-linked glycosylation is a covalent modification catalyzed by specific glycosyltransferases, at the side-chain nitrogen atom of asparagine residues. Similarly, O-linked glycosylation is a covalent oligosaccharide modification of the hydroxyl group of serine, threonine, or sometimes hydroxylysine. The complex glycosylation machinery involved in the formation of various oligosaccharide structures for protein glycosylation has been reviewed *(2–*

From: *Contemporary Neuroscience:*
Neurotransmitter Transporters: Structure, Function, and Regulation, 2nd Edition
Edited by: M. E. A. Reith © Humana Press Inc., Totowa, NJ

Oligosaccharide Attachment

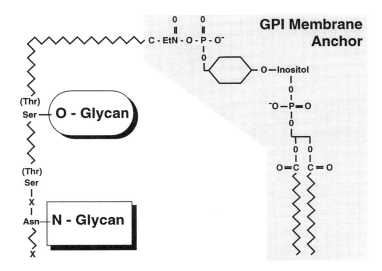

Fig. 1. The three main forms of eukaryotic protein glycosylation.

7). The initial core of oligosaccharide structure transferred onto the asparagine residue is made up of two N-acetylglucosamine (GlcNAc), nine mannose, and three glucose residues (Fig. 2). The assembly of the core oligosaccharide begins in the cytoplasm with a lipid carrier, dolichol pyrophosphate, and phosphorylated N-acetylglucosamine as the substrate catalyzed by N-acetylglucosamine phosphate transferase *(8).* Tunicamycin prevents glycosylation by inhibiting the N-acetylglucosamine phosphate transferase *(9).* The dolichol-PP-GlcNAc in the endoplasmic reticulum (ER) enters a cycle of elongation of the oligosaccharide in the presence of specific glycosyltransferases and substrates. The entire oligosaccharide core is transferred from a dolichol carrier to an asparagine on the emerging nascent polypeptide, a cotranslational modification. In the ER, post-translational modification of the oligosaccharide takes place with sequential removal of three glucose residues and a mannose residue. Further processing of the glycoprotein takes place in the golgi apparatus *(2).* In the ER and golgi, the precursor oligosaccharide core is processed, analogous to an assembly line with different stations modifying the core to achieve the final product.

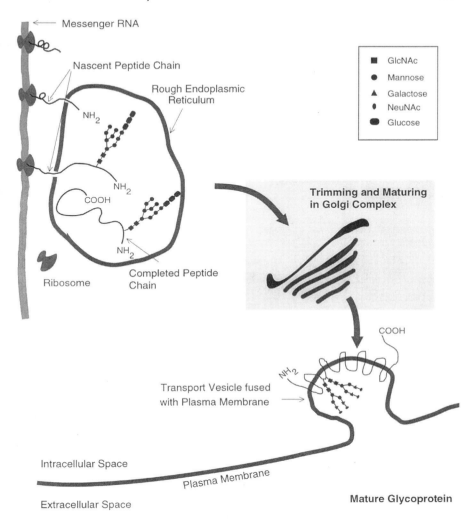

Fig. 2. Biosynthesis of N-linked oligosaccharides.

2.1. Potential Glycosylation Sites

The consensus sequence for potential N-glycosylation includes the asparagine residue followed by any amino acid except proline, followed in turn by either serine or threonine (Asn-Xaa-Ser/Thr). Some potential glycosylation sites are not glycosylated in the native state. Furthermore, the pattern of glycosylation of a protein may differ when it is expressed in foreign cell systems (*see* **Subheading 3.2.1.**). Glycosylation is affected by the presence of certain amino acids, disulfide bonds, protein-folding regions, and proximity to the amino or carboxy terminus *(5)*. The O-linked glycosylation sequence motif is not well-defined. The hydroxyl group of the serine and threonine can accept GlcNAc as the starting sugar residue in an oligosaccharide. Examples of the O-linked glycoproteins include human chorionic gonadotrophin hormone, mucins, and O-fucosylated blood coagulation factor proteins, in which the glycan is O-linked to hydroxylysine in the sequence gly-gly-OHlys-Xaa *(5,6)*.

To date, all of the cloned neurotransmitter-transporter cDNAs encode at least two or more potential N-glycosylation sites. One of the potential N-glycosylation sites is conserved in the dopamine, norepinephrine, and serotonin transporters. The transporter proteins are heavily glycosylated, with sugars representing 15–50% of the molecular mass.

2.2. Detection of Glycosylated Transporter Proteins

During the last decade, the study of glycoproteins has escalated with breakthroughs in analytical technology, detection of minute quantities of oligosaccharides, and the availability of contamination-free recombinant glycosidases. Carbohydrate detection kits are available to detect gross and specific oligosaccharides by chemical and enzymatic methods from commercial sources. Details of the detection and analysis of carbohydrates can be found in texts of carbohydrate analysis *(4,6,10)*. The glycosylation nature of a protein can also be assessed from the purification schemes of solubilized protein over a variety of commercially available lectin-affinity columns. Commonly used lectins are wheat-germ agglutinin (WGA), concanavalin A, and lentil lectin agarose. Purification afforded by lectin-affinity chromatography rarely exceeds 10%; however, it is a useful technique to enhance purification and separation of glycosylated peptides from nonglycosylated peptides. Lectin columns have been used in partial purification of dopamine, GABA, glutamate, and serotonin transporters *(11–15)*. [125I]-labeled lectins are also used as an overlay for proteins blotted onto membranes to detect glycosylated polypeptides. Specific glycosidases are used for the removal of terminal residues such as sialic acid, for complete N- and O-deglycosylation, and for the removal of the glycophosphatidylinositol (GPI) anchor with phosphatidylinositol-specific phospholipase-C *(16)*.

The complement of sugars attached to a potential glycosylation site on a single polypeptide may vary from site to site during the glycosylation process. This variance produces glycoforms with heterogeneous sugar composition, yielding a distinct, broad band in a denaturing gel, and multiple bands in two-dimensional gels. Upon complete deglycosylation, the broad band appears as a sharp band. However, for the dopamine and other neurotransmitter-transporter proteins, the enzymatic N-deglycosylated polypeptide is not a sharp band. This result may be caused by O-linked glycosylation, charged moieties on the polypeptide such as sulphates or phosphates, or N-glycosylation, which is resistant to N-deglycosylation enzymes. A chemical deglycosylation with trifluoromethanesulphonic acid that removes both N- and O-glycosylation and charged species from the polypeptide *(17)* may resolve this discrepancy.

Some proteins in denaturing polyacrylamide gels display anomalous mobilities—i.e., a discrepancy between predicted and observed molecular mass. The deglycosylated glucose transporter in sodium dodecyl sulfate polyacrylamide denaturing gels has anomalous mobility, in that the electrophoretic molecular mass is higher than that predicted from the amino-acid sequence *(18)*. Similar findings have been observed with the neurotransmitter transporters and other proteins including basic proteins, such as histones *(19)*. The difference in the reported molecular mass of 68–90 kDa for the dopamine transporter (DAT) results from both the source and concentration of acrylamide and the use of certain prestained protein mol-wt standards. Also, the mobility of the prestained protein standards varies with the difference in dye content between batches.

3. DIFFERENCES IN GLYCOSYLATION PATTERNS

3.1. Glycosylation Pattern of Native Neurotransmitter Transporters

3.1.1. Dopamine Transporter (DAT)

The DAT is a heavily glycosylated protein with a molecular mass of ~80 kDa on denaturing polyacrylamide gels *(20–25)*. A modest level of partial purification of DAT has been achieved with DEAE ion-exchange and wheat-germ lectin-affinity chromatography *(12)*. The DAT has been photoaffinity-labeled with GBR12935 derivatives *(20–25)*, the cocaine analog RTI-82 *(24,26)*, and the tropane analog GA II 34 *(27)*. The photoaffinity ligands have been valuable in the characterization of the glycosylation state of the transporter protein. The rat striatal DAT has a molecular mass of 80 kDa *(23,24,28,29)*; treatment with neuraminidase and N-glycanase suggested the presence of terminal sialic-acid residues and N-linked sugar cores, respectively *(20–22,25)*. The molecular mass of N-deglycosylated DAT polypeptides from the caudate of canine, human, and rat DAT expressed from a cDNA transfected into COS-7 cells is ~50 kDa *(25)*. The presence of O-linked glycosylation in the native rat DAT or rDAT cDNA expressed in COS-7 cells was undetectable *(25)*. The rDAT cDNA expressed in COS-7 cells did not seem to contain sialic-acid residues *(25)*. The differential glycosylation pattern of a DAT is dependent on the cell, tissue, species, and developmental state (*see* below).

3.1.2. Norepinephrine Transporter (NET)

The human norepinephrine transporter (hNET) is a glycoprotein, and its properties have been studied by stable expression of hNET cDNA in LLC-PK1, a porcine kidney epithelial-cell line *(30,31)*. A photoaffinity ligand for the NET is not yet available. However, hNET-specific antibody is available, and has been used to study glycosylated forms of the transporters. The bovine noradrenaline transporters (bNAT) expressed in HEK293 cells displays a molecular mass of 80 kDa (glycosylated) and 50 kDa (core protein) for bNAT1 and bNAT2, respectively *(32)*. It appears that bNAT2 is unglycosylated and has a truncated C-terminal tail (18 residues) compared to bNAT1 (C-terminal tail of 31 residues) which is glycosylated *(32)*. Nguyen and Amara *(33)* immunoprecipitated 90- and 60-kDa glycoforms and a 50-kDa core protein of hNET from HeLa-hNET cells.

3.1.3. Serotonin Transporter (SERT)

Launay et al. *(13)* purified the serotonin transporter (SERT) from human platelet membranes to homogeneity using affinity chromatography. The SERT protein in one- and two-dimensional denaturing polyacrylamide gels migrated with a molecular mass of 68 kDa and an acidic p*I* value of 5.2–6.2. The glycoprotein nature of the transporter protein was shown by its adsorption to a WGA agarose, and its specific elution with N-acetylglucosamine. A 68-kDa glycoprotein from human platelet membranes was also identified by specific photoaffinity labeling with paroxetine and cyanoimipramine *(13)*. In human platelet and rat-brain membranes, [^{3}H]2-nitroimipramine photolabeled a 30–35-kDa band *(34)*. The smaller size of this polypeptide could be the tricyclic-binding domain of the SERT or a degradation product. Qian et al. *(35)* reported cell-specific N-linked glycosylation of native SERT proteins from the rat brain (76 kDa), lung (80 kDa), platelet (94 kDa), and rSERT expressed from a cDNA transfected in HeLa cells (61 kDa). The human platelet SERT protein was sensitive to neuraminidase, decreasing [^{3}H]5HT binding to the desialated transporter. This may suggest a role for the sialic acid residues in substrate recognition.

3.1.4. GABA Transporter

The GABA transporter, an 80-kDa heavily glycosylated protein, was one of the first neurotransmitter transporters to be purified to homogeneity from the rat brain *(11,36)*. The transporter protein could be adsorbed and specifically eluted from a WGA agarose column. Enzymatic N-deglycosylation of the transporter protein yields a molecular mass of 60 kDa in a denaturing polyacrylamide gel. Peptide-sequence information from the purified protein resulted in successful cloning of the rat and human brain GABA (GAT1) transporters with predicted molecular masses of 67 kDa *(37,38)*.

3.1.5. Glutamate Transporter

The glutamate-transporter protein has been purified to homogeneity from the rat brain of glial origin. In denaturing polyacrylamide gel, the protein migrated with a molecular mass of 73–80 kDa and an acidic p*I* value of 6.2. Enzymatic N-deglycosylation of the purified protein yielded an apparent molecular mass of 63 kDa, with sugars representing 14% of the mass *(39–41)*. In the rat brain, two glycosylated glutamate transporters (GLAST-1) with molecular masses of 64 kDa and 70 kDa have been detected, which could be separated by lectin-specific affinity chromatography. The 64-kDa and the 70-kDa forms of the glycoproteins were separated by lentil- and wheat-germ agglutinin lectin columns, respectively *(15)*. As yet, splice variants of GLAST-1 have not been reported, and the transporter protein is probably differentially glycosylated. It would be of interest to determine whether GLAST-1 transporters with varying sugar components exhibit functional differences.

3.1.6. Glycine Transporter (GLYT)

Liu et. al. *(42,43)* has cloned two glycine transporter (GLYT) genes. GLYT1 is widely distributed in the central nervous system (CNS), and GLYT2 is mainly localized in the brain stem and spinal cord *(44)*. There are four potential N-linked glycosylation sites on the large EL between transmembrane domains (TMDs) 3 and 4, of which two sites are glycosylated. The protein may not be O-glycosylated, since enzymatic O-deglycosylation with O-glycanase in the presence of neuraminidase failed to alter the mobility of the band in a denaturing polyacrylamide gel. Neuraminidase also failed to alter the mobility of the transporter protein compared with untreated transporter protein, suggesting very low sialic-acid content *(14)*. GLYT2, a glycoprotein of 100 kDa, has been purified to homogeneity from pig-brain stems *(14,45)*. Aragon and Lopez-Corcuera *(45)* have characterized the biochemical and functional properties of purified GLYT2.

3.2. Cell, Tissue, and Species-Specific Glycosylation

3.2.1. Cell

Insight from the expression and the processing of the glycosylated proteins in foreign host cells is now readily available with the success of cloning and expression of candidate genes in amphibian, insect, and mammalian cell systems. Cloned transporter genes, transiently or stably expressed in foreign cell systems, are used to understand the biochemistry, physiology, and pharmacology of transport mechanisms, and as model systems to screen for potential therapeutic agents. The transporter-protein genes cloned from brain cells have been successfully expressed in foreign, non-neuronal cells.

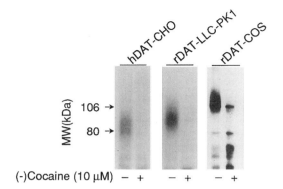

(-)Cocaine (10 µM) − +   − +   − +

Fig. 3. Cell-specific glycosylation of DATs. [^{125}I]DEEP-labeled DAT proteins from hDAT-CHO, rDAT-LLC-PK1, and rDAT-COS-7 are shown. The cells were labeled with [^{125}I]DEEP in the absence (−) and in the presence (+) of 10 µM (−)-cocaine, as described previously *(25)*. The cells were solubilized with SDS-PAGE buffer, resolved in 10% polyacrylamide gels, dried, and exposed to Kodak XAR-5 film. Prestained protein standards (Bio-Rad, Hercules, CA) were electrophoresed on the same gels to estimate the mass of DAT. Human DAT expressed in CHO cells has a molecular mass of 93 kDa. The rat DAT has a molecular mass of 100 and 110 kDa when expressed in LLC-PK1 and COS-7 cells, respectively.

The molecular mass of rat DAT expressed in LLC-PK1, a porcine kidney epithelial-cell line, is 100 kDa, whereas that expressed in the COS-7 monkey-kidney cell line is 110 kDa *(25,46)*. The human DAT expressed in transfected CHO (Chinese Hamster ovary) cells has a molecular mass of 93 kDa *(46;* Fig. 3), and human DAT expressed in HEK-293 cells has a mass of ~80 kDa *(47)*.

Studies indicate that some differences exist in the functional properties of the expressed DAT proteins in various mammalian cells. For example, the difference in K_m of [^{3}H]dopamine uptake could be attributed to the post-translational processing, particularly, the glycosylation of the transporter protein *(25)*. In fact, the K_m of DA uptake in HEK-293 cells expressing the human DAT was considerably reduced when two (N181 and N188) or three (N181, N188, and N205) of the consensus asparagine residues for N-glycosylation were removed *(47)*. It is not yet known whether the concomitant reduction observed for the DAT in V_{max} is mediated by functional changes associated with partial or non-glycosylation or by a reduction in surface-resident DATs.

Olivares et al. *(48)* have studied the properties of the glycine (GLYT1) transporter when expressed in COS-7 cells. The GLYT1 protein migrates in denaturing polyacrylamide gels as a broad band of 80–100 kDa and another band of 57 kDa. The glycine-transporter antipeptide antibodies also detected a single band of 47 kDa on Western blots from COS-7 cells expressing GLYT1 transporters grown in the presence of tunicamycin. A similar result was obtained by expressing a GLYT1 mutant with all four potential asparagine glycosylation sites conservatively replaced with glutamine. These results suggest that the 80–100-kDa band is fully glycosylated, the 57-kDa band corresponds to the partially glycosylated form, and the 47-kDa band is the nonglycosylated polypeptide *(48)*. Additional mutagenesis experiments of the glycosylation sites suggested that all four potential glycosylation sites are glycosylated in the African Green Monkey kidney (COS) cell system *(48)*. Site-directed mutagenesis experiments indicate that all four potential N-glycosylation sites are utilized in rDAT expressed in transfected COS-7 cells (Wang, Patel, and Uhl, unpublished data). In addition, all three potential N-glycosylation sites in human DAT are used upon expression in HEK-293 cells with incremental reductions in mol wt upon substitution of glutamine for asparagine in the single-mutant N181Q, the double-mutant N181,188Q, and the triple-mutant N181,188,205Q *(47)*.

Because there is a cell-specific difference in the glycosylation pattern, the usage of all four potential N-glycosylation sites in the rat brain or three sites in the human brain should not be assumed based on the COS-7 and HEK-293 cell results. The usage of potential N-glycosylation sites in the native state of the dopamine and other transporter proteins must be empirically determined. Culture conditions for mammalian and sf9 insect cells can influence the N-glycosylation process, leading to glycan heterogeneity *(49)*.

Cell specific differences in the glycosylation pattern of other neurotransmitter transporter proteins have also been observed. The cell-specific N-linked glycosylation of native SERT protein yields a molecular mass of 76, 80, and 94 kDa in rat brain, lung, and platelets, respectively *(35)*; in HeLa cells, rat and human glycosylated SERT proteins have a molecular mass of 61 kDa and the deglycosylated transporter protein has a molecular mass of 56 kDa *(35)*. The glycosylated SERT protein in sf9 insect cells has a molecular mass of 60 kDa, and the deglycosylated or unglycosylated transporters have a molecular mass of 54 kDa *(50)*.

The human norepinephrine transporter (hNET) properties have been studied by stable expression of hNET in LLC-PK1, porcine kidney epithelial-cell line *(30,31,50)*. The biosynthesis of the hNET protein was detected with anti-hNET antibodies following metabolic labeling and labeled membranes resolved on denaturing polyacrylamide gels. The antibodies reacted with two proteins with molecular masses of 54 and 80 kDa in the control cells and a 46 kDa polypeptide in cells grown in the presence of tunicamycin. The enzymatic N-deglycosylated hNET has a molecular mass of 46 kDa. The results show that the unglycosylated polypeptide has a molecular mass of 46 kDa. The 54-kDa band is the partially glycosylated intermediate product, and the 80-kDa band is the fully glycosylated polypeptide. Bruss et al. *(52)* detected hNET with molecular mass of 50 and 58 kDa using antipeptide antibodies in hNET-COS-7 cells, and Melikian et al. *(51)* reported species of 46 and 54 kDa in hNET-COS-1 cells. Nguyen and Amara *(33)*, using hNET antibodies detected three forms of hNET with a molecular mass of 90, 60, and unglycosylated 50 kDa bands in HeLa cells.

Conradt et al. *(15)* studied the glycosylation properties of rat glial GLAST-1 transporters expressed in Xenopus oocytes. The rat GLAST-1 expressed in Xenopus oocytes have a monomeric molecular mass of 65 kDa, reduced to 56 kDa on deglycosylation. The glutamate transporter formed homodimers in Xenopus oocytes with molecular mass of ~100 kDa *(15)*.

3.2.2. Tissue

Differences in tissue- and species-related processing of native DAT is shown in Figs. 4 and 5. Lew et. al. *(28)* had shown that the molecular mass of rat DAT in striatum was lower than that in nucleus accumbens. The difference was caused by core oligosaccharide and not the terminal sialic acid residues *(22)*. However, the labeling of human nucleus accumbens (Fig. 4) with [^{125}I]DEEP or [^{125}I]RTI-82 did not show any apparent difference in the human DAT molecular mass from caudate and nucleus accumbens (unpublished data). Also, no apparent difference in the molecular mass of the glycosylated human DAT from caudate and putamen was found (unpublished data and ref. *53*). Vaughan et. al. *(29)* has shown that DATs immunoprecipitated from rat striatum, nucleus accumbens, prefrontal cortex, and midbrain have a molecular mass of ~80 kDa. In addition, the N-linked carbohydrate and sialic-acid content is comparable with that of the rat striatal DAT.

Differences in the tissue-specific N-liked glycosylation of native SERT protein from the rat brain (76 kDa), lung (80 kDa), and platelet (94 kDa) have been reported *(35)*. Since there is only a single gene described for the SERT protein, the size variance is probably the result of differences in glycosylation, similar to those described for the DAT protein.

3.2.3. Species

A difference in the glycosylation of DATs from various species has been observed, with less variability between individuals of the same species *(25)*. The molecular mass of photoaffinity-labeled canine caudate DAT is 78 kDa, the human caudate 62–74 kDa, and the rat striatum 80 kDa *(25,53;* Fig. 5). Vaughan et. al. *(29)* also show DATs with molecular mass of ~80 kDa from the rat, mouse, dog, and human caudates. The deglycosylated size of the DAT from canine, human, and rat caudate/striatum is ~50 kDa in a denaturing polyacrylamide gel *(25)*.

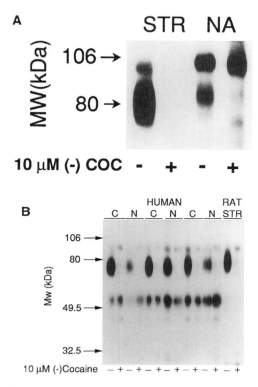

Fig. 4. (A) [^{125}I]DEEP-labeled DATs from rat nucleus accumbens and striatum. This figure shows differences in the molecular mass of rat DATs from nucleus accumbens and striatal membranes labeled with [^{125}I]DEEP. The membranes were labeled with [^{125}I]DEEP, as described in the legend to Fig. 3. **(B)** [^{125}I]DEEP-labeled human and rat DAT. The figure shows [^{125}I]DEEP-labeled caudate (C) and nucleus accumbens (N) membranes from human subjects with postmortem intervals (PMI) of 22, 11, and 36 h (from left to right) and rat striatal (RAT STR) membranes with PMI of 0 h. Molecular mass of DAT from the caudate and the nucleus accumbens from individuals with PMI of 2 h and 4 h was not different from those shown in this figure. The membranes were labeled as described in the legend to Fig. 3. The positions of the prestained protein markers are shown with an arrow.

3.3. Developmentally Regulated Glycosylation

The DAT protein in the rat brain undergoes differential glycosylation as determined by photoaffinity labeling during postnatal development *(54)*. The changes in the molecular mass of rat striatal DAT from postnatal d 4, 14, and 60 (adult) were significantly different (Fig. 6). The DAT molecular mass was slightly higher in older (24-mo-old) rats compared with adult rats, and there was no apparent difference in DAT molecular mass between postnatal

Patel and Reith

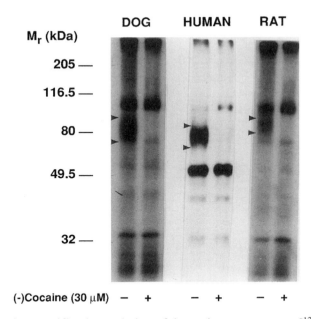

Fig. 5. Species-specific glycosylation of dopamine transporters. [125I]DEEP-labeled DAT proteins from the dog caudate, human caudate, and rat striatal membranes are shown. The membranes were labeled with [125I]DEEP in the absence (–) and in the presence (+) of 10 μM (–)-cocaine, as described in the legend to Fig. 3. Specifically labeled broad DAT bands are shown between arrow heads. The molecular mass of the dog, human, and rat DATs are ~78, 74, and 80 kDa, respectively.

d 0 (birth) and d 4 *(54)*. The difference in DAT molecular mass was caused by glycosylation especially, by the oligosaccharide core, and not caused by the terminal sialic-acid content or the protein backbone as determined by glycosidases and DAT antibodies, respectively *(54)*. The glutamate-transporter protein (GLAST), or proteins associated with the transporter that crossreacts with GLAST antibodies, may also be differentially glycosylated *(55)*. There are examples in the literature for the differential glycosylation of glycoproteins during development. The rat pituitary thyrotropin hormone *(56)*, human erythrocyte band 3 protein *(57)*, and chick embryo fibroblast glycopeptides *(58)* undergo developmentally related differential glycosylation.

In contrast, some glycosylated proteins show no changes in their glycosylation pattern during development. A 60-kDa rat-brain somatostatin receptor, a N-linked glycoprotein with terminal sialic acid residues *(59)* shows no glycosylation difference from embryonic d 13 through postnatal d 18 *(60)*. Similarly, glucose-transporter proteins GLUT1 and GLUT3 from

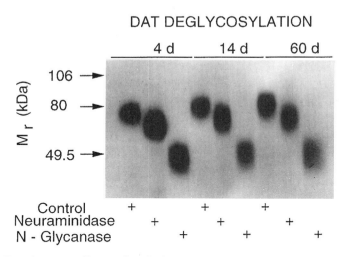

Fig. 6. Developmentally regulated glycosylation of rat striatal DATs. DAT protein from 4-, 14-, and 60-d-old rats were labeled with [^{125}I]DEEP and electrophoresed in SDS-polyacrylamide gels, as described in the legend to Fig. 3. This figure shows changes in molecular mass of control, after neuraminidase and N-glycanase enzyme treatments. The deglycosylation of the [^{125}I]deep-labeled, band was carried out as described previously *(54)*. The difference in molecular mass of DAT during early postnatal development persisted after desialation, but not after N-deglycosylation. (Figure taken from Patel et al. *[54]*).

various brain regions also show no changes in the glycosylation pattern from embryonic d 19 through postnatal d 30 and adult *(61)*. The differences in glycosylation patterns, or the lack thereof, can be explained in the availability of different glycosyltransferases during ontogenesis, as shown previously in the rat intestinal mucosa *(62)*. N-acetylglucosaminyltransferase-IV enzyme-mediated modification of glycoprotein processing may also account for the differences in glycosylation patterns *(63)*.

4. ROLE OF GLYCOSYLATION: FUNCTIONAL AND STRUCTURAL IMPLICATIONS

The roles of sugars associated with proteins and lipids are diverse in their contribution to structure, function, and disease states *(2–7,64,65)*. The cloned transporter genes have been expressed in amphibian, insect and mammalian cell systems. The expressed protein is cell-specifically processed, with varying glycosylation. The difference in glycosylation, however, has not appreciably affected the functioning of the SERT protein, which is expressed in sf9 insect cells compared with that expressed in the rat brain *(50)*. Uptake of

[^{3}H]5HT and the [^{125}I]RTI-55 ligand binding to the SERTs in sf9 insect cells do not require glycosylation for function. Similarly, the functional properties of hNET or GLYT1 transporters were not compromised by varying degrees in unglycosylated to fully glycosylated transporter proteins *(14,15,30,33)*. However, this conclusion may be affected by the cell type used for study. Thus, in COS cells, nonglycosylated and glycosylated hNET populated the surface membrane with comparable efficiency, but the mutant protein did not sustain as much DA uptake relative to wild-type *(51)*.

A C-terminal variant of the bovine norepinephrine transporter, bNAT2, when expressed in HEK293 cells, did not show desipramine-sensitive [^{3}H] norepinephrine uptake. Furthermore, bNAT2 was shown to lack glycosylation, and not targeted to the plasma membrane. In contrast, bNAT1 is glycosylated, and does show desipramine-sensitive [^{3}H] norepinephrine uptake in bNAT1-HEK293 cells *(32)*. Thus, the C-terminal portion of bNAT is important for membrane targeting. Could extending the C-terminal tail of bNAT2 restore function and targeting to the plasma membrane? Indeed, it was found that addition of a C-terminal segment of NAT1 restores activity to NAT2 (D. Christie, personal communication).

The DAT protein expressed in COS-7 cells is insensitive to neuraminidase, suggesting either a reduced number or the absence of sialic-acid residues *(25)*. Such differences in post-translational processing may account for the differences in K_m for [^{3}H]dopamine uptake in COS-7 and other mammalian-cell expression systems compared with that in the native state *(25) (see* also **Subheading 3.2.1.***)*. The removal of sialic-acid residues from the rat striatal DAT and human platelet SERT affected [^{3}H]dopamine uptake and [^{3}H]5HT binding, respectively *(13,66)*. However, the desialation of the rat striatal membranes did not affect the binding of [^{3}H]CFT to the DAT *(22)* or the functioning of the glycine, norepinephrine, and serotonin transporters *(14,30)*. The stability and trafficking/targeting of the glycine, norepinephrine, and serotonin transporters, were greatly dependent on the glycosylation state *(14,15,30,32,33,50,51)*. The turnover of the unglycosylated proteins is more rapid than that of the glycosylated proteins. This is because the oligosaccharide, N- and O-linked structures hinder access of cytosolic proteases and thereby decrease the turnover of proteins.

The glucose transporters GLUT1–GLUT5 *(67)* have a single potential N-linked glycosylation site in the EL between TMDs 1 and 2, except the GLUT2 isoform. The GLUT2 glucose transporter lacks the N-glycosylation motif. Furthermore, enzymes targeted against N-linked and O-linked glycosylation, and against sialic-acid linkages, failed to affect the mobility of GLUT2 polypeptide in denaturing polyacrylamide gels *(68)*. However, the transport function of GLUT2 appears uncompromised, and the transport activity of the deglycosylated red-blood-cell glucose transporter is also unaffected *(69)*. In contrast, Asano et al. *(70)*, showed that glycosylation is important for the GLUT1 glucose-transporter protein expressed in CHO cells for intracellular targeting and protein stability. Similarly, N-glycosylation in human erythrocyte band 3 anion-transport protein is not required for anion transport function, but plays a role in correct folding of the protein *(71)*.

5. OTHER POST-TRANSLATIONAL MODIFICATIONS

5.1. Fatty-Acid Acylation

The regulatory proteins, particularly the guanine-nucleotide-binding proteins and the receptors interacting with them, are modified largely by myristoylation, palmitoylation, and prenylation (for a review covering literature up to 1995, *see* ref. *72*). The fatty-acyl modification plays an important role in signal transduction and regulation of cell activity. The acylation-deacylation process is analogous to the well-described role of phosphorylation-dephosphorylation in the regulation of cellular activity. Modification of cysteine residues by palmitate is a post-translational modification *(73)* described for cellular proteins, including receptors *(74,75)*. Such fatty-acid acylation of neurotransmitter transporters has not yet been described. However, there is a recent report of palmitoylation of glucose transporters and its significance to glucose-transport regulation in blood-brain barrier capillaries *(76)*.

5.1.1. Myristoylation

Fatty-acid acylation as a form of post-translational modification of a variety of cellular proteins has been described *(72)*. The enzyme responsible for myristoylation is N-myristoyltransferase, which catalyzes the transfer of myristic acid from myristoyl-CoA. In regulatory guanine-nucleotide binding protein (G-protein), myristoylation of the N-terminal glycine of alpha subunits of G-proteins (Gi, Go, and Gq) has been shown *(77)*. The function of this acylation of soluble cellular proteins is to allow their anchoring to the cell membranes in close proximity to receptors or effectors with which they interact.

5.1.2. Palmitoylation

Pouliot and Beliveau *(76)* have shown glucose transporters (GLUT1) in brain barrier capillaries to be palmitoylated. Interestingly, preparations from diabetic or hyperglycemic rats showed increased palmitoylation, suggesting a functional role for palmitoylation in the regulation of GLUT1, At present, it is unknown whether the increased palmitoylation indicates that a greater number of free cysteines is available for reaction with [³H]palmitate, which would indicate decreased in vivo palmitoylation. A variety of cellular proteins are palmitoylated and involved in signal transduction and membrane trafficking *(74,75,78)*. The well-known regulatory α-G-protein subunits, several nonreceptor tyrosine-kinase family members and receptors interacting with G-proteins are palmitoylated *(74,75)*. Are palmitoylation sites theoretically predictable, as are PKC or PKA phosphorylation sites or glycosylation sites? No consensus sites for palmitoylation have been definatively established, but suggestions have been made. First, a flanking motif containing several positively charged amino acids, which surrounds the palmitoylated cysteine of several G-protein coupled receptors *(74,75)* and the GluR6 kainate receptor *(79)* has been discussed by Pickering et al. *(79)*. Second, Bouvier et al. *(74)* have proposed a consensus sequence of F X X L/I L/I $(X)_n$ C^P (n = 0 to 4) based on results for G-protein-coupled receptors. Thirdly, a more generalized sequence of F/Y $(X)_{n1}$ B $(X)_{n2}$ C^P (B = hydrophobic residue and n1 or n2 = 0 to 4) accommodates palmitoylation of most G-protein-coupled receptors *(74)*. The significance of myristoylation or palmitoylation of receptors and G-proteins, apart from the anchoring of the subunits to the plasma membrane, is found in protein-protein interactions, either allowing or preventing access to phosphorylation sites by specific kinases *(74,75)*.

5.1.3. Prenylation

The prenylation of G-protein γ-subunits plays a crucial role in a receptor-specific signal transduction *(72,80)*. The various isoforms of G-protein γ-subunits are isoform-specific acylated with different chain lengths of fatty acids, isoprenyl, farnesyl, and geranylgeranyl. The acylation is at the C-terminal cystine residues with a motif CAAX (C = cystine, A = aliphatic amino acid and X = any amino acid) forming a thioester bond. The acylated gamma subunits can form membrane attachment with the receptor and other G-protein subunits *(80)*. For a review of protein prenylation, see ref. *81*.

5.2. Phosphorylation

Phosphorylation is one of the most widely known forms of post-translational reversible modification of many proteins regulating a variety of physiological functions. Specific sequences within the polypeptide have been identified as potential phosphorylation sites for substrates for specific kinases. The consensus sequence motifs for many protein kinase and phosphatase substrates have been described *(82,83)*. The regulation of neurotransmitter-transporter protein activity by phosphorylation has been reviewed *(84,85)* and is covered in detail in Chapter 1. Briefly, [³H]dopamine uptake by the DAT can be inhibited through activation of protein kinase C (PKC) enzyme in rat synaptosomes *(86,87)*, rDAT-COS-7 cells, and rDAT-LLC-PK1 cells *(88,89)*. However, [^{125}I]RTI-55 binding to the DAT in rDAT-LLC-PK1 cells was not affected by PKC activator phorbol 12-myristate 13-acetate (PMA) (unpublished data). Zhang and Reith *(90)* have shown that the DAT function can be modulated by arachidonic acid and PKC by separate mechanisms.

Over the last five years, a considerable amount of research has focused on understanding the functional regulation of transporters by phosphorylation. For a review of biogenic amine-transporter regulation and trafficking over this period, *see* ref. *85* and Chapter 1. Melikian and Buckley *(91)* have shown that PKC-activated hDAT (stably expressed in PC12 cells) undergoes downregulation and sequestration in endosomes. They further propose that the endocytosed hDAT may be recycled to the plasma membrane. Unlike the DAT endocytosis-recycling pathway proposed by Melikian and Buckley *(91)*, PKC-activated GFP-DAT (green fluorescent protein tagged-hDAT expressed in MDCK cells) undergoes an endosomal/lysosomal pathway leading to complete transporter degradation and not recycling of the transporter to the plasma membrane *(92)*. Clearly, there is a difference in processing hDAT, either at steady-state, or in PKC activation in these cell lines. Like the differential glycosylation of transporters, it appears that the processing of the transporter from the cell surface to endosomes for recycling, or to lysosomes for degradation, is cell-dependent. There is also a difference among transporters for substrate-dependent redistribution of transporters; serotonin modulates SERT trafficking *(93)*, whereas dopamine fails to modulate DAT trafficking *(92)*.

Perhaps the neurotransmitter transporters share similar mechanisms and pathways to those of the G-protein-coupled receptors and other transporters, such as glucose (GLUT1 and GLUT4) transporters.

5.3. Ubiquitination

Ubiquitin is a 76-amino-acid-polypeptide with a sequence highly conserved in evolution *(94)*. Its name, which signifies ubiquitous immunopoietic polypeptide, is probably a misnomer because the original reports of its stimulatory effect on the differentiation of lymphocytes and on adenylate cyclase have not been replicated, perhaps because of a contamination in the original preparation (for citations, *see* ref. *94*). Ubiquitin can become conjugated with proteins in an ATP-dependent process. Such ubiquitination can play an important role in protein degradation and in protein modification *(95)*. The former role has been studied extensively during the past 15 yr and represents a pathway for breakdown of cytosolic as well as integral membrane proteins by the 26S proteasome, an ATP-dependent multicatalytic protein-ase complex *(96–98).* In this process, ubiquitin is activated in an ATP-dependent fashion by ubiquitin-activating enzyme; in steps involving a carrier protein, ubiquitin is then linked to substrate protein by the covalent formation of an isopeptide linkage between the C-terminal Gly of ubiquitin and ε-amino groups of Lys residues of the substrate protein. Polyubiquitin chains also form post-translationally—for instance, through ubiquitin-Lys48-ubiquitin linkages *(96),* and are in fact required for degradation of some proteins targeted for breakdown by the ubiquitin pathway *(99).* Examples of membrane proteins targeted in this manner are the human cys-tic fibrosis transmembrane conductance regulator (CFTR) protein, a mem-ber of the ABC transporter family *(100),* and Sec61, a yeast protein involved in protein translocation *(101).* Much attention has been focused on the implication of ubiquitination in extra-lysosomal protein degradation, but there is also substantial evidence for targeting of proteins by ubiquitin for lysosomal protein catabolism *(102).*

Studies also indicate that transporters can be regulated by ubiquitination. Thus, the yeast multidrug transporter Pdr5 *(103),* and the Ste6 α-factor pheromone transporter *(97),* both members of the ATP-binding cassette (ABC) multidrug transporter family, appear to be prepared by ubiquitination for endocytic delivery to the vacuole for proteolytic turnover. In these cases, ubiquitination may serve as a signal for protein trafficking rather than a sig-nal for protein degradation itself. The group of Hicke *(104)* has provided more information on ubiquitin as a signal for endocytosis of the plasma-membrane protein Ste2p, the mating pheromone α-factor receptor. Unlike the ubiquitin proteasome recognition signal, the internalization signal does not require polyubiquitin formation through Lys18, but rather relies on monoubiquitination on one lysine residue of the Ste2p. In contrast, Hicke *(104)* reviews evidence that uracil permease, another plasma-membrane pro-tein, probably requires monoubiquitination of two lysine residues for endocytosis, whereas linkage of these ubiquitins through Lys63 facilitates permease internalization. The epithelial sodium channel is also a candidate for ubiquitination mediating downregulation *(104),* which would be impor-tant for the control of hypertension.

In contrast to the situation with yeast transporters, very little information is available on the potential role of ubiquitination of neurotransmitter trans-porters. Two intriguing studies have been reported by Meyer and colleagues *(105,106),* who examined the effect of affinity-purified antibodies directed against ubiquitin (anti-Ub) on the high-affinity uptake of various transmit-ters into rat cerebrocortical synaptosomes. Anti-Ub inhibited the Na^+-dependent uptake of [^{3}H]choline, [^{3}H]GABA, [^{3}H]glutamate, [^{3}H]aspartate,

[^{3}H]norepinephrine, and [^{3}H]serotonin. There was no effect on Na$^+$-independent [^{3}H]choline uptake *(105)*. Furthermore, non-neuronal Na$^+$-dependent uptake of [^{3}H]proline or [^{3}H]glucose into small-intestine brush-border membrane vesicles was unaffected by anti-Ub, suggesting neuronal specificity *(106)*. Cerebrocortical plasma-membrane vesicles allowed the measurement of [^{3}H]GABA and [^{3}H]glutamate uptake driven by investigator-applied ion gradients, and the effectiveness of anti-Ub in this system suggests that GABA and glutamate transporters themselves or proteins in close proximity to them are affected by the antibody *(106)*. Although the study did not include measurements of [^{3}H]dopamine uptake into striatal synaptosomes, a range of Na$^+$-dependent neuronal transporters appears to be susceptible to anti-Ub, suggesting interaction with a common element in these proteins or closely associated proteins.

The presence of ubiquitin in the brain has been recognized for a number of years, since the finding of anti-Ub reactive deposits in inclusions characteristic of neurodegenerative diseases such as Alzheimer's disease *(107)* or Parkinson's disease *(108)* and the identification of a ubiquitin- and ATP-dependent protein-breakdown pathway in the rat cerebral cortex *(109)*. Some synaptic proteins which are conjugated with ubiquitin play a role in the stability of the synapses, and are also component of occlusion bodies in Alzheimer's and Parkinson's diseases *(110)*. Furthermore, the formation of ubiquitin-protein conjugates in the brain has been reported, especially in structures that contain neuronal cell bodies *(111)*, although it is important to note that this study focused on cytosolic fractions only.

6. CONCLUDING COMMENTS

Glycosylation of the neurotransmitter-transporter proteins is cell-, tissue-, and species-specific, and varies during development and aging. The results obtained with different transporters expressed in various host-cell systems show that glycosylation is important for the stability and targeting of the transporter to the plasma membrane. Partial or complete N-deglycosylation has less effect on the transporter function. Because post-translational modification differences are observed in non-neuronal cells, post-translational processes must also be studied in neuronal cells to understand the processing of the transporters in the brain.

Considerable progress has been made in understanding the role of phosphorylation in the regulation and trafficking of transporters. The role of fatty-acid acylation or ubiquitination of transporters needs further consideration. Exciting work lies ahead in elucidating the pathways of downregulation/internalization and the trafficking of transporters. The neurotransmitter transporters may share mechanisms and pathways similar to those of the G-protein-coupled receptors and other transporters, such as glucose transporters.

ACKNOWLEDGMENTS

The author thanks Glyko, Inc. for providing Figs. 1 and 2.

REFERENCES

1. Amara, S. G. and Kuhar, M. J. (1993) Neurotransmitter transporters, Recent progress. *Annu. Rev. Neurosci.* **16,** 73–93.
2. Hubbard, S. C. and Ivatt, R. J. (1981) Synthesis and processing of asparagine-linked oligosaccharides. *Ann. Rev. Biochem.* **50,** 555–583.
3. Kornfeld, R. and Kornfeld, S. (1985) Assembly of asparagine-linked oligosaccharides. *Ann. Rev. Biochem.* **54,** 631–664.
4. Kobata, A. (1992) Structures and function of the sugar chains of glycoproteins. *Eur. J. Biochem.* **209,** 483–501.
5. Opdenakker, G., Rudd, P. M., Ponting, C. P., and Dwek, R. A. (1993) Concepts and principles of glycobiology. *FASEB J.* **7,** 1330–1337.
6. Lis, H. and Sharon, N. (1993) Protein glycosylation, structural and functional aspects. *Eur. J. Biochem.* **218,** 1–27
7. Dwek, R. A. (1995) Glycobiology: towards understanding the function of sugars. *Biochem. Soc. Trans.* **23,** 1–25.
8. Hirschberg, C. B. and Snider, M. D. (1987) Topography of glycosylation in the rough endoplasmic reticulum and Golgi apparatus. *Ann. Rev. Biochem.* **56,** 63–87.
9. Lehrman, M. A. (1991) Biosynthesis of N-acetylglucosamine-P-P-dolichol, the committed step of asparagine-linked oligosaccharide assembly. *Glycobiology* **1,** 553–562.
10. Caplin, M. F. and Kennedy, J. F. (eds). (1986) *Carbohydrate Analysis, A Practical Approach.* IRL Press, Oxford, England.
11. Radian, R., Bendahan, A., and Kanner, B. I. (1986) Purification and identification of the functional sodium- and chloride-coupled g-aminobutyric acid transport glycoprotein from rat brain. *J. Biol. Chem.* **261,** 15,437–15,441.
12. Simantov, R., Vaughan, R., Lew, R., Wilson, A., and Kuhar, M. J. (1991) Dopamine transporter cocaine receptor characterization and purification. *Adv. Biosci.* **82,** 151–154.
13. Launay, J.-M., Geoffroy, C., Mutel, V., Buckel, M., Cesura, A., Alouf, J. E., and Da Prada, M. (1992) One-step purification of the serotonin transporter located at the human platelet plasma membrane. *J. Biol. Chem.* **267,** 11,344–11,351.
14. Nunez, E. and Aragon, C. (1994) Structural analysis and functional role of the carbohydrate component of glycine transporter. *J. Biol. Chem.* **269,** 16,920–16,924.
15. Conradt, M., Storck, T., and Stoffel, W. (1995) Localization of N-glycosylation sites and functional role of the carbohydrate units of GLAST-1, a cloned rat brain L-glutamate/L-aspartate transporter. *Eur. J. Biochem.* **229,** 682–687.
16. Low, M. G. (1989) Glycosyl-phosphatidylinostiol, a versatile anchor for cell surface proteins. *FASEB J.* **3,** 1600–1608.

17. Edge, A. S. B., Faltynek, C. R., Hof, L., Reichert, Jr., L. E., and Weber, P. (1981) Deglycosylation of glycoprotein by trifluoromethanesulfonic acid. *Analyt. Biochem.* **118,** 131–137.

18. Haspel, H. C., Revillame, J., and Rosen, O. M. (1988) Structure, biosynthesis, and function of the hexose transporter in Chinese hamster ovary cells deficient in N-acetylglucosaminyl transferase 1 activity. *J. Cell Physiol.* **136,** 361–366.

19. Hames, B. D. (1981) An introduction to polyacrylamide gel electrophoresis, in *Gel Electrophoresis of Protein: A Practical Approach*, (Hames, B. D. and Rickwood, D., eds.), IRL Press, pp. 1–91.

20. Grigoriadis, D. E., Wilson, A. A., Lew, R., Sharkey, J. S., and Kuhar, M. J. (1989) Dopamine transporter sites selectively labeled by a novel photoaffinity probe, [^{125}I]DEEP. *J. Neurosci.* **9,** 2664–2670.

21. Sallee, F. R., Fogel, E. L., Schwartz, E., Choi, S. M., Curran, D. P., and Niznik, H. B. (1989) Photoaffinity labeling of the mammalian dopamine transporter. *FEBS Lett.* **256,** 219–224.

22. Lew, R., Patel, A., Vaughan, R. A., Wilson, A., and Kuhar, M. J. (1992) Microheterogeneity of dopamine transporters in rat striatum and nucleus accumbens. *Brain Res.* **584,** 266–271.

23. Berger, P., Martenson, R., Laing, P., Thurcauf, A., DeCosta, B., Rice, K. C., and Paul, S. M. (1991) Photoaffinity labeling of the dopamine reuptake carrier protein with 3-azido[3H]GBR-12935. *Mol. Pharmacol.* **39,** 429–435.

24. Patel, A., Boja, J. W., Lever, J., Lew, R., Simantov, R., Carroll, F. I., Lewin, A. H., Phillip, A., Gao, Y., and Kuhar, M. J. (1991) A cocaine analog and a GBR analog label the same protein in rat striatal membranes. *Brain Res.* **576,** 173,174.

25. Patel, A., Uhl, G., and Kuhar, M. J. (1993) Species differences in dopamine transporters, Postmortem changes and glycosylation differences. *J. Neurochem.* **61,** 496–500.

26. Vaughan, R. A. (1998) Cocaine and GBR photoaffinity labels as probes of dopamine transporter structure. *Methods Enzymol.* 296–230.

27. Agoston, G. E., Vaughan, R., Lever, J. R., Izenwasser, S., Terry, P. D., and Newman, A. H. (1997) A novel photoaffinity label for the dopamine transporter based on N-substituted 3"-[bis(4'-fluorophenyl) methoxy] tropane. *Bioorg. Med. Chem. Lett.* **7,** 3027–3032.

28. Lew, R., Vaughan, R., Simantov, R., Wilson, A., and Kuhar, M. J. (1991) Dopamine transporters in the nucleus accumbens and the striatum have different apparent molecular weights. *Synapse* **8,** 152,153.

29. Vaughan, R. A., Brown, V. L., McCoy, M. T., and Kuhar, M. J. (1996) Species- and brain region-specific dopamine transporters, Immunological and glycosylation characteristics. *J. Neurochem.* **66,** 2146–2152.

30. Melikian, H. E., McDonald, J. K., Gu, H., Rudnick, G., Moore, K. R., and Blakely, R. D. (1994) Human norepinephrine transporter, biosynthetic studies using a site-directed polyclonal antibody. *J. Biol. Chem.* **269,** 12,290–12,297.

31. Blakely, R. D., De Felice, L. J., and Hartzell, H. C. (1994) Molecular physiology of norepinephrine and serotonin transporters. *J. Exp. Biol.* **196,** 263–281.

32. Burton, L. D., Kippenberger, A. G., Lingen, B., Bruss, M., Bonisch, D., and Christie, D. (1998) A variant of the bovine noradrenaline transporter reveals the importance of the C-terminal region for correct targeting to the membrane and functional expression. *Biochem. J.* **330,** 909–914.

33. Nguyen, T. T. and Amara, S. G. (1996) N-Linked oligosaccharides are required for cell surface expression of the norepinephrine transporter but do not influence substrate or inhibitor recognition. *J. Neurochem.* **67,** 645–655.

34. Wennogle, L. P., Ashton, R. A., Schuster, D. I., Murphy, R. B., and Meyerson, L. R. (1985) 2-Nitroimipramine, a photoaffinity probe for the serotonin uptake/tricyclic binding site complex. *EMBO J.* **4,** 971–977.

35. Qian, Y., Melikian, H. E., Rye, D. B., Levey, A. I., and Blakely, R. D. (1995) Identification and characterization of antidepressant-sensitive serotonin transporter proteins using site-specific antibodies. *J. Neurosci.* **15,** 1261–1274.

36. Kanner, B. I. (1994) Sodium-coupled neurotransmitter transport, structure, function and regulation. *J. Exp. Biol.* **196,** 237–249.

37. Guastella, J. Nelson, N., Nelson, H., Czyzyk, L., Keynan, S., Miedel, M. C., et al. (1990) Cloning and expression of a rat brain GABA transporter. *Science* **249,** 1303–1306.

38. Nelson, H., Mandiyan, S., and Nelson, N. (1990) Cloning of the human brain GABA transporter. *FEBS Lett.* **269,** 181–184.

39. Danbolt, N. C., Pines, G., and Kanner, B. I. (1990) Purification and reconstitution of the sodium- and potassium-coupled glutamate transport glycoprotein from rat brain. *Biochemistry* **29,** 6734–6740.

40. Danbolt, N. C., Storm-Mathisen, J., and Kanner, B. I. (1992) A [Na$^+$-K$^+$]coupled L-glutamate transporter purified from rat brain is located in glial cell processes. *Neuroscience* **51,** 295–310.

41. Danbolt, N. C. (1994) The high affinity uptake system for excitatory amino acids in the brain. *Prog. Neurobiol.* **44,** 377–396.

42. Liu, Q. R., Nelson, H., Mandiyan, S., Lopez-Corcuera, B., and Nelson, N. (1992) Cloning and expression of a glycine transporter from mouse brain. *FEBS Lett.* **305,** 110–114.

43. Liu, Q. R., Lopez-Corcuera, B., Mandiyan, S., Nelson, H., and Nelson, N. (1993) Cloning and expression of a spinal cord- and brain-specific glycine transporter with novel structural features. *J. Biol. Chem.* **269,** 22,802–22,808.

44. Zafra, F., Gomeza, J., Olivaries, L., Aragon, C., and Gimenez, C. (1995) Regional distribution and developmental variation of the glycine transporters GLYT1 and GLYT2 in the CNS. *Eur. J. Neurosci.* **7,** 1342–1352.

45. Aragon, C. and Lopez-Corcuera, B. Purification, hydrodynamic properties, and glycosylation analysis of glycine transporters (1998) *Methods Enzymol.* **296,** 3–17.

46. Patel, A. P., Martel, J.-C., Vandenbergh, D. J., Uhl, G. R., and Kuhar, M. J. (1995) Cell lines expressing human and rat dopamine transporter cDNAs, Different ligands yield different radiolabeling patterns. *Soc. Neurosci.* **21,** 781.

47. Li, L.-B., Wang, L. C., Chen, N. and Reith, M. E. A. (2000) Removal of potential consensus sites for N-linked glycosylation of the human dopamine transporter. *Soc. Neurosci. Abstr.* **26,** 1168.

48. Olivares, L., Aragon, C., Gimenez, C., and Zafra, F. (1995) The role of N-glycosylation in the targeting and activity of the GLYT1 glycine transporter. *J. Biol. Chem.* **270,** 9437–9442.

49. Jenkins, N. (1995) Monitoring and control of recombinant glycoprotein heterogeneity in animal cell cultures. *Biochem. Soc. Trans.* **23,** 171–175.

50. Tate, C. G. and Blakely, R. D. (1994) The effect of N-linked glycosylation on activity of the Na^+ and Cl^- dependent serotonin transporter expressed using recombinant baculovirus in insect cells. *J. Biol. Chem.* **269,** 26,303–26,310.

51. Melikian, H. E., Ramamoorthy, S., Tate, C. G. and Blakely, R. D. (1996) Inability to N-glycosylate the human norepinephrine transporter reduces protein stability, surface trafficking, and transport activity but not ligand recognition, *Mol. Pharmacol.* **50,** 266–276.

52. Bruss, M., Hammermann, R., Brimijoin, S., and Bonisch, H. (1995) Antipeptide antibodies confirm the topology of the human norepinephrine transporter. *J. Biol. Chem.* **270,** 9197–9201.

53. Niznik, H. B., Fogel, E. F., Fasso, F. F., and Seeman, P. (1991) The dopamine transporter is absent in Parkinsonian putamen and reduced in the caudate nucleus. *J. Neurochem.* **56,** 192–198.

54. Patel, A. P., Cerruti, C., Vaughan, R. A., and Kuhar, M. J. (1994) Developmentally regulated glycosylation of dopamine transporter. *Dev. Brain Res.* **83,** 53–58.

55. Furuta, A., Rothstein, J. D., and Martin, L. J. (1997) Glutamate transporter protein subtypes are expressed differentially during rat CNS development. *J. Neurosci.* **17,** 8363–8375.

56. Gyves, P. W., Gesundheit, N., Stannard, B. S., DeCherney, G. S., and Weintraub, B. D. (1989) Alterations in the glycosylation of secreted thyrotropin during ontogenesis. *J. Biol. Chem.* **264,** 6104–6110.

57. Fukuda, M., Fukuda, M. N., and Hakomori, S. (1979) Developmental change and defect in the carbohydrate structure of band 3 glycoprotein of human erythrocyte membrane. *J. Biol. Chem.* **254,** 3700–3703.

58. Codogno, P., Botti, J., Font, J., and Aubery, M. (1985) Modification of the N-linked oligosaccharides in cell surface glycoproteins during chick embryo development. *Eur. J. Biochem.* **149,** 453–460.

59. Rens-Domiano, S. and Reisine, T. (1991) Structural analysis and functional role of the carbohydrate component of somatostatin receptors. *J. Biol. Chem.* **266,** 20,094–20,102.

60. Theveniau, M. and Reisine, T. (1993) Developmental changes in expression of a 60 kDa somatostatin receptor immunoreactivity in the rat brain. *J. Neurochem.* **60,** 1870–1875.

61. Vannucci, S. (1994) Developmental expression of GLUT1 and GLUT3 glucose transporters in rat brain. *J. Neurochem.* **62,** 240–246.

62. Biol, M. C., Martin, A., Richard, M., and Louisot, P. (1987) Developmental changes in intestinal glycosyl-transferase activities. *Pediatric Res.* **22,** 250–256.

63. Goldberg, D. E. and Kornfeld, S. (1983) Evidence for extensive subcellular organization of asparagine-linked oligosaccharide processing and lysosomal enzyme phosphorylation. *J. Biol. Chem.* **258,** 3159–3165.

64. Paulson, J. C. (1989) Glycoproteins: What are the sugars chains for? *Trends Biochem. Sci.* **14,** 272–276.
65. Dwek, R. A. (1995) Glycobiology, More functions for oligosaccharides. *Science* **269,** 1234,1235.
66. Zaleska, M. M. and Erecinska, M. (1987) Involvement of sialic acid in high-affinity uptake of dopamine by synaptosomes from rat brain. *Neurosci. Lett.* **82,** 107–112.
67. Pessin, J. E. and Bell, G. I. (1992) Mammalian facilitative glucose transporter family, structure and molecular regulation. *Annu. Rev. Physiol.* **54,** 911–930.
68. Brant, A. M., Gibbs, M. E., and Gould, G. W. (1992) Examination of the glycosidation state of five members of the human facilitative glucose transporter family. *Biochem Soc. Trans.* **20,** 235S.
69. Wheeler, T. J. and Hinkel, P. C. (1981) Kinetic properties of the reconstituted glucose transporter from human erythrocytes. *J. Biol. Chem.* **256,** 8907–8914.
71. Groves, J. D. and Tanner, M. J. (1994) Role of N-glycosylation in the expression of human band 3-mediated anion transport. *Mol. Membr. Biol.* **11,** 31–38.
70. Asano, T., Takata, K., Katagiri, H, Ishihara, H., Inukai, K., Anai, M., Hirano, H., Yazaki, Y., and Oka, Y. (1993) The role of N-glycosylation in the targeting and stability of GLUT1 glucose transporter. *FEBS Lett.* **324,** 258–261.
72. Casey, P. J. and Buss, J. E. (eds.) (1995) Lipid modifications of proteins. *Meth. Enzymol.* **250.**
73. Saltiel, A. R., Ravetch, A. R., and Aderem, A. A. (1991) Functional consequences of lipid-mediated protein-membrane interactions. *Biochem. Pharmacol.* **42,** 1–11.
74. Bouvier, M., Loisel, T. P. and Hebert, T. (1995) Dynamic regulation of G-protein coupled receptor palmitoylation, potential role in receptor function. *Biochem. Soc. Trans.* **23,** 577–581.
75. Milligan, G., Parenti, M., and Magee, A. I. (1995) The dynamic role of palmitoylation in signal transduction. *Trends Biochem. Sci.* **20,** 181–186.
76. Pouliot, J.-F. and Beliveau, R. (1994) Palmitoylation of the glucose transporter in blood-brain barrier capillaries. *Biochimica et Biophysica Acta.* **1234,** 191–196.
77. Buss, J. E., Mumby, S. M., Casey, P. J., Gilman, A. G., and Sefton, B. M. (1987) Myristoylated alpha subunits of guanine nucleotide-binding regulatory proteins. *Proc. Natl. Acad. Sci. USA* **84,** 7493–7497.
78. Mundy, D. I. (1995) Protein palmitoylation in membrane trafficking. *Biochem. Soc. Trans.* **23,** 572–576.
79. Pickering, D. S., Taverna, F. A., Salter, M. W., and Hampson, D. R. (1995) Palmitoylation of the GluR6 kainate receptor. *Proc. Natl. Acad. Sci. USA* **92,** 12,090–12,094.
80. Muller, S. and Lohse, M. J. (1995) The role of *bg* subunits in signal transduction. *Biochem. Soc. Trans.* **23,** 141–148.
81. Sinensky, M. (2000) Recent advances in the study of prenylated proteins. *Biochimica et Biophysica Acta* **1484,** 93–106.

82. Kemp, B. E. and Pearson, R. B. (1990) Protein kinase recognition sequence motifs. *Trends Biochem. Sci.* **15**, 342–346.

83. Kennely, P. J. and Krebs, E. G. (1991) Consensus sequences as substrate specificity determinants for protein kinases and protein phosphatases. *J. Biol. Chem.* **266**, 15,555–15,558.

84. Boja, J. B., Vaughan, R. A., Patel, A., Shaya, E., and Kuhar, M. J. (1994) The dopamine transporter, in *Dopamine Receptors and Transporters,* (Niznik, H. B., ed.), Marcel Dekker, New York, pp. 611–644.

85. Blakely, R. D. and Bauman, A. L. (2000) Biogenic amine transporters, regulation in flux. *Curr. Opin. Neurobiol.* **10**, 328–336.

86. Copeland, B. J., Neff, N. H., and Hadjiconstantinou, M. (1995) Protein kinase C activators decrease dopamine uptake into striatal synaptosomes. *Soc. Neurosci.* **21**, 1381.

87. Vaughan, R. A., Huff, R. A., Uhl, G. R., and Kuhar, M. J. (1997) Protein kinase C-mediated phosphorylation and functional regulation of dopamine transporters in striatal synaptosomes. *J. Biol. Chem.* **272**, 15,541–15,546.

88. Kitayama, S., Dohi, T., and Uhl, G. R. (1994) Phorbol esters alter functions of the expressed dopamine transporter. *Eur. J. Pharmacol.* **268**, 115–119.

89. Huff, R. A., Vaughan, R. A., Kuhar, M. J., and Uhl, G. R. (1995) Protein kinase activity modulates dopamine transporter function. *Soc. Neurosci.* **21**, 1380.

90. Zhang, L. and Reith, M. E. A. (1996) Regulation of the functional activity of the human dopamine transporter by the arachidonic acid pathway. *Eur. J. Pharmacol.* **315**, 345–354.

91. Melikian, H. E., Buckley, K. M. (1999) Membrane trafficking regulates the activity of the human dopamine transporter. *J. Neurosci.* **19**, 7699–7710.

92. Daniels, G. M. and Amara, S. G. (1999) Regulated trafficking of the human dopamine transporter. Clathrin-mediated internalization and lysosomal degradation in response to phorbol esters. *J. Biol. Chem.* **274**, 35,794–35,801.

93. Ramamoorthy, S. and Blakely, R. D. (1999) Phosphorylation and sequestration of serotonin transporters differentially modulated by psychostimulants. *Science* **285**, 763–766.

94. Hershko, A. and Ciechanover, A. (1982) Mechanisms of intracellular protein breakdown. *Annu. Rev. Biochem.* **51**, 335–364.

95. Hicke, L. (1999) Gettin' down with ubiquitin, turning off cell-surface receptors, transporters and channels. *Trends Cell Biol.* **9**, 107–112.

96. Hershko, A. and Ciechanover, A. (1992) The ubiquitin system for protein degradation. *Annu. Rev. Biochem.* **61**, 761–807.

97. Kolling, R. and Losko, S. (1997) The linker region of the ABC-transporter Ste6 mediates ubiquitination and fast turnover of the protein, *EMBO J.* **16**, 2251–2261.

98. Rubin, D. M. and Finley, D. (1995) Proteolysis. The proteasome, a protein-degrading organelle? *Curr. Biol.* **5**, 854–858.

99. Tobias, J. W. and Varshavsky, A. (1991) Cloning and functional analysis of the ubiquitin-specific protease gene UBP1 of Saccharomyces cerevisiae, *J. Biol. Chem.* **266**, 12,021–12,028.

100. Jensen, T. J., Loo, M. A., Pind, S., Williams, D. B., Goldberg, A. L. and Riordan, J. R. (1995) Multiple proteolytic systems, including the proteasome, contribute to CFTR processing, *Cell* **83,** 129–135.

101. Biederer, T., Volkwein, C. and Sommer, T. (1996) Degradation of subunits of the Sec61p complex, an integral component of the ER membrane, by the ubiquitin-proteasome pathway, *EMBO J.* **15,** 2069–2076.

102. Mayer, R. J., Arnold, J., Laszlo, L., Landon, M. and Lowe, J. (1991) Ubiquitin in health and disease, *Biochim. Biophys. Acta,* **1089,** 141–157.

103. Egner, R. and Kuchler, K. (1996) The yeast multidrug transporter Pdr5 of the plasma membrane is ubiquitinated prior to endocytosis and degradation in the vacuole, *FEBS Lett.* **378,** 177–181.

104. Hicke, L. (1997) Ubiquitin-dependent internalization and down-regulation of plasma membrane proteins, *FASEB J.* **11,** 1215–1226.

105. Meyer, E. M., West, C. M., and Chau, V. (1986) Antibodies directed against ubiqintin inhibit high affinity [^{3}H]choline uptake in rat cerebral cortical synaptosomes. *J. Biol. Chem.* **261,** 14,365–14,368.

106. Meyer, E. M., West, C. M., Stevens, B. R., Chau, V., Nguyen, M-t., and Judkins, J. H. (1987) Ubiquitin-directed antibodies inhibit neuronal transporters in rat brain synaptosomes. *J. Neurochem.* **49,** 1815–1819.

107. Mori, H., Kondo, J. and Ihara, Y. (1987) Ubiquitin is a component of paired helical filaments in Alzheimer's disease, *Science* **235,** 1641–1644.

108. Torack, R. M. and Miller, J. W. (1994) Immunoreactive changes resulting from dopaminergic denervation of the dentate gyrus of the rat hippocampal formation, *Neurosci. Lett.* **169,** 9–12.

109. Okada, M., Ishikawa, M., and Mizushima, Y. (1991) Identification of a ubiquitin- and ATP-dependent protein degradation pathway in rat cerebral cortex, *Biochim. Biophys. Acta,* **1073,** 514–520.

110. Beesley, P. W., Mummery, R., Tibaldi, J., Chapman, A. P., Smith, S. J., and Rider, C. C. (1995) The post-synaptic density, putative involvement in synapse stabilization via cadherins and covalent modification by ubiquitination. *Biochem. Soc. Trans.* **23,** 59–64.

111. Adamo, A. M., Moreno, M. B., Soto, E. F. and Pasquini, J. M. (1994) Ubiquitin-protein conjugates in different structures of the central nervous system of the rat, *J. Neurosci. Res.* **38,** 358–364.

11

Dopamine-Transporter Uptake Blockers

Structure-Activity Relationships

F. Ivy Carroll, Anita H. Lewin, and S. Wayne Mascarella

1. INTRODUCTION

The dopamine transporter (DAT), a protein located on presynaptic nerve terminals *(1–3)*, plays a major role in the reuptake of released dopamine. Uptake of DA is sodium- and chloride-ion- as well as temperature- and time-dependent, and is inhibited by a variety of compounds, including cocaine. Even though cocaine binds to several sites in the brain, cocaine abuse has been shown to be related to binding at the DAT site. Thus, analysis of the binding potencies of a series of cocaine analogs demonstrates that, in animal models, the reinforcing properties of these analogs correlate only with binding potencies at the DAT site. The DAT site has been called a cocaine receptor *(4,5)*; it may be the initial site responsible for producing cocaine's drug reinforcement. The cDNA for the DAT has been cloned from rat *(6–8)*, mouse *(9,10)*, bovine *(11)*, and human *(12)* brains, and these clones exhibit 92, 93, and 88% homology to human DAT (hDAT), respectively. A transporter with 92% homology and with neuronal DAT properties has also been characterized from African Green Monkey kidney (COS-7) cell lines *(13)*. The hydrophobicity profile of the DAT indicates 12 possible membrane-spanning regions with the amino and carboxy termini located intracellularly. The protein from human and rat brains contains three and four extracellular *N*-glycosylation sites, respectively (Fig. 1). A detailed review of structural information about the DAT has recently been published *(14)*.

Since changes in DA neuron density, DAT sites, or both are involved in a number of neurological disorders, as well as in drug abuse, probes for the DAT may be useful as markers for dopaminergic terminals in some diseases (*see* Chapters 12 and 13). For example, cocaine congeners have been used as binding ligands to detect reduced densities of the DAT in Parkinson's-

From: *Contemporary Neuroscience:*
Neurotransmitter Transporters: Structure, Function, and Regulation, 2nd Edition
Edited by: M. E. A. Reith © Humana Press Inc., Totowa, NJ

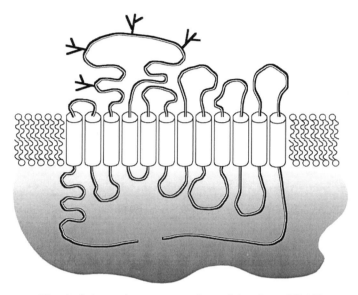

Fig. 1. Schematic representation of the cloned DAT.

diseased brains. In order to develop potent and selective probes for the DAT, an understanding of the structure-activity relationship (SAR) of the binding site(s) is required. This may be accomplished experimentally by correlation of structural variations with inhibition of dopaminé uptake, or, more conveniently, with the inhibition of radioligand binding at the DAT. The latter approach also has the potential for identifying structures capable of blocking the binding of cocaine without inhibiting dopamine transport. Such compounds are considered "dopamine-transport-sparing cocaine antagonists" and have not yet been identified (*see* **Subheading 3.**).

Compounds capable of inhibiting DA transport and of binding at the DAT display remarkable variations in structure. The compound classes for which SAR are discussed in this chapter are: tropane analogs (**Subheading 2.1.**), 1,4-dialkylpiperazine analogs (**Subheading 2.2.**), mazindol analogs (**Subheading 2.3.**), phencyclidine analogs (**Subheading 2.4.**), and methylphenidate analogs (**Subheading 2.5.**). The SARs discussed in this chapter are based on the inhibition of radioligand binding at the DAT and/or on inhibition of [³H]dopamine uptake.

1.1. Assay Conditions

To conduct a SAR study for a binding site requires a radio-ligand that shows specific, saturable binding that is linear with increasing tissue con-

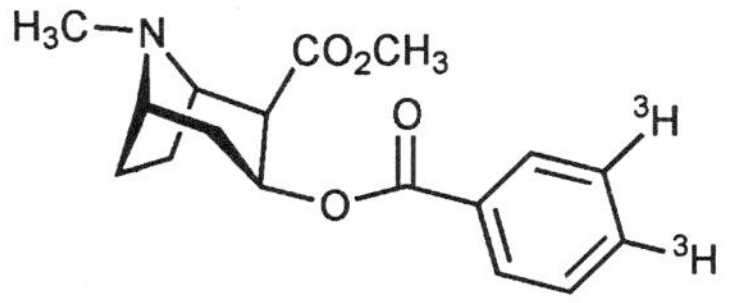

[³H]Cocaine

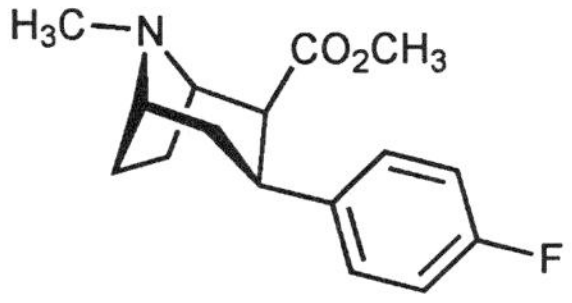

[³H]WIN 35,428

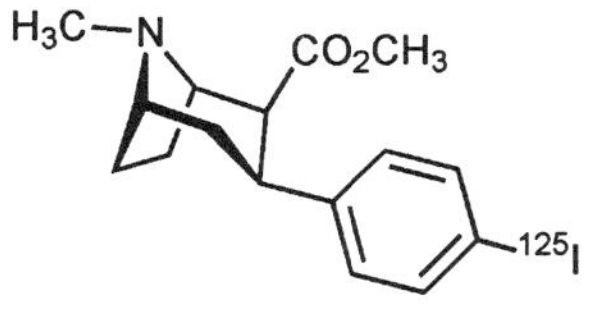

[¹²⁵ I] RTI-55

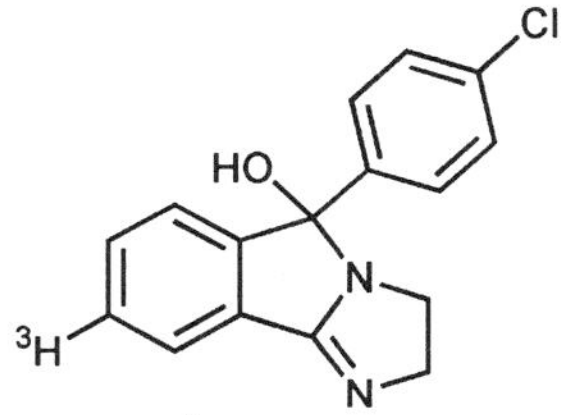

[³H]Mazindol

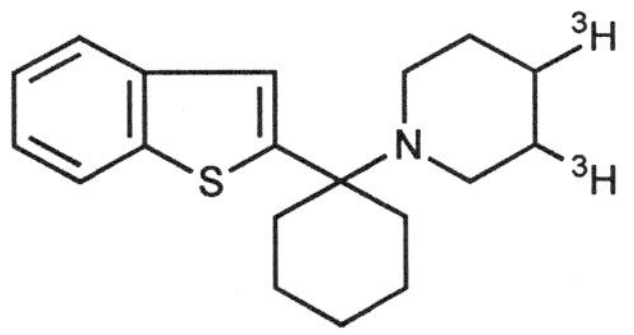

[³H]BTCP

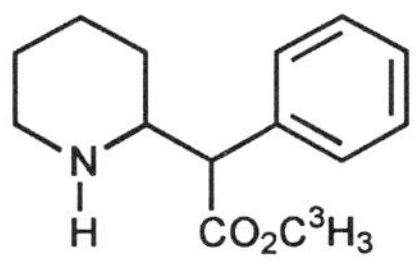

[³H]Methylphenidate

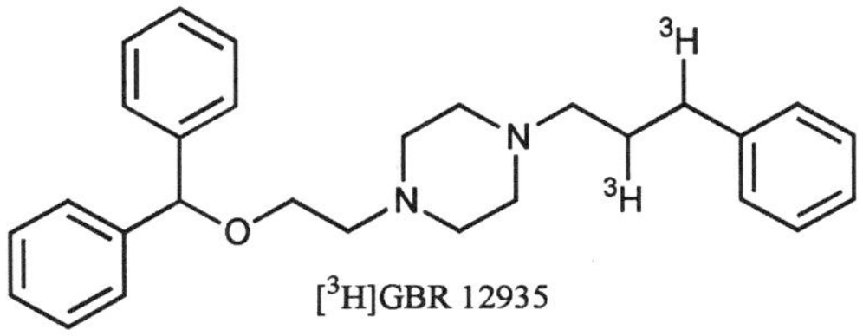

[³H]GBR 12935

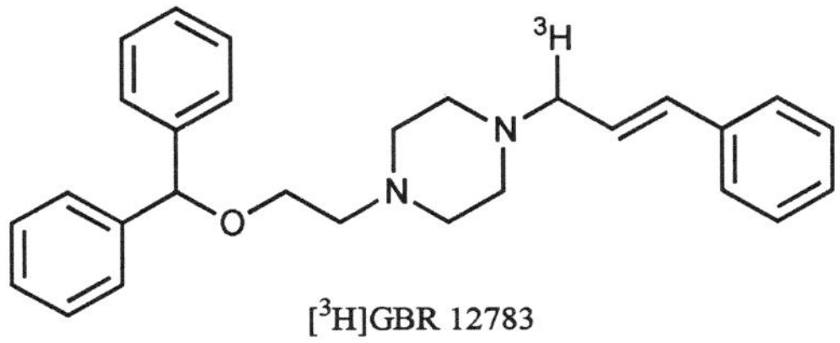

[³H]GBR 12783

Scheme 1. Radioligans used for inhibition of binding to the DAT.

tent. Several suitable radioligands for the DAT have been developed. Scheme 1 shows the radioligands that have been used for the SAR studies presented in this chapter. Some of the ligands bind only one site, while others, like [³H]cocaine, bind a high- and a low-affinity site. There are conflicting reports as to whether the various radioligands are binding to the same site(s) or different site(s), or possibly to the same site in structurally and/or biochemically different ways. Most SAR studies have been conducted using rat brain striata; however, some studies have used monkey caudate or putamen. In either case, the buffer used, as well as the incubation temperature and time, can have effects on the assay results. Nevertheless, there is evidence for mutually exclusive binding of the ligands *(15)*, and there is remarkable consistency in most of the reported SAR studies. Thus, unless there are gross differences, no distinctions will be made in this chapter as to the exact assay conditions.

Recently, a few SAR studies have been reported using cloned rat or human DATs. Due to the limited amount of data on cloned DATs and possible post-translational differences, these studies are not included.

2. CLASSES OF COMPOUNDS

2.1. Tropane Class

Tropane (8-methyl-8-azabicyclo[**3.2.1**]octane) provides the backbone for three distinct classes of compounds whose SARs for binding the DAT have been investigated. For the purpose of this chapter, these classes have been termed cocaine analogs, WIN 35,065–2 analogs, and benztropine analogs. Cocaine analogs are characterized by possession of a benzoyloxy or similar group at the 3-position of the tropane ring. The WIN 35,065–2 and benztropine analogs have an aryl and diphenylmethylether group, respectively, at this position.

2.1.1. Cocaine Analogs

Structurally, natural (–)-cocaine (**2.1.1a**) is [1*R*-(exo,exo)]-3-benzoyloxy-8-methyl-8-azabicyclo[3.2.1]octane-2-carboxylic acid methyl ester. Since this structure has three sites of asymmetry, there are eight possible isomers. Only (–)-cocaine (**2.1.1a**), with an IC_{50} value of 89–150 n*M*, has appreciable affinity for the DAT; other isomers are 60 to 600× weaker *(16)*. SAR studies of cocaine have included variations in the 3-benzoyloxy, 2-carbomethoxy, and N-methyl groups, and in the tropane ring. Each will be addressed separately. Since stereochemistry plays an important part in the potency of cocaine analogs, results from substrates that are mixtures of isomers are not included in this chapter.

2.1.1.1. MODIFICATION OF THE 3β-BENZOYLOXY GROUP

Substitution of the aromatic ring with electron-releasing, as well as electron-donating groups in the 3'- and 4'-positions (**2.1.1b**), results in a small to large loss of affinity for the DAT *(17–20)*. However, some 2'-substituted analogs—for example, **2.1.1c**, Z = OH, X = Y = H, show greater affinity for the DAT than cocaine *(21)*. Large losses in affinity result from changing the 3β-benzoyloxy group to a cinnamoyloxy (**2.1.2a**) or to a 1- or 2-naphthoyloxy (**2.1.2b–c**) group *(22,23)*. Replacement of the 3β-benzoyloxy group of cocaine with an *m*-nitrophenylcarbamoyloxy (*see* **2.1.3a**) group has little effect on binding potency; however, the *m*-amino, as well as *p*-nitro and *p*-amino analogs (**2.1.3b–d**) have reduced potency *(24)*.

2.1.1.2. MODIFICATION OF THE 2-POSITION

Early SAR studies showed that the presence and nature of a 2β-substituent contributes substantially to binding affinities *(16,17,25,26)*. The substantial reduction in binding potency observed on replacement of the 2β-carbomethoxy group in cocaine (**2.1.1a**) by a hydrogen atom (**2.1.4a**), or by a methylenehydroxy group (**2.1.4b**), and the partial restoration of binding potency with a methyleneacetoxy group (**2.1.4c**) suggests that one or more heteroatoms, and/or an electron-rich moiety, may be required for high potency and that some type of electrostatic interaction with the DAT may be involved *(25)*. However, since the 2β-vinyl analog **2.1.4d** possesses good affinity for the DAT, a heteroatom in the 2β-group does not appear to be an essential feature for high-potency binding *(27)*.

A number of 2β-alkyl and -aryl ester analogs of cocaine have been evaluated *(25,28–30)*. In general, only minor effects on binding affinity at the DAT were observed *(25)* even when the 2β-carbomethoxy methyl group in cocaine was replaced by very large groups. For example, compound **2.1.4e**, which possesses a 2β-[*p*-(ethylsuccinamido)phenyl]ethyl ester group, has an IC_{50} value of 86 n*M*, which is essentially identical to that of cocaine *(25)*. The most interesting finding in this group of compounds is that the phenyl and isopropyl esters **2.1.4f** and **2.1.4g**, respectively, are similar to cocaine in their affinity for the DAT, but show much lower potencies at serotonin (5-HT) and norepinephrine (NE) transporters *(28–30)*. In contrast, the benzyl, phenylethyl, and phenypropyl esters (**2.1.4j–l**) all show increased affinity for the 5-HT transporter *(30)*.

There are a number of 2β-substituted analogs of cocaine with wide variation in size, shape, hydrophobicity, and other physical parameters which show binding affinity equal to, or slightly less than, that of cocaine. However, there are only a few compounds in this class of analogs, such as the 2β-amide **2.1.4h** *(30)*, the 2β-carbethoxyvinyl analog **2.1.4i** *(31)*, and the 2β-isoxazole **2.1.5** *(31)*, that possess binding potencies exceeding that of cocaine.

2.1.1.3. MODIFICATION OF THE *N*-METHYL GROUP

The syntheses and binding affinities of a variety of *N*-nor-*N*-substituted cocaine analogs have been described *(26,32–34)*. Replacement of the *N*-

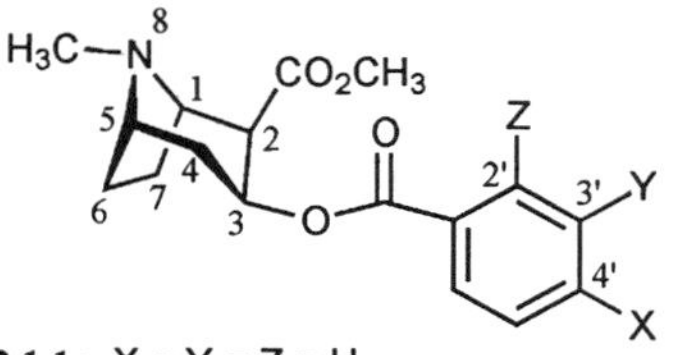

2.1.1a, X = Y = Z = H
  b, X or Y = substituent, Z = H
  c, Z = substituent or OH, X = Y = H

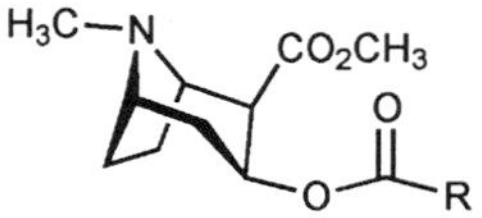

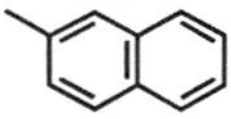

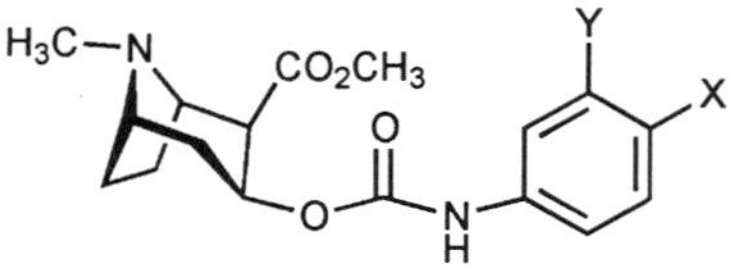

2.1.3a, X = H, Y = NO₂
  b, X = H, Y = NH₂
  c, X = NO₂, Y = H
  d, X = NH₂, Y = H

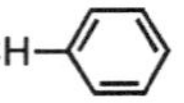

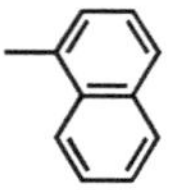

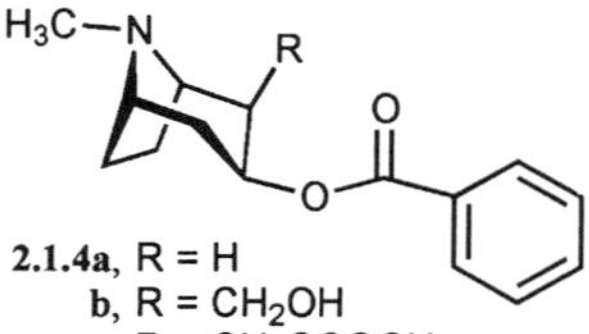

2.1.5

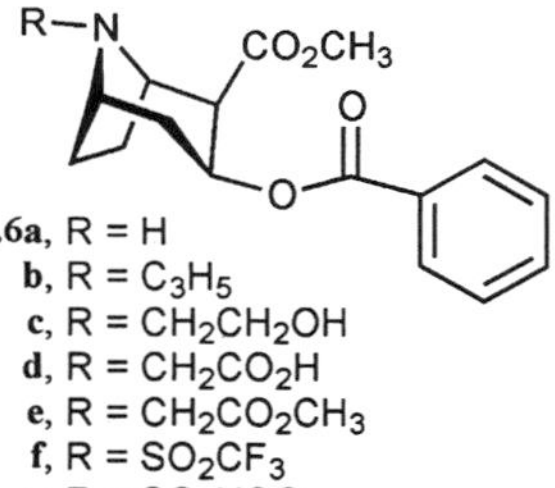

2.1.6a, R = H
  b, R = C₃H₅
  c, R = CH₂CH₂OH
  d, R = CH₂CO₂H
  e, R = CH₂CO₂CH₃
  f, R = SO₂CF₃
  g, R = SO₂NCO
  h, R = CH₂Ph
  i, R = COCH₃
  j, R = NO₂
  k, R = NH₂

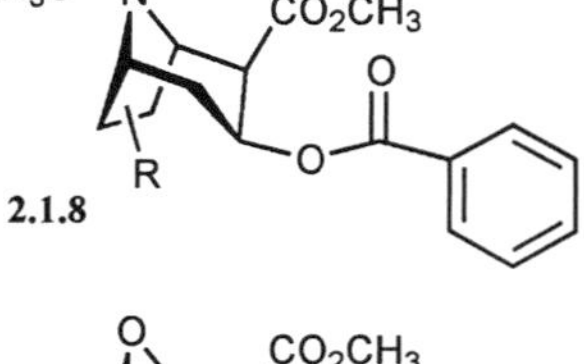

2.1.2a, R = CH=CH—

b, R =

c, R =

2.1.4a, R = H
  b, R = CH₂OH
  c, R = CH₂OCOCH₃
  d, R = CH = CH₂
  e, R = CO₂(CH₂)₂PhNHCO(CH₂)₂CO₂C₂H₅
  f, R = CO₂Ph
  g, R = CO₂CH(CH₃)₂
  h, R = CON(OCH₃)CH₃
  i, R = CH=CHCO₂Et
  j, R = CO₂CH₂Ph
  k, R = CO₂(CH₂)₂Ph
  l, R = CO₂(CH₂)₃Ph

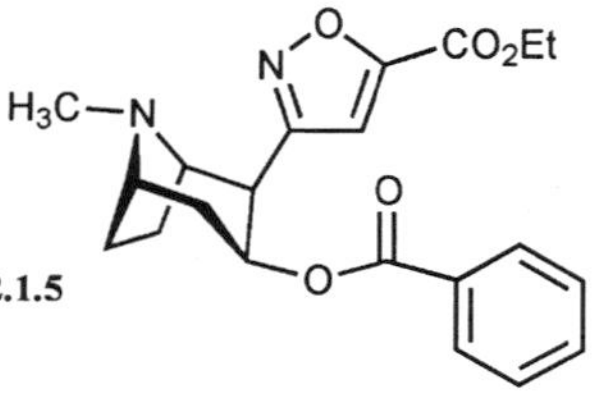

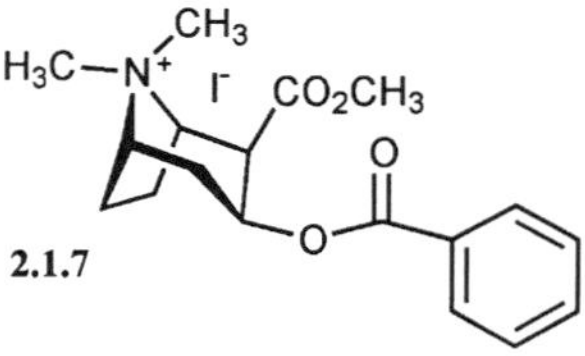

2.1.7

2.1.8

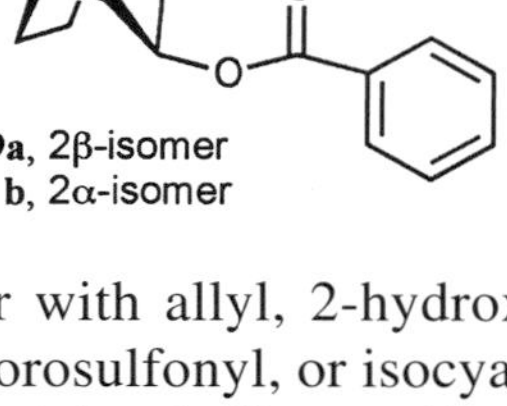

2.1.9a, 2β-isomer
  b, 2α-isomer

methyl group with a hydrogen atom, or with allyl, 2-hydroxyethyl, car-boxymethyl, carbomethoxymethyl, trifluorosulfonyl, or isocyanatosulfonyl groups (compounds **2.1.6a–g**) has only a minimal effect on binding potency. Changing the *N*-methyl group to an *N*-benzyl (**2.1.6h**) group results in a 7-fold loss in affinity. In contrast to these small effects, replacement of the *N*-

methyl group by an acetyl moiety to yield the amide **2.1.6i**, or addition of a methyl group to create the quaternary salt **2.1.7**, reduces binding potency by factors of 33 and 111, respectively. Replacement of the *N*-methyl with an electron-withdrawing nitro group or electron-releasing amino group (**2.1.6j** and **2.1.6k**, respectively) results in an even larger loss of affinity.

2.1.1.4. TROPANE RING MODIFICATION

The addition of substituents to the 6 or 7 (**2.1.8**) position of the tropane ring results in large losses of affinity for binding the DAT *(35)*. Replacement of the nortropane NH group with the bioisosteric oxygen to give structure **2.1.9a** also results in an 8-fold loss in affinity for the DAT *(36)*. In contrast to the cocaine series, the pseudoisomer **2.1.9b** possesses essentially the same affinity at the DAT as **2.1.9a**. This suggests that these analogs may bind the DAT in a mode different from that of cocaine.

2.1.2. WIN 35,065-2 Analogs

The structure of WIN 35,065-2 (**2.1.10**) differs from that of cocaine (**2.1.1a**) by having an aromatic ring (with β-stereochemistry) connected directly to the 3-position of the tropane ring. Analogs in this series possess potencies to bind the DAT that are more than 100× that of cocaine and its analogs. Like cocaine (**2.1.1a**), WIN 35,065-2 (**2.1.10**) has three sites of asymmetry and eight possible isomers. The 2α-isomer (WIN 35,140, **2.1.11**) and the enantiomer (WIN 35,065-3, **2.1.12**) both have a much lower affinity for the DAT than WIN 35,065-2 (**2.1.10**), showing that the orientation at the 2-position and the absolute stereochemistry are important contributors to high potency *(17,26,37)*. Interestingly, the 2β,3α-isomer (**2.1.13a**) of WIN 35,065-2 and 4'-substituted analogs **2.1.13b–d**, which were shown to exist in the boat conformation, all possess affinity at the DAT comparable to that of the corresponding 2β,3β-isomers *(38)*. Binding properties for the other isomers have not been reported. SAR studies of WIN 35,065-2 analogs follow a pattern much like that described for cocaine and include variation in the 3β-phenyl, 2β-carbomethoxy, and *N*-methyl groups. In **Subheading 2.1.2.1.**, the effect of changes in the 3β-phenyl ring is addressed. For the most part, the SAR studies, addressing changes in the 2β-carbomethoxy and *N*-methyl group have been carried out on 3β-(substituted phenyl) WIN 35,065-2 analogs. For the purpose of this chapter, we have assumed that the SAR observed for changes in the 2β-carbomethoxy and *N*-methyl groups are not significantly affected by the substituent on the 3β-phenyl group.

2.1.2.1. MODIFICATION OF THE 3β-PHENYL GROUP

The parent compound in this series, WIN 35,065-2, demonstrates a five-fold higher affinity for the DAT than cocaine *(39)*. SAR studies from several laboratories have shown that, unlike for cocaine, binding affinity in this

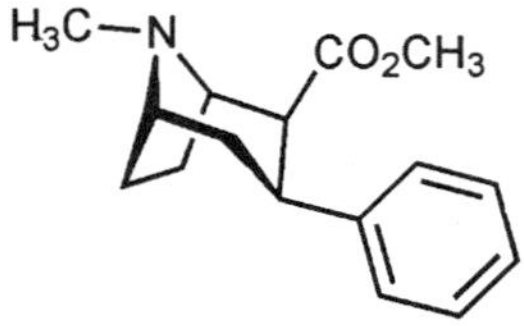

2.1.10, WIN 35,065-2

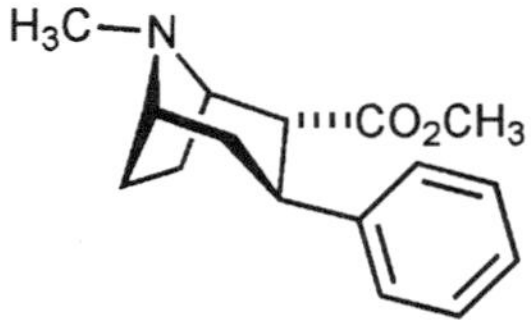

2.1.11, WIN 35,140

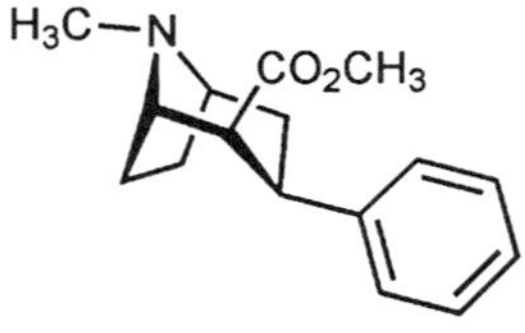

2.1.12, WIN 35,065-3

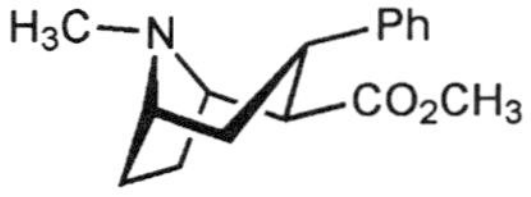

2.1.13a, X = H
    b, X = F
    c, X = Cl
    d, X = I

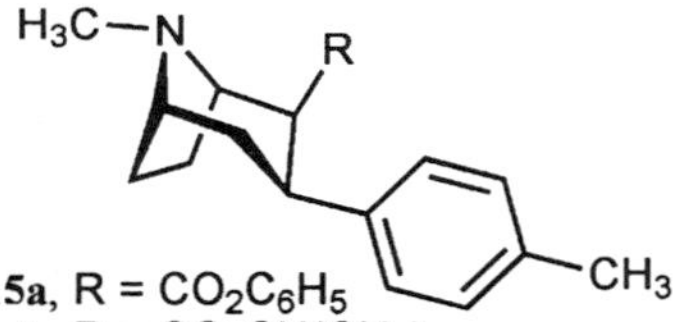

2.1.15a, R = $CO_2C_6H_5$
    b, R = $CO_2CH(CH_3)_2$
    c, R = Ph

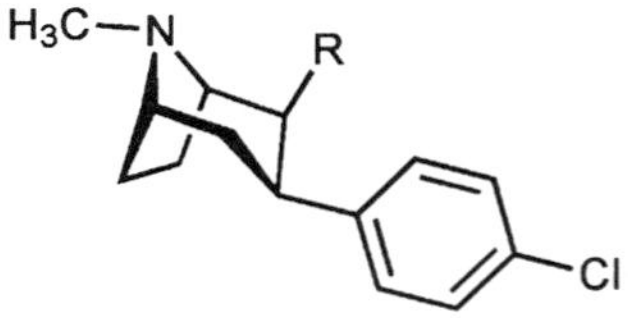

2.1.16a, R = CO_2Ph
    b, R = $CO_2CH(CH_3)_2$
    c, R = $CO_2(CH_2)_2C_6H_4$-4-NH_2

    d, R = CON⟨O⟩

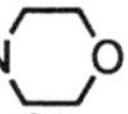

    e, R = CON⟨⟩

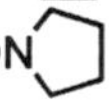

    f, R = $CH=CH_2$
    g, R = $(CH_2)_2CH_3$
    h, R = CH=CHCl
    i, R = CH=CHPh
    j, R = CH_2CH_2Ph

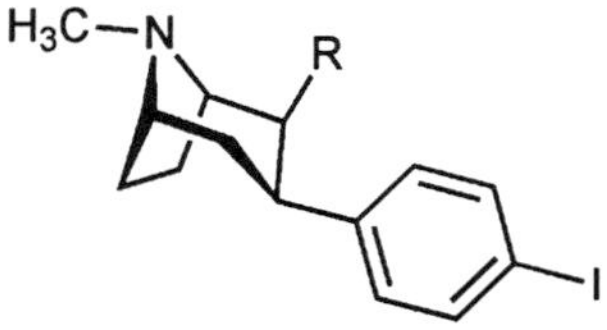

2.1.17a, R = CO_2Ph (RTI-122)
    b, R = $CO_2CH(CH_3)_2$ (RTI-121)

    c, R = CON⟨⟩

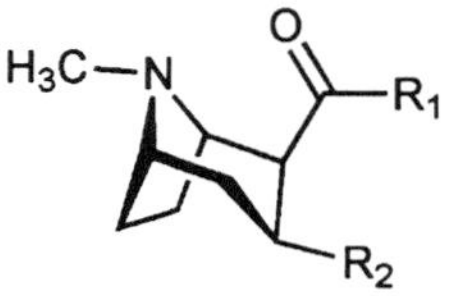

2.1.18a, R_1 = C_2H_5, R_2 = 2-naphthyl
    b, R_1 = CH_3, R_2 = 2-naphthyl

series is dependent upon the substituents and their positions on the 3β-phenyl ring *(17,23,26,29,37,40–45)*. The results from some of these studies

(Table 2.1) reveal that compounds possessing many different types of substituents have IC_{50} values less than 12 nM. The 4'-methyl, 4'-chloro, 4'-bromo, and 4'-iodo (**2.1.14a–d**) and the 4'-amino-3'-iodo, 3',4'-dichloro, 4'-chloro-3'-methyl, 4'-vinyl, and 4'-acetylenic analogs (**2.1.14a–i**) possess IC_{50} values of less than 2 nM, as does the 3β-naphthyl analog **2.1.14j** *(22)*. Strikingly, **2.1.14l**, with an electron-withdrawing 4'-nitro substituent, and **2.1.14m–o**, which possess a 4'-amino, a 4'-hydroxy, and a 4'-methoxy group, respectively, all have IC_{50} values in the same range. Compounds **2.1.14dd–kk** with large substituents such as 4'-acetylamino, 4'-propionylamino, 4'-ethoxycarbonylamino, and 4'-trimethylstannyl possess relatively low affinity for the DAT. However, even these compounds have IC_{50} values in the range of cocaine's. It is surprising that the potency of the 4'-ethyl analog **2.1.14hh**, with an IC_{50} value of 55 nM, is so different from that of the 4'-methyl derivative **2.1.14a** *(42)*. Note, however, that the 4'-vinyl (**2.1.14h**) and the 4'-acetylenic (**2.1.14i**) analogs both have good affinity for the DAT. Substituents in the 3'-position show somewhat inconsistent results. The 3'-chloro, 3'-bromo, and 3'-iodo analogs (**2.1.14p**, **2.1.14q**, and **2.1.14aa**) have lower affinity than the corresponding 4'-substituted analogs, **2.1.14b–d**, the 3'- and 4'-fluoro analogs (**2.1.14x** and **2.1.14y**, respectively) fall into the same IC_{50} range. It is interesting to note that addition of a 3'-chloro or 3'-methyl group to analog **2.1.14b** to give the 3',4'-dichloro and 4'-chloro-3'-methyl analogs, **2.1.14f** and **2.1.14g**, respectively, has only a minor effect on potency. By contrast, the addition of a 3'-methyl group to the 4'-fluoro analog **2.1.14y** or of a 3'-iodo substituent to the 4'-amino analog **2.1.14m** to give **2.1.14s** and **2.1.14e**, respectively, increases potency relative to the 4'-substituted analogs.

Quantitative structure-activity relationship (QSAR) and comparative molecular-field analysis (CoMFA) models were initially reported for 12 *(40)*, and then for 25 *(44)* of the compounds listed in Table 2.1. The classical QSAR models suggest that distribution properties (hydrophobicity) are important contributors to binding at the DAT. The CoMFA models show that some steric bulk extending from and above the 4'-position contributes to enhanced potency, while excessive bulk leads to reduced potency. In addition, the model suggests that electrostatic forces account for approximately one-quarter of the binding affinity, and thus may make a significant contribution to potency.

2.1.2.2. MODIFICATION OF THE 2-POSITION

Since WIN 35,065-2 (**2.1.10**) is considerably more potent than its 2α-isomer WIN 35,140 (**2.1.11**), it appears that, like for cocaine (**2.1.1a**), a 2β-substituent is required for high affinity at the DAT *(17,26,37)*. A number of

DAT Binding Potency of 3β-(Substituted Phenyl)tropane-2β-carboxylic Acid Methyl Esters

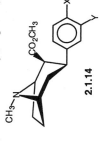

2.1.14

IC_{50} Values

Compound	< 2 nM X, Y	Compound	2 to 12 nM X, Y	Compound	12 to 30 nM X, Y	Compound	30 to 200 nM X, Y
a	CH_3, H	k	N_3, H	**2.1.10**	H, H	dd	$NHCOCH_3$, H
b	Cl, H	l	NO_2, H	x	H, F	ee	$NHCOC_2H_5$, H
c	Br, H	m	NH_2, H	y	F, H	ff	$NHCO_2C_2H_5$, H
d	I, H	n	OH, H	z	CF_3, H	gg	$Sn(CH_3)_3$, H
e	NH_2, I	o	OCH_3, H	aa	H, I	hh	CH_2CH_3, H
f	Cl, Cl	p	H, Cl	bb	$-C(=CH_2)CH_3$, H	ii	C_3H_7, H
g	Cl, CH_3	q	H, Br	cc	c-$CH=CHCH_3$, H	jj	$CH(CH_3)_2$, H
h	$-CH=CH_2$, H	r	N_3, I			kk	$CH_2CH=CH_2$, H
i	$-C≡CH$, H	s	F, CH_3				
j	(C6H5 ring, attached)	t	NH_2, Br				
		u	t-$CH_2=CHCH_3$, H				
		v	$-C≡CCH_3$, H				
		w	C_6H_5, H				

2β-alkyl and aryl ester analogs of **2.1.14a**, **2.1.14b**, and **2.1.14c** have been studied *(28–30)*. As in the cocaine series, the phenyl and isopropyl esters (**2.1.15a**, **2.1.15b**, **2.1.16a**, **2.1.16b**, **2.1.17a**, and **2.1.17b**) are highly potent and more selective for the DAT relative to binding at the 5-HT and NE transporters *(29,30,46)*. Also, similar to the cocaine series, replacement of the methyl group of **2.1.14b** with a large group such as a 2-(4'-aminophenyl)ethyl group (**2.1.16c**) has only small effects on the IC_{50} value for binding at the DAT *(47)*.

Results obtained with amide analogs of **2.1.14a**, **2.1.14b**, and **2.1.14d** *(30,46)* demonstrate tertiary amides to be more potent than primary and secondary amide analogs. In particular, 3β-(4'-chlorophenyl)tropane-2β-*N*-morpholinocarboxamide (**2.1.16d**), as well as 3β-(4'-chlorophenyl)- and 3β-(4'-iodophenyl)-tropane-2β-*N*-pyrrolidinocarboxamides (**2.1.16e** and **2.1.17c**), possess high affinity and selectivity for the DAT.

The binding affinity of the 3β-aryl-, 2β-acetyl- and 2β-propanoyl tropanes has been reported *(48–51)*. The most potent compounds in this series are 3β-(2-naphthyl)-2β-acetyltropane (**2.1.18a**), and 2β-(2'-naphthyl)-2β-propanoyltropane (**2.1.18b**).

Among other 2-substituted analogs of **2.1.14b**, compounds with the 2β-substituent equal to vinyl (**2.1.16f**), propyl (**2.1.16g**), chlorovinyl (**2.1.16h**), phenylvinyl (**2.1.16i**), and phenylethyl (**2.1.16j**) show strong affinity for the DAT. In addition, the 2β-propyl (**2.1.19a**), carbomethoxyvinyl (**2.1.19b**), hydroxy-methylvinyl (**2.1.19c**), 2-carbomethoxyethyl (**2.1.19d**), and 3-hydroxypropyl (**2.1.19e**) all exhibit potencies similar to their parent 2β-carbomethoxy compound WIN 35,065-2 *(27,52,53)*. The SAR of this set of compounds has been interpreted to suggest a hydrophobic pocket for the 2β-substituent *(27,52,53)*. However, no correlation with the hydrophobicity of the various 2β-substituents was reported *(27,52,53)*.

In another study *(54,55)*, the DAT affinity of the 3β-(4'-chlorophenyl)-2β-heterocyclic tropanes (**2.1.20a–i**) along with reference compounds was evaluated to determine possible relationships between the molecular electrostatic (MEP), hydrophobic (calculated log P), and steric (substituent volume) properties of the 2β-heterocyclic analogs **2.1.20a–i**. A good correlation was found between the relative MEP minima adjacent to one of the heteroatoms and the affinities at the DAT. Neither calculated log P nor the substituent volume provided a good correlation. This molecular modeling study is consistent with a predominantly electrostatic interaction for these 2β-heterocyclic analogs. A possible explanation for the wide range of substituents that can be accommodated at the C-2 position of the 3-phenyltropanes is the existence of more than one binding mode for groups

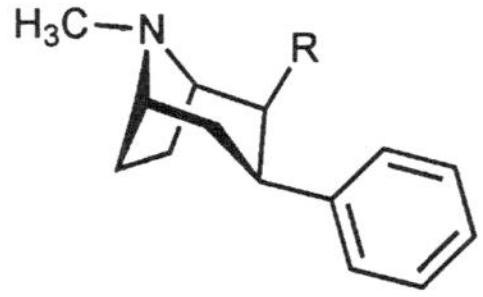

2.1.19a, R = (CH₂)₂CH₃
b, R = CH=CHCO₂CH₃
c, R = CH=CHCH₂OH
d, R = CH₂CH₂CO₂CH₃
e, R = CH₂CH₂CH₂OH

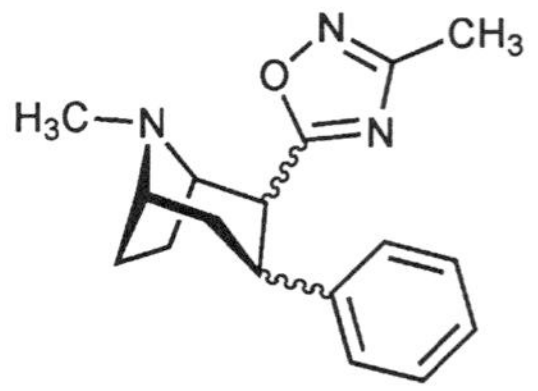

2.1.21a, 2β,3β-isomer
b, 2α,3β-isomer
c, 2β,3α-isomer
d, 2α,3α-isomer

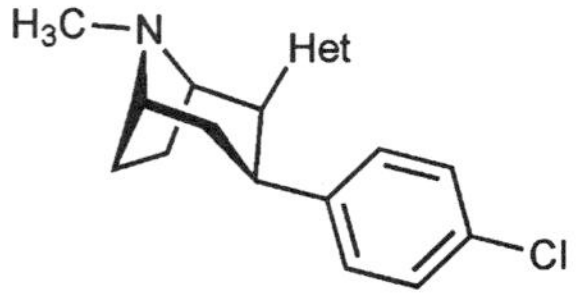

2.1.20

	Het		Het
a	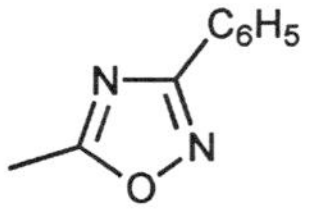	f	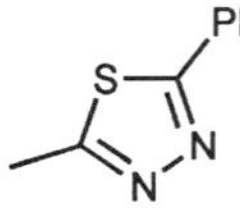
b	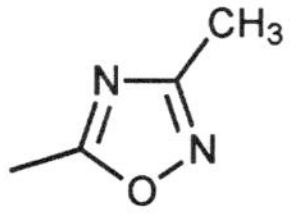	g	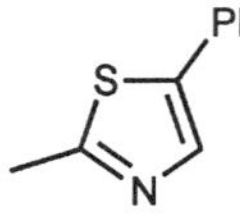
c	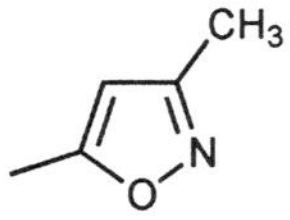	h	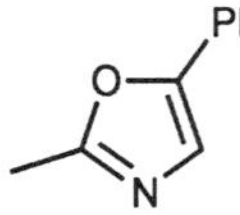
d	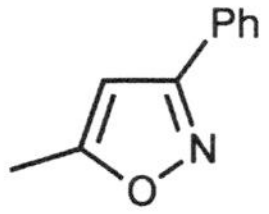	i	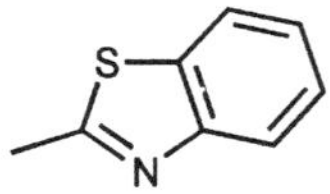
e	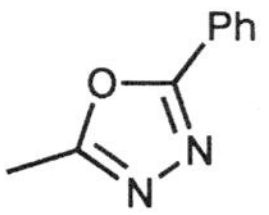		

at this position. The 3β-(4-chlorophenyl)-2β-heterocyclic tropanes **2.1.20c**, **2.1.20d**, **2.1.20g**, and **2.1.20i** have high affinity and selectivity for the DAT.

Compound **2.1.20d** (RTI-177) is in preclinical development as a pharmaco-therapy for cocaine addiction.

The binding properties of all four stereoisomers of the (1*R*,5*S*)-3-phenyl-2-(3'-methyl-1',2',4'-oxadiazol-5'-yl)tropane (**2.1.21a–d**) bioisosteres of WIN 35,065-2 (**2.1.10**) have been reported *(56)*. As expected, the 2β,3β-isomer, which has the WIN 35,065-2 (**2.1.10**) stereochemistry, possesses the greatest potency for the DAT. However, the IC$_{50}$ values of the 2β,3α- and 2α,3α-isomers are only slightly higher than those of the 2β,3β-isomer.

2.1.2.3. MODIFICATION OF THE *N*-METHYL GROUP

A number of *N*-nor-*N*-substituted 3-phenyltropane analogs have been synthesized—many as potential positron emission tomography (PET) or single photon emission computed tomography (SPECT) brain imaging agents, or as intermediates to prepare the imaging agents. Removal of the *N*-methyl group from the 2β-carbomethoxy or 2β-propanoyl tropanes to give the corresponding *N*-nor analogs (**2.1.22a–j** and **2.1.23a–b**) has little effect on potency at the DAT, but increases potency at the 5-HT transporter *(17,41,45,51,57–59)*. Replacement of the *N*-methyl group of **2.1.14y** with allyl, propyl, phenylethyl, phenylpropyl, or iodopropenyl groups (**2.1.24a–e**) or of the *N*-methyl group of **2.1.14d** with ethyl, propyl, butyl, fluoroethyl, or fluoropropyl groups (**2.1.24f–j**) has minor effects on binding potency *(22,57,58,60,61)*. Even compounds with the large *N*-phthalimidoalkyl group (**2.1.24k–l**) show good affinity for the DAT *(62)*.

2.1.2.4. TROPANE RING MODIFICATION

Similar to the cocaine analogs, the addition of substituents to the 6- or 7-positions (**2.1.25**) in this series results in large losses of affinity for the DAT *(63)*. A number of bioisosteric analogs of nortropane was studied where the NH at the tropane 8-position was replaced by an oxygen or methylene group *(64,65)*. Surprisingly, some of the analogs (**2.1.26a–b**) showed only a five-to 10-fold loss in affinity relative to the parent nortropane analog. Some compounds, which possess an unsaturated substituent at the 2-position, showed reasonable affinity for the DAT. For example, the 3-naphthyl analog **2.1.27** (RTI-328) has an IC$_{50}$ value of 5.7 nm at the DAT *(22)*.

Introduction of a 2- or 3-carbon bridge between the 8-nitrogen and the 2-position (**2.1.28a–b**), or a 3-carbon bridge between the 8-nitrogen and the 3-position (**2.1.29**), results in a large loss of affinity relative to the parent 2β-carbomethoxy analog *(22,66,67)*. Compounds **2.1.30** and **2.1.31**, which possess a 2-carbon bridge between the 8-nitrogen and the 6- and 7-positions on the tropane ring, show only an eight- to ninefold loss in affinity relative to the parent 3β-(4'-methylphenyl)-2β-butyltropane.

2.1.22a, X = Y = H (WIN 35,981)
b, X = F, Y = H
c, X = Cl, Y = H
d, X = I, Y = H
e, X = C_2H_5, Y = H
f, X = Y = Cl
g, X = CH=CH_2, Y = H
h, X = C≡CH, Y = H
i, X = C≡CCH_3, Y = H
j, X = C(=CH_2)CH_3, Y = H

2.1.23a, X = CH=CH_2
b, X = C(=CH_2)CH_3

2.1.24a, X = F, R = CH_2CH=CH_2
b, X = F, R = $CH_2CH_2CH_3$
c, X = F, R = $(CH_2)_3$Ph
d, X = F, R = $(CH_2)_5$Ph
e, X = F, R = CH_2CH=CHI
f, X = I, R = CH_2CH_3
g, X = I, R = $CH_2CH_2CH_3$
h, X = I, R = $CH_2(CH_2)_2CH_3$
i, X = I, R = CH_2CH_2F
j, X = I, R = $CH_2CH_2CH_2$F
k, X = I, R = $(CH_2)_4$Pht
l, X = I, R = $(CH_2)_8$Pht

Larger, 3-phenyl-9-azabicyclo[3.3.1]nonane **2.1.32**, and smaller, 3-phenyl-7-azabicyclo[2.2.1]heptane (**2.1.33**), analogs of WIN 35,065-2 show large reductions in affinity at the DAT relative to the parent compound.

2.1.3. Benztropine Analogs

Benztropine (**2.1.34a**) binds the DAT with potency of about one-half to one-third that of cocaine *(68–71)*. Since the structure of **2.1.34a** contains a tropane and a diphenylmethoxy moiety, it resembles both cocaine and the 1,4-

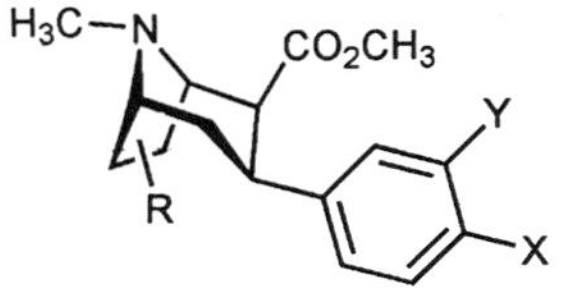

2.1.25

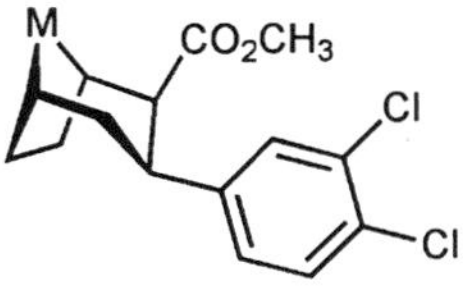

2.1.26a, M = O
b, M = CH$_2$

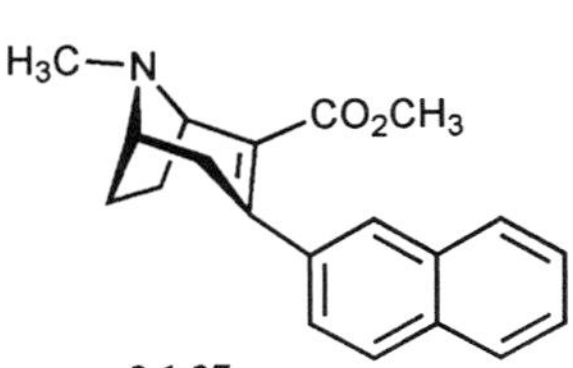

2.1.27

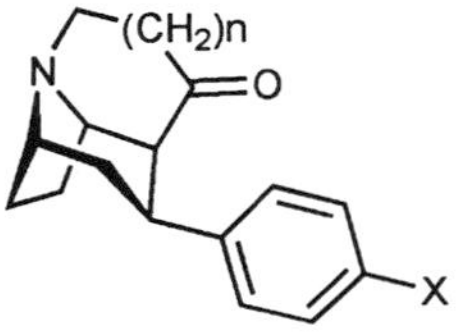

2.1.28a, n = 1, X = CH$_3$
b, n = 2, X = Cl

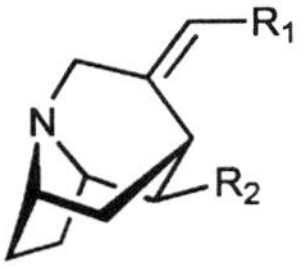

2.1.29, R$_1$ = 2-thienyl
R$_2$ = CO$_2$CH$_3$

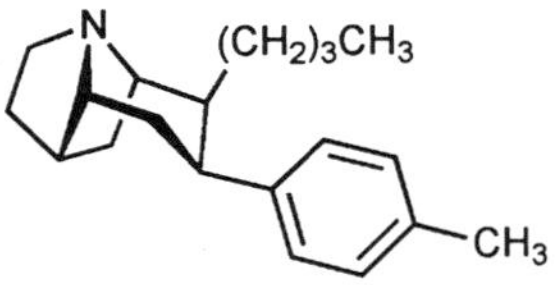

2.1.30

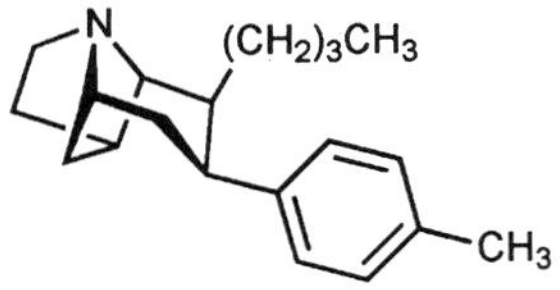

2.1.31

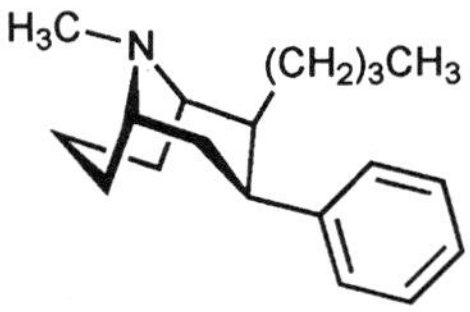

2.1.32

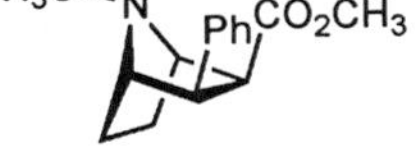

2.1.33

dialkylpiperazine (GBR) series of dopamine inhibitors (*see* **Subheading 2.2.**). Photoaffinity labeling of the DAT with the benztropine analog [^{123}I] GA II **2.1.35a** ([^{125}I]N-[n-butyl-4-(4'''-azido-3'''-iodophenyl)]-4',4''-difluoro-3α-(diphenylmethoxy)tropane) has shown that it interacts with the same regions as the dialkylpiperazine class of compounds (*72*).

2.1.3.1. MODIFICATION TO THE PHENYL RINGS

The binding affinity of a number of 4'- and 4',4"-substituted analogs of **2.1.34a** has been reported *(68,69,73,74)*. The 4',4"-difluoro analog (**2.1.34b**), which is 10× more potent than **2.1.34a**, is the most potent analog. The ring substituted analogs **2.1.34b–m** have binding potencies at least 3× greater than **2.1.34a**. A CoMFA study has been developed to predict DAT potencies of this class of compounds at the DAT *(74)*. While the addition of one or two halogens to the aromatic system either has no effect or increases potency relative to the parent **2.1.34a**, addition of most other groups decreases potency. In particular, *bis*(4-methoxy) (**2.1.34n**), 4-*t*-butyl (**2.1.34o**), and 2-amino (**2.1.34p**) analogs are 10-fold less potent than the parent compound **2.1.34a**. Overall, the effects suggest that the steric bulk and the orientation of the phenyl rings affect potency. All of the analogs, including **2.1.34a**, show weak affinity at the 5-HT and NE transporters. Benztropine (**2.1.34a**) shows high affinity for both the m_1 and m_2 muscarinic receptors. The analogs **2.1.34b–h** show potency at the m_1 receptor similar to those at the DAT, but are weaker at the m_2 receptor as compared to **2.1.34a**.

2.1.3.2. ADDITIONS AND MODIFICATIONS OF THE TROPANE RING

The addition of a 2-carbomethoxy group to the benztropine structure creates three centers of asymmetry and, like for cocaine, eight possible diastereoisomers. An SAR study of the eight possible diastereoisomers of 2-carbomethoxy-4',4"-difluorobenztropine reveals that only the *S*-isomer **2.1.36a** with potency 28.6× greater than **2.1.34a** possesses high affinity for the DAT *(70)*. An SAR study of other 4',4"-disubstituted analogs of **2.1.36a** reveals that analogs **2.1.36b–i** have potencies at least 3× greater than **2.1.34a**.

2.1.3.3. MODIFICATION OF THE N-METHYL GROUP

A number of *N*-substituted analogs of **2.1.34b** have been prepared and evaluated *(75)*. Replacement of the *N*-methyl group with a hydrogen (**2.1.35b**) or with a large 4'-phenylbutyl group (**2.1.35c**) has very little effect on DAT affinity. In contrast, replacement of the *N*-methyl group with the acetyl group **2.1.35d** results in a large loss of affinity for the DAT. A CoMFA model was derived, and was used to correlate DAT affinity for a number of *N*-substituted analogs of **2.1.34b** *(76)*. Somewhat similar to the finding with **2.1.34b**, replacement of the N-methyl group of **2.1.36a** with a hydrogen (**2.1.37a**) or by a large 3'-phenylpropyl group (**2.1.37b**), results in only a 2-fold loss of affinity at the DAT *(71)*. On the other hand, 10- to 20-fold decreases in potency result from replacement of the *N*-methyl group by a benzyl group when the benzhydryl substituents are 3'-Cl, 4'-Cl, or 4',4"-*bis*Cl (**2.1.38a**, **2.1.38b**, and **2.1.38c**) *(77)*.

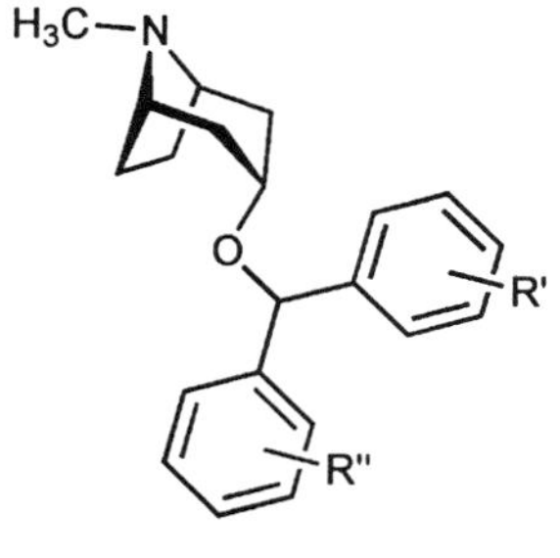

2.1.34a, R' = R" = H
 b, R' = R" = 4-F
 c, R' = R" = 4-Cl
 d, R' = 4-F, R" = H
 e, R' = 4-Cl, R" = H
 f, R' = 4-Br, R" = H
 g, R' = 4-F, R" = 3-Cl
 h, R' = 4-F, R" = 3,4-Cl$_2$
 i, R' = H, R" = 3,4-F$_2$
 j, R' = H, R" = 3,4-Cl$_2$
 k, R' = 4-Br, R" = 3-Br
 l, R' = 4-F, R" = 3-Cl
 m, R' = 4-F, R" = 3-CH$_3$
 n, R' = R" = 4-OCH$_3$
 o, R' = 4-t-Bu, R"=H
 p, R' = 2-NH$_2$, R" = H

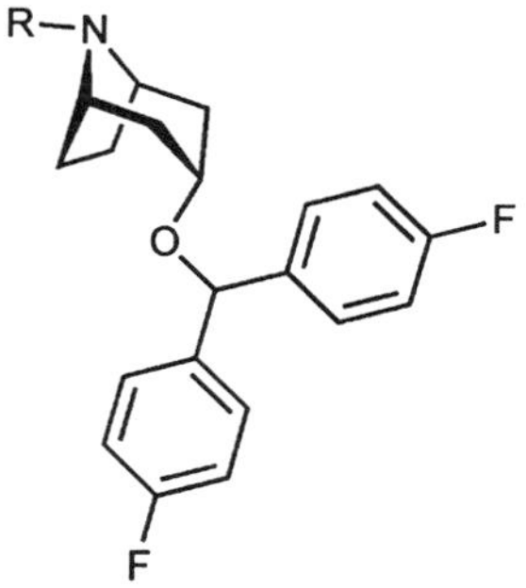

2.1.35a, R = (CH$_2$)$_4$(4-N$_3$-3-^{125}IPh)
 b, R = H
 c, R = (CH$_2$)$_4$Ph
 d, R = COCH$_3$

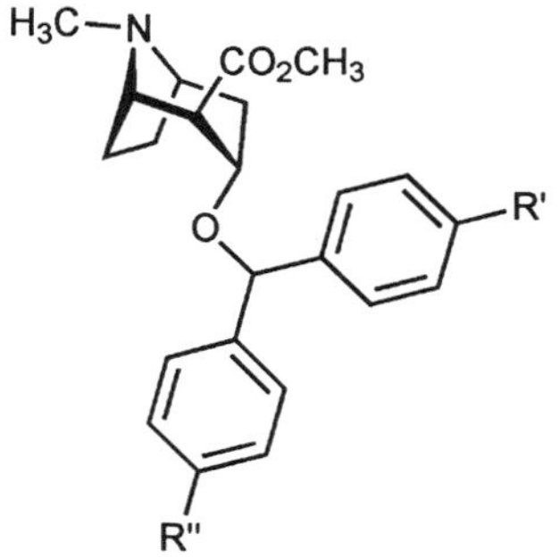

2.1.36a, R' = R" = F
 b, R' = F, R" = H
 c, R' = Cl, R" = H
 d, R' = CH$_3$, R" = H
 e, R' = I, R" = H
 f, R' = R" = H
 g, R' = R" = Cl
 h, R' = Br, R" = H
 i, R' = R" = Br

2.2. 1,4-Dialkylpiperazine Class

The 1,4-dialk(en)ylpiperazines had been found to be potent and selective inhibitors of DA transport *(78)* and to label the DAT *(79–81)*. However, none of the compounds in this class are as potent as those in the phenyltropane class (*see* **Subheading 2.1.**). In addition, although the binding of cocaine and analogs in the tropane class (*see* **Subheading 2.1.**) has two components, the compounds in the 1,4-dialk(en)ylpiperazine class label only one component *(37)*. Furthermore, ligands in the 1,4-dialk (en)ylpiperazine class do not fully displace ligands in the tropane class from

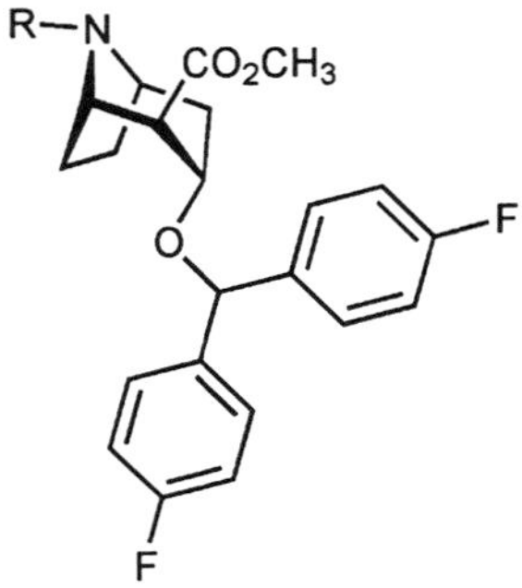

2.1.37a, R = H
b, R = (CH$_2$)$_3$Ph

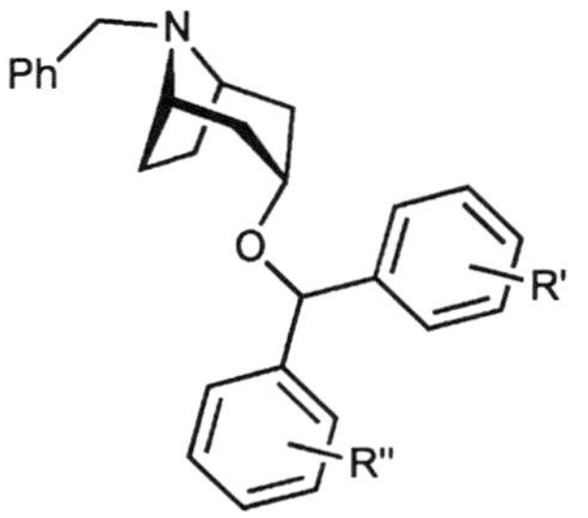

2.1.38a, R' = 3-Cl, R'' = H
b, R' = 4-Cl, R'' = H
c, R' = R'' = 4-Cl

binding at the DAT and vice versa *(37,82)*, suggesting that the binding sites for these two classes of compounds may be discrete. Photoaffinity labeling results support this conclusion. Thus, the 1,4-dialkylpiperazine analog DEEP ([^{125}I]1-2-(diphenylmethoxy) ethyl-4-[2-(4-azido-3-iodophenyl)ethyl] piperazine) labels the DAT in a region including the transmembrane domains (TMDs) 1 and 2 and the 6-amino linker *(83,84)* while the cocaine analog [^{125}I]4'-azido-3'-iodophenylethyl ester RTI-82 labels a region of TMDs 4–7 *(83,84)*. Finally, in addition to binding the DAT, 1,4-dialk(en)ylpiperazines bind a "piperazine-acceptor site" *(81)*.

Structural changes have been made in the benzhydryl ether moiety, in the ethylene tether between the benzhydryl ether oxygen and the proximal nitrogen, in the 1,2-diaminoethylene fragment, and in the 4-substituent.

2.2.1. Modification of the Benzhydryl Ether Moiety

Many modifications to the benzhydryl group have been carried out. Greatest potencies to inhibit DA uptake are achieved with the parent compound (e.g., GBR 12783, **2.2.1a**) or with *bis*(4-fluoro)-substitution (e.g., GBR 12789, **2.2.1e**), which usually serves to enhance potency *(78,85–94)*. A surprisingly large (2 orders of magnitude) increase in potency has been found to be associated with *bis*(4-fluoro)-substitution of the [3.2.1]bicyclic analog **2.2.2** (**2.2.2b** vs **2.2.2a**) *(87)*; effects of other aromatic substituents in this system have not been investigated. Symmetrically or unsymmetrically, 3- and 4-substituted analogs of GBR 12783 exhibit small decreases in potency to inhibit DA uptake *(78,95,96)*. Larger substituent effects on potency to inhibit binding at DAT are seen in analogs in which the proximal amino group was replaced by a methine, particularly when the distal amino group bears a phenethyl substituent (**2.2.3a**) *(90)*. *Bis*(4-methyl) (**2.2.3b**) *(90)*, *bis*(4-bromo) *(90)*, (**2.2.3c**) *(90)*, and *bis*(4-chloro) (**2.2.3d**) *(90)* substitution leads to 10- to 40-fold decreases in binding potency at DAT. In this series of compounds, replacement of one of the benzhydryl phenyl groups by a bioisosteric 2-thiophene (**2.2.4**) *(92)* has no effect on potency to inhibit binding at DAT, but replacement by a methyl group to give the α-methylbenzyl analog **2.2.5a** *(97)* decreases binding potency by 3 orders of magnitude. This observation is consistent with the proposal that both phenyl rings are involved in binding *(98)*.

Large decreases in potency to inhibit DA uptake (100- to 400-fold) are observed by introduction of a 3-isothiocyanato group (**2.2.1b**) *(96,98)*, of a 4-maleimide group (**2.2.1c**) *(96,98)*, or of 2,4-dichloro-4'-fluoro substituents (**2.2.1d**) *(78)* onto a GBR.12783 skeleton (**2.2.1a**). Binding inhibition data obtained for the isothiocyanato- (**2.2.1b**) and maleimide- (**2.2.1c**) substituted compounds have been interpreted to indicate that the domain recognized by the benzhydryl portion of the ligand may be a small cleft which is sensitive to steric effects *(98)*. Whether these data relate to inhibition of DA uptake is not clear, since the potencies of 3- and 4-isothiocyanato-substituted analogs to inhibit [³H]methylphenidate binding, are identical, but the potency of the 3-isothiocyanato analog to inhibit DA uptake is fourfold greater than that of the 4-isothiocyanato analog.

2.2.2. Modification of the 1,2-Diaminoethylene Group

Introduction of methyl groups, *trans* to each other, at positions 2 and 5 of the six-membered ring containing the 1,2-diaminoethylene group (piperazine ring) of the parent compound GBR 12783 (**2.2.1a**) or of the *bis*(4-fluoro) compound GBR 12789 (**2.2.1e**) affords the chiral compounds **2.2.6a** and **2.2.6b** *(85)*. Analogous modification of the piperazine ring of GBR 12935 (**2.2.1f**), GBR 12909 (**2.2.1g**), and GBR 11513 (**2.2.1h**) yields the chiral compounds **2.2.6c** *(85)*, **2.2.6d** *(88)*, and **2.2.6e** *(88)*. While no improvements in potency are obtained by this modification, the enantioselectivity observed for the 2*S*,5*R* enantiomers suggests the operation of a receptor-mediated mechanism for the inhibition of DA reuptake by this class of compounds *(99)*. This is also supported by the recent results obtained for analogs in which the 1,2-diaminoethylene fragment is contained in bicyclic ring systems. In particular, the similar potencies to inhibit DA reuptake observed for the 2*R*,5*S*-enantiomer of **2.2.6d** *(88)* and for the [2.2.1]bicyclic analog **2.2.7b** *(87)* support the requirement of 2*S*,5*R* configuration for high potency.

Imposition of conformational constraints on the piperazine ring by introduction of an ethylene bridge to create the [3.2.1]- and [4.2.1]bicyclic series of analogs **2.2.2** and **2.2.8** significantly reduces potency to inhibit DA uptake in the absence of aromatic substituents (**2.2.2a**, **2.2.8a**, **2.2.8c**) *(87)*. But potency is restored in the *bis*(4-fluoro) analogs **2.2.2b**, **2.2.8b**, and **2.2.8d** *(87)*.

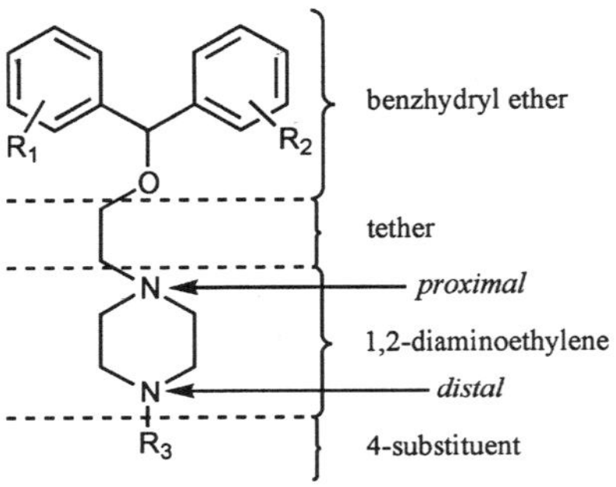

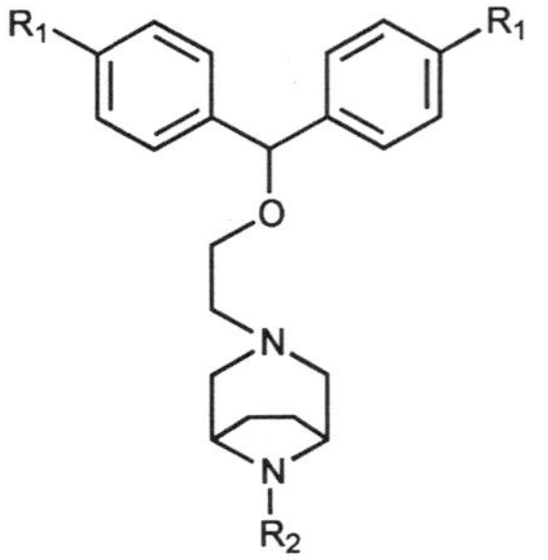

benzhydryl ether

tether

proximal

1,2-diaminoethylene

distal

4-substituent

2.2.1a, $R_1 = R_2 = H$, $R_3 = CH_2CH=CHPh$ (GBR 12783)
  b, $R_1 = 3\text{-NCS}$, $R_2 = H$, $R_3 = CH_2CH=CHPh$

c, $R_1 = 4\text{-}N$ [maleimide] $R_2 = H$ $R_3 = CH_2CH=CHPh$

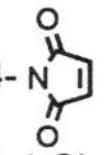

d, $R_1 = 2,4\text{-}Cl_2$, $R_2 = 4'\text{-}F$, $R_3 = CH_2CH=CHPh$
e, $R_1 = R_2 = 4\text{-}F$, $R_3 = CH_2CH=CHPh$ (GBR 12879)
f, $R_1 = R_2 = H$, $R_3 = (CH_2)_3Ph$ (GBR 12935)
g, $R_1 = R_2 = 4\text{-}F$, $R_3 = (CH_2)_3Ph$ (GBR 12909)
h, $R_1 = R_2 = 4\text{-}F$, $R_3 = H$ (GBR 11513)
i, $R_1 = R_2 = 4\text{-}F$, $R_3 = CH_3$ (NNC 12-0722)

2.2.2a, $R_1 = H$, $R_2 = (CH_2)_3Ph$
  b, $R_1 = F$, $R_2 = (CH_2)_3Ph$
  c, $R_1 = F$, $R_2 = CH_2CH=CHPy$

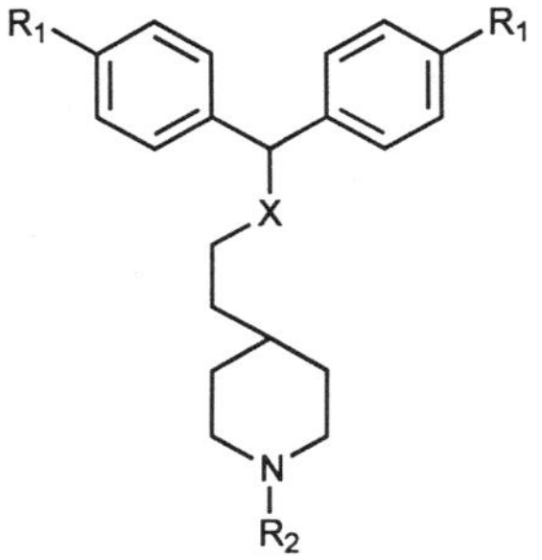

2.2.3a, $X = O$, $R_1 = H$, $R_2 = (CH_2)_2Ph$
  b, $X = O$, $R_1 = CH_3$, $R_2 = (CH_2)_2Ph$
  c, $X = O$, $R_1 = Br$, $R_2 = (CH_2)_2Ph$
  d, $X = O$, $R_1 = Cl$, $R_2 = (CH_2)_2Ph$
  e, $X = O$, $R_1 = F$, $R_2 = (CH_2)_2Ph$
  f, $X = NH$, $R_1 = H$, $R_2 = CH_2Ph$
  g, $X = NH$, $R_1 = F$, $R_2 = CH_2Ph$
  h, $X = NH$, $R_1 = H$, $R_2 = CH_2(4\text{-FPh})$
  i, $X = NH$, $R_1 = H$, $R_2 = CH_2(3,4\text{-}F_2Ph)$

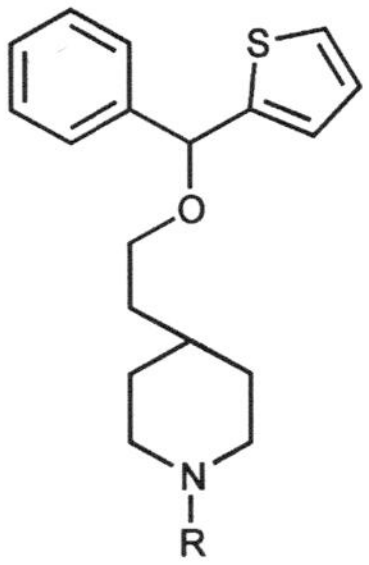

2.2.4a, $R_1 = H$, $R_2 = (CH_2)_3Ph$
  b, $R_1 = F$, $R_2 = (CH_2)_3Ph$
  c, $R_1 = H$, $R_2 = CH_2Ph$
  d, $R_1 = H$, $R_2 = CH_2(4\text{-FPh})$
  e, $R_1 = H$, $R_2 = CH_2(3\text{-Py})$

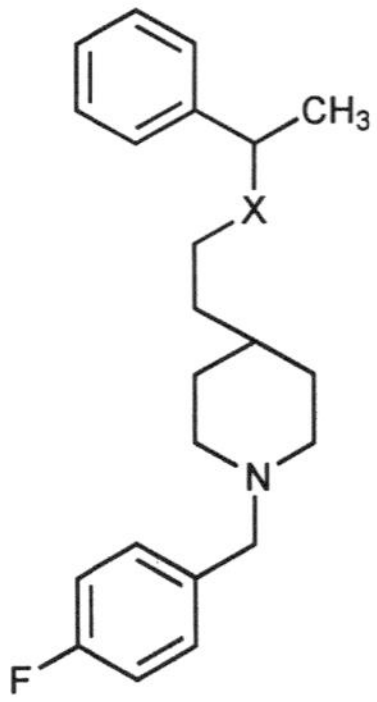

2.2.5a, $X = O$
  b, $X = NH$

The effects of allowing the 1,2-diaminoethylene fragment to be within 7- or 8-membered heterocycles, to be partially exocyclic, or to exist as an open chain, have been evaluated in analogs of GBR 12783 (**2.2.1a**) and GBR 12935 (**2.2.1f**). Most of these changes do not significantly affect the potency of the compounds to inhibit DA uptake. Two exceptions are (a) replacement of the tertiary amino groups by secondary amino groups in open-chain analogs and (b) replacement of the piperazine by a *gem*-dimethyl 7-membered,

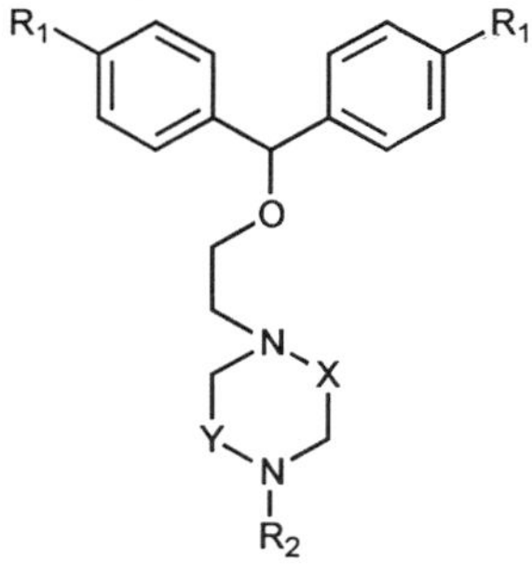

2.2.6a, R₁ = H, R₂ = CH₂CH=CHPh, X = Y = CHCH₃
b, R₁ = F, R₂ = CH₂CH=CHPh, X = Y = CHCH₃
c, R₁ = H, R₂ = (CH₂)₃Ph, X = Y = CHCH₃
d, R₁ = F, R₂ = (CH₂)₃Ph, X = Y = CHCH₃
e, R₁ = F, R₂ = H, X = Y = CHCH₃
f, R₁ = H, R₂ = (CH₂)₃Ph, X = CH₂, Y = C(O)
g, R₁ = F, R₂ = (CH₂)₂Ph, X = C(O), Y = CH₂

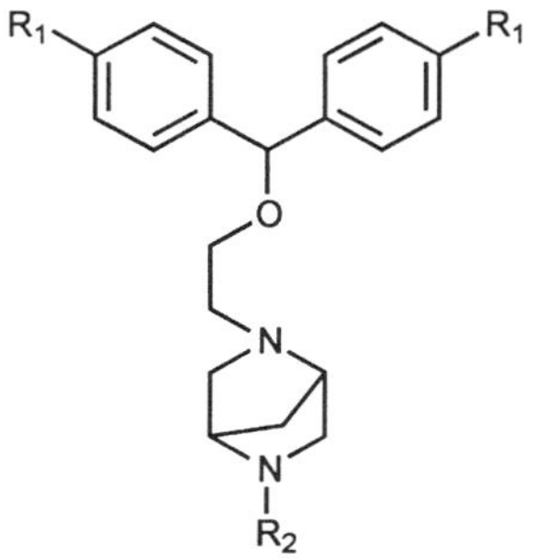

2.2.7a, R₁ = H, R₂ = (CH₂)₃Ph
b, R₁ = F, R₂ = (CH₂)₃Ph
c, R₁ = H, R₂ = CH₃
d, R₁ = F, R₂ = CH₃

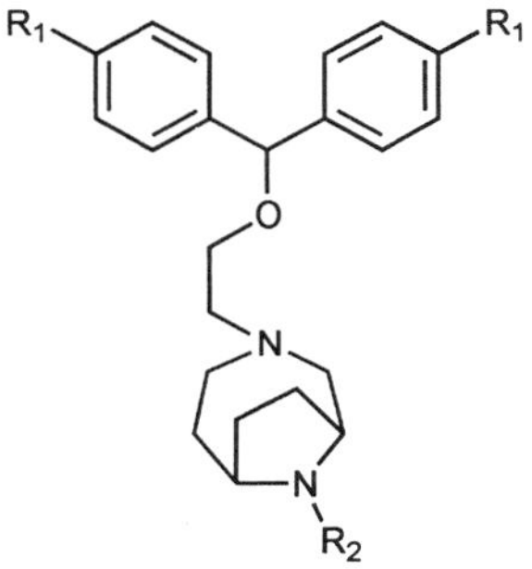

2.2.8a, R₁ = H, R₂ = CH₃
b, R₁ = F, R₂ = CH₃
c, R₁ = H, R₂ = (CH₂)₃Ph
d, R₁ = F, R₂ = (CH₂)₃Ph

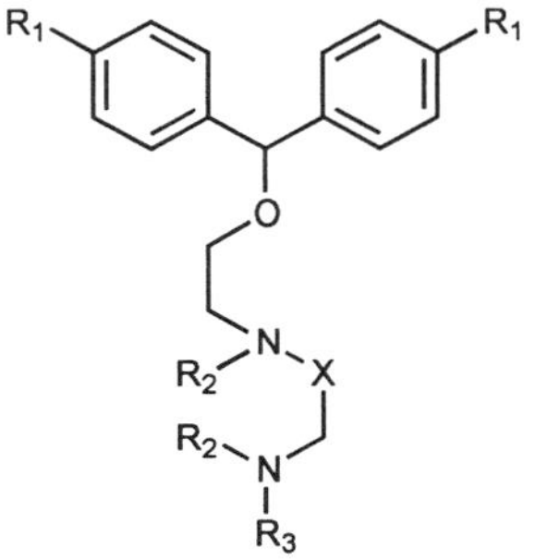

2.2.9a, R₁ = H, R₂ = CH₃, R₃ = (CH₂)₃Ph, X = CH₂
b, R₁ = F, R₂ = CH₃, R₃ = (CH₂)₃Ph, X = CH₂
c, R₁ = R₂ = H, R₃ = (CH₂)₃Ph, X = CH₂
d, R₁ = F, R₂ = H, R₃ = (CH₂)₃Ph, X = CH₂
e, R₁ = H, R₂ = CH₃, R₃ = (CH₂)₂Ph, X = (CH₂)₂

or by an 8-membered ring. Thus, although the open chain analogs of GBR 12935 (**2.2.1f**) and GBR 12909 (**2.2.1g**) are only slightly less potent than the parent compounds, provided that the amino groups are tertiary (**2.2.9a** *[85]* and **2.29b** *[91]*), the secondary amine analogs (**2.2.9c** and **2.2.9d** *[91]*) are devoid of potency. Similarly, while the seven-membered ring analogs of GBR 12783 (**2.2.1a**) and GBR 12935 (**2.2.1f**), compounds **2.2.10a** and **2.2.10b**, and their *bis*(4-fluoro)analogs **2.2.10c** and **2.2.10d**, have similar potencies to inhibit DA reuptake, the 6,6-dimethyl 7-membered analog, as well as the 8-membered analog of GBR 12935 (**2.2.10e** and **2.2.10f**, respectively) are almost 2 orders of magnitude less potent. These structural modi-

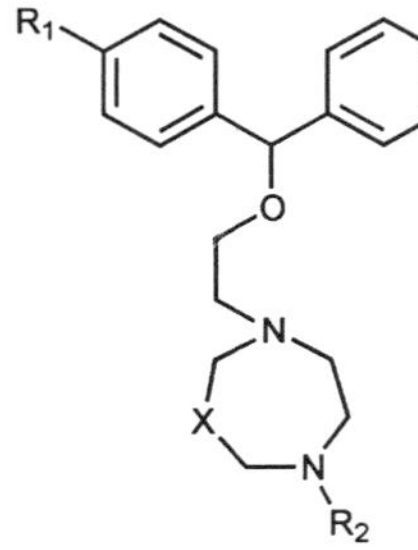

2.2.10a, R$_1$ = H, R$_2$ = CH$_2$CH=CHPh, X = CH$_2$
b, R$_1$ = H, R$_2$ = (CH$_2$)$_3$Ph, X = CH$_2$
c, R$_1$ = F, R$_2$ = CH$_2$CH=CHPh, X = CH$_2$
d, R$_1$ = F, R$_2$ = (CH$_2$)$_3$Ph, X = CH$_2$
e, R$_1$ = H, R$_2$ = (CH$_2$)$_3$Ph, X = C(CH$_3$)$_2$
f, R$_1$ = H, R$_2$ = (CH$_2$)$_3$Ph, X = (CH$_2$)$_2$

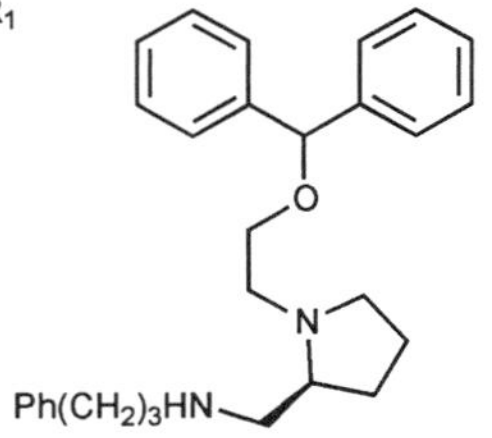

2.2.11

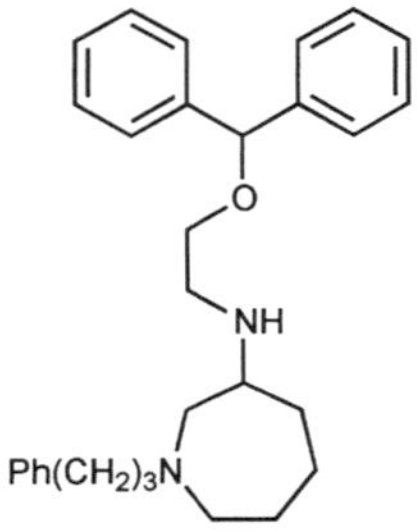

2.2.12

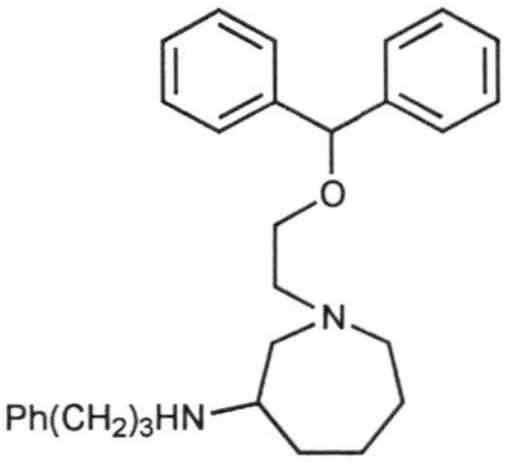

2.2.13

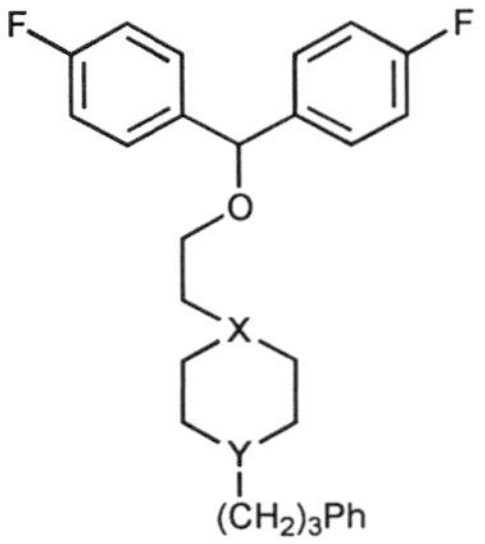

2.2.14a, X = N, Y = CH
b, X = CH, Y = N

fications indicate that steric constraints exist, and support a receptor-based mechanism for the inhibition of DA uptake. Further support for a receptor-mediated mechanism is derived from the selectivity of the chiral pyrrolidine analogs of GBR 12935 (**2.2.1f**) in which the distal amino group is exocyclic [(+)- and (−)-**2.2.11**]; in this structure the *S*-enantiomer is 150-fold more potent than the *R*-analog. Conformational factors may also be involved in the decrease in potency observed for the analog in which the proximal amino group is exocyclic to an azepine ring (**2.2.12**); no effects are seen with an exocyclic distal amino group (**2.2.13**).

Increasing the separation between amino groups by modification of the ethylene chain (**2.2.9d**) to a propylene chain (**2.2.9e**) decreases potency to inhibit DA uptake by a factor of 15, suggesting an important role for the position of the distal amino group. Elimination of one of the amino groups by replacement of the piperazine ring of 1-[2-*bis*(4-fluorophenyl)-methoxy]-ethyl-4-(3-propylphenyl)-piperazine (GBR 12909, **2.2.1g**) by a

2.2.15a, X = O, Y = NH
b, X = CH$_2$, Y = NH
c, X = CH$_2$, Y = NCH$_3$

2.2.16a, R$_1$ = H,F, R$_2$ = (CH$_2$)$_3$Ph
b, R$_1$ = H,F, R$_2$ = (CH$_2$)$_2$Ph
c, R$_1$ = H,F, R$_2$ = CH$_2$Ph

piperidine ring with a nitrogen atom either proximal (**2.2.14a**) or distal (**2.2.14b**) to the benzhydryl moiety *(100)* indicates, based on the relative potencies of these analogs to displace [^{3}H]WIN 35,428 and [^{3}H]GBR 12935 binding, that only the distal nitrogen (**2.2.14b**) is required for binding the DAT. Although the piperidine analogs bear structural resemblance to the parent piperazines, recent data indicate that they may possess a different binding profile. Analysis of the binding isotherm of the *N*-benzyl analog **2.2.15a** in monkey striatum is consistent with two component binding *(97)*; GBR 12909 binding under the same conditions indicates the presence of only one binding site *(101)*. The loss of potency to inhibit DA uptake by modification of the distal piperazine nitrogen in GBR 12935 (**2.2.1f**) to an endocyclic amide (**2.2.6f**) *(102)* is also consistent with a key role for the distal amino group. However, the observation that modification of the proximal nitrogen to an amide (**2.2.6g**) *(102)* also leads to similar reduction in potency to inhibit DA uptake suggests that the effect of converting an amino to an amide group may be unique.

2.2.3. Modification of the Oxapropylene Tether

Because analogs in which the proximal amino group was replaced by a methine were found to be more selective than GBR 12909 (**2.2.1g**) for binding at DAT relative to the binding at the "piperazine-acceptor site," *(100)* analogs of **2.2.3a** *(90,93)* were used to explore the effects of modifications of the oxapropylene tether. Replacement of the ether by a secondary amine (**2.2.3f** *[90,93]*, **2.2.3g** *[97]*, **2.2.3h** *[89]*, and **2.2.3i** *[89,97]*) reduces potency to inhibit DA uptake, but the effects are all within one order of magnitude. Interestingly, replacement of the oxygen of the α-methylbenzyl analog **2.2.5a** *(97)* by a nitrogen to give **2.2.5b** *(97)* increases potency to inhibit binding at DAT 2.5-fold. Replacement of the 2-position methylene group by a secondary amino group to yield **2.2.15a** *(97)* only slightly reduces potency to inhibit DA uptake, but replacement of the ether oxygen by a methylene group (**2.2.15b**) *(97)* results in significant reduction in potency. *N*-methylation to give **2.2.15c** *(97)* reduces binding potency at DAT relative to the secondary amine **2.2.15b**.

Shortening of the tether to produce **2.2.16** *(90,93)* has no significant effects on potency to inhibit binding at DAT. Effects on potency to inhibit DA uptake have not been reported.

2.2.4. Modification of the 4-Substituent

Many modifications of the 4-substituent have been examined for their effects on potency to inhibit DA uptake. Highest potencies are obtained when the 4-substituent is 3-phenyl-2-propenyl or 3-phenylpropyl. Interestingly, while replacement of the 4-position hydrogen in GBR 11513 (**2.2.1h**) by a 3-phenylpropyl substituent to give GBR 12909 (**2.2.1g**) increases potency to inhibit DA uptake by a factor of 70 *(78,95)*, the analogous replacement of the 4-position hydrogen in the *trans*-2,5-dimethylpiperazine analog **2.2.6e** by a 3-phenylpropyl group to give **2.2.6d** only increases potency by a factor of 2.4 in both the racemate and the more potent enantiomer *(99)*; a decrease in potency is observed for the less active enantiomer *(99)*. Replacement of the 4-position hydrogen (or methyl) by a 3-phenylpropyl group in the bicyclic analogs **2.2.7**, **2.2.2**, and **2.2.8** *(87)* increases potency to inhibit DA uptake two- to fourfold only. Substituents on the phenyl portion of the 3-phenylpropenyl group of GBR 12789 (**2.2.1e**) do not lead to significant changes in potency, although electron-donating substituents have a slightly adverse effect on potency *(78)*. An important compound in this series is the 4-methyl analog **2.2.1i** (NNC12-0722) *(103)*. Although not highly potent, this compound could be a useful imaging ligand for the DAT when labeled with carbon-11. Overall, the site is fairly tolerant to substitution. Furanyl (**2.2.17a**), thiophenyl (**2.2.17b**), indolyl (**2.2.17c**), benimidazoyl (**2.2.17d**) and pyridylprop(en)yl (**2.2.17e**) *(86)* analogs are less potent than the analogous phenyl compounds (**2.2.1a**, **2.2.1e**, **2.2.1f**, and **2.2.1g**), but not significantly so. The largest decreases in potency are obtained with 2-(α-naphthyl)ethyl (**2.2.18**), 2-quinolylmethyl (**2.2.19**), and 7-quinolylmethyl (**2.2.20**) *(86)* substitution. A large effect is also observed for the pyridylpropenyl [**3.2.1.**] bicyclic analog **2.2.2c**, which exhibits 200-fold decrease in potency relative to its 3-phenylpropyl analog **2.2.2b** *(92)*.

Similar effects are seen in the inhibition of binding at DAT for analogs in which the proximal amino group was replaced by a methine (**2.2.3a**) *(90,83)*. The largest decreases in potency are seen when the 4-substituent is attached to nitrogen by a methylene group and is bulky. Thus, there is virtually no effect on potency when the 4-substituent is varied from phenylpropyl (**2.2.14b** *[93]* and **2.2.21a** *[90,83]*) to phenethyl (**2.2.3a–e** *[90,83]*) to benzyl (**2.2.21b** *[90,83]*), but 4-substitution with 2-benzthiophenylmethyl (**2.2.21c** *[93,104]*) and *p*-aminobenzyl (**2.2.21d** *[93,104]*), decreases potency 10- to 20-fold. Particularly interesting is the effect of "shortening" the 4-substituent in the series in which the ethylene tether is shortened to a methylene group (**2.2.16** *[90,83]*). In this series potency of the ligand with a benzyl 4-substituent (**2.2.16c**) is 50-fold reduced over potency of the analogous phenylpropyl compound **2.2.16a**; the phenethyl compound **2.2.16b** has potency to inhibit binding at DAT between **2.2.16a** and **2.2.16c** *(90,83)*.

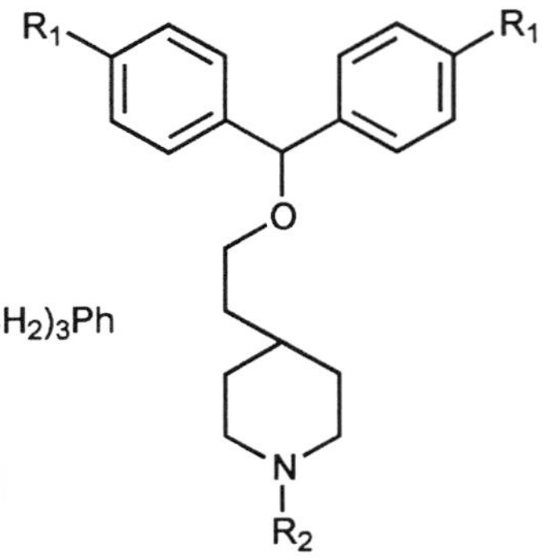

2.2.21a, R₁ = H, Cl, Br, OMe, R₂ = (CH₂)₃Ph

b, R₁ = H, F, R₂ = CH₂Ph

c, R₁ = H, R₂ = CH₂─[benzothiophene]

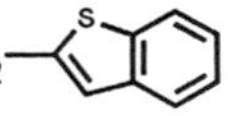

d, R₁ = H, R₂ = CH₂(4-NH₂Ph)

2.3. Methylphenidate

Methylphenidate (Ritalin®, **2.3.1**) is a drug used for the treatment of hyperactive children *(105)*. Since **2.3.1** has two centers of asymmetry, four isomers are possible. Methylphenidate is the (±)-*threo*-isomer (**2.3.1a**). It possesses an IC_{50} value of 83 to 390 nM for radioligand displacement at the DAT, depending on the radioligand used in the assay *(4,106,107)*. In a comparison of the potency of the stereoisomers to displace [³H]methylphenidate binding, it was found that the *threo* diastereomer (**2.3.1a**) is approx 80× more potent than the *erythro* diastereomer (**2.3.1b**) *(107)*; within the more active *threo* diastereomer, the (+)-enantiomer is 13.6 and 2.4× more potent than the (−)-enantiomer and the racemate, respectively *(107)*. Similar enantioselectivity was observed for the 4'-substituted analog of methylphenidate *(108)*.

2.3.1. Modification of the Carbomethoxy Group

Replacement of the carbomethoxy methyl group by a hydrogen to give **2.3.2a** leads to total loss of potency *(107)*. Similarly, alkyl groups other than methyl afford inactive compounds (**2.3.2b**); losses in potency appear to be associated with increasing mass of the alkyl group, but not necessarily with steric bulk *(107)*.

2.3.2. Modification of the Phenyl Ring

Investigation of the effect of aromatic substitution on binding potency in (±)-*threo*-**2.3.1** revealed all 2-substituted analogs to have greatly reduced potencies, yet electron withdrawing substituents in positions 3- and 4- are linked to increased potency, and electron-donating groups in the 3- and 4-positions have little effect, or decrease potency *(106)*. The most potent compounds are the 3-chloro, 3- and 4-bromo, and 3,4-dichloro analogs **2.3.3a–d**, which are 12 to 20× more potent than **2.3.1a**. Analogs **2.3.3e** and **2.3.3f**,

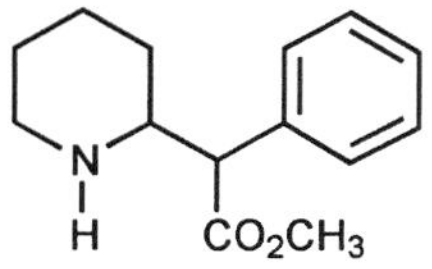

2.3.1a, (±)-*threo*
b, (±)-*erythro*

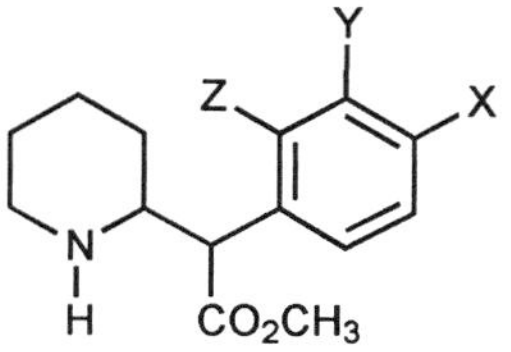

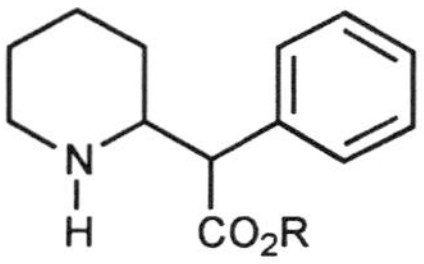

2.3.2a, R = H
b, R = group larger
than CH₃

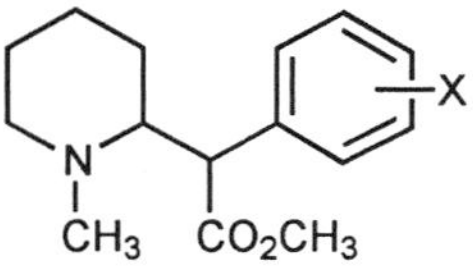

2.3.4

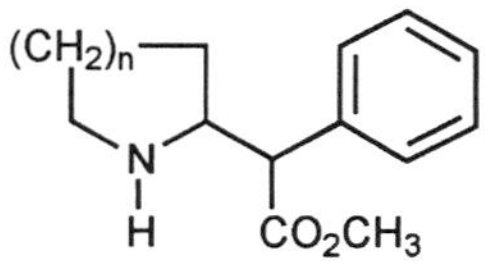

2.3.6

2.3.3a, Y = Cl, X = Z = H
b, Y = Br, X = Z = H
c, X = Br, Y = Z = H
d, X = Y = Cl, Z = H
e, X = OH, Y = Z = H
f, X = NH₂, Y = Z = H
g, X = Z = H, Y = OH
h, X = Z = H, Y = NH₂
i, X = Y = H, Z = CH₃O
j, X = Y = H, Z = F

k, X, Y = ⟨⟩, Z = H

l, X = H, Y, Z = ⟨⟩

2.3.5a, n = 1
b, n = 3
c, n = 4

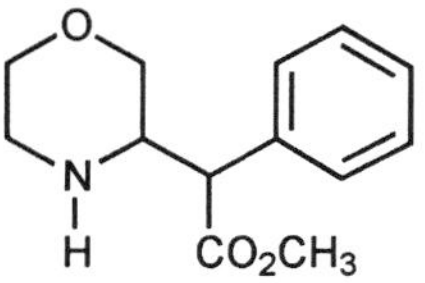

which are 4-hydroxy- and 4-amino-substituted, have slightly reduced and enhanced potencies, respectively, whereas the 3-hydroxy- and 3-amino analogs **2.3.3g** and **2.3.3h** show three- to fourfold decreases in potency relative to **2.3.1a**. The most deactivating substituent is a methoxy group in the 2-position (1200-fold lower potency, **2.3.3i**), but even the 2-fluoro analog **2.3.3j** shows a 17-fold decrease in potency. The 2-naphthyl analog **2.3.3k** is approx 2.5-fold more potent than the 1-naphthyl analog **2.3.3l**, and is also more potent than methylphenidate *(109)*.

2.3.3. Modification of the Piperidine Ring

Methylphenidate analogs that possess an *N*-methyl group (**2.3.4**), a smaller or larger ring (**2.3.5a**, **2.3.5b**, and **2.3.5c**), as well as a ring containing an oxygen (**2.3.6**), all show reduced affinity for the DAT *(109,110)*.

2.4. Phencyclidine Class

PCP (**2.4.1a**) has been observed to induce generalization to cocaine in rats trained to discriminate cocaine solutions from saline *(111)*. In addition, the PCP class of compounds had been found to inhibit synaptosomal dopamine uptake *(112)*, and it has been suggested that the complex nature of the behaviors associated with PCP abuse is a result of the interaction of PCP with both the *N*-methyl-D-aspartate acid (NMDA)-receptor complex and the DA transporter *(113)*. Since the reinforcing properties of cocaine have been correlated with its ability to increase dopamine levels in the synapse by virtue of its effectiveness in inhibiting dopamine uptake *(114)*, the interaction of PCP (**2.4.1a**) and its congeners with the DAT *(115)* may involve the cocaine binding site.

Replacement of the phenyl group in PCP (**2.4.1a**) clearly demonstrates that different factors are involved in the inhibition of ligand binding at DAT and DA uptake. Substitution by a *m*-hydroxyphenyl (to give **2.4.1b**), or by a bioisosteric 2-thienyl group (to yield TCP, **2.4.1c**), increases binding affinity but decreases potency to inhibit DA uptake *(113)*. Conversely, replacement of the phenyl group in PCP (**2.4.1a**) by a 2-benzothienyl group (to give BTCP, **2.4.2a**) decreases potency to displace [^{3}H]PCP but increases potency to inhibit DA uptake *(113,116)*. Further selectivity is associated with chirality of the ligands. Thus, the homochiral (+)-*(R)*-3-methyl analog (**2.4.3**) of TCP is selective for the PCP site, and the homochiral (−)-*(S)* analog 2.4.2c of BTCP is selective for the DAT-receptor site *(117)*. Direct displacement of binding of the radiolabeled form of **2.4.2a** [^{3}H]BTCP, by cocaine *(118)* and identification of two [^{3}H]BTCP-binding sites, with affinities of 0.9 and 20 n*M* and B_{max} values of 3.5 and 7.5 pmol/mg protein, respectively *(113)*, supports similarity to the binding of cocaine.

Replacement of the phenyl group in PCP (**2.4.1a**) by a 2-naphthyl group to give **2.4.4** has essentially the same effect on potency to displace binding at DAT as replacement by a 2-benzothienyl group to give BTCP (**2.4.2a**) *(119)*. However, replacement of the sulfur atom in BTCP by an oxygen to give the 2-benzofuranyl analog **2.4.5** reduces potency at DAT by a factor of 5 *(120)*.

Structural modifications of BTCP (**2.4.2a**) have involved changes in the cyclohexyl and piperidine moieties. The observed effects on potency to inhibit DA uptake appear to be stereospecific, supporting a receptor-based mechanism. Structural variations designed to explore correlations of potency to inhibit DA uptake with lipophilicity have not produced clear results *(121)*.

2.4.1. Modification of the Carbocyclic Ring

Analogs and derivatives of BTCP with modified carbocyclic rings generally have lower potency to inhibit DA uptake than the parent structure. Decreasing the size of the cyclohexyl ring to cyclopentyl, (as in **2.4.2d–f**),

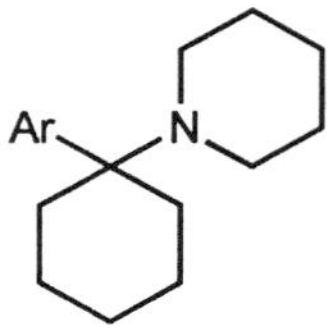

2.4.1a, Ar = Ph
b, Ar = 3-OHPh
c, Ar = 2-thienyl

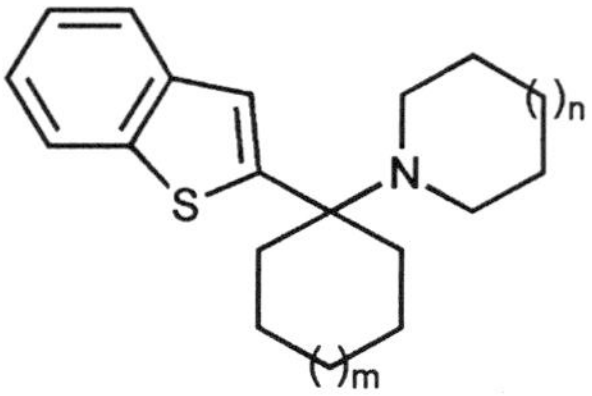

2.4.2a, m = n = 1
b, m = 1, n = 0
c, m = 1, n = 2
d, m = 0, n = 1
e, m = n = 0
f, m = 0, n = 2
g, m = 2, n = 1
h, m = 2, n = 0
i, m = n = 2

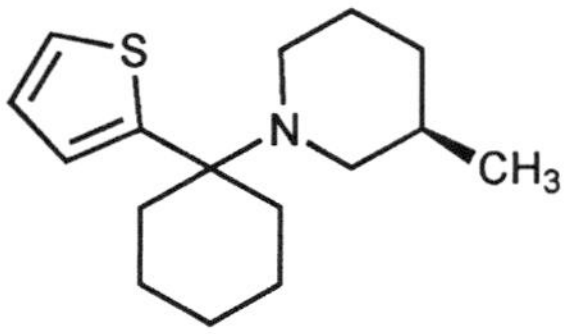

2.4.3

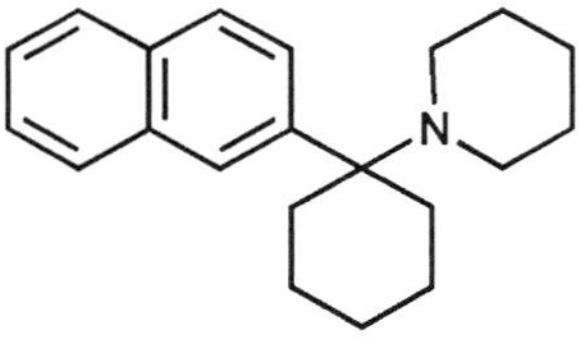

2.4.4

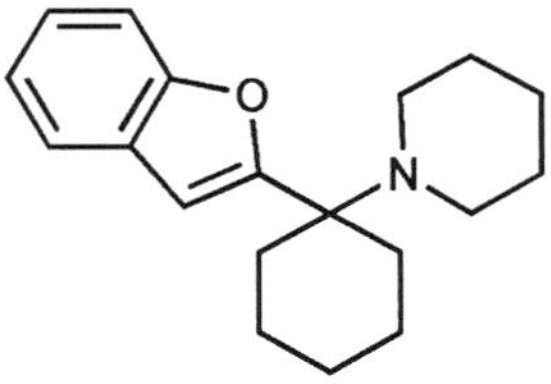

2.4.5

or increasing it to cycloheptyl, as in **2.4.2g–i**, leads to potencies 2–3× lower than those of the parent compound *(116)*. While the introduction of a 2-, or a 4-methyl substituent *cis* to the piperidine ring (**2.4.6a**, **2.4.6b**) has no effect on potency to inhibit DA uptake, or to inhibit binding at DAT, *(122)* steric effects do appear to play a role. Thus, a *trans*-2-methyl substituent (**2.4.7a**) decreases binding potency at DAT by a factor of 170, but a *trans*-4-methyl substituent (**2.4.7b**) only decreases potency by a factor of 2 *(122)*. Similarly, the potency of the *cis*-4–hydroxy analog **2.4.6d** to inhibit binding at DAT is fivefold lower than that of BTCP, but the potency of the *trans* isomer **2.4.7d** is reduced by a factor of 15 *(123)*. Very large decreases in potency to inhibit DA uptake are observed with 4-*t*-butyl substitution: 100-fold for the cis isomer **2.4.6e** and 10,000-fold for the *trans* isomer **2.4.7e** *(122)*. The effects of 3-methyl substitution are large, and are stereospecific as well: the potency of the *cis*-3-methyl-substituted analog **2.4.6c** is reduced by a factor of 70, and that of the *trans* isomer **2.4.6c** is reduced by a factor of 40 *(122)*. Striking stereospecificity is observed by the addition of a 3-methyl substituent to the cyclohexyl ring of the 3-methylpiperidine BTCP analog **2.4.8a** to give **2.4.9**. The binding potency of (*RS* and *R*) -**2.4.9** and of (*RS* and *S*) -**2.4.9** at DAT is 20-fold lower than that of BTCP, while the binding of the enantiomeric (*SRS*) -**2.4.9** and *(SRR)* -**2.4.9** is only reduced by a factor of 3 *(124)*. Effects on potency to inhibit DA uptake have not been reported.

Modification of the BTCP carbocycle to a tetrahydrothiapyran to give **2.4.10a** has virtually no effect on potency to inhibit [³H]BTCP binding *(119)*, but modification to a tetrahydropyran (**2.4.10b**) reduces potency by a factor of 10 *(121,125)*. Effects on DA uptake have not been reported.

2.4.2. Modification of the Amino Group

Modifications in the size of the nitrogen heterocycle have led to modest increases in potency to inhibit DA uptake with the largest (threefold) observed for the homopiperidine analog (**2.4.2c**) of BTCP *(116)*. A 10-fold decrease in potency is observed when both the carbocyclic and the heterocyclic rings are 5-membered (**2.4.2e**), and a 20-fold decrease is observed when both rings are 7-membered (**2.4.2i**). The combined effects of bulk and lipophilicity are also apparent from a study in which numerous replacements of the amino group were investigated *(126)*. In general, analogs with secondary amino groups (**2.4.8b**) are more potent than the primary parent (**2.4.8c**) or their tertiary amine counterparts (**2.4.8d**). Secondary amines with three- and four-carbon substituents (*R* = Pr, Bu, allyl, cyclopropylmethyl, and cyclobutylmethyl) exhibit the highest potency to inhibit DA uptake. The effects of piperidine ring substitution are relatively small, but are stereoselective. Introduction of a 2'-hydroxyl substituent produces a modest decrease in potency, but a 4'-hydroxyl substituent increases potency by a factor of 6. Similarly, introduction of a methyl substituent at the 3'-position of the piperidine ring of **2.4.2a**, to give **2.4.8a**, has been found to increase potency by a factor of 2.5 *(117,119)*. This effect has been found to be enantioselective with a threefold greater increase in potency for the (–)-(*S*)-1-[1-(2-benzo[*b*]thiophenyl)cyclohexyl]-3-methylpiperidine over the (+)-(*R*)-enantiomer *(117)*. A similar effect is observed when both the carbocycle and the piperidine ring are methyl-substituted (*see* **2.4.1**): the potency of the diastereomer in which the configuration of the methyl-substituted piperidine is *R* to inhibit binding at DAT exceeds that of the isomer with the *S* configuration by a factor of 2 [(*SRS*)- **2.4.9** vs *(SRR)*- **2.4.9**] *(124)*.

Replacement of the amino group by an imidazoline-2-one (**2.4.8e**) leads to a total loss of potency *(126)*, demonstrating the need for an amino functionality. Replacement by a β-aminoacetamido group (**2.4.8f**) restores some

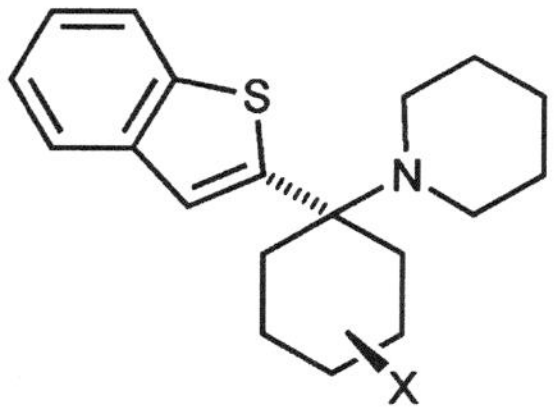

2.4.6a, X = 2-CH₃
b, X = 4-CH₃
c, X = 3-CH₃
d, X = 4-OH
e, X = 4-C(CH₃)₃

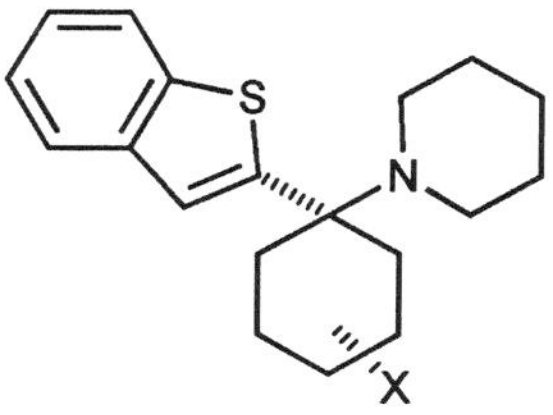

2.4.7a, X = 2-CH₃
b, X = 4-CH₃
c, X = 3-CH₃
d, X = 4-OH
e, X = 4-C(CH₃)₃

potency, but reduction of the carbonyl group to a methylene, i.e., replacement of the amino group by an ethylenediamine group (**2.4.8g**), increases potency by a factor of 5 *(126)*. Replacement by a cyclic diamine—i.e., by a piperazino group (**2.4.8h**)—improves potency by an additional factor of 2 *(127)*. Modification of the piperidine ring to a morpholine in the BTCP analog in which the cyclohexyl has been replaced by a thiatetrahydropyran to give **2.4.10c** reduces binding potency by a factor of 6 *(117)*. Potencies to inhibit DA uptake are not known.

2.5. Mazindol Class

Mazindol (**2.5a**) (Sanorex® or Mazanor®) *(128–130)* is a clinically utilized appetite suppressant reported to be free of abuse potential *(131)*. It potently inhibits monoamine uptake *(132)* and reduces craving for cocaine in human subjects *(133)*. [³H]Mazindol labels the DAT in the striatum and the NE transporter in the cerebral cortex with K_d values of 18 and 4 nM, respectively *(132)*. Mazindol has been reported to exist exclusively in the enol form in the solid state *(134)*. However, it can be present in either the enol or keto form in solution, depending on the pH *(129)*. While there is no direct evidence to indicate which mazindol tautomer is responsible for the affinity at the DAT the relatively low affinity of mazindol analogs that exist solely in the keto form suggests that the enol tautomer plays a significant role in high-affinity binding *(135)*. For convenience, all mazindol analogs in this chapter are represented as the enol tautomer.

The DAT binding and uptake properties of only a limited number of mazindol analogs have been reported *(135–140)*. Replacement of the 4'-chlorophenyl ring (ring D) with either methyl, *t*-butyl, or 2-furyl groups (**2.5b–d**) causes a large loss of binding affinity, indicating that this phenyl ring is required for potent binding at the DAT *(139–141)*.

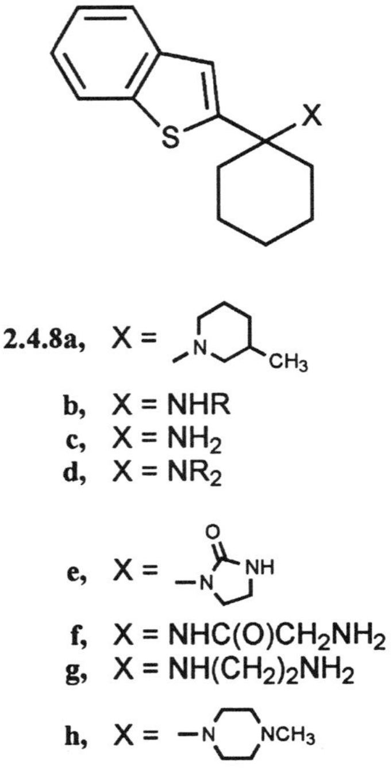

2.4.8a, X = (piperidine with CH₃)

b, X = NHR
c, X = NH₂
d, X = NR₂

e, X = (imidazolidinone)

f, X = NHC(O)CH₂NH₂
g, X = NH(CH₂)₂NH₂

h, X = (N-methylpiperazine)

Within a series of 4'-substituted analogs, the order of binding potency is Br > Cl > F > I >> H (**2.5e**, **2.5a**, and **2.5f–h**, respectively), with the 4'-Br analog (**2.5e**) exhibiting a threefold increase in affinity relative to mazindol. The 4'-F (**2.5f**) and 4'-I (**2.5g**) analogs are twofold less potent and the unsubstituted (**2.5h**) analog is eightfold less potent than mazindol *(135)*. In a separate study of 3'-monosubstituted analogs, the order of binding potency was found to be Cl ≈ Br > I > F (**2.5i–l**). The 3'-Cl and 3'-Br analogs were found to have approximately the same potency as mazindol, and the 3'-F and 3'-I analogs (**2.5k** and **2.5.l**) were four- to sevenfold less potent *(136)*. Other studies have reported the 3'-Cl analog to be twofold more potent than mazindol *(135)*. Substitution in the 2'-position uniformly lowers binding potency at the DAT. The 2'-Cl and 2'-Br monosubstituted analogs (**2.5m** and **2.5n**) bind 36- to >150-fold less potently to the DAT, while adding a 2'-Cl substituent to mazindol to produce the 2',4'-dichloro analog (**2.5o**) results in a ninefold loss of DAT affinity. This is in contrast to the 3',4'-dichloro analog (**2.5p**) for which there is a threefold increase in DAT affinity *(135)*. Substitution of the phenyl ring C can result in a very slight increase in binding potency (4'-Cl, 7-F analog, **2.5q**) or a twofold loss of potency (4', 7, 8-trichloro analog, **2.5r**). The 4'-F, 7-Cl analog **2.5s** is sevenfold less potent

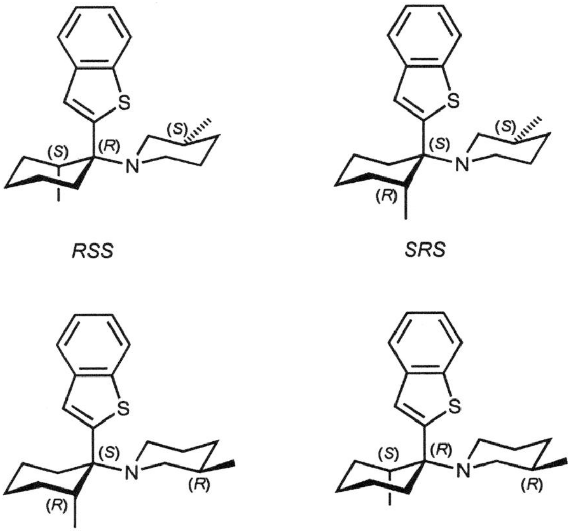

RSS

SRS

SRR

RSR

2.4.9

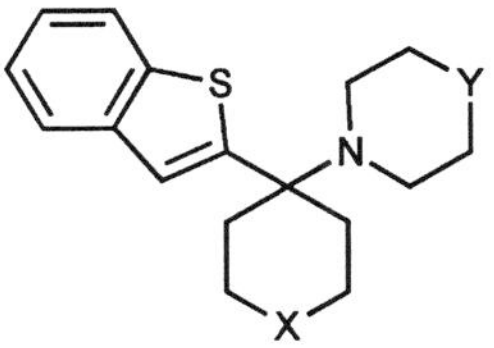

2.4.10a, X = S, Y = CH$_2$
b, X = O, Y = CH$_2$
c, X = S, Y = O

than the 4'-Cl, 7-F analog *(135)*. Derivatives bearing halogen substituents on the C-ring alone are 4- to 10-fold less potent than mazindol (**2.5t and 2.5u**) *(135)*.

Modification of the imidazolo ring A to produce the *gem*-dimethyl analog **2.5v** results in a 6-fold loss of potency *(135)*. Expansion of the imidazolo ring to a pyrimidino- or diazepino-ring system (**2.5w** and **2.5x**, respectively) yields analogs that are 5× more potent than mazindol *(135)*. The ability of mazindol to form a tricyclic enol tautomer appears to be important for high affinity binding to the DAT, since replacement of the 5-position with groups that prevent formation of the enol tautomer, such as CH$_2$OH, CH$_2$, oxygen, sulfur, and sulfone (**2.5y–cc**), results in a 24- to 2,050-fold loss of potency *(139)*.

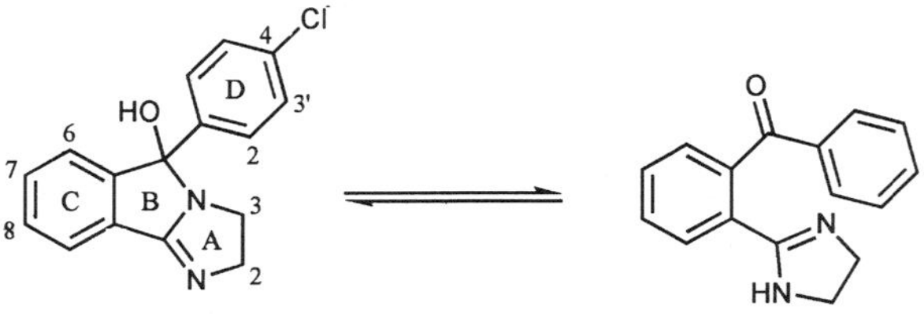

Enol tautomer   *Keto tautomer*

2.5a, Mazindol

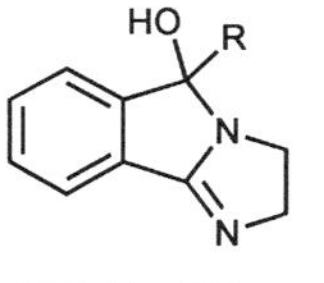

2.5b, R = CH₃
    c, R = C(CH₃)₃
    d, R = 2-furyl

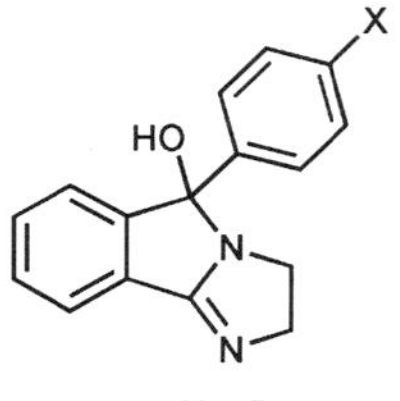

2.5e, X = Br
    f, X = F
    g, X = I
    h, X = H

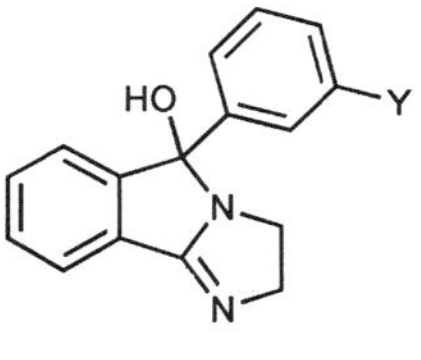

2.5i, Y = Cl
    j, Y = Br
    k, Y = F
    l, Y = I

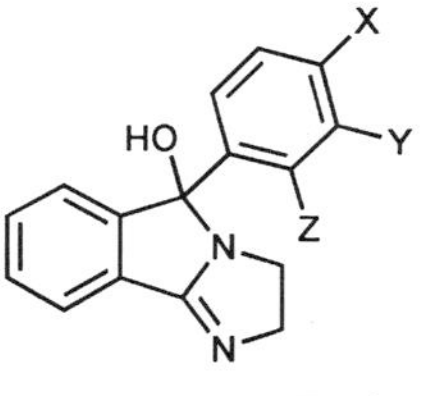

2.5m, X = Y = H, Z = Cl
    n, X = Y = H, Z = Br
    o, X = Cl, Y = H, Z = Cl
    p, X = Y = Cl, Z = H

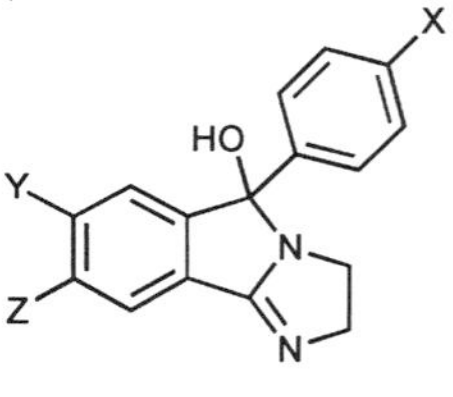

2.5q, X = Cl, Y = F, Z = H
    r, X = Y = Z = Cl
    s, X = F, Y = Cl, Z = H
    t, X = H, Y = F, Z = H
    u, X = H, Y = Cl, Z = H

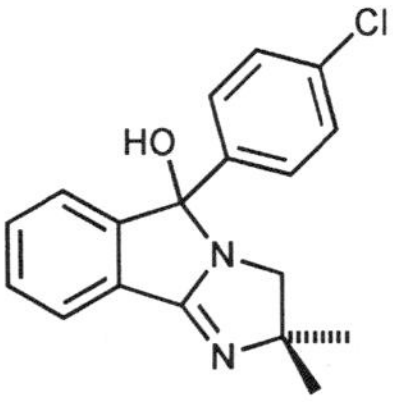

2.5v

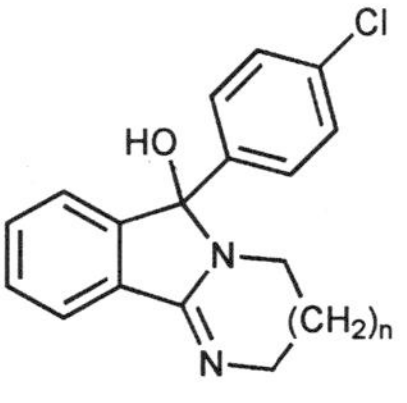

2.5w, n = 1
    x, n = 2

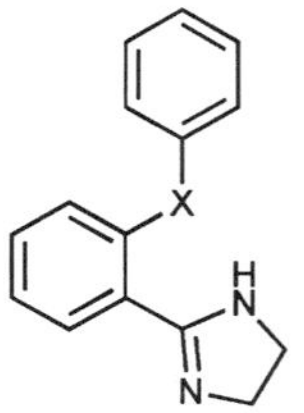

2.5y, X = CH₂OH
    z, X = CH₂
    aa, X = O
    bb, X = S
    cc, X = SO₂

3. COCAINE ANTAGONISTS

The function of the DAT is to transport (move) the neurotransmitter dopamine from the synaptic gap back into the cytoplasm of the presynaptic neuron. The uptake of dopamine is coupled to the flow of sodium and chloride ions, and, as indicated in this chapter, is inhibited by many structurally different compounds. Considerable effort has been devoted to the biochemical characterization of the site or sites where these inhibitors bind *(142)*. Cocaine has taken center stage in many studies, because the TMDs 5–8 appear to be involved in both substrate translocation and recognition of cocaine and tropane analogs *(14)*. Some studies suggest a competitive mechanism for the inhibition of DA uptake by cocaine and other inhibitors, while other studies suggest an allosteric, or more complex, mechanism *(142)*. Uncompetitive inhibition of DA uptake by cocaine was observed in a multisubstrate study of analogs of dopamine, which used rotating-disk voltametry to measure DA uptake in striatal suspension *(see* Fig. 2); in the same study, cocaine was found to competitively inhibit the binding of both sodium and chloride ions *(143)*. Furthermore, based on the differential effects on [^{3}H]DA uptake and [^{3}H]WIN 35,428 binding in mutants prepared by site-directed mutagenesis of cloned DAT(s), as compared to the wild-type clones, it was concluded that aspartate 79, in the first hydrophobic region, and serine 356 and 359, in the seventh hydrophobic region, are more instrumental in DA transport than in WIN 35,428 binding *(144)* (also *see* Chapter 3). These reports suggest that the DA and the WIN 35,428 binding sites may not be identical *(144)*. Dissociation of DA uptake and cocaine binding has also been concluded from investigations of chimeras of DAT and NET *(145)*. Other investigations, which have demonstrated that cocaine is substantially better than DA in protecting the DAT to reaction with sulfhydryl reagents, also suggest that the sites are different *(146)*. Thus, the data from several studies, which show that there are differences between the cocaine-binding site(s) on the DAT and the site(s) for DA uptake, suggest that some compounds might inhibit cocaine binding without inhibiting the translocation of DA. A recent review of the structure and function of the DAT has concluded that, despite the apparent overlap between the TMDs involved in DAT and cocaine binding, discrete domains of the transporter influence dopamine translocation and ligand recognition *(14)*. In particular, mutation of phenylalanine 361, located on TMD 7 and close to serines implicated in dopamine translocation in rat DAT, has shown that it is possible to inhibit the binding of cocaine analogs without affecting DA transport *(147)*. Therefore, it should be possible, in principle, to identify a compound that might be referred to as a "cocaine antagonist" or, perhaps preferably, as a "dopamine-transport-sparing cocaine antagonist" *(144)* *(see* Fig. 3). Results obtained with non-amine cocaine analogs have led to the conclusion that the development of such a drug should be feasible, because molecular volume, as well

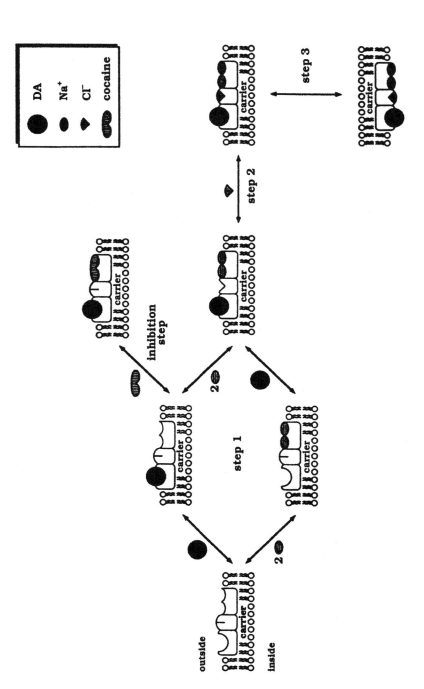

Fig. 2. Schematic illustration of the partially random, sequential multisubstrate mechanism of dopamine uptake in striatal tissue and how it may be influenced by cocaine. The sequence proceeds from left to right. In step 1, DA or Na$^+$ binds to the uptake carrier in random order, followed by Cl$^-$ binding in step 2. In step 2, the carrier transports DA across the membrane. If cocaine is present, the process is inhibited by cocaine binding at the Na$^+$-binding site as illustrated in the inhibition step. **Note:** the illustration is highly schematic and is not intended to represent the structure of the carrier protein and its mechanism of translocation.

416

as functionality, is essential for DAT inhibition *(148)*. Experimentally it has been found that, when measured under identical conditions, the potencies of a group of compounds to inhibit [^{3}H]DA uptake are 2.3× greater than to inhibit [^{3}H]WIN 35, 428 binding *(149)*. In other words, only one-half of the binding sites must be occupied in order to completely inhibit DA transport. Cases in which potency to inhibit radioligand binding to the cocaine site exceeds the potency to inhibit DA uptake by four- to ninefold, when evaluated under the usual binding and uptake assay conditions, have been reported *(30,50,68,96,106,150)*. However, it should be noted that when the assays for these compounds were repeated under nearly identical conditions, these differences disappeared *(50,149,151,152)*. One compound, the tricyclic depressant desipramine, has been found to fit the profile of a "dopamine-sparing-cocaine-antagonist" since its potency to inhibit cocaine binding is 11.5-fold its potency to inhibit DA uptake, when the assays are carried out under identical conditions in CHO cells expressing hDAT *(153)* and 5.6-fold in HEK-293 cells that express hDAT *(154)*. In the clinic, patients treated with desipramine for cocaine abuse were found to have longer periods of abstinence than the placebo controls. Unfortunately, there was no difference in the likelihood of dropping out of treatment between the groups *(155)*.

4. SUMMARY

The crucial role of the DAT in several neurological disorders and in cocaine abuse has prompted investigations aimed at a better understanding of the mechanism of DA uptake and of its inhibition. The short-range goal of identifying ligands with high potency and selectivity for the DAT has largely been met. Thus, compounds with potency to inhibit radioligand binding at the "cocaine site" at the DAT <2 nM and with selectivities greater than two orders of magnitude for binding at the DAT relative to the norepinephrine or serotonin transporter have been identified (*see* **Subheading 2.1.2.**). Some of these compounds have been used successfully as biological markers, particularly in imaging.

In addition, some progress has been made in the identification of specific molecular features associated with potency and selectivity at the DAT; most of this has been associated with compounds in the tropane class (**Subheading 2.1.**). On the other hand, attempts to correlate molecular features of compounds in different compound classes have met with complete failure. In fact, the only important conclusion that can be drawn from examination of the various compound classes that bind the so-called "cocaine receptor" at the DAT is that the protein may be quite adaptable to the several compound classes, assuming different conformations and possibly utilizing different amino-acid residues to bind compounds in the different compound classes.

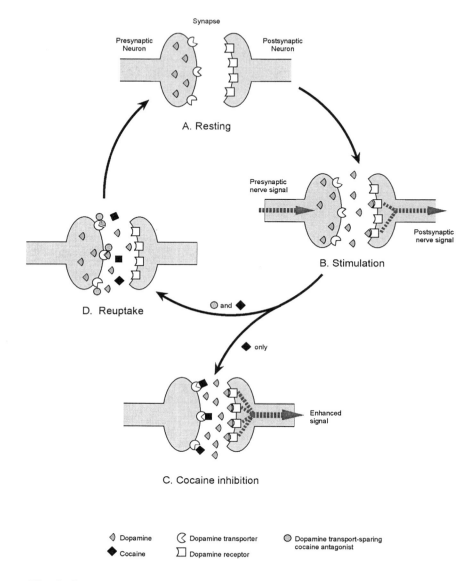

Fig. 3. Cartoon representation of a dopamine-transport-sparing cocaine antagonist within the context of the DA hypothesis *(112)*. Stimulation of a resting neuron **(A)** releases DA (◁) from the presynaptic terminal into the synapse, where it may bind DA receptors (▢) on the postsynaptic terminal, resulting in transmission of the signal **(B)**. Reuptake of DA (◁) by its transporter (ᘓ) is inhibited by cocaine (◆) binding to its recognition site **(C)** but a dopamine-transport-sparing cocaine antagonist (◎) compromises the ability of cocaine (◆) to bind to its recognition site without inhibiting DA (◁) reuptake **(D)**.

The differences in binding properties (e.g., B_{max}; number of binding sites, Hill coefficients) associated with the various compound classes may also be accounted for in this manner. This hypothesis may be confirmed experimentally by investigations with cloned wild-type receptors and cloned mutants (*see* **Subheading 3.**). The observations that, in the tropane class, high-potency ligands can be obtained when the 2β-substituent is capable of hydrogen bonding, or of electrostatic interaction, with the binding site, but also when it is incapable of such interactions, may similarly be accommodated because the protein is so readily adaptable. These problems have also prevented the definition of a specific pharmacophore for the "cocaine site" at the DAT. Good progress has been made within subsets of the tropane class of compounds, but it appears that definition of a generalized pharmacophore may not be a realistic goal.

A very important goal which has not been met thus far is the determination as to whether "dopamine-transporter-sparing cocaine antagonism" may be feasible. Since, in general, it appears that occupancy of only one-half of the cocaine-receptor sites at the DAT is required for complete inhibition of DA transport, it may be possible that DAT oligomers (dimers?) may be involved. In principle, it would seem that the state of aggregation and the nature of the oligomers (e.g., homo- or heterodimers) may be related to the state of occupancy of the receptor, and, if the protein is indeed very adaptable, DA transport may or may not be inhibited by occupancy of the receptor site *(156)*. Moreover, it has been recently pointed out that mutations in hDAT which completely abolish tropane binding retain high-potency inhibition of DA transport, suggesting that recognition by the carrier may not be reflected by determination of classical equilibrium binding *(14)*. These issues present interesting research challenges. Since crucial research tools (e.g., potent and selective ligands, irreversible ligands, radio-isotopically labeled ligands, as well as wild-type cloned DAT protein and mutagens thereof) are now available, the goal of identifying a ligand capable of blocking cocaine binding yet sparing DA transport may be reached.

ACKNOWLEDGMENT

The authors express their thanks to Dr. Maarten Reith for his encouragement and patience, as well as for sharing prepublication data.

REFERENCES

1. Kuhar, M. J. (1973) Neurotransmitter uptake: A tool in identifying neurotransmitter-specific pathways. *Life Sci.* **13,** 1623–1634.
2. Horn, A. S. (1978) Characteristics of neuronal dopamine uptake, in *Advances*

in Biochemistry and Psychopharmacology, Vol. 19, (Roberts, P. J., Woodruff, G. N., and Iversen, L. L. eds.), Raven Press, New York, NY, pp. 25–34.

3. Horn, A. S. (1990) Dopamine uptake: A review of progress in the last decade. *Prog. Neurobiol.* **34**, 387–400.

4. Ritz, M. C., Lamb, R. J., Goldberg, S. R., and Kuhar, M. J. (1987) Cocaine receptors on dopamine transporters are related to self-administration of cocaine. *Science* **237**, 1219–1223.

5. Bergman, J., Madras, B. K., Johnson, S. E., and Spealman, R. D. (1989) Effects of cocaine and related drugs in nonhuman primates. III. Self-administration by squirrel monkeys. *J. Pharmacol. Exp. Ther.* **251**, 150–155.

6. Giros, B., el Mestikawy, S., Bertrand, L., and Caron, M. G. (1991) Cloning and functional characterization of a cocaine-sensitive dopamine transporter. *FEBS Lett.* **295**, 149–154.

7. Kilty, J. E., Lorang, D. B., and Amara, S. G. (1991) Cloning and expression of a cocaine-sensitive rat dopamine transporter. *Science* **254**, 578–579.

8. Shimada, S., Kitayama, S., Lin, C.-L., Patel, A., Nanthakumar, E., Gregor, P., et al. (1991) Kuhar, M., and Uhl, G. Cloning and expression of a cocaine-sensitive dopamine transporter complementary DNA. *Science* **254**, 576–578.

9. Brüss, M., Wieland, A., and Bönisch, H. (1999) Molecular cloning and functional expression of the mouse dopamine transporter. *J Neural Transm* **106**, 657–662.

10. Wu, X. and Gu, H. H. (1999) Molecular cloning of the mouse dopamine transporter and pharmacological comparison with the human homologue. *Gene* **233**, 163–170.

11. Usdin, T. B., Mezey, E., Chen, C., Brownstein, M. J., and Hoffman, B. J. (1991) Cloning of the cocaine-sensitive bovine dopamine transporter. *Proc. Natl. Acad. Sci. USA* **88**, 11,168–11,171.

12. Vandenbergh, D. J., Persico, A. M., Hawkins, A. L., Griffin, C. A., Li, X., Jabs, E. W., et al. (1992) Human dopamine transporter gene (DAT1) maps to chromosome 5p15.3 and displays a VNTR. *Geonomics* **14**, 1104–1106.

13. Sugamori, K. S., Lee, F. J. S., Pristupa, Z. B., and Niznik, H. B. (1999) A Cognate dopamine transporter-like activity endogenously expressed in a COS-7 kidney-derived cell line. *FEBS Lett* **451**, 169–174.

14. Chen, N., and Reith, M. E. A. (2000) Structure and function of the dopamine transporter. *Eur. J. Pharmacol.* **405**, 329–339.

15. Reith, M. E. A., de Costa, B., Rice, K. C., and Jacobson, A. E. (1992) Evidence for mutually exclusive binding of cocaine, BTCP, GBR 12935, and dopamine to the dopamine transporter. *Eur. J. Pharmacol.—Mol. Pharmacol. Sec.* **227**, 417–425.

16. Carroll, F. I., Lewin, A. H., Abraham, P., Parham, K., Boja, J. W., and Kuhar, M. J. (1991) Synthesis and ligand binding of cocaine isomers at the cocaine receptor. *J. Med. Chem.* **34**, 883–886.

17. Ritz, M. C., Cone, E. J., and Kuhar, M. J. (1990) Cocaine inhibition of ligand binding at dopamine, norepinephrine and serotonin transporters: a structure-activity study. *Life Sci.* **46**, 635–645.

18. Gatley, S. J., Yu, D.-W., Fowler, J. S., MacGregor, R. R., Schlyer, D. J., Dewey, S. L., et al. (1994) Studies with differentially labeled [^{11}C]cocaine, [^{11}C]norcocaine, [^{11}C]benzoylecgonine, and [^{11}C]- and 4'-[^{18}F]fluorococaine to probe the extent to which [^{11}C]cocaine metabolites contribute to PET images of the baboon brain. *J. Neurochem.* **62,** 1154–1162.

19. Yu, D. W., Gatley, S. J., Wolf, A. P., MacGregor, R. R., Dewey, S. L., Fowler, J. S., et al. (1992) Synthesis of carbon-11 labeled iodinated cocaine derivatives and their distribution in baboon brain measured using positron emission tomography. *J. Med. Chem.* **35,** 2178–2183.

20. Metwally, S. A. M., Gatley, S. J., Wolf, A. P., and Yu, D.-W. (1992) Synthesis and binding to striatal membranes of no carrier added I-123 labeled 4'-iodococaine. *J. Label. Compd. Radiopharm.* **XXXI,** 219–225.

21. El-Moselhy, T. F., Avor, K. S., and Basmadjian, G. P. (2000) Synthesis and dopamine transporter binding of 2'-substituted cocaine analogs. *Med. Chem. Res.* **10,** 50–57.

22. Carroll, F. I., et al. (unpublished).

23. Lieske, S. F., Yang, B., Eldefrawi, M. E., MacKerell, J., Alexander D., and Wright, J. (1998) (–)-3b-Substituted ecgonine methyl esters as inhibitors for cocaine binding and dopamine uptake. *J. Med. Chem.* **41,** 864–876.

24. Kline, R. H., Jr., Wright, J., Eshleman, A. J., Fox, K. M., and Eldefrawi, M. E. (1991) Synthesis of 3-carbamoylecgonine methyl ester analogues as inhibitors of cocaine binding and dopamine uptake. *J. Med. Chem.* **34,** 702–705.

25. Lewin, A. H., Gao, Y., Abraham, P., Boja, J. W., Kuhar, M. J., and Carroll, F. I. (1992) 2b-Substituted analogues of cocaine. Synthesis and inhibition of binding to the cocaine receptor. *J. Med. Chem.* **35,** 135–140.

26. Reith, M. E. A., Meisler, B. E., Sershen, H., and Lajtha, A. (1986) Structural requirements for cocaine congeners to interact with dopamine and serotonin uptake sites in mouse brain and to induce stereotyped behavior. *Biochem. Pharmacol.* **35,** 1123–1129.

27. Kozikowski, A. P., Roberti, M., Xiang, L., Bergmann, J. S., Callahan, P. M., Cunningham, K. A., et al. (1992) Structure-activity relationship studies of cocaine: replacement of the C-2 ester group by vinyl argues against H-bonding and provides an esterase-resistant, high-affinity cocaine analogue. *J. Med. Chem.* **35,** 4764–4766.

28. Boja, J. W., McNeill, R. M., Lewin, A. H., Abraham, P., Carroll, F. I., and Kuhar, M. J. (1992) Selective dopamine transporter inhibition by cocaine analogs. *NeuroReport* **3,** 984–986.

29. Carroll, F. I., Abraham, P., Lewin, A. H., Parham, K. A., Boja, J. W., and Kuhar, M. J. (1992) Isopropyl and phenyl esters of 3b-(4-substituted phenyl)tropan-2b-carboxylic acids. Potent and selective compounds for the dopamine transporter. *J. Med. Chem.* **35,** 2497–2500.

30. Carroll, F. I., Kotian, P., Dehghani, A., Gray, J. L., Kuzemko, M. A., Parham, K. A., et al. (1995) Cocaine and 3b-(4'-substituted phenyl)tropane-2b-carboxylic acid ester and amide analogues. New high-affinity and selective compounds for the dopamine transporter. *J. Med. Chem.* **38,** 379-388.

31. Kozikowski, A. P., Xiang, L., Tanaka, J., Bergmann, J. S., and Johnson, K.

M. (1991) Use of nitrile oxide cycloaddition (*NOC*) chemistry in the synthesis of cocaine analogues: mazindol binding and dopamine uptake studies. *Med. Chem. Res.* **1**, 312-321.

32. Abraham, P., Pitner, J. B., Lewin, A. H., Boja, J. W., Kuhar, M. J., and Carroll, F. I. (1992) N-Modified analogues of cocaine: synthesis and inhibition of binding to the cocaine receptor. *J. Med. Chem.* **35**, 141–144.

33. Stoelwinder, J., Roberti, M., Kozikowski, A. P., Johnson, K. M., Bergmsnn, J. S. Differential binding and dopamine uptake activity of cocaine analogues modified at nitrogen. (1994) *Bioorg. Med. Chem. Lett.* 4303-4308.

34. Kozikowski, A. P., Saiah, M. K. E., Bergmann, J. S., and Johnson, K. M. (1994) Structure–activity relationship studies of *N*-sulfonyl analogs of cocaine: role of ionic interaction in cocaine binding. *J. Med. Chem.* **37**, 3440–3442.

35. Simoni, D., Roberti, M., Andrisano, V., Manferdini, M., Rondanin, R., and Invidiata, F. P. (1999) Two-carbon bridge substituted cocaines: enantioselective synthesis, attribution of the absolute configuration and biological activity of novel 6- and 7-methoxylated cocaines. *Il Farmaco* **54**, 275–287.

36. Kozikowski, A. P., Simoni, D., Roberti, M., Rondanin, R., Wang, S., Du, P., et al. (1999) Synthesis of 8-Oxa analogues of norcocaine endowed with interesting cocaine-like activity. *Bioorg. Med. Chem. Lett.* **9**, 1831–1836.

37. Madras, B. K., Spealman, R. D., Fahey, M. A., Neumeyer, J. L., Saha, J. K., and Milius, R. A. (1989) Cocaine receptors labeled by [^{3}H]2β-carbomethoxy-3β-(4-fluorophenyl)tropane. *Mol. Pharmacol.* **36**, 518–524.

38. Holmquist, C. R., Keverline-Frantz, K. I., Abraham, P., Boja, J. W., Kuhar, M. J. K., and Carroll, F. I. (1996) 3α-(4'-Substituted phenyl)tropane-2β-carboxylic acid methyl esters: novel ligands with high affinity and selectivity at the dopamine transporter. *J. Med. Chem.* **39**, 4139-4141.

39. Clarke, R. L., Daum, S. J., Gambino, A. J., Aceto, M. D., Pearl, J., Levitt, M., et al. (1973) Compounds affecting the central nervous system. 4. 3β-Phenyltropane-2-carboxylic esters and analogs. *J. Med. Chem.* **16**, 1260–1267.

40. Carroll, F. I., Gao, Y., Rahman, M. A., Abraham, P., Lewin, A. H., Boja, J. W., et al. (1991) Synthesis, ligand binding, QSAR, and CoMFA study of 3β-(p-substituted phenyl)tropan-2β-carboxylic acid methyl esters. *J. Med. Chem.* **34**, 2719–2927.

41. Meltzer, P. C., Liang, A. Y., Brownell, A. -L., Elmaleh, D. R., Madras, B. K. (1993) Substituted 3-phenyltropane analogs of cocaine: Synthesis, inhibition of binding at cocaine recognition sites, and positron emission tomography imaging. *J. Med. Chem.* **36**, 855–862.

42. Kline, R. H., Jr., Wright, J., Fox, K. M., and Eldefrawi, M. E. (1990) Synthesis of 3-arylecgonine analogues as inhibitors of cocaine binding and dopamine uptake. *J. Med. Chem.* **33**, 2024–2027.

43. Carroll, F. I., Kuzemko, M. A., Gao, Y., Abraham, P., Lewin, A. H., Boja, J. W., et al. (1992) Synthesis and ligand binding of 3β-(3-substituted phenyl)- and 3β-(3,4-disubstituted phenyl)tropane-2β-carboxylic acid methyl esters. *Med. Chem. Res.* **1**, 382–387.

44. Carroll, F. I., Mascarella, S. W., Kuzemko, M. A., Gao, Y., Abraham, P., Lewin, A. H., et al. (1994) Synthesis, ligand binding, and QSAR (CoMFA

and classical) study of 3β-(3'-substituted phenyl)-, 3β-(4'-substituted phenyl)-, and 3β-(3',4'-disubstituted phenyl)tropane-2b-carboxylic acid methyl esters. *J. Med. Chem.* **37,** 2865–2873.

45. Blough, B. E., Abraham, P., Lewin, A. H., Kuhar, M. J., Boja, J. W., and Carroll, F. I. (1996) Synthesis and transporter binding properties of 3β-(4'-Alkyl-, 4'-alkenyl-, and 4'-alkynylphenyl)nortropane-2β-carboxylic acid methyl esters: Serotonin transporter selective analogs. *J. Med. Chem.* **39,** 4027–4035.

46. Carroll, F. I., Kotian, P., Gray, J. L., Abraham, P., Kuzemko, M. A., Lewin, A. H., et al. (1993) 3β-(4'-Chlorophenyl)tropan-2β-carboxamides and cocaine amide analogues: new high affinity and selective compounds for the dopamine transporter. *Med. Chem. Res.* **3,** 468–472.

47. Carroll, F. I., Gao, Y., Abraham, P., Lewin, A. H., Lew, R., Patel, A., et al. (1992) Probes for the cocaine receptor. Potentially irreversible ligands for the dopamine transporter. *J. Med. Chem.* **35,** 1813–1817.

48. Davies, H. M. L., Saikali, E., Sexton, T., and Childers, S. R. (1993) Novel 2-substituted cocaine analogs: binding properties at dopamine transport sites in rat striatum. *Eur. J. Pharmacol.-Mol. Pharmacol. Sec.* **244,** 93–97.

49. Davies, H. M. L., Saikali, E., Huby, N. J. S., Gilliat, V. J., Matasi, J. J., Sexton, T., et al. (1994) Synthesis of 2β-Acyl-3β-aryl-8-azabicyclo[3.2.1]octanes and their binding affinities at dopamine and serotonin transport sites in rat striatum and frontal cortex. *J. Med. Chem.* **37,** 1262–1268.

50. Bennett, B. A., Wichems, C. H., Hollingsworth, C. K., Davies, H. M. L., Thornley, C., Sexton, T., et al. (1995) Novel 2-substituted cocaine analogs: uptake and ligand binding studies at dopamine, serotonin and norepinephrine transport sites in the rat brain. *J. Pharmacol. Exp. Ther.* **272,** 1176–1186.

51. Davies, H. M. L., Kuhn, L. A., Thornley, C., Matasi, J. J., Sexton, T., and Childers, S. R. C. (1996) Synthesis of 3β-Aryl-8-azabicyclo[3.2.1]octanes with high binding affinities and selectivities for the serotonin transporter site. *J. Med. Chem.* **39,** 2554–2558.

52. Kozikowski, A. P., Saiah, M. K. E., Johnson, K. M., and Bergmann, J. S. (1995) Chemistry and biology of the 2β-alkyl-3β-phenyl analogues of cocaine: Subnanomolar affinity ligands that suggest a new pharmacophore model at the C-2 position. *J. Med. Chem.* **38,** 3086=3093.

53. Kelkar, S. V., Izenwasser, S., Katz, J. L., Klein, C. L., Zhu, N., and Trudell, M. L. (1994) Synthesis, cocaine receptor affinity, and dopamine uptake inhibition of several new 2β-substituted 3β-phenyltropanes. *J. Med. Chem.* **37,** 3875–3877.

54. Kotian, P., Abraham, P., Lewin, A. H., Mascarella, S. W., Boja, J. W., Kuhar, M. J., et al. (1995) Synthesis and ligand binding study of 3β-(4'-substituted phenyl)-2β-(heterocyclic)tropanes. *J. Med. Chem.* **38,** 3451–3453.

55. Kotian, P., Mascarella, S. W., Abraham, P., Lewin, A. H., Boja , J. W., Kuhar, M. J., et al. (1996) Synthesis, ligand binding, and quantitative structure–activity relationship study of 3β-(4'-substituted phenyl)-2β-(heterocyclic)tropanes: Evidence for an electrostatic interaction at the 2β-position. *J. Med. Chem.* **39,** 2753–2763.

56. Carroll, F. I., Gray, J. L., Abraham, P., Kuzemko, M. A., Lewin, A. H., Boja, J. W., et al. (1993) 3-Aryl-2-(3'-substituted-1',2',4'-oxadiazol-5'-yl)tropane analogues of cocaine: affinities at the cocaine binding site at the dopamine, serotonin, and norepinephrine transporters. *J. Med. Chem.* **36,** 2886–2890.

57. Madras, B. K., Kamien, J. B., Fahey, M. A., Canfield, D. R., Milius, R. A., Saha, J. K., et al. (1990) N-Modified fluorophenyltropane analogs of cocaine with high affinity for cocaine receptors. *Pharmacol. Biochem. Behav.* **35,** 949–953.

58. Milius, R. A., Saha, J. K., Madras, B. K., and Neumeyer, J. L. (1991) Synthesis and receptor binding of N-substituted tropane derivatives. High-affinity ligands for the cocaine receptor. *J. Med. Chem.* **34,** 1728–1731.

59. Boja, J. W., Kuhar, M. J., Kopajtic, T., Yang, E., Abraham, P., Lewin, A. H., et al. (1994) Secondary amine analogues of 3β-(4'-substituted phenyl)-tropane-2β-carboxylic acid esters and *N*-norcocaine exhibit enhanced affinity for serotonin and norepinephrine transporters. *J. Med. Chem.* **37,** 1220–1223.

60. Neumeyer, J. L., Wang, S., Gao, Y., Milius, R. A., Kula, N. S., Campbell, A., et al. (1994) N-w-Fluoroalkyl analogs of (1R)-2β-carbomethoxy-3β-(4-iodophenyl)-tropane (β-CIT): radiotracers for positron emission tomography and single photon emission computed tomography imaging of dopamine transporters. *J. Med. Chem.* **37,** 1558–1561.

61. Elmaleh, D. R., Fischman, A. J., Shoup, T. M., Byon, C., Hanson, R. N., Liang, A. Y., Meltzer, P. C., Madras, B. K. (1996) Preparation and biological evaluation of iodine-125-IACFT: A selective SPECT agent for imaging dopamine transporter sites. *J. Nucl. Med.* **37,** 1197–1202.

62. Neumeyer, J. L., Tamagnan, G., Wang, S., Gao, Y., Milius, R. A., Kula, N. S., et al. (1996) N-substituted analogs of 2β-carbomethoxy-3β-(4'-iodophenyl)tropane (β-CIT) with selective affinity to dopamine or serotonin transporters in rat forebrain. *J. Med. Chem.* **39,** 543–548.

63. Zhao, L., Johnson, K. M., Zhang, M., Flippen-Anderson, J., and Kozikowski, A. P. (2000) Chemical synthesis and pharmacology of 6- and 7-hydroxylated 2-carbomethoxy-3-(p-tolyl)tropanes: antagonism of cocaine's locomotor stimulant effects. *J. Med. Chem.* **43,** 3283–3294.

64. Meltzer, P. C., Blundell, P., Yong, Y. F., Chen, Z., George, C., Gonzalez, M. D., et al. (2000) 2-Carbomethoxy-3-aryl-8-bicyclo[3.2.1]octanes: potent non-nitrogen inhibitors of monoamine transporters. *J. Med. Chem.* **43,** 2982–2991.

65. Meltzer, P. C., Liang, A. Y., Blundell, P., Gonzalez, M. D., Chen, Z., George, C., et al. (1997) 2-Carbomethoxy-3-aryl-8-oxabicyclo[3.2.1]octanes: potent non-nitrogen inhibitors of monoamine transporters. *J. Med. Chem.* **40,** 2661–2673.

66. Hoepping, A., Johnson, K. M., George, C., Flippen-Anderson, J., and Kozikowski, A. P. (2000) Novel conformationally constrained tropane analogues by 6-endo-trig radical cyclization and stille coupling-switch of activity toward the serotonin and/or norepinephrine transporter. *J. Med. Chem.* **43,** 2064–2071.

67. Tamiz, A. P., Smith, M. P., and Kozikowski, A. P. (2000) Design, synthesis and biological evaluation of 7-azatricyclodecanes: analogues of cocaine. *Bioorg. Med. Chem. Lett.* **10,** 297–300.

68. Newman, A. H., Allen, A. C., Izenwasser, S., and Katz, J. L. (1994) Novel 3a-(diphenylmethoxy)tropane analogs: Potent dopamine uptake inhibitors without cocaine-like behavioral profiles. *J. Med. Chem.* **37,** 2258–2261.

69. Newman, A. H., Kline, R. H., Allen, A. C., Izenwasser, S., George, C., and Katz, J. L. (1995) Novel 4'- and 4',4"-substituted-3α-(diphenylmethoxy)-tropane analogs are potent and selective dopamine uptake inhibitors. *J. Med. Chem.* **38,** 3933-3940.

70. Meltzer, P. C., Liang, A. Y., and Madras, B. K. (1994) The discovery of an unusually selective and novel cocaine analog: Difluoropine. Synthesis and Inhibition of binding at cocaine recognition sites. *J. Med. Chem.* **37,** 2001–2010.

71. Meltzer, P. C., Liang, A. Y., and Madras, B. K. (1996) 2-Carbomethoxy-3-(diarylmethoxy)-1aH,5aH-tropane analogs: Synthesis and inhibition of binding at the dopamine transporter and comparison with piperazines of the GBR series. *J. Med. Chem.* **39,** 371–379.

72. Vaughan, R. A., Agoston, G. E., Lever, J. R., and Newman, A. H. (1999) Differential binding of tropane-based photoaffinity ligands on the dopamine transporter. *J. Neurosci.* **19.**

73. Katz, J. L., Izenwasser, S., Kline, R. H., Allen, A. C., and Newman, A. H. (1999) Novel 3α-diphenylmethoxytropane analogs: selective dopamine uptake inhibitors with behavioral effects distinct from those of cocaine. *J. Pharmacol. Exp. Ther.* **288,** 302–315.

74. Newman, A. H., Izenwasser, S., Robarge, M. J., and Kline, R. H. (1999) CoMFA study of novel phenyl ring-substituted 3α-(diphenylmethoxy)-tropane analogues at the dopamine transporter. *J. Med. Chem.* **42,** 3502–3509.

75. Agoston, G. E., Wu, J. H., Izenwasser, S., George, C., Katz, J., Kline, R. H., et al. (1997) Novel N-substituted 3α-[bis(4'-fluorophenyl)methoxy]tropane analogues: selective ligands for the dopamine transporter. *J. Med. Chem.* **40,** 4329–4339.

76. Robarge, M. J., Agoston, G. E., Izenwasser, S., Kopajtic, T., George, C., Katz, J. L., et al. (2000) Highly selective chiral N-substituted 3α-[bis(4'-fluorophenyl)methoxy]tropane analogues for the dopamine transporter: synthesis and comparative molecular field analysis. *J. Med. Chem.* **43,** 1085–1093.

77. Newman, A. H., Robarge, M. J., Howard, I. M., Wittkopp, S. L., George, C., Kopajtic, T., et al. (2001) Structure–activity relationships at monoamine transporters and muscarinic receptors for N-substituted-3α-(3'-chloro-, 4'-chloro-, and 4',4"-dichloro-substituted-diphenyl)methoxytropanes. *J. Med. Chem.* **44,** 633–640.

78. Van der Zee, P., Koger, H. S., Gootjes, J., and Hespe, W. (1980) Aryl 1,4-dialk(en)ylpiperazines as selective and very potent inhibitors of dopamine uptake. *Eur. J. Med. Chem.* **15,** 363–370.

79. Berger, P., Janowsky, A., Vocci, F., Skolnick, P., Schweri, M. M., and Paul, S. M. (1985) [³H]GBR-12935: a specific high affinity ligand for labeling the dopamine transport complex. *Eur. J. Pharmacol.* **107,** 289–290.

80. Bonnet, J.-J., Protais, P., Chagraoui, A., and Costentin, J. (1986) High-affinity [³H]GBR 12783 binding to a specific site associated with the neuronal

dopamine uptake complex in the central nervous system. *Eur. J. Pharmacol.* **126,** 211–222.

81. Anderson, P. H. (1987) Biochemical and pharmacological characterization of [³H]GBR-12935 binding in vitro to rat striatal membranes: labeling of the dopamine uptake complex. *J. Neurochem.* **48,** 1887–1896.

82. Fahey, M. A., Canfield, D. A., Spealman, R. D., and Madras, B. K. (1989) Comparison of [³H]GBR 12935 and [³H]cocaine binding sites in monkey brain. *Soc. Neurosci. Abst.* **515,** 252.

83. Vaughan, R. A. (1995) Photoaffinity-labeled ligand binding domains on dopamine transporters identified by peptide mapping. *Mol. Pharmacol.* **47,** 956–964.

84. Vaughan, R. A., Brown, V. I., McCoy, M. T., and Kuhar, M. J. (1996) Species- and brain region-specifc dopamine transporters: immunological and glycosylation characteristics. *J. Neurochem.* **66,** 2146–2152.

85. Matecka, D., Rothman, R. B., Radesca, L., de Costa, B. R., Dersch, C. M., partilla, J. S., et al. (1996) Development of novel, potent, and selective dopamine reuptake inhibitors through alteration of the piperazine ring of 1-[2-(diphenylmethoxy)ethyl]- and 1-[2-[bis(4-fluorophenyl)methoxy]ethyl]-4-(3-phenylpropyl)piperazines (GBR 12935 and GBR 12909). *J. Med. Chem.* **39,** 4704–4716.

86. Matecka, D., Lewis, D., Rothman, R. B., Dersch, C. M., Wojnicki, F. H., Glowa, J. R., et al. (1997) Heteroaromatic analogs of 1-[2-(diphenylmethoxy)ethyl]- and 1-[2-[bis(4-fluorophenyl)methoxy]ethyl]-4-(3-phenylpropyl)piperaz ines (GBR 12935 and GBR 12909) as high-affinity dopamine reuptake inhibitors. *J. Med. Chem.* **40,** 705–716.

87. Zhang, Y., Rothman, R. B., Dersch, C. M., de Costa, B. R., Jacobson, A. E., and Rice, K. C. (2000) Synthesis and transporter binding properties of bridged piperazine analogues of 1-2-[Bis(4-fluorophenyl)methoxy]ethyl-4-(3-phenylpropyl)piperazine (GBR 12909) [In Process Citation]. *J. Med. Chem.* **43,** 4840–4849.

88. Matecka, D., Rice, K. C., Rothman, R. B., de Costa, B. R., Glowa, J. R., Wojnicki, F. H., et al. (1995) Synthesis and absolute configuration of chiral piperazines related to GBR 12909 as dopamine reuptake inhibitors. *Med. Chem. Res.* **5,** 43–53.

89. Dutta, A., Xu, C., and Reith, M. E. A. (1998) Tolerance in the replacement of the benzhydrylic O atom in 4-[2-(Diphenylmethoxy)ethyl]-1-benzylpiperidine derivatives by an N atom: development of new-generation potent and selective N-analogue molecules for the dopamine transporter. *J. Med. Chem.* **41,** 3293–3297.

90. Dutta, A. K., Xu, C., and Reith, M. E. (1996) Structure-activity relationship studies of novel 4-[2-[bis(4-fluorophenyl)methoxy]ethyl]-1-(3-phenylpropyl)piperidine analogs: synthesis and biological evaluation at the dopamine and serotonin transporter sites. *J. Med. Chem.* **39,** 749–756.

91. Elmaleh, D. R., Hanson, R. N., and C., S.-W., (1999) WO9912893A1: diagnostic and therapeutic alkylenediamine compounds and process. *WO 9912893,* pp. 63.

92. Dutta, A. K., Coffey, L. L., and Reith, M. E. A. (1998) Potent and selective ligands for the dopamine transporter (DAT): structure-activity relationship studies of novel 4-[2-(diphenylmethoxy)ethyl]-1-(3-phenylpropyl)piperidien analogues. *J. Med. Chem.* **41,** 699–705.

93. Madras, B. K., Reith, M. E., Meltzer, P. C., and Dutta, A. K. (1994) O-526, a piperidine analog of GBR 12909, retains high affinity for the dopamine transporter in monkey caudate-putamen. *Eur. J. Pharmacol.* **267,** 167–173.

94. Smith, M. P., Johnson, K. M., Zhang, M., Flippen-Anderson, J. L., and Kozikowski, A. P. (1998) Tuning the selectivity of monoamine transporter inhibitors by the stereochemistry of the nitrogen lone pair. *J. Am. Chem. Soc.* **120,** 9072–9073.

95. van der Zee, P. and Hespe, W. (1985) Interaction between substituted 1-[2-(diphenylmethoxy)ethyl] piperazines and dopamine receptors. *Neuropharmacology* **24,** 1171–1174.

96. Deutsch, H. M., and Schweri, M. M. (1994) Can stimulant binding and dopamine transport be differentiated? studies with GBR 12783 derivatives. *Life Sci.* **55,** 115–120.

97. Dutta, A. K., Fei, X. S., Beardsley, P. M., Newman, J. L., Reith, M. E. A. (2001) Structure activity relationshop studies of 4-[2[(diphenylmethoxyl) ethyl]-1-benzylpiperidine derivatives and their N-analogues: evaluation of behavioral activity of O- and N-analogues and their binding to monoamine transporters. *J. Med. Chem.* **44,** 937–948.

98. Deutsch, H. M., Schweri, M., M., Culbertson, C. T., and Zalkow, L. H. (1992) Synthesis and pharmacology of irreversible affinity labels as potential cocaine antagonists: aryl 1,4-dialkylpiperazines related to GBR-12783. *Eur. J. Pharmacol.* **220,** 173–180.

99. Matecka, D., Rice, K. C., Rothman, R. B., de Costa, B. R., Glowa, J. R., Wojnicki, F. H., et al. (1994) Synthesis and absolute configuration of chiral piperazines related to GBR 12909 as dopamine uptake inhibitors. *Med. Chem. Res.* **5,** 43–53.

100. Madras, B. K., Reith, M. E. A., Meltzer, P. C., and Dutta, A. K. (1993) O-526, a piperidine analog of GBR 12909, retains high affinity for the dopamine transporter in monkey caudate-putamen. *Eur. J. Pharmacol.-Mol. Pharmacol. Sec.* **267,** 167–173.

101. Madras, B. K., Fahey, M. A., Bergman, J., Canfield, D. R., and Spealman, R. D. (1989) Effects of cocaine and related drugs in nonhuman primates. I. [^{3}H]Cocaine binding sites in caudate-putamen. *J. Pharmacol. Exp. Ther.* **251,** 131–141.

102. Choi, S. W., Elmaleh, D. R., Hanson, R. N., and Fischman, A. J. (1999) Design, synthesis, and biological evaluation of novel non-piperazine analogues of 1-[2-(Diphenylmethoxy)ethyl]- and 1-[2-[Bis(4- fluorophenyl)-methoxy]ethyl]-4-(3-phenylpropyl)piperazines as dopamine transporter inhibitors. *J Med Chem* **42,** 3647–3656.

103. Muller, L., Halldin, C., Farde, L., Karlsson, P., Hall, H., Swahn, C. G., et al. (1993) Neumeyer, J., Gao, Y., and Milius, R. [11C]b-CIT, a cocaine analogue. Preparation, autoradiography and preliminary PET investigations. *Nucl. Med. Biol.* **20,** 249–255.

104. Dutta, A., Coffey, L. L., and Reith, M. E. A. (1997) Highly selective, novel analogs of 4-[2-(diphenylmethoxy)ethyl]-1-benzylpiperidine for the dopamine transporter: Effect of different aromatic substitutions on their affinity and selectivity. *J. Med. Chem.* **40,** 35–43.

105. *Physicians' Desk Reference*, 49th ed., Medical Economics Data, (1995), Montvale, NJ, pp. 309, 897.

106. Deutsch, H. M., Shi, Q., Gruszecka-Kowalik, E., and Schweri, M. M. (1996) Synthesis and pharmacology of potential cocaine antagonists. 2. Structure-activity relationship studies of aromatic ring-substituted methylphenidate analogs. *J. Med. Chem.* **39,** 1201–1209.

107. Schweri, M. M., Skolnick, P., Rafferty, M. F., Rice, K. C., Janowsky, A. J., and Paul, S. M. (1985) [^{3}H]Threo-(±)-methylphenidate binding to 3,4-dihydroxyphenylethylamine uptake sites in corpus striatum correlation with the stimulant properties of ritalinic acid esters. *J. Neurochem.* **45,** 1062–1070.

108. Thai, D. L., Sapko, M. T., Reiter, C. T., Bierer, D. E., and Perel, J. M. (1998) Asymmetric synthesis and pharmacology of methylphenidate and its para-substituted derivatives. *J. Med. Chem.* **41,** 591–601.

109. Axten, J., Krim, L., Kung, H. F., and Winkler, J. D. (1998) A stereoselective synthesis of *dl-threo*-methylphenidate: preparation and biological evaluation of novel analogues. *J. Org. Chem.* **63,** 9628–9629.

110. Froimowitz, M., Deutsch, H. M., Shi, Q., Wu, K. M., Glasser, R., Adin, I., et al. (1997) *Bioorg. Med. Chem. Lett.* **7,** 1213.

111. Colpaert, F. C., Niemegeers, C. J. E., and Janssen, P. A. J. (1978) Discriminative stimulus properties of cocaine. Neuropharmacological characteristics as derived from stimulus generalization experiments. *Pharmacol. Biochem. Behav.* **10,** 535–546.

112. Vignon, J. and Lazdunski, M. (1984) Structure-function relationships in the inhibition of synaptosomal dopamine uptake by phencyclidine and analogues: potential correlation with binding sites identified with [^{3}H]phencyclidine. *Biochem. Pharmacol.* **33,** 700–702.

113. Vignon, J., Cerruti, C., Chaudieu, I., Pinet, V., Chicheportiche, M., Kamenka, J.-M., et al. (1988) Interaction of molecules in the phencyclidine series with the dopamine uptake system. Correlation with their binding properties to the phencyclidine receptor. Binding properties of ^{3}H-BTCP, a new PCP analog, to the dopamine uptake complex, in *Sigma and Phencyclidine-Like Compounds as Molecular Probes in Biology*, (Domino, E. F. and Kamenka J.-M. eds.), NPP Books, Ann Arbor, MI, pp. 199–208.

114. Kuhar, M. J., Ritz, M. C., and Boja, J. W. (1991) The dopamine hypothesis of the reinforcing properties of cocaine. *Trends Neurosci.* **14,** 299–302.

115. Koek, W., Colpaert, F. C., Woods, J. H., and Kamenka, J.-M. (1989) The phencyclidine (PCP) analog N-[1-(2-benzo(*B*)thiophenyl)cyclohexyl]-piperidine shares cocaine-like but not other characteristic behavioral effects with PCP, ketamine and MK-801. *J. Pharmacol. Exp. Ther.* **250,** 1019–1027.

116. He, X.-S., Raymon, L. P., Mattson, M. V., Eldefrawi, M. E., and de Costa, B. R. (1993) Synthesis and biological evaluation of 1-[1-(2-benzo[b]-

thienyl)cyclohexyl] piperidine homologues at dopamine-uptake and phencyclidine-and s-binding sites. *J. Med. Chem.* **36,** 1188–1193.

117. Coderc, E., Cerruti, P., Vignon, J., Rouayrenc, J. F., and Kamenka, J.-M. (1995) PCP receptor and dopamine uptake sites are discriminated by chiral TCP and BTCP derivatives of opposite configuration. *Eur. J. Med. Chem.* **30,** 463–470.

118. Vignon, J., Pinet, V., Cerruti, C., Kamenka, J.-M., and Chicheportiche, R. (1988) [³H]N-[1-(2-Benzo(b)thiophenyl)cyclohexyl]piperidine ([³H]BTCP): a new phencyclidine analog selective for the dopamine uptake complex. *Eur. J. Pharmacol.* **148,** 427–436.

119. Duterte-Boucher, D., Vaugeois, J.-M., Costentin, J., Ilagouma, A., Maurice, T., Vignon, J., et al. (1992) In vivo and in vitro effects of BTCP and derivatives related to dopamine and norepinephrine neuronal uptake inhibition, in *Multiple Sigma and PCP Receptor Ligands*, (Kamenka, J.-M. and Domino, E. F., eds.), NPP Books, Ann Arbor, MI, pp. 435–443.

120. Billaud, G., Menard, J. F., Marcellin, N., Kamenka, J. M., Costentin, J., and Bonnet, J. J. (1994) Thermodynamics of the binding of BTCP (GK 13) and related derivatives on the dopamine neuronal carrier. *Eur J Pharmacol* **268,** 357–363.

121. Hamon, J., Vignon, J., and Kamenka, J. M. (1996) Effect of lowered lipophilicity on the affinity of PCP analogues for the PCP receptor and the dopamine transporter. *Eur. J. Med. Chem.* **31,** 489–495.

122. Ilagouma, A. T., Maurice, T., Duterte-Boucher, D., Coderc, E., Vignon, J., Costentin, J., et al. (1993) Arylcyclohexylamines derived from BTCP are potent indirect catecholamine agonists. *Eur. J. Med. Chem.* **28,** 377–385.

123. Deleuze-Masquefa, C., Michaud, M., Vignon, J., and Kamenka, J. M. (1997) 1-[1-2-Benzo[b]thiopheneyl)cyclohexyl]piperidine hydrochloride (BTCP) yields two active primary metabolites in vitro: synthesis, identification from rat liver microsome extracts, and affinity for the neuronal dopamine transporter. *J Med Chem* **40,** 4019–4025.

124. Coderc, E., Cerruti, P., Vignon, J., Rouayrenc, J. F., and Kamenka, J.-M. (1997) Sensitivity of the PCP receptor and the dopamine transporter to ligands bearing multiple asymmetric centers. *Eur. J. Med. Chem.* **32,** 263–271.

125. Saunders, J., Cassidy, M., Freedman, S. B., Harley, E. A., Iversen, L. L., Kneen, C., et al. (1990) Novel quinuclidine-based ligands for the muscarinic cholinergic receptor. *J. Med. Chem.* **33,** 1128–1138.

126. He, X.-S., Raymon, L. P., Mattson, M. V., Eldefrawi, M. E., and de Costa, B. R. (1993) Further studies of the structure–activity relationships of 1-[1-(2-benzo[*b*]thienyl)cyclohexyl]piperidine. Synthesis and evaluation of 1-(2-benzo[*b*]thienyl)-*N,N*-dialkylcyclohexylamines at dopamine uptake and phencyclidine binding sites. *J. Med. Chem.* **36,** 4075–4081.

127. Coderc, E., Martin-Fardon, R., Vignon, J., and Kamenka, J.-M. (1993) New compounds resulting from structural and biochemical similarities between GBR 12783 and BTCP, two potent inhibitors of dopamine uptake. *Eur. J. Med. Chem.–Chim. Thera.* **28,** 893–898.

128. *Physicians' Desk Reference*, 49th ed., Medical Economics Data, (1995), Montvale, NJ, pp. 2190–2191, 2269.

129. Aeberli, P., Eden, P., Gogerty, J. H., Houlihan, W. J., and Penberthy, C. (1975) 5-Aryl-2,3-dihydro-5*H*-imidazo[2-1-*a*]isoindol-5-ols. A novel class of anorectic agents. *J. Med. Chem.* **18**, 177–182.

130. Aeberli, P., Eden, P., Gogerty, J. H., Houlihan, W. J., and Penberthy, C. (1975) Anorectic agents. 2. Structural analogs of 5-(*p*-chlorophenyl)-2,3-dihydro-5*H*-imdazo[2,1-*a*]isoindol-5-ol. *J. Med. Chem.* **18**, 182–185.

131. Chait, L. D., Uhlenhuth, E. H., and Johansen, C. E. (1987) Reinforcing and subjective effects of several anorectics in normal human volunteers. *J. Pharmacol. Exp. Ther.* **242**, 777–783.

132. Javitch, J. A., Blaustein, R. O., and Snyder, S. H. (1984) [^{3}H]Mazindol binding associated with neuronal dopamine and norepinephrine uptake sites. *Mol. Pharmacol.* **26**, 35–44.

133. Berger, P., Gawin, F., and Koster, T. R. (1989) Treatment of cocaine abuse with mazindol. *Lancet II*, 283.

134. Barcza, S., and Houlihan, W. J. (1975) Structure determination of the anorexic agent Mazindol. *J. Pharm. Sci.* **64**, 829–831.

135. Houlihan, W. J., Boja, J. W., Parrino, V. A., Kopajtic, T. A., and Kuhar, M. J. (1996) Halogenated mazindol analogs as potential inhibitors of the cocaine binding site at the dopamine transporter. *J. Med. Chem.* **39**, 4935–4941.

136. Galinier, E., Garreau, L., Dognon, A. M., Ombetta-Goka, J. E., Frangin, Y., Chalon, S., et al. (1993) Synthesis of halogenated analogs of 5-(4-chlorophenyl)-2,3-dihydro-5-hydroxy-5H-imidazo[2,1-a]-isoindole or mazindol for exploration of the dopmaine transporter. *Eur. J. Med. Chem.* **28**, 927–933.

137. Heikkila, R. E., Manzino, L., and Cabbat, F. S. (1981) Stereospecific effects of cocaine derivatives on ^{3}H-dopamine uptake: Correlations with behavioral effects. *Sub. Alcohol Actions/Misuse* **2**, 115–121.

138. Houlihan, W. J., Boja, J. W., Kopajtic, T. A., Kuhar, M. J., Degrado, S. J., and Toledo, L. (1998) Positional isomers and analogs of mazindol as potential inhibitors of the cocaine binding site of the dopmaine transporter site. *Med. Chem. Res.* **8**, 77–90.

139. Houlihan, W. J., Boja, J. W., Parrino, V. A., Kuhar, M. J., and Kopajtic, T. A., (1994) Halogenated mazindol analogs as potential inhibitors of the cocaine binding site at the dopamine transporter. *208th Am. Chem. Soc. Natl. Meeting, Vol. Abstr. 173*, Washington, DC, August 21–25.

140. Houlihan, W. J., Boja, J. W., Kuhar, M. J., Kopajtic, T. A., and Parrino, V. A. (1995) Mazindol analogs as potential inhibitors of the cocaine binding site. Structure activity relationships. *Abst. College Prob. Drug Depend.* Scottsdale, AZ, June 10–15, p. 66.

141. Heikkila, R. E., Babington, R. G., and Houlihan, W. J. (1981) Pharmacological studies with several analogs of mazindol: correlation between effects on dopamine uptake and various in vivo responses. *Eur. J. Pharmacol.* **71**, 277–286.

142. Carroll, F. I., Lewin, A. H., Boja, J. W., and Kuhar, M. J. (1992) Cocaine receptor: Biochemical characterization and structure-activity relationships for the dopamine transporter. *J. Med. Chem.* **35**, 969–981.

143. McElvain, J. S., and Schenk, J. O. (1992) A multisubstrate mechanism of

striatal dopamine uptake and its inhibition by cocaine. *Biochem. Pharmacol.* **43**, 2189–2199.

144. Kitayama, S., Shimada, S., Xu, H., Markham, L., Donovan, D. M., and Uhl, G. R. (1992) Dopamine transporter site-directed mutations differentially alter substrate transport and cocaine binding. *Proc. Natl. Acad. Sci. USA* **89**, 7782–7785.

145. Giros, B., Wang, Y.-M., Suter, S., McLeskey, S. B., Pifl, C., and Caron, M. G. (1994) Delineation of discrete domains for substrate, cocaine, and tricyclic antidepressant interactions using chimeric dopamine-norepinephrine transporters. *J. Biol. Chem.* **269**, 15,985–15,988.

146. Johnson, K. M., Bergman, J. S., and Kozikowski, A. P. (1992) Cocaine and dopamine differentially protect [³H]mazinol binding sites from alkylations by N-ethylmaleimide. *Eur. J. Pharmacol.* **227**, 411–415.

147. Lin, Z., Wang, W., Kopajtic, T., Revay, R. S., and Uhl, G. R. (1999) Dopamine transporter: transmembrane phenylalanine mutations can selectively influence dopamine uptake and cocaine analog recognition. *Mol. Pharmacol.* **56**, 434–447.

148. Meltzer, P. C., Blundell, P., and Madras, B. K. (1998) Structure activity relationships of inhibition of the dopamine transporter by 3-arylbicyclo-[3,2,1]octanes. *Med. Chem. Res.* **8**, 12–34.

149. Xu, C., Coffey, L. L., and Reith, M. E. A. (1995) Translocation of dopamine and binding of 2β-carbomethoxy-3β-(4-fluorophenyl)tropane (WIN 35,428) measured under identical conditions in rat striatal synaptosomal preparations. Inhibition by various blockers. *Biochem. Pharmacol.* **49**, 339–350.

150. Simoni, D., Stoelwinder, J., Kozikowski, A. P., Johnson, K. M., Bergmann, J. S., and Ball, R. G. (1993) Methoxylation of cocaine reduces binding affinity and produces compounds of differential binding and dopamine uptake inhibitory activity: discovery of a weak cocaine "antagonist". *J. Med. Chem.* **36**, 3975–3977.

151. Rothman, R. B., Becketts, K. M., Radesca, L. R., de Costa, B. R., Rice, K. C., Carroll, F. I., et al. (1993) Studies of the biogenic amine transporters. II. A brief study on the use of [³H]DA-uptake-inhibition to transporter-binding-inhibition ratios for the in vitro evaluation of putative cocaine antagonists. *Life Sci.* **53**, PL267–PL272.

152. Reith, M. E. A., Xu, C., Carroll, F. I., and Chen, N.-H. (1998) Inhibition of [3H]dopamine translocation and [3H] cocaine analog binding a potential screening device for cocaine antagonists, in *Methods in Enzymology, Vol. 296, Neurotransmitter Transporters* (Amara, S. G. ed.), Academic Press, San Diego, CA, pp. 248–259.

153. Slusher, B. S., Tiffany, C. W., Olkowski, J. L., and Jackson, P. F. (1997) Use of identical assay conditions for cocaine analog binding and dopamine uptake to identify potential antagonists. *Drug Alcohol Depend.* **48**, 43–50.

154. Eshleman, A. J., Carmolli, M., Cumbay, M., Martens, C. R., Neve, K. A., and Janowsky, A. (1999) Characteristics of drug interactions with recombinant giogenic amine transporters expressed in the same cell type. *J. Pharmacol. Exp. Ther.* **289**, 877–885.

155. Warner, E. A., Kosten, T. R., and O'Connor, P. G. (1997) Pharmacotherapy for Opioid and Cocaine Abuse. *Med. Clin. North Am.* **81,** 909-925.
156. Wang,L. C., Berfield, J. L., Kuhar, M. J., Carroll, F. I., and Reith, M. E. A. (2000) RTI-76, an isothiocyanate derivative of a phenyltropane cocaine analog, as a tool for irreversibly inactivating dopamine transporter function in vitro. *Naunyn-Schmiedeberg's Arch. Pharmacol.* **362,** 238–247.

12
Imaging the Brain Dopamine Transporter Using PET and SPECT

S. John Gatley, Nora D. Volkow, Joanna S. Fowler,
Yu-Shin Ding, Jean Logan, Gene-Jack Wang,
Christoph Felder, Frank W. Telang,
and Andrew N. Gifford

1. INTRODUCTION

The corresponding chapter in the first edition of this book *(1)* reviewed progress in imaging of the dopamine transporter (DAT), and related topics, up to about the end of 1995. There have been considerable advances in this area in the last five years, including the development of novel radioligands as well as the use of more established radioligands to evaluate changes in DAT-density in various disease states, and to estimate the degree of occupancy of the DAT associated with drug effects.

2. PET AND SPECT

Positron emission tomography (PET) is an imaging method used to track the regional distribution and kinetics of chemical compounds labeled with short-lived positron-emitting isotopes in the living body *(2)*. It was the first technology that enabled direct measurement of components of the neurochemical systems in the living human brain *(3)*.

Current commercially available PET technology for human brain scanning offers spatial resolution of 3–4 mm in studies that encompass the entire brain *(4)*. This resolution is sufficient to tomographically isolate the putamen and caudate nucleus, but not to provide a pure signal from smaller dopaminergic brain regions such as the nucleus accumbens or substantia nigra (SN). PET is an expensive technology because of the cost of producing positron-emitting radiotracers as well as the cost of the tomograph. Because of the short half-life of ^{11}C (20 min), this must be produced using a

From: *Contemporary Neuroscience:*
Neurotransmitter Transporters: Structure, Function, and Regulation, 2nd Edition
Edited by: M. E. A. Reith © Humana Press Inc., Totowa, NJ

cyclotron and radiochemical laboratory in close proximity to the PET laboratory. The half-life of ^{18}F (110 min) allows transport of radiotracers over only limited distances. Both the radiotracer production and imaging phases of the PET studies demand highly skilled technical assistance. SPECT (single-photon emission computed tomography) methodology can also be used to measure some of the same components of the DA system as PET. SPECT is inferior to PET in terms of spatial resolution (approx 6 mm), sensitivity, and quantitation. However, this methodology is less expensive than PET, and far more widely available because of its advantages in clinical nuclear medicine. These include the use of radionuclides of longer half-life, such as ^{123}I (13 h) and ^{99m}Tc (6 h), that do not necessitate PET's tight coupling between radiotracer preparation and imaging study. Although their radiosynthesis is typically less involved, it is often more difficult to design SPECT tracers than PET tracers. Table 1 provides key properties of the radionuclides discussed in this chapter.

3. THE DOPAMINE SYSTEM

The dopamine (DA) system REF is involved in the regulation of brain regions that subserve motor, cognitive, and motivational behaviors. Disruptions of DA function have been implicated in neurological and psychiatric illnesses including substance abuse, as well as in some of the deficits associated with aging of the human brain. This has made the DA system an important topic of research in the neurosciences and an important molecular target for drug development. DA is released into the synapse in response to an action potential, and interacts with postsynaptic DA receptors. The concentration of DA in and around the synapse is regulated primarily by its reuptake by the DAT, to maintain low (nanomolar) steady-state concentrations. The DAT is thus a critical element in the control of (extra-)synaptic DA *(5)*.

4. PET AND SPECT RADIOTRACERS OF THE DA SYSTEM

Although DA was labeled with ^{11}C 30 yr ago *(6)*, it does not cross the blood-brain barrier, so that direct imaging studies of DA reuptake in the living brain are not possible. However, PET investigation can be conducted with radiotracers that are DA precursors or that have high affinity for DA receptors or DA-metabolizing enzymes as well as the DAT. Tracers that provide information on regional brain-glucose metabolism and cerebral blood flow have also made possible assessment of the functional consequences of changes in brain DA activity. We have recently reviewed progress in this area *(7)*.

Table 1
Nuclear Decay Properties of Pertinent Radionuclides

Nuclide	Half-life	Decay mode
$^{11}C^a$	20 min	Positron (511 keV annihilation photon)
$^{18}F^a$	110 min	Positron (511 keV annihilation photon)
$^{99m}Tc^b$	6 h	Gamma, 140 keV
$^{123}I^a$	13.2 h	Electron capture (gamma, 159 keV)
$^{3}H^b$	12.6 yr	Beta

[a]Nuclides produced using cyclotrons
[b]Nuclides produced using nuclear reactors (^{99m}Tc is the metastable daughter of ^{99}Mo (half-life = 66 h). It is obtained inexpensively in the radiopharmacy from "generators" containing ^{99}Mo. ^{99m}Tc has favorable properties for clinical nuclear imaging.)

Interest in imaging of the DAT has been stimulated, in part, by the fact that it constitutes the main target site for the reinforcing properties of stimulant drugs such as cocaine and methylphenidate. Because transporters are localized on the presynaptic terminal, they may serve as markers of DA neurons in studies of neurodegenerative diseases. Many radioligands have been evaluated for their suitability as PET and SPECT probes of the DAT. These compounds differ with respect to their affinities for the DAT, their specific to nonspecific binding ratios, and their specificity for the DAT as well as their kinetics. The mass of labeled drug injected in tracer-level PET or SPECT experiments is routinely less than 50 μg (i.e., <1 μg/kg) so that a very small fraction of the transporters is occupied by the radioligand. Thus, no pharmacological or toxic effects are expected, even from very potent DA reuptake blockers.

Table 2 summarizes information for six DAT radioligands which have been examined in reasonable detail in human subjects. These include [^{11}C]cocaine and three structurally related compounds containing the tropane moiety: WIN 35,428 (also known as CFT), RTI-55 (also known as β-CIT) and TRODAT-1. A modified form of this table appeared in an earlier review *(7)*, where data for DAT radioligands were compared. Data for TRODAT-1 are from more recent articles of Kung and colleagues *(8–10)*. Table 3 shows the structures of these radioligands together with the structures of several other radiolabeled tropanes mentioned in this chapter. The initial uptake in brain for each of the radioligands in Table 2 is high; for the PET tracers it corresponds to approx 7–10% of the injected dose, and for RTI-55 and TRODAT-1 it is about half this level. As a whole the data form a self-consistent set in that in vivo binding parameters are clearly related to

Table 2
Radioligands for DATs

Radioligand	K_i in vitro[a] (nM)	Peak ST[g] (nCi/cc/mCi)	Human data ST/CB[f]	Time max[b] (min)	B_{max}/K_d'[c] (mL/g)	ST/Pl[d]	Rodent data ST/CB[f]	Time max[b] (min)
RTI-55	1.5	130 (500')	12+	>1200	6.7[i]	250	10+	120
WIN 35 428	12	150 (60')	4+	>120	5	?	4	60
dtMP	40	100	2.3	30–40	1.6	25	2	15
Nomifensine	45	85[h]	1.9	20–30	0.9[e]	10[e]	–	–
Cocaine	100	85	1.7	5–7	0.6	4	1.5	5
TRODAT-1	14	50	2.4	200–300	1.3	?	2.7	60

[a]Rat striatal membranes.

[b]Time of maximum uptake in striatum.

[c]Calculated using the ratio of distribution volume in striatum to that in reference tissue, obtained using graphical analysis for cocaine and dtMP; B_{max}/K_d from 2-compartment model, for nomifensine; calculated as k_3/k_4 from 3-compartment model, for WIN 35,428, RTI-55, and TRODAT-1.

[d]At time of maximal striatal binding; not corrected for plasma protein binding.

[e]Metabolite correction based on plasma/erythrocyte partitioning.

[f]Maximum value observed; at end of measurement period if followed by "+".

[g]Representative radioactivity concentration in striatum (nCi/cc/mCi injected) at peak or at indicated time after injection.

[h]Estimated from MBq (g tissue)-1/MBq (g body wt)-1.

[i]Termed V3", or "equilibrium partition coefficient" (251).

[j]Mixture of diastereomers.

Table 3
Structures of Some DAT Radioligands

Compound	R1	R2	R3
Cocaine	CH$_3$	COOCH$_3$	Benzoyl
WIN 35,428	CH$_3$	COOCH3	4-F-phenyl
RTI-55	CH$_3$	COOCH$_3$	4-I-phenyl
TRODAT-1	See figure	COOCH$_3$	4-Cl-phenyl
Technepine	CH$_3$	See figure	4-F-phenyl
PE2I	3-iodoprop-2-enyl	COOCH	4-Me-phenyl
FP-CIT	3-fluoropropyl	COOCH$_3$	4-I-phenyl
(+)-FCT	CH$_3$	COCH$_3$	4-F-phenyl
IPT	3-iodoprop-2-enyl	COOCH	4-Cl-phenyl
IACFT	Iodoallyl	COOCH	4-F-phenyl

Technepine   TRODAT-1

437

the affinity of the radioligand for the DAT. The peak binding, the time to peak binding, and the clearance half-time in striatum are inversely correlated with in vitro K_i or K_d value. The times to maximum uptake of the six radioligands vary by a factor of over 200, that is between about 6 min for cocaine, and over 1200 min for the high-affinity cocaine analog RTI-55. The choice of PET radioligand may depend on the purpose of the study: visualizing a gross abnormality in DAT distribution; quantifying altered regional DAT concentrations, or measuring the degree of occupancy of the DAT by an abused or therapeutic drug. For example, the low affinity of [¹¹C]cocaine results in fast-clearance kinetics, which limits the statistical quality of PET images compared with the other radioligands. However, low affinity also causes rapid equilibration, which gives [¹¹C]cocaine superior properties for measurements of occupancy of the DAT by other drugs *(11)*.

[¹¹C]Nomifensine was the first DAT radioligand developed for PET *(12)*. Its uptake in the brain is relatively rapid and peak concentrations are achieved approx 20 min after its administration, which allows for proper modeling and quantification. Although this ligand binds more tightly to the norepinephrine transporter (NET) than to the DAT, this does not appear to be a problem in PET studies. The specific binding of [¹¹C]*d-threo* meth-ylphenidate in brain is similar to that of nomifensine *(13,14)*. Striatal kinetics are rapid enough to allow for proper quantification, but slow enough to permit appropriate counting statistics. An advantage in developing [¹¹C]*d-threo* methylphenidate was that the pharmacology of methylphenidate in humans was well-studied. This simplified and accelerated the development process; for example, unlabeled methylphenidate could be administered to allow the assessment of nonspecific binding *(15)*. The cocaine analogs WIN 35,428 and RTI-55, bind more strongly to the DAT than S-(+)-nomifensine or *d-threo*-methylphenidate *(16–19)*. They have affinities approx 10 and 100× higher than those of cocaine, respectively, and exhibit correspondingly slower in vivo kinetics.

[¹¹C]WIN 35,428 does not achieve a maximum value of striatal binding in humans within the time constraints imposed by the short half-life of ¹¹C; the decay-corrected concentration of radioactivity in striatum is still rising after more than 5 half-lives for ¹¹C have passed (110 min) *(20)*. The extent to which this failure to reach equilibrium may compromise its ability to detect small decreases in DAT concentration is presently unclear. The potential problem under these conditions is that tissues with some threshold concentration of binding sites may trap essentially every molecule of radiotracer which is delivered. This is expected to cause radiotracer binding to become very sensitive to alterations in tissue perfusion, which proportionately alters delivery of the radioligand. It is suggestive that studies with

Table 4
PET and SPECT Studies of the Effect of Aging
on the DAT

Investigators	Results	#	Tracer
(20)	No effect	10	[^{11}C]WIN 35
(59)	Decrease	24	[^{11}C]Cocaine
(60)	Decrease 8%/decade	28	[^{123}I]RTI-55
(61)	Decrease 5-10%/decade	15	[^{18}F]FP-CIT
(62)	Nonlinear decrease	18	[^{123}I]IPT
(69)	Decrease 7%/decade	23	[^{11}C]dtMP
(64)	Decrease	10	[^{123}I]RTI-55
(65)	No effect	15	[^{123}I]RTI-55
(48)	Decrease 7%/decade	7	[^{18}F]FP-CIT
(31)	Nonlinear decrease	55	[^{99m}Tc]TRODAT-1
(66)	Decrease 4%/decade	16	[^{123}I]PE2I
(67)	Decrease 8%/decade	35	[^{123}I]RTI-55
(68)	Decrease	45	[^{18}F]FP-CIT

#, Number of subjects in the study.

[^{11}C]WIN 35,428, unlike those using other radioligands (Table 4) failed to demonstrate an age-related decline in DAT *(20)*. Yet under conditions of decreased DAT availability ligands with high affinities such as [^{11}C]WIN 35,428 may be more desirable than a ligand with relatively lower affinity, such as [^{11}C]*d-threo* methylphenidate.

RTI-55 labeled with ^{123}I has highly desirable properties for SPECT (i.e., very high affinity for the DAT and a high specific-to-nonspecific binding ratio) *(21,22)*. The SPECT measurements with [^{123}I]RTI-55 are made when the radioligand distribution corresponds to an equilibrium situation, which is believed to be the case after approx 18 h. Although RTI-55 has been labeled with ^{11}C as well as ^{123}I, PET measurements with [^{11}C]RTI-55 (half-life = 20 min) must be conducted very far from equilibrium *(23)*. Almost 200 papers have been published that involve RTI-55 or close congeners since its first synthesis in 1991. Its in vitro and in vivo binding, metabolism, and behavioral pharmacology have been well-characterized, and tracer kinetic modeling approaches have been explored. Much of this work has been conducted by Innis and colleagues *(24–29)*.

The DAT radioligand [^{99m}Tc]TRODAT-1 was the first ^{99m}Tc-labeled compound that binds to its neuroreceptor target in vivo. It was introduced by Kung and colleagues only in 1996 *(30)*, but has already had extensive use in human subjects, especially outside the United States *(8,31–34)*. Its popularity stems from the favorable economics of ^{99m}Tc radiopharmaceuticals. The radiolabel is available inexpensively from generator systems in hospital radiopharmacies, and can be incorporated into the final product radiotracer using fairly simple procedures *(35–37)*. TRODAT-1 has advantages for clinical SPECT studies over ^{123}I-labeled tracers, but it is possible that it may eventually be supplanted by a ^{99m}Tc tracer with improved characteristics, such as greater brain penetration. A similar compound, Technepine, which incorporates a very similar Tc-containing chelate attached to the 2β position of a phenyltropane, was developed by Madras and Meltzer and their colleagues contemporaneously with the development of TRODAT-1 *(38,39)*.

5. OTHER DAT RADIOTRACERS

Many positron- and ^{123}I-labeled DAT radioligands have been described in the nuclear medicine literature. Many of these (Table 3) have been aryltropane derivatives. Various labeled groups have been placed on the bridgehead N-atom, the 2β-position, or the aromatic ring *(40–44)*. However, as far as ^{11}C is concerned, it seems unlikely that superior agents to cocaine and d-threo-methylphenidate can be developed for studies of the DAT in striatum. Cocaine has good properties for measuring DAT occupancy by other drugs—although its in vivo binding is relatively weak, it is sufficiently strong for measurement of specific binding and it achieves equilibrium rapidly. The binding affinity of d-threomethylphenidate is higher than that of cocaine, resulting in a higher striatum-to-cerebellum ratio. However, as discussed here, ^{11}C-labeled compounds with higher affinities than methylphenidate, such as WIN 35,428, are likely to be flow-limited in terms of in vivo striatal binding, with reduced sensitivity to alterations in DAT densities. Furthermore, the use of any novel ^{11}C DAT radioligand in human subjects would first require extensive preclinical studies of its pharmacology and toxicity. An expectation of considerably improved properties relative to cocaine or methylphenidate would be required in order to make this worthwhile. A similar consideration applies to SPECT studies with ^{123}I-labeled DAT radioligands, because of the extensive validation studies that have been conducted using RTI-55. To repeat these studies with an alternative radioligand would require a compelling rationale that is presently unavailable. The fact that brain uptake of RTI-55 reflects binding to serotonin as well as DATs does not present practical difficulties for clinical SPECT studies of striatal DAT, because the striatum is almost devoid of serotonin transporters (SERTs). Related compounds such as RTI-121, which is more specific for DATs than RTI-55, have been labeled with ^{123}I, but have been little used for SPECT studies *(45)*. As already mentioned, it is too early to tell whether TRODAT-1 is likely to remain the most popular ^{99m}Tc-labeled DAT radioligand.

If a compelling clinical need for quantitative (i.e., PET) scanning of the DAT develops, an ^{18}F-labeled radioligand would be useful. This is because the half-life of ^{18}F, unlike that of ^{11}C, is long enough to allow distribution of tracers from central production sites located within moderate travel times to hospitals without their own cyclotron/radiochemistry facilities. A commercial network for distribution of [^{18}F]FDG (2-deoxy-2-fluoro-glucose)—which is increasingly used in nuclear oncology—now exists, which would facilitate the dissemination of other ^{18}F tracers. WIN 35,428 contains a F-atom, and is thus an obvious candidate because its binding kinetics would be better suited to the 110-min half-life of ^{18}F than the 20-min half-life of [11]C. [^{18}F]WIN 35,428 has been synthesized, though from a route providing lower radiochemical yields and material of lower specific radioactivity than those available for many other ^{18}F radiotracers, and has been used in several studies *(46,47)*. Among the several related compounds, [^{18}F]FP-CIT—in which the N-methyl group of RTI-55 is replaced by a fluoropropyl group to allow ^{18}F labeling, and the methyl ester is replaced by an isopropyl ester to enhance DAT/SERT selectivity—has the advantage that the behavior of the same compound labeled with ^{123}I is well-understood *(48)*. Another DAT radioligand that appears to be well-validated is [^{18}F](+)-FCT, developed by Mach and colleagues *(43)*.

6. VALIDATION OF [^{11}C]COCAINE AS A PET DAT RADIOLIGAND

Confidence in the results of neuroreceptor PET imaging experiments requires a demonstration that radioactivity in the region(s) of interest predominantly reflects radioligand bound to the target receptor, as opposed to that bound to other receptors and to "nonspecific" sites. Correspondence between PET images and receptor distributions determined using in vitro techniques is an important validation. Another question is whether changes in the level of endogenous ligands for the binding site significantly alter radioligand binding, because this would be a confounding factor in measures of binding-site density. The possible contributions of radioactive metabolites to PET images must also be considered.

Initial PET experiments with [^{11}C]cocaine demonstrated selective binding of [^{11}C]cocaine in the basal ganglia of human and baboon brains *(49)*. Binding of [^{11}C]cocaine in the basal ganglia was maximal at 5 min, and cleared with a half-time of 20 min. This time-course was similar to that of the "high" previously reported by cocaine abusers, suggesting that the subjective effects of cocaine were associated with the binding in striatum (Fig. 1).

Pretreatment of subjects with the DA reuptake blocker nomifensine reduced [^{11}C]cocaine binding in the striatum, but not in the cerebellum—

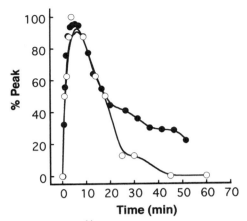

Fig. 1. Time activity curves for [¹¹C]cocaine in striatum (**filled circles**) along-side the temporal course for cocaine-induced high (**empty circles**) for a single subject. For comparison purposes, values are expressed as percent from peak. Notice the good correspondence between the kinetics of [¹¹C]cocaine in striatum and the duration of cocaine-induced high for the initial uptake and peak concentration of cocaine in striatum. The high dissipates more rapidly than the levels of [¹¹C]cocaine. Reproduced with permission from ref. *49*.

which lacks DAT, consistent with the notion that the radioactive cocaine is binding to the DAT *(49)*. Many in vitro studies show binding of cocaine to the serotonin and NETs as well as to the DAT (e.g., ref. *50*), but as a PET ligand it shows specificity of binding to DAT. This is probably because the concentration of the other monoamine transporters is much lower, so that their in vivo specific binding with [¹¹C]cocaine is negligible compared with the tracer's nonspecific binding. Radioactivity in some peripheral tissues— such as the heart, kidneys, and adrenals—after administration of [¹¹C]cocaine probably represents in part binding to NETs *(51)*.

The susceptibility of cocaine to butyrylcholinesterase (e.g., ref. *52*) and other esterases highlighted the question of what contribution labeled metabolites of (–)-cocaine may make to PET images of the brain. This issue was addressed by labeling the cocaine molecule in three different positions *(53)*. In addition to [N-¹¹CH₃]cocaine, which is the form routinely employed, [O-¹¹CH₃]cocaine and 4'-[¹⁸F]fluorococaine were prepared. Although 4'-fluorococaine is a distinct compound, in vitro assays showed that cocaine and 4'-fluorococaine were equipotent in terms of binding to the DAT. PET baboon brain scans, as well as regional brain kinetics and plasma time-activity curves corrected for the presence of labeled metabolites, were nearly identical for all three radioactive compounds, as well as for 4'-[N-

[11]CH$_3$]fluorococaine. The rationale behind these experiments was that the patterns of labeled metabolites for the labeled compounds are different. The nearly identical brain kinetics for the labeled forms of cocaine indicate that ^{11}C metabolites do not significantly affect PET images of the brain *(53)*.

The issue of competition between DA and [^{11}C]cocaine was addressed in baboon striatum after treatment animals with drugs expected to deplete synaptic DA. The increases in binding were not statistically significant, and were smaller than the changes found in corresponding experiments with the D2-receptor radioligand, [^{11}C]raclopride *(54)*. Subsequent experiments using d-threo-methylphenidate, WIN 35,428, RTI-55, and TRODAT-1 have failed to find significant effects of altered DA on radioligand binding in rodent and in vitro models, suggesting that PET measures of DAT availability have very low sensitivity to this factor *(29,55,56)*.

Early work with [^{11}C]cocaine resulted in the development of a convenient graphical data analysis method, the "Logan plot" *(57)*, which has since been extensively applied to many "reversibly binding" PET radiotracers *(58)*.

7. PET AND SPECT STUDIES OF DAT DENSITIES

Many studies in humans using DAT radioligands have shown age-related decreases in binding *(20,31,48,59–68)*. Taken as a whole (Table 4), the studies are consistent with the reduction in DAT and in DA cells documented in postmortem studies of approx 5% per decade of life. The decrease appears to be regionally variable (putamen > caudate) and nonlinear, with a greater rate of decrease in early adulthood than in later life. Recent studies also suggest an effect of gender on DAT densities, with women exhibiting higher densities than men *(68)*.

DAT radioligands have also been used in many studies to assess DA-cell degeneration in subjects with Parkinson's disease. These subjects show marked reductions in DAT when compared with healthy age-matched controls (Table 5) *(32,41,65,70–84,101)*.

In vivo DAT radioligand binding has been evaluated in several patient populations, as shown in Table 6.

DAT levels were reported to be unchanged in impulsive violent behavior, idiopathic cervical dystonia and schizophrenia *(88,90,98,99)*.

Decreases in DAT were documented in Lesch-Nyhan syndrome, diffuse Lewy body disease, social phobia, Wilson's disease, dominantly inherited fronto-temporal dementia, Machado-Joseph disease, traumatic brain injury, juvenile neuronal ceroid lipofuscinosis, REM sleep behavior disorder, detached personality disorder, and sporadic olivopontocerebellar atrophy *(33,75,86,87,89,93–97,100,102)*.

Table 5
PET and SPECT Studies of the DAT in Parkinsonism

Investigators	Condition	#	Tracer
(70)	Hemi-parkinsonism	8	RTI-55
(41)	PD	18	FP-CIT
(71)	PD	17	IPT
(72)	Drug-naive PD	11	RTI-32
(73)	Drug-naive PD	33	RTI-55
(74)	PD	47	RTI-55
(75)	PD	7	RTI-55
(76)	PD and treated schizophrenia	2	IPT
(77)	Drug-naive PD	16	RTI-55
(78)	Drug-naive PD	15	RTI-55
(65)	PD	20	RTI-55
(79)	PD	27	$[^{18}F]$WIN35,428
(80)	Drug-naive PD	10	$[^{11}C]$WIN35,428
(81)	PD	11	RTI-55
(82)	PD	10	RTI-55
(83)	PD	20	RTI-55
(32)	PD	42	TRODAT-1
(84)	PD	41	FP-CIT

#, Number of subjects in the study.

Increases in DAT have been reported in Tourette's syndrome and major depression *(85,92)*. A recent, preliminary SPECT study *(91)* has also indicated that the striatal DAT density is elevated in adult attention-deficit hyperactivity disorder (ADHD), yet another SPECT study with TRODAT-1 *(34)* suggests that chronic treatment with methylphenidate in these patients may lead to a decrease in DAT radioligand binding. These studies are significant because they are consistent with the idea that the symptoms of ADHD result from a subnormal level of synaptic DA caused by an excessive rate of reuptake. Thus, they appear to provide an explanation for the therapeutic effect of the drugs that are used to treat ADHD, since these are DA reuptake blockers or DA releasers that may elevate synaptic DA into the normal range. Confirmatory PET studies in adult ADHD patients are urgently required.

Three studies have reported decreased DAT in nonviolent alcoholic subjects *(103–105)*. One of these studies found slightly increased DAT in habitually violent alcoholics *(105)*. Another study found no changes in DAT levels in alcoholics *(106)*. PET studies of abused drugs in both humans and nonhuman primates are summarized in Table 7.

Table 6
PET and SPECT Studies of the DAT in Psychiatric Conditions

Investigators	Condition	#	Change	Tracer
(85)	Tourette's syndrome	5	Up	RTI-55
(86)	Lesch-Nyhan disease	6	Down	[^{11}C]WIN 35
(87)	Diffuse Lewy body disease	7	Down	RTI-55
(88)	Impulsive violent behavior	21	Unch	RTI-55
(89)	Social phobia	11	Down	RTI-55
(75)	Wilson's Disease	6	Down	RTI-55
(90)	Idiopathic cervical dystonia	10	Unch	RTI-55
(91)	ADHD	6	Up	Technepine
(92)	Major depression		Up	RTI-55
(93)	Dominantly inherited fronto-temporal dementia and parkinsonism	2	Down	RTI-55
(33)	Machado-Joseph disease	10	Down	TRODAT-1
(94)	Traumatic brain injury	10	Down	RTI-55
(95)	Juvenile neuronal ceroid lipofuscinosis	17	Down	RTI-55
(96)	Sporadic olivopontocerebellar atrophy	9	Down	RTI-55
(34)	ADHD			TRODAT-1
(97)	Detached personality disorder	18		[^{18}F]WIN 35, 428
(98)	Drug naïve schizophrenia		Unch	[^{18}F]WIN 35, 428
(99)	Schizophrenia		Unch	RTI-55
(100)	REM sleep behavior disorder	5	Down	IPT

#, Number of subjects in the study. Unch, unchanged.

A state of intermittently increased synaptic DA, as is believed to exist in cocaine abusers, may be expected to lead to a compensatory increase in the DAT level so that clearance of dopamine from the synaptic cleft would be increased. A SPECT study of acutely abstinent cocaine abusers using [^{123}I]RTI-55 was consistent with this view, and was suggestive of a statistical relationship between cocaine-related depression and DAT levels *(112)*. PET studies with [^{11}C]cocaine of a group of 12 detoxified cocaine abusers showed decreased uptake of cocaine in the brain. However, radioactivity decreased in both the striatum and cerebellum and subjects showed no differences from controls in DAT availability when the data were subjected to graphical analysis *(111,116)*. A subsequent study with 8 additional cocaine abusers and 20 controls also found no group differences in DAT levels *(117)*. However, this study found that cocaine abusers did not show the age-related decline in DAT radioligand binding seen in control subjects, raising the intriguing possibility that cocaine use may protect against age-related decline in DAT *(117)*.

Table 7
PET and SPECT Studies of the DAT with Abused Drugs

Investigators	Drug/subjects	#	Change	Tracer
(16)	MPTP monkeys		Down	[¹¹C]WIN 35
(107)	MPTP rhesus monkeys		Down	IACFT
(108)	Methamphetamine baboons			[¹¹C]WIN 35
(109)	MPTP macaques		Down	RTI-55
(110)	Methamphetamine vervets			[¹¹C]WIN 35
(111)	Detoxified cocaine abusers	12	Unch*	[¹¹C]Cocaine
(112)	Cocaine abstinence	28	Up	RTI-55
(113)	Methamphetamine abusers	10	Down	[¹¹C]WIN 35
(104)	Late onset alcoholism	9	Down	PE2I
(114)	alcoholism	27	Down	RTI-55
(115)	alcoholism	14	Unch	RTI-55

#, Number of subjects in the study. Unch, unchanged.
See text and refs. *116* and *117*.

Several postmortem studies of DAT levels in cocaine abusers vs control subjects have also been inconsistent, with indications of increased, decreased, and unchanged DAT levels. Little and colleagues have reported that striatal DAT radioligand binding is increased in cocaine abusers, although midbrain DAT mRNA levels paradoxically were decreased and striatal-nerve-terminal densities estimated using [³H]dihydrotetrabenazine were also decreased *(118–121)*. Another postmortem study found unchanged nigral mRNA levels in cocaine-overdose victims and decreased mRNA in subjects with agitated cocaine delirium *(122)*. The results of Wilson et al. *(123)* suggested decreased DAT in some autopsy subjects, and another paper from the same group found that chronic cocaine decreased DAT in a rat study *(124)*.

The differences between studies may relate to factors such as time after last drug use, and to the radioligand and other methodology used for quantification of transporters, as well as to variability in choice of subject and control groups.

8. STUDIES OF DAT OCCUPANCY

The BNL group has conducted extensive studies in which [¹¹C]cocaine is administered in conjunction with pharmacologically active doses of cocaine *(125–128)*. The reduction in striatal binding of ¹¹C compared with that seen when tracer [¹¹C]cocaine is given alone allows one to estimate the degree of occupancy of the DAT by the non-radioactive drug (Fig. 2) *(116)*. Com-

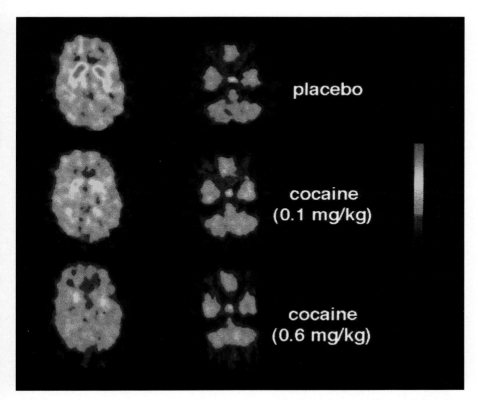

Fig. 2. Brain images at the level of striatum **(left)** and cerebellum **(right)** in a representative subject studied three times with tracer [^{11}C]cocaine only **(top)**, [^{11}C]cocaine plus 0.1 mg/kg cocaine **(middle)** and [^{11}C]cocaine plus 0.6 mg/kg cocaine **(bottom)**. (Reprinted with permission from ref. *129*.

bined PET/neuropsychological experiments have demonstrated that occupation of 50% or a higher fraction of the transporters is necessary for a cocaine abuser to experience a "high" (Fig. 3) *(129)*. Similar results were obtained in normal controls with intravenously administered methylphenidate, as would be predicted by the similar neurochemical properties of cocaine and methylphenidate, although not all subjects found intravenous (iv) methylphenidate pleasurable *(130)*. Methylphenidate given orally in clinically used doses is also associated with blockade of approx 50% of the DAT *(130)*. The fact that a "high" is not induced by oral methylphenidate is presumably a consequence of the slower delivery of drug to the brain with the oral route compared to the iv route *(131)*. Many studies have indicated that the speed of delivery of abused drugs is crucial to their rewarding

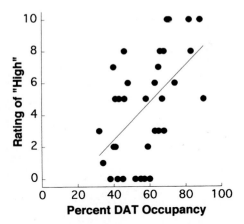

Fig. 3. Correlation between DAT occupancy by cocaine and subjects' reports of high. Reproduced with permission from ref. *129*.

effects. Presumably, the "high" is related to the rate of increase of occupancy of the DAT (and thus the rate of activation of DA receptors) rather than to the degree of occupancy *per se*. This view is supported by studies with [^{11}C]d-threomethylphenidate, which show that the "high" decreases while the DAT is still occupied by the drug *(127)*. Additional, dramatic support was given by experiments involving sequential iv administration of two doses of methylphenidate, 40 min apart, to human subjects. Although the [^{11}C]cocaine measurements at the time of the second injection of methylphenidate showed that the DAT was still over 50% occupied from the first injection, an identical "high" was reported *(127)*.

In contrast to the results of PET experiments with [^{11}C]cocaine, SPECT studies using [^{123}I]RTI-55 as the DAT radioligand suggested that a pharmacologically active dose of cocaine occupied only 6% of the transporters *(132)*. Similar experiments in mice also indicated low occupancy of the DAT by cocaine when [^{123}I]RTI-55 was used to probe unoccupied DAT *(133)*. The reason for the discrepancy is probably that the net tissue-dissociation-rate constant of the very high-affinity ligand RTI-55 is too low to allow [^{123}I]RTI-55 to come to equilibrium during the period that the weaker ligand cocaine is available for binding. An erroneously low-degree DAT occupancy by cocaine is thus estimated. An indication that the discrepancy is not caused by a problem with radiolabeled cocaine as the DAT radioligand is given by recent results in which DAT occupancy by nonradioactive RTI-55 was estimated in mice using [^{3}H]cocaine *(134)*. The time-course of occupancy over the period 1–12 h was very similar to the time-courses of inhibition of DA transport and of elevation of locomotor activity. These observations suggest that the measurement of DAT blockade by labeled cocaine is accurately tracking the consequences of the blockade by RTI-55.

Experiments using PET and [^{11}C]cocaine to study the clearance of nonradioactive RTI-55 from baboon striatum in vivo also support the validity of using labeled cocaine in occupancy measurements because the calculated clearance half-times of RTI-55 were very similar to those measured with [^{123}I]RTI-55 and SPECT *(135)*. One can conclude from these studies with cocaine and its high-affinity analog RTI-55 that the most accurate estimates of DAT occupancy will be obtained by using a radioligand of equal or lower affinity than that of the drug being tested. Baboon PET studies in which [^{11}C]cocaine and [^{11}C]d-threo-methylphenidate were each used to explore DAT occupancy by nonradioactive cocaine and methylphenidate also support this conclusion, since slightly lower occupancies were indicated using [^{11}C]d-threo-methylphenidate than using [^{11}C]cocaine *(11)*.

9. IMAGING STUDIES OF OTHER MONOAMINE TRANSPORTERS

The neuronal membrane DAT has been by far the largest target of development work in this area. However, the SERT has also received a significant amount of attention, consistent with its important role in antidepressant drug action. Radioligands include [^{123}I]RTI-55, previously discussed as a DAT radioligand; it can also be used to image SERTs, although in different brain regions, where there are few or no DATs, such as the midbrain/hypothalamus and the cerebral cortex *(136–139)*. SPECT studies using [^{123}I]RTI-55 have demonstrated decreased cortical SERT levels in MDMA abusers, and decreased midbrain SERT levels in subjects suffering from impulsive, violent behavior *(88)*, alcoholism *(106)*, major depression *(140)*, and acute abstinence during cocaine dependency *(141)*. A significant age-related decrease in the binding of [^{123}I]RTI-55 has also been reported *(67,142)*. Studies have failed to find altered SERT levels in parkinsonism or schizophrenia *(99,143)*.

A number of other SPECT radioligands for the SERT have been described, including nor-methyl [^{123}I]RTI-55, and [^{123}I]5-iodo-6-nitroquipazine *(144–148)*. Weak binding of [^{99m}Tc]TRODAT-1 to SERTs during SPECT studies in non-human primates has also been documented *(149)*, suggesting that improved ^{99m}Tc-labeled SERT radioligands may become available. PET studies of the SERT have been conducted using [^{11}C]McN 5652 *(150,151)*, and several other compounds have been evaluated *(152–155)*.

The brain vesicular monoamine transporter, VMAT2, has also been a target for radioligand development. The best-characterized compound is [^{11}C]dihydrotetrabenazine *(156–158)*. An advantage of the vesicular transporter as a nerve-terminal marker is that it does not appear to up or downregulate in response to treatment with drugs such as L-DOPA or deprenyl *(159)*, and may thus be less sensitive to the confounding effects of medication. The striatal binding of [¹¹C]dihydrotetrabenazine is decreased by Parkinson's disease and in normal aging *(160,161)*. PET studies have shown that VMAT2 is decreased in Huntington's disease, severe chronic alcoholism, and olivopontocerebellar atrophy or multiple systems atrophy *(162–165)*, but is not altered in schizophrenia or Tourette's syndrome *(166,167)*. VMAT2 binding was increased in a study of bipolar I disorder *(168)*.

Advances in medicinal chemistry are needed to provide useful PET radioligands for all transporters in the brain. Even for the brain NET there are no satisfactory radioligands for in vivo studies. Nisoxetine, which is used in tritium-labeled form for in vitro studies, is not satisfactory as an in vivo radioligand *(169)*. Several radiolabeled compounds, including [¹⁸F]fluorocatecholamines and [¹²³I]iodobenzylguanidine are substrates for the NET and accumulate inside sympathetic nerve terminals in peripheral tissues *(170–172)*. However, these compounds do not cross the blood-brain barrier.

10. CONCLUSIONS

Brain imaging results with several PET and SPECT radioligands for the DAT have provided two strong indications that such studies can yield valid measures of DAT availability in striatum in the living human brain. Firstly, in vivo tracer kinetics for individual tracers are well-correlated with tracer affinity for the DAT measured in vitro. Secondly, 11 of 13 studies using 7 different tracers found decreased DAT ligand binding with age, and the magnitude of the decrease was consistent with the decreases documented in in vitro studies of human brain.

It appears unlikely that the PET tracers developed for research studies of the striatal DAT will be superior to those already available. However, it is possible that ⁹⁹ᵐTc-labeled tracers for clinical SPECT studies will continue to be refined. For example, current TRODAT-1 studies are conducted using a mixture of diastereomers with somewhat different brain penetration and transporter-binding affinity. Eventually, it is probable that a diastereomerically pure compound will be used. Also, a ⁹⁹ᵐTc-labeled compound with higher brain uptake and perhaps a higher binding affinity may allow lower amounts of radioactivity to be injected, reducing the radiation dose to the patient.

In vitro studies have shown that DATs are present in many brain regions, although in much lower concentrations than in the striatum. It may be possible to study extrastriatal DAT in individual subjects using a radioligand

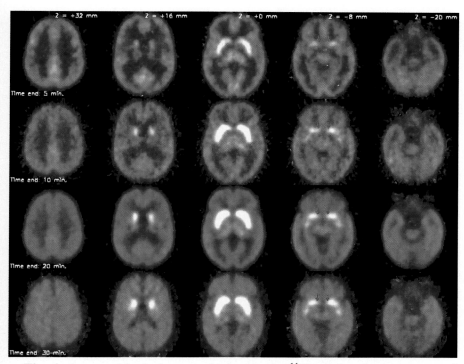

Fig. 4. Averaged brain images obtained with [¹¹C]cocaine. Images averaged from studies with 19 subjects are shown at four different levels obtained at 5, 10, 20, and 30 min after radiotracer administration. Images have been normalized to highest activity in a given time frame. PET studies were carried out with a CTI 931 tomograph (6 × 6 × 6.5-mm full width half maximum). Emission scans were started immediately after injection of 6.5–10 mCi of [¹¹C] cocaine (specific activity >0.2 Ci/mmol at time of injection). Image Registration (normalization) was performed with SPM implemented in PRO-MATLAB (Mathworks, Sherborn, MA). After identifying the intercommissural line, images were reoriented, linearly rescaled in three dimensions, and resliced to correspond to the stereotactic space in a brain atlas. These stereotactically normalized images had 31 planes to cover the full brain anatomy. Voxels were 2 × 2 × 4 mm. The images were smoothed by convolution in three dimensions with a Gaussian filter (full width at half maximum of 16 mm) to account for the normal variations in gyral anatomy between individuals and to increase the signal-to-noise ratio. Dynamically acquired images were averaged over the first seven time frames before being processed by SPM (averaging over more time frames did not improve registration). Each dynamic image was then processed individually using the parameters of the averaged image. (Reprinted with permission from ref. *173.*

with subnanomolar affinity. Such a radioligand would have to be very selective and possess high affinity in order to avoid contaminating signals from

SERTs and NETs outside the striatum. An alternative strategy to visualizing or quantifying extrastriatal DATs might be to average PET scans from groups of subjects. An example of such a study is shown in Fig. 4, which shows averaged scans from 19 normal subjects injected with [^{11}C]cocaine. This approach involved acquiring an MRI scan as well as a PET scan for each individual, using an algorithm that adjusted each individual's MRI scan to fit a common template, and then averaging PET data for each region of the template across the subjects.

REFERENCES

1. Kuhar, M. J., Carroll, F. I., Lewin, A. H., Boja, J. W., Scheffel, U., and Wong, D. F. (1997) in *Neurotransmitter Transporters: Structure, Function and Regulation* (Reith, M. E. A., ed), 1st ed., Humana Press, Totowa, NJ.
2. Wolf, A. P. and Fowler, J. S. (1995) Positron emission tomography. Biomedical research and clinical application. *Neuroimaging Clin. N. Am.* **5,** 87–101.
3. Fowler, J. S. and Wolf, A. P. (1990) The heritage of radiotracers for positron emission tomography. *Acta Radiol. Suppl.* **374,** 13–16.
4. Budinger, T. F. (1998) PET instrumentation: what are the limits? *Semin. Nucl. Med.* **28,** 247–267.
5. Giros, B., Jaber, M., Jones, S. R., Wightman, R. M., and Caron, M. G. (1996) Hyperlocomotion and indifference to cocaine and amphetamine in mice lacking the dopamine transporter. *Nature* **379,** 606–612.
6. Christman, D. R., Hoyte, R. M., and Wolf, A. P. (1970) Organic radiopharmaceuticals labeled with isotopes of short half-life. I. 11C-1-dopamine hydrochloride. *J. Nucl. Med.* **11,** 474–478.
7. Volkow, N. D., Fowler, J. S., Gatley, S. J., Logan, J., Wang, G. J., Ding, Y. S., and Dewey, S. (1996) PET evaluation of the dopamine system of the human brain. *J. Nucl. Med.* **37,** 1242–1256.
8. Kushner, S. A., McElgin, W. T., Kung, M. P., Mozley, P. D., Plossl, K., Meegalla, S. K., et al. (1999) Kinetic modeling of [^{99m}Tc]TRODAT-1: a dopamine transporter imaging agent. *J. Nucl. Med.* **40,** 150–158.
9. Meegalla, S. K., Plossl, K., Kung, M. P., Stevenson, D. A., Mu, M., Kushner, S., et al. (1998) Specificity of diastereomers of [^{99m}Tc]TRODAT-1 as dopamine transporter imaging agents. *J. Med. Chem.* **41,** 428–436.
10. Acton, P. D., Meyer, P. T., Mozley, P. D., Plössl, K., and Kung, H. (2000) Simplified quantification of dopamine transporters in humans using [^{99m}Tc]TRODAT-1 and single photon emission tomography. *Eur. J. Nucl. Med.,* **27,** 1714–1718.
11. Fowler, J. S., Volkow, N. D., Logan, J., Gatley, S. J., Pappas, N., King, P., et al. (1998) Measuring dopamine transporter occupancy by cocaine in vivo: radiotracer considerations. *Synapse* **28,** 111–116.
12. Aquilonius, S. M., Bergstrom, K., Eckernas, S. A., Hartvig, P., Leenders, K. L., Lundquist, H., et al. (1987) In vivo evaluation of striatal dopamine reuptake sites using 11C-nomifensine and positron emission tomography. *Acta Neurol. Scand.* **76,** 283–287.

13. Ding, Y. S., Fowler, J. S., Volkow, N. D., Gatley, S. J., Logan, J., Dewey, S. L., et al. (1994) Pharmacokinetics and in vivo specificity of [^{11}C]dl-threo-methylphenidate for the presynaptic dopaminergic neuron. *Synapse* **18**, 152–160.

14. Ding, Y. S., Fowler, J. S., Volkow, N. D., Logan, J., Gatley, S. J., and Sugano, Y. (1995) Carbon-11-d-threo-methylphenidate binding to dopamine transporter in baboon brain. *J. Nucl. Med.* **36**, 2298–2305.

15. Volkow, N. D., Ding, Y. S., Fowler, J. S., Wang, G. J., Logan, J., Gatley, S. J., Schlyer, D. J., and Pappas, N. (1995) A new PET ligand for the dopamine transporter: studies in the human brain. *J. Nucl. Med.* **36**, 2162–2168.

16. Hantraye, P., Brownell, A. L., Elmaleh, D., Spealman, R. D., Wullner, U., Brownell, G. L., et al. (1992) Dopamine fiber detection by [^{11}C]-CFT and PET in a primate model of parkinsonism. *Neuroreporter* **3**, 265–268.

17. Cline, E. J., Scheffel, U., Boja, J. W., Mitchell, W. M., Carroll, F. I., Abraham, P., et al. (1992) In vivo binding of [125I]RTI-55 to dopamine transporters: pharmacology and regional distribution with autoradiography. *Synapse* **12**, 37–46.

18. Davies, H. M., Saikali, E., Sexton, T., and Childers, S. R. (1993) Novel 2-substituted cocaine analogs: binding properties at dopamine transport sites in rat striatum. *Eur. J. Pharmacol.* **244**, 93–97.

19. Frost, J. J., Rosier, A. J., Reich, S. G., Smith, J. S., Ehlers, M. D., Snyder, S. H., et al. (1993) Positron emission tomographic imaging of the dopamine transporter with 11C-WIN 35,428 reveals marked declines in mild Parkinson's disease [see comments]. *Ann. Neurol.* **34**, 423–431.

20. Wong, D. F., Yung, B., Dannals, R. F., Shaya, E. K., Ravert, H. T., Chen, C. A., et al. (1993) In vivo imaging of baboon and human dopamine transporters by positron emission tomography using [^{11}C]WIN 35,428. *Synapse* **15**, 130–142.

21. Innis, R. B., Seibyl, J. P., Scanley, B. E., Laruelle, M., Abi-Dargham, A., Wallace, E., et al. (1993) Single photon emission computed tomographic imaging demonstrates loss of striatal dopamine transporters in Parkinson disease. *Proc. Natl. Acad. Sci. USA* **90**, 11,965–11,969.

22. Laruelle, M., Wallace, E., Seibyl, J. P., Baldwin, R. M., Zea-Ponce, Y., Zoghbi, S. S., et al. (1994) Graphical, kinetic, and equilibrium analyses of in vivo [^{123}I] beta-CIT binding to dopamine transporters in healthy human subjects. *J. Cereb. Blood Flow Metab.* **14**, 982–994.

23. Farde, L., Halldin, C., Muller, L., Suhara, T., Karlsson, P., and Hall, H. (1994) PET study of [^{11}C]beta-CIT binding to monoamine transporters in the monkey and human brain. *Synapse* **16**, 93–103.

24. Laruelle, M., Giddings, S. S., Zea-Ponce, Y., Charney, D. S., Neumeyer, J. L., et al. (1994) Methyl 3 beta-(4-[125I]iodophenyl)tropane-2 beta-carboxylate in vitro binding to dopamine and serotonin transporters under "physiological" conditions. *J. Neurochem.* **62**, 978–986.

25. Zea-Ponce, Y., Baldwin, R. M., Laruelle, M., Wang, S., Neumeyer, J. L., and Innis, R. B. (1995) Simplified multidose preparation of iodine-123-beta-CIT: a marker for dopamine transporters. *J. Nucl. Med.* **36**, 525–529.

26. Seibyl, J. P., Laruelle, M., van Dyck, C. H., Wallace, E., Baldwin, R. M., Zoghbi, S., et al. (1996) Reproducibility of iodine-123-beta-CIT SPECT brain measurement of dopamine transporters. *J. Nucl. Med.* **37,** 222–228.

27. Abi-Dargham, A., Innis, R. B., Wisniewski, G., Baldwin, R. M., Neumeyer, J. L., and Seibyl, J. P. (1997) Human biodistribution and dosimetry of iodine-123-fluoroalkyl analogs of beta-CIT. *Eur. J. Nucl. Med.* **24,** 1422–1425.

28. Innis, R. B. (1998) Single photon emission computed tomography imaging of dopaminergic function: presynaptic transporter, postsynaptic receptor, and "intrasynaptic" transmitter. *Adv. Pharmacol.* **42,** 215–219.

29. Innis, R. B., Marek, K. L., Sheff, K., Zoghbi, S., Castronuovo, J., Feigin, A., and Seibyl, J. P. (1999) Effect of treatment with L-dopa/carbidopa or L-selegiline on striatal dopamine transporter SPECT imaging with [^{123}I]beta-CIT. *Mov. Disord.* **14,** 436–442.

30. Kung, H. F., Kim, H. J., Kung, M. P., Meegalla, S. K., Plossl, K., and Lee, H. K. (1996) Imaging of dopamine transporters in humans with technetium-99m TRODAT-1 [see comments]. *Eur. J. Nucl. Med.* **23,** 1527–1530.

31. Mozley, P. D., Acton, P. D., Barraclough, E. D., Plossl, K., Gur, R. C., Alavi, A., et al. (1999) Effects of age on dopamine transporters in healthy humans. *J. Nucl. Med.* **40,** 1812–1817.

32. Mozley, P. D., Schneider, J. S., Acton, P. D., Plossl, K., Stern, M. B., Siderowf, A., et al. (2000) Binding of [^{99m}Tc]TRODAT-1 to dopamine transporters in patients with Parkinson's disease and in healthy volunteers. *J. Nucl. Med.* **41,** 584–589.

33. Yen, T. C., Lu, C. S., Tzen, K. Y., Wey, S. P., Chou, Y. H., Weng, Y. H., et al. (2000) Decreased dopamine transporter binding in Machado-Joseph disease. *J. Nucl. Med.* **41,** 994–998.

34. Krause, K. H., Dresel, S. H., Krause, J., Kung, H. F., and Tatsch, K. (2000) Increased striatal dopamine transporter in adult patients with attention deficit hyperactivity disorder: effects of methylphenidate as measured by single photon emission computed tomography. *Neurosci. Lett.* **285,** 107–110.

35. Choi, S. R., Kung, M. P., Plossl, K., Meegalla, S., and Kung, H. F. (1999) An improved kit formulation of a dopamine transporter imaging agent: [Tc-99m]TRODAT-1. *Nucl. Med. Biol.* **26,** 461–466.

36. Mu, M., Kung, M. P., Plossl, K., Acton, P. D., Mozley, P. D., and Kung, H. F. (1999) A simplified method to determine [^{99m}Tc]TRODAT-1 in human plasma. *Nucl. Med. Biol.* **26,** 821–825.

37. Fang, P., Wu, C. Y., Liu, Z. G., Wan, W. X., Wang, T. S., Chen, S. D., et al. (2000) The preclinical pharmacologic study of dopamine transporter imaging agent [^{99m}Tc]TRODAT-1. *Nucl. Med. Biol.* **27,** 69–75.

38. Madras, B. K., Jones, A. G., Mahmood, A., Zimmerman, R. E., Garada, B., Holman, B. L., et al. (1996) Technepine: a high-affinity 99m-technetium probe to label the dopamine transporter in brain by SPECT imaging. *Synapse* **22,** 239–246.

39. Meltzer, P. C., Blundell, P., Jones, A. G., Mahmood, A., Garada, B., Zimmerman, R. E., et al. (1997) A technetium-99m SPECT imaging agent which targets the dopamine transporter in primate brain. *J. Med. Chem.* **40,** 1835–1844.

40. Guilloteau, D., Emond, P., Baulieu, J. L., Garreau, L., Frangin, Y., Pourcelot, L., et al. (1998) Exploration of the dopamine transporter: in vitro and in vivo characterization of a high-affinity and high-specificity iodinated tropane derivative (E)-N-(3-iodoprop-2-enyl)-2beta-carbomethoxy-3beta-(4'-methylph enyl)nortropane (PE2I). *Nucl. Med. Biol.* **25,** 331–337.

41. Booij, J., Tissingh, G., Boer, G. J., Speelman, J. D., Stoof, J. C., Janssen, A. G., et al. (1997) [^{123}I]FP-CIT SPECT shows a pronounced decline of striatal dopamine transporter labelling in early and advanced Parkinson's disease. *J. Neurol. Neurosurg. Psychiatry* **62,** 133–140.

42. Mozley, P. D., Stubbs, J. B., Kim, H. J., McElgin, W., Kung, M. P., Meegalla, S., and Kung, H. F. (1996) Dosimetry of an iodine-123-labeled tropane to image dopamine transporters. *J. Nucl. Med.* **37,** 151–159.

43. Mach, R. H., Nader, M. A., Ehrenkaufer, R. L., Gage, H. D., Childers, S. R., Hodges, L. M., et al. (2000) Fluorine-18-labeled tropane analogs for PET imaging studies of the dopamine transporter. *Synapse* **37,** 109–117.

44. Elmaleh, D. R., Fischman, A. J., Shoup, T. M., Byon, C., Hanson, R. N., Liang, A. Y., et al. (1996) Preparation and biological evaluation of iodine-125-IACFT: a selective SPECT agent for imaging dopamine transporter sites. *J. Nucl. Med.* **37,** 1197–1202.

45. Hume, S. P., Luthra, S. K., Brown, D. J., Opacka-Juffry, J., Osman, S., Ashworth, S., et al. (1996) Evaluation of [^{11}C]RTI-121 as a selective radioligand for PET studies of the dopamine transporter. *Nucl. Med. Biol.* **23,** 377–384.

46. Haaparanta, M., Bergman, J., Laakso, A., Hietala, J., and Solin, O. (1996) [^{18}F]CFT ([^{18}F]WIN 35,428), a radioligand to study the dopamine transporter with PET: biodistribution in rats. *Synapse* **23,** 321–327.

47. Laakso, A., Bergman, J., Haaparanta, M., Vilkman, H., Solin, O., and Hietala, J. (1998) [^{18}F]CFT [(18F)WIN 35,428], a radioligand to study the dopamine transporter with PET: characterization in human subjects. *Synapse* **28,** 244–250.

48. Kazumata, K., Dhawan, V., Chaly, T., Antonini, A., Margouleff, C., Belakhlef, A., et al. (1998) Dopamine transporter imaging with fluorine-18-FPCIT and PET. *J. Nucl. Med.* **39,** 1521–1530.

49. Fowler, J. S., Volkow, N. D., Wolf, A. P., Dewey, S. L., Schlyer, D. J., Macgregor, R. R., et al. (1989) Mapping cocaine binding sites in human and baboon brain in vivo. *Synapse* **4,** 371–377.

50. Gatley, S. J., Pan, D., Chen, R., Chaturvedi, G., and Ding, Y. S. (1996) Affinities of methylphenidate derivatives for dopamine, norepinephrine and serotonin transporters. *Life Sci.* **58,** 231–239.

51. Fowler, J. S., Ding, Y. S., Volkow, N. D., Martin, T., MacGregor, R. R., Dewey, S., et al. (1994) PET studies of cocaine inhibition of myocardial nore-pinephrine uptake. *Synapse* **16,** 312–317.

52. Gatley, S. J., MacGregor, R. R., Fowler, J. S., Wolf, A. P., Dewey, S. L., and Schlyer, D. J. (1990) Rapid stereoselective hydrolysis of (+)-cocaine in baboon plasma prevents its uptake in the brain: implications for behavioral studies. *J. Neurochem.* **54,** 720–723.

53. Gatley, S. J., Yu, D. W., Fowler, J. S., MacGregor, R. R., Schlyer, D. J., Dewey, S. L., et al. (1994) Studies with differentially labeled [^{11}C]cocaine, [^{11}C]norcocaine, [^{11}C]benzoylecgonine, and [^{11}C]- and 4'-[^{18}F]fluorococaine to probe the extent to which [^{11}C]cocaine metabolites contribute to PET images of the baboon brain. *J. Neurochem.* **62**, 1154–1162.

54. Gatley, S. J., Volkow, N. D., Fowler, J. S., Dewey, S. L., and Logan, J. (1995) Sensitivity of striatal [^{11}C]cocaine binding to decreases in synaptic dopamine. *Synapse* **20**, 137–144.

55. Gatley, S. J., Ding, Y. S., Volkow, N. D., Chen, R., Sugano, Y., and Fowler, J. S. (1995) Binding of d-threo-[^{11}C]methylphenidate to the dopamine transporter in vivo: insensitivity to synaptic dopamine. *Eur. J. Pharmacol.* **281**, 141–149.

56. Dresel, S. H., Kung, M. P., Plossl, K., Meegalla, S. K., and Kung, H. F. (1998) Pharmacological effects of dopaminergic drugs on in vivo binding of [^{99m}Tc]TRODAT-1 to the central dopamine transporters in rats. *Eur. J. Nucl. Med.* **25**, 31–39.

57. Logan, J., Fowler, J. S., Volkow, N. D., Wolf, A. P., Dewey, S. L., Schlyer, D. J., et al. (1990) Graphical analysis of reversible radioligand binding from time-activity measurements applied to [N-11C-methyl]-(-)-cocaine PET studies in human subjects. *J. Cereb Blood Flow Metab.* **10**, 740–747.

58. Logan, J. (2000) Graphical analysis of PET data applied to reversible and irreversible tracers. *Nucl. Med. Biol.* **27**, 661–670.

59. Volkow, N. D., Fowler, J. S., Wang, G. J., Logan, J., Schlyer, D., MacGregor, R., et al. (1994) Decreased dopamine transporters with age in health human subjects. *Ann. Neurol.* **36**, 237–239.

60. van Dyck, C. H., Seibyl, J. P., Malison, R. T., Laruelle, M., Wallace, E., Zoghbi, S. S., et al. (1995) Age-related decline in striatal dopamine transporter binding with iodine-123-beta-CITSPECT [see comments]. *J. Nucl. Med.* **36**, 1175–1181.

61. Ishikawa, T., Dhawan, V., Kazumata, K., Chaly, T., Mandel, F., Neumeyer, J., et al. (1996) Comparative nigrostriatal dopaminergic imaging with iodine-123-beta CIT-FP/SPECT and fluorine-18-FDOPA/PET. *J. Nucl. Med.* **37**, 1760–1765.

62. Mozley, P. D., Kim, H. J., Gur, R. C., Tatsch, K., Muenz, L. R., McElgin, W. T., et al. (1996) Iodine-123-IPT SPECT imaging of CNS dopamine transporters: nonlinear effects of normal aging on striatal uptake values. *J. Nucl. Med.* **37**, 1965–1970.

63. Volkow, N. D., Ding, Y. S., Fowler, J. S., Wang, G. J., Logan, J., Gatley, S. J., et al. (1996) Dopamine transporters decrease with age. *J. Nucl. Med.* **37**, 554–559.

64. Tissingh, G., Bergmans, P., Booij, J., Winogrodzka, A., Stoof, J. C., Wolters, E. C., et al. (1997) [^{123}I]beta-CIT single-photon emission tomography in Parkinson's disease reveals a smaller decline in dopamine transporters with age than in controls. *Eur. J. Nucl. Med.* **24**, 1171–1174.

65. Muller, T., Farahati, J., Kuhn, W., Eising, E. G., Przuntek, H., Reiners, C., et al. (1998) [^{123}I]beta-CIT SPECT visualizes dopamine transporter loss in de novo parkinsonian patients. *Eur. Neurol.* **39**, 44–48.

66. Kuikka, J. T., Tupala, E., Bergstrom, K. A., Hiltunen, J., and Tiihonen, J. (1999) Iodine-123 labelled PE2I for dopamine transporter imaging: influence of age in healthy subjects. *Eur. J. Nucl. Med.* **26,** 1486–1488.

67. Pirker, W., Asenbaum, S., Hauk, M., Kandlhofer, S., Tauscher, J., Willeit, M., et al. (2000) Imaging serotonin and dopamine transporters with 123I-beta-CIT SPECT: binding kinetics and effects of normal aging. *J. Nucl. Med.* **41,** 36–44.

68. Lavalaye, J., Booij, J., Reneman, L., Habraken, J. B., and van Royen, E. A. (2000) Effect of age and gender on dopamine transporter imaging with [^{123}I]FP-CIT SPET in healthy volunteers [In Process Citation]. *Eur. J. Nucl. Med.* **27,** 867–869.

69. Volkow, N. D., Wang, G. J., Fowler, J. S., Logan, J., Hitzemann, R., Ding, Y. S., et al. (1996) Decreases in dopamine receptors but not in dopamine transporters in alcoholics. *Alcohol Clin. Exp. Res.* **20,** 1594–1598.

70. Marek, K. L., Seibyl, J. P., Zoghbi, S. S., Zea-Ponce, Y., Baldwin, R. M., Fussell, B., et al. (1996) [^{123}I] beta-CIT/SPECT imaging demonstrates bilateral loss of dopamine transporters in hemi-Parkinson's disease. *Neurology* **46,** 231–237.

71. Kim, H. J., Im, J. H., Yang, S. O., Moon, D. H., Ryu, J. S., Bong, J. K., et al. (1997) Imaging and quantitation of dopamine transporters with iodine-123-IPT in normal and Parkinson's disease subjects. *J. Nucl. Med.* **38,** 1703–1711.

72. Guttman, M., Burkholder, J., Kish, S. J., Hussey, D., Wilson, A., DaSilva, J., et al. (1997) [^{11}C]RTI-32 PET studies of the dopamine transporter in early dopa-naive Parkinson's disease: implications for the symptomatic threshold. *Neurology* **48,** 1578–1583.

73. Eising, E. G., Muller, T. T., Zander, C., Kuhn, W., Farahati, J., Reiners, C., and Coenen, H. H. (1997) SPECT-evaluation of the monoamine uptake site ligand [^{123}I](1R)-2-beta-carbomethoxy-3-beta-(4-iodophenyl)-tropane ([^{123}I]beta-CIT) in untreated patients with suspicion of Parkinson disease. *J. Investig. Med.* **45,** 448–452.

74. Asenbaum, S., Brucke, T., Pirker, W., Podreka, I., Angelberger, P., Wenger, S., et al. (1997) Imaging of dopamine transporters with iodine-123-beta-CIT and SPECT in Parkinson's disease. *J. Nucl. Med.* **38,** 1–6.

75. Jeon, B., Kim, J. M., Jeong, J. M., Kim, K. M., Chang, Y. S., Lee, D. S., et al. (1998) Dopamine transporter imaging with [^{123}I]-beta-CIT demonstrates presynaptic nigrostriatal dopaminergic damage in Wilson's disease. *J. Neurol. Neurosurg. Psychiatry* **65,** 60–64.

76. Schwarz, J., Scherer, J., Trenkwalder, C., Mozley, P. D., and Tatsch, K. (1998) Reduced striatal dopaminergic innervation shown by IPT and SPECT in patients under neuroleptic treatment: need for levodopa therapy? *Psychiatry Res.* **83,** 23–28.

77. Tissingh, G., Bergmans, P., Booij, J., Winogrodzka, A., van Royen, E. A., Stoof, J. C., and Wolters, E. C. (1998) Drug-naive patients with Parkinson's disease in Hoehn and Yahr stages I and II show a bilateral decrease in striatal dopamine transporters as revealed by [^{123}I]beta-CIT SPECT. *J. Neurol.* **245,** 14–20.

78. Wenning, G. K., Donnemiller, E., Granata, R., Riccabona, G., and Poewe, W. (1998) 123I-beta-CIT and 123I-IBZM-SPECT scanning in levodopa-naive Parkinson's disease. *Mov. Disord.* **13,** 438–445.

79. Rinne, J. O., Ruottinen, H., Bergman, J., Haaparanta, M., Sonninen, P., and Solin, O. (1999) Usefulness of a dopamine transporter PET ligand [(18)F]beta-CFT in assessing disability in Parkinson's disease. *J. Neurol. Neurosurg. Psychiatry* **67,** 737–741.

80. Ouchi, Y., Kanno, T., Okada, H., Yoshikawa, E., Futatsubashi, M., Nobezawa, S., Torizuka, T., and Sakamoto, M. (1999) Presynaptic and postsynaptic dopaminergic binding densities in the nigrostriatal and mesocortical systems in early Parkinson's disease: a double-tracer positron emission tomography study. *Ann. Neurol.* **46,** 723–731.

81. Ichise, M., Kim, Y. J., Ballinger, J. R., Vines, D., Erami, S. S., Tanaka, F., et al. (1999) SPECT imaging of pre- and postsynaptic dopaminergic alterations in L-dopa-untreated PD. *Neurology* **52,** 1206–1214.

82. Hamano, T., Tsuchida, T., Hirayama, M., Fujiyama, J., Mutoh, T., Yonekura, Y., et al. (2000) Dopamine transporter SPECT in patients with Parkinson's disease. *Kaku Igaku* **37,** 125–129.

83. Muller, U., Wachter, T., Barthel, H., Reuter, M., and von Cramon, D. Y. (2000) Striatal [^{123}I]beta-CIT SPECT and prefrontal cognitive functions in Parkinson's disease. *J. Neural. Transm.* **107,** 303–319.

84. Benamer, H. T., Patterson, J., Wyper, D. J., Hadley, D. M., Macphee, G. J., and Grosset, D. G. (2000) Correlation of Parkinson's disease severity and duration with 123I-FP-CIT SPECT striatal uptake [In Process Citation]. *Mov. Disord.* **15,** 692–698.

85. Malison, R. T., McDougle, C. J., van Dyck, C. H., Scahill, L., Baldwin, R. M., Seibyl, J. P., et al. (1995) [^{123}I]beta-CIT SPECT imaging of striatal dopamine transporter binding in Tourette's disorder. *Am. J. Psychiatry* **152,** 1359–1361.

86. Wong, D. F., Harris, J. C., Naidu, S., Yokoi, F., Marenco, S., Dannals, R. F., et al. (1996) Dopamine transporters are markedly reduced in Lesch-Nyhan disease in vivo. *Proc. Natl. Acad. Sci. USA* **93,** 5539–5543.

87. Donnemiller, E., Heilmann, J., Wenning, G. K., Berger, W., Decristoforo, C., Moncayo, R., et al. (1997) Brain perfusion scintigraphy with 99mTc-HMPAO or 99mTc-ECD and 123I-beta-CIT single-photon emission tomography in dementia of the Alzheimer-type and diffuse Lewy body disease. *Eur. J. Nucl. Med.* **24,** 320–325.

88. Tiihonen, J., Kuikka, J. T., Bergstrom, K. A., Karhu, J., Viinamaki, H., Lehtonen, J., et al. (1997) Single-photon emission tomography imaging of monoamine transporters in impulsive violent behaviour. *Eur. J. Nucl. Med.* **24,** 1253–1260.

89. Tiihonen, J., Kuikka, J., Bergstrom, K., Lepola, U., Koponen, H., and Leinonen, E. (1997) Dopamine reuptake site densities in patients with social phobia. *Am. J. Psychiatry* **154,** 239–242.

90. Naumann, M., Pirker, W., Reiners, K., Lange, K. W., Becker, G., and Brucke, T. (1998) Imaging the pre- and postsynaptic side of striatal dopaminergic

synapses in idiopathic cervical dystonia: a SPECT study using [^{123}I] epidepride and [^{123}I] beta-CIT. *Mov. Disord.* **13**, 319–323.

91. Dougherty, D. D., Bonab, A. A., Spencer, T. J., Rauch, S. L., Madras, B. K., and Fischman, A. J. (1999) Dopamine transporter density in patients with attention deficit hyperactivity disorder [letter] [see comments]. *Lancet* **354**, 2132–2133.

92. Laasonen-Balk, T., Kuikka, J., Viinamaki, H., Husso-Saastamoinen, M., Lehtonen, J., and Tiihonen, J. (1999) Striatal dopamine transporter density in major depression. *Psychopharmacology (Berl)* **144**, 282–285.

93. Sperfeld, A. D., Collatz, M. B., Baier, H., Palmbach, M., Storch, A., Schwarz, J., et al. (1999) FTDP-17: an early-onset phenotype with parkinsonism and epileptic seizures caused by a novel mutation [see comments]. *Ann. Neurol.* **46**, 708–715.

94. Donnemiller, E., Brenneis, C., Wissel, J., Scherfler, C., Poewe, W., Riccabona, G., and Wenning, G. K. (2000) Impaired dopaminergic neurotransmission in patients with traumatic brain injury: a SPECT study using 123I-beta-CIT and 123I-IBZM [In Process Citation]. *Eur. J. Nucl. Med.* **27**, 1410–1414.

95. Aberg, L., Liewendahl, K., Nikkinen, P., Autti, T., Rinne, J. O., and Santavuori, P. (2000) Decreased striatal dopamine transporter density in JNCL patients with parkinsonian symptoms. *Neurology* **54**, 1069–1074.

96. Kim, G. M., Kim, S. E., and Lee, W. Y. (2000) Preclinical impairment of the striatal dopamine transporter system in sporadic olivopontocerebellar atrophy: studied with [(123)I]beta-CIT and SPECT. *Eur. Neurol.* **43**, 23–29.

97. Laakso, A., Vilkman, H., Kajander, J., Bergman, J., Paranta, M., Solin, O., and Hietala, J. (2000) Prediction of detached personality in healthy subjects by low dopamine transporter binding. *Am. J. Psychiatry* **157**, 290–292.

98. Laakso, A., Vilkman, H., Alakare, B., Haaparanta, M., Bergman, J., Solin, O., et al. (2000) Striatal dopamine transporter binding in neuroleptic-naive patients with schizophrenia studied with positron emission tomography. *Am. J. Psychiatry* **157**, 269–271.

99. Laruelle, M., Abi-Dargham, A., van Dyck, C., Gil, R., D'Souza, D. C., Krystal, J., et al. (2000) Dopamine and serotonin transporters in patients with schizophrenia: an imaging study with [(123)I]beta-CIT. *Biol. Psychiatry* **47**, 371–379.

100. Eisensehr, I., Linke, R., Noachtar, S., Schwarz, J., Gildehaus, F. J., and Tatsch, K. (2000) Reduced striatal dopamine transporters in idiopathic rapid eye movement sleep behaviour disorder. Comparison with Parkinson's disease and controls. *Brain* **123**, 1155–1160.

101. Ouchi, Y., Yoshikawa, E., Okada, H., Futatsubashi, M., Sekine, Y., Iyo, M., and Sakamoto, M. (1999) Alterations in binding site density of dopamine transporter in the striatum, orbitofrontal cortex, and amygdala in early Parkinson's disease: compartment analysis for beta-CFT binding with positron emission tomography. *Ann. Neurol.* **45**, 601–610.

102. Jeon, B. S., Jeong, J. M., Park, S. S., Kim, J. M., Chang, Y. S., Song, H. C., (1998) Dopamine transporter density measured by [^{123}I]beta-CIT single-pho-

ton emission computed tomography is normal in dopa-responsive dystonia. *Ann. Neurol.* **43**, 792–800.

103. Laine, T. P., Ahonen, A., Rasanen, P., and Tiihonen, J. (1999) Dopamine transporter availability and depressive symptoms during alcohol withdrawal. *Psychiatry Res.* **90**, 153–157.

104. Repo, E., Kuikka, J. T., Bergstrom, K. A., Karhu, J., Hiltunen, J., and Tiihonen, J. (1999) Dopamine transporter and D2-receptor density in late-onset alcoholism. *Psychopharmacology (Berl)* **147**, 314–318.

105. Tiihonen, J., Kuikka, J., Bergstrom, K., Hakola, P., Karhu, J., Ryynanen, O. P., and Fohr, J. (1995) Altered striatal dopamine re-uptake site densities in habitually violent and non-violent alcoholics [see comments]. *Nat. Med.* **1**, 654–657.

106. Heinz, A., Ragan, P., Jones, D. W., Hommer, D., Williams, W., Knable, M. B., et al. (1998) Reduced central serotonin transporters in alcoholism. *Am. J. Psychiatry* **155**, 1544–1549.

107. Fischman, A. J., Babich, J. W., Elmaleh, D. R., Barrow, S. A., Meltzer, P., Hanson, R. N., et al. (1997) SPECT imaging of dopamine transporter sites in normal and MPTP-Treated rhesus monkeys. *J. Nucl. Med.* **38**, 144–150.

108. Villemagne, V., Yuan, J., Wong, D. F., Dannals, R. F., Hatzidimitriou, G., Mathews, W. B., et al. (1998) Brain dopamine neurotoxicity in baboons treated with doses of methamphetamine comparable to those recreationally abused by humans: evidence from [^{11}C]WIN-35,428 positron emission tomography studies and direct in vitro determinations. *J. Neurosci.* **18**, 419–427.

109. Eberling, J. L., Bankiewicz, K. S., Pivirotto, P., Bringas, J., Chen, K., Nowotnik, D. P., et al. (1999) Dopamine transporter loss and clinical changes in MPTP-lesioned primates. *Brain Res.* **832**, 184–187.

110. Harvey, D. C., Lacan, G., Tanious, S. P., and Melega, W. P. (2000) Recovery from methamphetamine induced long-term nigrostriatal dopaminergic deficits without substantia nigra cell loss. *Brain Res.* **871**, 259–270.

111. Volkow, N. D., Wang, G. J., Fowler, J. S., Logan, J., Hitzemannn, R., Gatley, S. J., et al. (1996) Cocaine uptake is decreased in the brain of detoxified cocaine abusers. *Neuropsychopharmacology* **14**, 159–168.

112. Malison, R. T., Best, S. E., van Dyck, C. H., McCance, E. F., Wallace, E. A., Laruelle, M., et al. (1998) Elevated striatal dopamine transporters during acute cocaine abstinence as measured by [^{123}I] beta-CIT SPECT. *Am. J. Psychiatry* **155**, 832–834.

113. McCann, U. D., Wong, D. F., Yokoi, F., Villemagne, V., Dannals, R. F., and Ricaurte, G. A. (1998) Reduced striatal dopamine transporter density in abstinent methamphetamine and methcathinone users: evidence from positron emission tomography studies with [^{11}C]WIN-35,428. *J. Neurosci.* **18**, 8417–8422.

114. Laine, T. P., Ahonen, A., Torniainen, P., Heikkila, J., Pyhtinen, J., Rasanen, P., (1999) Dopamine transporters increase in human brain after alcohol withdrawal. *Mol. Psychiatry* **4**, 104–185, 189–191.

115. Heinz, A., Goldman, D., Jones, D. W., Palmour, R., Hommer, D., Gorey, J. G., et al. (2000) Genotype influences in vivo dopamine transporter availability in human striatum. *Neuropsychopharmacology* **22**, 133–139.

116. Logan, J., Volkow, N. D., Fowler, J. S., Wang, G. J., Fischman, M. W., Foltin, R. W., et al. (1997) Concentration and occupancy of dopamine transporters in cocaine abusers with [^{11}C]cocaine and PET [see comments]. *Synapse* **27,** 347–356.

117. Wang, G. J., Volkow, N. D., Fowler, J. S., Fischman, M., Foltin, R., Abumrad, N. N., et al. (1997) Cocaine abusers do not show loss of dopamine transporters with age. *Life Sci.* **61,** 1059–1065.

118. Little, K. Y., Carroll, F. I., and Butts, J. D. (1998) Striatal [125I]RTI-55 binding sites in cocaine-abusing humans. *Prog. Neuropsychopharmacol. Biol. Psychiatry* **22,** 455–466.

119. Little, K. Y., Kirkman, J. A., Carroll, F. I., Clark, T. B., and Duncan, G. E. (1993) Cocaine use increases [^{3}H]WIN 35428 binding sites in human striatum. *Brain Res.* **628,** 17–25.

120. Little, K. Y., McLaughlin, D. P., Zhang, L., McFinton, P. R., Dalack, G. W., Cook, E. H., Jr., et al. (1998) Brain dopamine transporter messenger RNA and binding sites in cocaine users: a postmortem study. *Arch. Gen. Psychiatry* **55,** 793–799.

121. Little, K. Y., Zhang, L., Desmond, T., Frey, K. A., Dalack, G. W., and Cassin, B. J. (1999) Striatal dopaminergic abnormalities in human cocaine users. *Am. J. Psychiatry* **156,** 238–245.

122. Chen, L., Segal, D. M., Moraes, C. T., and Mash, D. C. (1999) Dopamine transporter mRNA in autopsy studies of chronic cocaine users. *Brain Res. Mol. Brain Res.* **73,** 181–185.

123. Wilson, J. M., Levey, A. I., Bergeron, C., Kalasinsky, K., Ang, L., Peretti, F., et al. (1996) Striatal dopamine, dopamine transporter, and vesicular monoamine transporter in chronic cocaine users. *Ann. Neurol.* **40,** 428–439.

124. Wilson, J. M. and Kish, S. J. (1996) The vesicular monoamine transporter, in contrast to the dopamine transporter, is not altered by chronic cocaine self-administration in the rat. *J. Neurosci.* **16,** 3507–3510.

125. Volkow, N. D., Fowler, J. S., Logan, J., Gatley, S. J., Dewey, S. L., MacGregor, R. R., et al. (1995) Carbon-11-cocaine binding compared at subpharmacological and pharmacological doses: a PET study [see comments]. *J. Nucl. Med.* **36,** 1289–1297.

126. Volkow, N. D., Gatley, S. J., Fowler, J. S., Logan, J., Fischman, M., Gifford, A. N., et al. (1996) Cocaine doses equivalent to those abused by humans occupy most of the dopamine transporters. *Synapse* **24,** 399–402.

127. Volkow, N. D., Wang, G. J., Fowler, J. S., Gatley, S. J., Ding, Y. S., Logan, J., et al. (1996) Relationship between psychostimulant-induced "high" and dopamine transporter occupancy. *Proc. Natl. Acad. Sci. USA* **93,** 10,388–10,392.

128. Volkow, N. D., Wang, G. J., and Fowler, J. S. (1997) Imaging studies of cocaine in the human brain and studies of the cocaine addict. *Ann. NY Acad. Sci.* **820,** 41–54; discussion 54–45.

129. Volkow, N. D., Wang, G. J., Fischman, M. W., Foltin, R. W., Fowler, J. S., Abumrad, N. N., et al. (1997) Relationship between subjective effects of cocaine and dopamine transporter occupancy. *Nature* **386,** 827–830.

130. Volkow, N. D., Wang, G. J., Fowler, J. S., Gatley, S. J., Logan, J., Ding, Y. S., et al. (1998) Dopamine transporter occupancies in the human brain induced by therapeutic doses of oral methylphenidate. *Am. J. Psychiatry* **155,** 1325–1331.

131. Volkow, N. D., Wang, G. J., Fischman, M. W., Foltin, R., Fowler, J. S., Franceschi, D., et al. (2000) Effects of route of administration on cocaine induced dopamine transporter blockade in the human brain. *Life Sci.* **67,** 1507–1515.

132. Malison, R. T., Best, S. E., Wallace, E. A., McCance, E., Laruelle, M., Zoghbi, S. S., et al. (1995) Euphorigenic doses of cocaine reduce [^{123}I]beta-CIT SPECT measures of dopamine transporter availability in human cocaine addicts. *Psychopharmacology (Berl)* **122,** 358–362.

133. Gatley, S. J., Volkow, N. D., Chen, R., Fowler, J. S., Carroll, F. I., and Kuhar, M. J. (1996) Displacement of RTI-55 from the dopamine transporter by cocaine. *Eur. J. Pharmacol.* **296,** 145–151.

134. Gatley, S. J., Gifford, A. N., Carroll, F. I., and Volkow, N. D. (2000) Sensitivity of binding of high-affinity dopamine receptor radioligands to increased synaptic dopamine [In Process Citation]. *Synapse* **38,** 483–488.

135. Volkow, N. D., Gatley, S. J., Fowler, J. S., Chen, R., Logan, J., Dewey, S. L., et al. (1995) Long-lasting inhibition of in vivo cocaine binding to dopamine transporters by 3 beta-(4-iodophenyl)tropane-2-carboxylic acid methyl ester: RTI-55 or beta CIT. *Synapse* **19,** 206–211.

136. Brucke, T., Kornhuber, J., Angelberger, P., Asenbaum, S., Frassine, H., and Podreka, I. (1993) SPECT imaging of dopamine and serotonin transporters with [^{123}I]beta-CIT. Binding kinetics in the human brain. *J. Neural. Transm. Gen. Sect.* **94,** 137–146.

137. Bergstrom, K. A., Kuikka, J. T., Ahonen, A., and Vanninen, E. (1994) [^{123}I] beta-CIT, a tracer for dopamine and serotonin re-uptake sites: preparation and preliminary SPECT studies in humans. *J. Nucl. Biol. Med.* **38,** 128–131.

138. Kuikka, J. T., Tiihonen, J., Bergstrom, K. A., Karhu, J., Hartikainen, P., Viinamaki, H., et al. (1995) Imaging of serotonin and dopamine transporters in the living human brain. *Eur. J. Nucl. Med.* **22,** 346–350.

139. Fujita, M., Takatoku, K., Matoba, Y., Nishiura, M., Kobayashi, K., Inoue, O., et al. (1996) Differential kinetics of [^{123}I]beta-CIT binding to dopamine and serotonin transporters. *Eur. J. Nucl. Med.* **23,** 431–436.

140. Malison, R. T., Price, L. H., Berman, R., van Dyck, C. H., Pelton, G. H., Carpenter, L., et al. (1998) Reduced brain serotonin transporter availability in major depression as measured by [^{123}I]-2 beta-carbomethoxy-3 beta-(4-iodophenyl)tropane and single photon emission computed tomography [see comments]. *Biol. Psychiatry* **44,** 1090–1098.

141. Jacobsen, L. K., Staley, J. K., Malison, R. T., Zoghbi, S. S., Seibyl, J. P., Kosten, T. R., et al. (2000) Elevated central serotonin transporter binding availability in acutely abstinent cocaine-dependent patients. *Am. J. Psychiatry* **157,** 1134–1140.

142. van Dyck, C. H., Malison, R. T., Seibyl, J. P., Laruelle, M., Klumpp, H., Zoghbi, S. S., et al. (2000) Age-related decline in central serotonin trans-

porter availability with [(123)I]beta-CIT SPECT. *Neurobiol. Aging* **21,** 497–501.

143. Kim, S. E., Lee, W. Y., Choe, Y. S., and Kim, J. H. (1999) SPECT measurement of iodine-123-beta-CIT binding to dopamine and serotonin transporters in Parkinson's disease: correlation with symptom severity. *Neurol. Res.* **21,** 255–261.

144. Jagust, W. J., Eberling, J. L., Biegon, A., Taylor, S. E., VanBrocklin, H. F., Jordan, S., et al. (1996) Iodine-123-5-iodo-6-nitroquipazine: SPECT radiotracer to image the serotonin transporter. *J. Nucl. Med.* **37,** 1207–1214.

145. Bergstrom, K. A., Halldin, C., Hall, H., Lundkvist, C., Ginovart, N., Swahn, C. G., and Farde, L. (1997) In vitro and in vivo characterisation of nor-beta-CIT: a potential radioligand for visualisation of the serotonin transporter in the brain. *Eur. J. Nucl. Med.* **24,** 596–601.

146. Hiltunen, J., Akerman, K. K., Kuikka, J. T., Bergstrom, K. A., Halldin, C., Nikula, T., et al. (1998) Iodine-123 labeled nor-beta-CIT as a potential tracer for serotonin transporter imaging in the human brain with single-photon emission tomography. *Eur. J. Nucl. Med.* **25,** 19–23.

147. Oya, S., Choi, S. R., Hou, C., Mu, M., Kung, M. P., Acton, P. D., et al. (2000) 2-((2-((dimethylamino)methyl)phenyl)thio)-5-iodophenylamine (ADAM): an improved serotonin transporter ligand. *Nucl. Med. Biol.* **27,** 249–254.

148. Zhuang, Z., Choi, S., Hou, C., Mu, M., Kung, M., Acton, P. D., et al. (2000) A novel serotonin transporter ligand: (5-iodo-2-(2-dimethylaminomethyl-phenoxy)-benzyl alcohol. *Nucl. Med. Biol.* **27,** 169–175.

149. Dresel, S. H., Kung, M. P., Huang, X., Plossl, K., Hou, C., Shiue, C. Y., et al. (1999) In vivo imaging of serotonin transporters with [^{99m}Tc]TRODAT-1 in nonhuman primates. *Eur. J. Nucl. Med.* **26,** 342–347.

150. Szabo, Z., Kao, P. F., Scheffel, U., Suehiro, M., Mathews, W. B., Ravert, H. T., et al. (1995) Positron emission tomography imaging of serotonin transporters in the human brain using [^{11}C](+)McN5652. *Synapse* **20,** 37–43.

151. Parsey, R. V., Kegeles, L. S., Hwang, D. R., Simpson, N., Abi-Dargham, A., Mawlawi, O., et al. (2000) In vivo quantification of brain serotonin transporters in humans using [^{11}C]McN 5652. *J. Nucl. Med.* **4,1** 1465–1477.

152. Wilson, A. A., Ginovart, N., Schmidt, M., Meyer, J. H., Threlkeld, P. G., and Houle, S. (2000) Novel radiotracers for imaging the serotonin transporter by positron emission tomography: synthesis, radiosynthesis, and in vitro and ex vivo evaluation of (11)C-labeled 2-(phenylthio)araalkylamines. *J. Med. Chem.* **43,** 3103–3110.

153. Lundkvist, C., Loc'h, C., Halldin, C., Bottlaender, M., Ottaviani, M., Coulon, C., et al. (1999) Characterization of bromine-76-labelled 5-bromo-6-nitroquipazine for PET studies of the serotonin transporter. *Nucl. Med. Biol.* **26,** 501–507.

154. Helfenbein, J., Loc'h, C., Bottlaender, M., Emond, P., Coulon, C., Ottaviani, M., et al. (1999) A selective radiobrominated cocaine analogue for imaging of dopamine uptake sites: pharmacological evaluation and PET experiments. *Life Sci.* **65,** 2715–2726.

155. Helfenbein, J., Sandell, J., Halldin, C., Chalon, S., Emond, P., Okubo, Y., et al. (1999) PET examination of three potent cocaine derivatives as specific radioligands for the serotonin transporter. *Nucl. Med. Biol.* **26,** 491–499.
156. DaSilva, J. N., Kilbourn, M. R., and Mangner, T. J. (1993) Synthesis of a [^{11}C]methoxy derivative of alpha-dihydrotetrabenazine: a radioligand for studying the vesicular monoamine transporter. *Appl. Radiat. Isot.* **44,** 1487–1489.
157. Koeppe, R. A., Frey, K. A., Kuhl, D. E., and Kilbourn, M. R. (1999) Assessment of extrastriatal vesicular monoamine transporter binding site density using stereoisomers of [^{11}C]dihydrotetrabenazine. *J. Cereb. Blood Flow Metab.* **19,** 1376–1384.
158. Koeppe, R. A., Frey, K. A., Vander Borght, T. M., Karlamangla, A., Jewett, D. M., Lee, L. C., et al. (1996) Kinetic evaluation of [^{11}C]dihydrotetrabenazine by dynamic PET: measurement of vesicular monoamine transporter. *J. Cereb. Blood Flow Metab.* **16,** 1288–1299.
159. Kilbourn, M. R., Frey, K. A., Vander Borght, T., and Sherman, P. S. (1996) Effects of dopaminergic drug treatments on in vivo radioligand binding to brain vesicular monoamine transporters. *Nucl. Med. Biol.* **23,** 467–471.
160. Frey, K. A., Koeppe, R. A., Kilbourn, M. R., Vander Borght, T. M., Albin, R. L., Gilman, S., et al. (1996) Presynaptic monoaminergic vesicles in Parkinson's disease and normal aging. *Ann. Neurol.* **40,** 873–884.
161. Lee, C. S., Samii, A., Sossi, V., Ruth, T. J., Schulzer, M., Holden, J. E., Wudel, J., et al. (2000) In vivo positron emission tomographic evidence for compensatory changes in presynaptic dopaminergic nerve terminals in Parkinson's disease. *Ann. Neurol.* **47,** 493–503.
162. Gilman, S., Frey, K. A., Koeppe, R. A., Junck, L., Little, R., Vander Borght, T. M., et al. (1996) Decreased striatal monoaminergic terminals in olivopontocerebellar atrophy and multiple system atrophy demonstrated with positron emission tomography. *Ann. Neurol.* **40,** 885–892.
163. Gilman, S., Koeppe, R. A., Adams, K. M., Junck, L., Kluin, K. J., Johnson-Greene, D., Martorello, S., et al. (1998) Decreased striatal monoaminergic terminals in severe chronic alcoholism demonstrated with (+)[^{11}C]dihydrotetrabenazine and positron emission tomography. *Ann. Neurol.* **44,** 326–333.
164. Gilman, S., Koeppe, R. A., Junck, L., Little, R., Kluin, K. J., Heumann, M., et al. (1999) Decreased striatal monoaminergic terminals in multiple system atrophy detected with positron emission tomography. *Ann. Neurol.* **45,** 769–777.
165. Bohnen, N. I., Koeppe, R. A., Meyer, P., Ficaro, E., Wernette, K., Kilbourn, M. R., et al. (2000) Decreased striatal monoaminergic terminals in Huntington disease. *Neurology* **54,** 1753–1759.
166. Taylor, S. F., Koeppe, R. A., Tandon, R., Zubieta, J., and Frey, K. A. (2000) In Vivo Measurement of the Vesicular Monoamine Transporter in Schizophrenia. *Neuropsychopharmacology* **23,** 667–675.
167. Meyer, P., Bohnen, N. I., Minoshima, S., Koeppe, R. A., Wernette, K., Kilbourn, M. R., et al. (1999) Striatal presynaptic monoaminergic vesicles are not increased in Tourette's syndrome. *Neurology* **53,** 371–374.
168. Zubieta, J. K., Huguelet, P., Ohl, L. E., Koeppe, R. A., Kilbourn, M. R., Carr, J. M., et al. (2000) High vesicular monoamine transporter binding in asymp-

tomatic bipolar I disorder: sex differences and cognitive correlates [In Process Citation]. *Am. J. Psychiatry* **157,** 1619–1628.

169. Haka, M. S. and Kilbourn, M. R. (1989) Synthesis and regional mouse brain distribution of [^{11}C]nisoxetine, a norepinephrine uptake inhibitor. *Intl. J. Rad. Appl. Instrum. B* **16,** 771–774.

170. Ding, Y. S., Fowler, J. S., Dewey, S. L., Logan, J., Schlyer, D. J., Gatley, S. J., et al. (1993) Comparison of high specific activity (-) and (+)-6-[^{18}F]fluoronorepinephrine and 6-[^{18}F]fluorodopamine in baboons: heart uptake, metabolism and the effect of desipramine. *J. Nucl. Med.* **34,** 619–629.

171. Ding, Y. S., Fowler, J. S., Gatley, S. J., Logan, J., Volkow, N. D., and Shea, C. (1995) Mechanistic positron emission tomography studies of 6-[^{18}F]fluorodopamine in living baboon heart: selective imaging and control of radiotracer metabolism using the deuterium isotope effect. *J. Neurochem.* **65,** 682–690.

172. Degrado, T. R., Zalutsky, M. R., and Vaidyanathan, G. (1995) Uptake mechanisms of meta-[^{123}I]iodobenzylguanidine in isolated rat heart. *Nucl. Med. Biol.* **22,** 1–12.

173. Telang, F. W., Volkow, N. D., Levy, A., Logan, J., Fowler, J. S., Felder, C., Wong, C., and Wang, G. J. (1999) Distribution of tracer levels of cocaine in the human brain as assessed with averaged [11C]cocaine images. *Synapse* **31,** 290–296.

13

In Vitro and In Vivo Imaging of the Human Dopamine Transporter in Cocaine Abusers

Deborah C. Mash and Julie K. Staley

1. INTRODUCTION

Deaths involving psychoactive drugs stem not only from overdose, but also from drug-induced mental states that may lead to serious injuries *(1)*. Mortality data have revealed the virulence of the cocaine epidemic, although other indicators including crime, drug-exposed neonates, drug-related traffic accidents, and drug use by workers provide a fuller view of the nature and extent of the problem of cocaine abuse. The arrival of inexpensive "crack" cocaine has radically changed the nature of the epidemic, and has revealed the serious addictive potential of this drug.

The epidemic of cocaine abuse has prompted the search for drugs to block cocaine's euphoriant effects or substitute for cocaine with lower abuse liability. The reinforcing and additive properties of cocaine arise from its ability to block reuptake of the neurotransmitter dopamine (DA) by binding to recognition sites on the DA transport carrier *(2–4)*. DA uptake is inhibited with a distinct pharmacological profile by a variety of structurally diverse compounds *(see* also Chapter 3). Identification and cloning of the gene for the DA transporter (DAT) has provided insight into the molecular mechanism of DA reuptake inhibition by cocaine binding to the transport carrier *(5–7)*. Rothman and colleagues *(8)* have suggested that DA transport inhibitors can be divided into two groups: Type 1 DA reuptake inhibitors such as cocaine, which produce euphoria and addiction in humans, and Type 2 inhibitors such as mazindol, bupropion, and GBR 12909, which do not produce euphoria and have low abuse liability. The underlying assumption is that Type 2 inhibitors interact differently than cocaine at sites on the DAT. Molecular biological and pharmacological studies have delineated discrete

From: *Contemporary Neuroscience:*
Neurotransmitter Transporters: Structure, Function, and Regulation, 2nd Edition
Edited by: M. E. A. Reith © Humana Press Inc., Totowa, NJ

domains within the structure of the DAT protein for substrate, cocaine, and antidepressant interactions *(9–11)* *(see* also Chapter 3), raising the possibility that it may be feasible to design cocaine antagonists that are devoid of uptake blockade for the clinical management of cocaine addiction.

The neuronal DAT has been mapped and characterized in the human brain postmortem by radioligand binding techniques using novel cocaine congeners and transport inhibitors. In vitro binding and autoradiography of radioligands specific for the DAT afford systematic visualization of the status of cocaine-binding sites on the DAT. Studies in the postmortem human brain serve as an important counterpart to the rapidly developing noninvasive techniques of positron emission tomography (PET) and single photon emission tomography (SPECT) imaging, and are important for establishing quantitative and regional neurochemical adaptations resulting from cocaine abuse. Equally important are studies aimed at determining the long-term neurobiological consequences of chronic cocaine abuse on the DAT in aging populations that have abused psychostimulants. Determining how cocaine exposure affects the human DAT may reveal the mechanisms responsible for tolerance and sensitization (reverse tolerance) to cocaine actions, and provide a basis for designing effective medications for treating cocaine dependence.

2. LIGAND-BINDING STUDIES OF THE HUMAN DAT

The DAT is a primary recognition site for cocaine that is related to drug abuse *(3,4)*. DAT functions to rapidly control the removal of transmitter molecules from the synaptic cleft. Cocaine potentiates DAergic neurotransmission by blocking the reuptake of DA, leading to marked elevations in the levels of neurotransmitter *(6,12)*. Radioligand binding to the DAT has been best characterized with the cocaine congeners [^{3}H]WIN 35,428, [^{125}I]RTI-55, and [^{125}I]RTI-121 (for review, *see* ref. *13*). In contrast to the classic DAT inhibitors ([^{3}H]mazindol, [^{3}H]GBR 12935, and [^{3}H]nomifensine), the cocaine congeners ([^{3}H]WIN 35,428, [^{125}I]RTI-55, and [^{125}I]RTI-121) label multiple sites with a pharmacological profile characteristic of the DAT in primate and human brains *(7,14–16)*. Radioligand binding to African Green Monkey kidney (COS) cells transfected with the cloned cocaine-sensitive DAT demonstrates that [^{3}H]WIN 35,428 identified two sites for binding to the protein expressed from a single cDNA *(17)*. Pharmacological studies have demonstrated a lack of correspondence between DAT function and ligand binding *(7)*. These findings provide additional support for the incomplete correspondence of pharmacologically overlapping sites for [^{3}H]WIN 35,428, [^{3}H]GBR 12,935, and [^{3}H]mazindol labeling of the DAT in native membrane preparations from rat brain *(18,19)*. Pharmacological heterogeneity of the cloned and native human DA transporter was suggested further by the dissociation of [^{3}H]WIN 35,428 and [^{3}H]GBR 12,935 binding sites *(7)*. Interestingly, the proportion of observed high- and low-affinity [^{3}H]WIN 35,428 binding sites differs across studies using cloned *(5,7,17)* or native membranes *(14,18,19)*. The binding of [^{3}H]WIN 35,428 to the cloned human DAT demonstrates the recognition of multiple sites with only one that is functionally correlated with that of the cloned DA uptake process *(6,17)*.

Saturation binding of the cocaine congeners [^{3}H]WIN 35,428, *(14,23)*, [^{125}I]RTI-55 *(15,21)*, and [^{125}I]RTI-121 *(16)* is biphasic, indicating the presence of multiple cocaine recognition sites on the DAT. In contrast, the binding of [^{3}H]GBR 12935 *(22)* and [^{3}H]mazindol (Staley and Mash, unpublished observations) to the human striatum is monophasic, suggesting that these radioligands may interact differently with the DAT. The binding-site densities in human striatum estimated from saturation isotherms for the cocaine congeners ([^{3}H]WIN 35,428 *(23)*, [^{125}I]RTI-55 *(15)*, and [^{125}I]RTI-121 *(16)* (approx 150–200 pmol/g original tissue) are in agreement with that observed for the noncocaine-like transport inhibitors [^{3}H]mazindol (Fig. 1) and [^{3}H]GBR 12935 *(22)*. The total binding densities corresponding to both the high- and low-affinity cocaine recognition sites were comparable to noncocaine-like transport inhibitors which recognize a single class of sites associated with the DAT (Fig. 1).

Multiple cocaine recognition sites associated with the DAT have not been consistently reported in human striatum *(7,24*; Kish, S., personal communication). The binding density corresponding to a single high affinity site is significantly lower than that comprised by both high- and low-affinity cocaine recognition sites in the human brain *(7,15,16,21,23,24)*. In some instances, the existence of only a single component for WIN 35,428 *(7,24)* or RTI-121 *(16)* may be caused by different conditions of the binding assay. For example, in the human striatum the binding of RTI-121 is biphasic, depending on the assay buffer *(16)*. RTI-121 saturation curves performed in sucrose-phosphate buffer reveal high- and low-affinity cocaine recognition sites with a total density approximately equal to 200 pmol/g tissue. When saturation-binding assays are conducted in a high-salt-containing buffer, a significant decrease in the overall density of binding (50 pmol/g tissue) is observed *(16)*. Similar total density values have been reported for [^{125}I]RTI-55 and [^{3}H]WIN 35,428 binding to human striatal membranes in Tris buffers *(21,24)*. Rosenthal plots of RTI-121 binding obtained in the presence of high sodium-containing buffers demonstrate a selective decrease in the density of the low-affinity binding component *(16)*. These findings suggest that the low-affinity cocaine-recognition site may overlap with sites on the transporter that confer ionic dependence for DAT. The

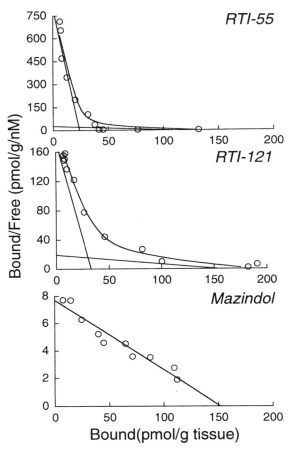

Fig. 1. Rosenthal plots of DA transport-inhibitor binding in human striatum. Saturation-binding studies of [^{125}I]RTI-55, [^{125}I]RTI-121, and [^{3}H]mazindol binding to human striatal membranes were conducted in 10 mM sodium phosphate buffer (pH 7.4) containing 0.32 M sucrose as previously described *(15,16)*. Data were analyzed using the iterative nonlinear curve-fitting program EBDA/LIGAND (Elsevier, Biosoft). In these representative experiments, the cocaine congeners RTI-55 and RTI-121 recognized high- and low-affinity binding sites with affinity values of 0.1 and 4.1 nM for RTI-55 and affinity values of 0.2 and 8.0 nM for RTI-121. The density values corresponding to the high- and low-affinity binding sites were 23.8 and 108.9 pmol/g for RTI-55 and 33.1 and 153.6 pmol/g tissue for RTI-121. In contrast, the noncocaine-like DA transport inhibitor [^{3}H]mazindol labeled a single binding site with a K_d value of 19.2 nM and B_{max} value of 148.8 pmol/g tissue.

heterogeneity of radioligand binding suggests that DA and cocaine interact with more than one class of recognition sites on the DAT.

The functional significance of multiple cocaine-recognition sites associated with the DAT remains unclear. The two cocaine-recognition sites may reflect binding interactions with distinct domains of a single DAT polypeptide or with different conformations of the DAT that are recognized with equal affinities by noncocaine-like DAT inhibitors. The function of the DAT is to translocate DA together with Na^+ and Cl^- across the presynaptic membrane into the cytoplasm. Re-orientation of the DAT from inside to outside or structural folding of the protein, which favors outward- vs inward-directed residues, may account for the pharmacological heterogeneity seen with different classes of radioligands that bind to the DAT. Posttranslational modification of the DAT protein that signal trafficking and membrane insertion of the protein may affect apparent radioligand-binding characteristics.

The recent cloning and expression of the human DAT gene has provided extensive information on the pharmacological signature of the DAT protein *(5,6,11)*. The cloned DAT cDNA encodes a single polypeptide strand of 620 amino acids that corresponds to a protein of 68,517 daltons *(6)*. The cDNA encodes consensus sites for glycosylation and phosphorylation, suggesting that secondary processing may contribute to the regulation of the transport protein (*see* Chapters 2 and 11). Like other neurotransmitter carriers, the predicted structure of the DAT based on hydropathicity analysis suggests the presence of 12 transmembrane domains with an intracellular N-terminus and C-terminus *(6)*. Expression of the cloned DAT in COS-7, mouse fibroblast, or glioma-cell lines *(5–7)* afforded biochemical characterization of saturable, Na-dependent DAT, which was blocked by psychostimulants (cocaine, amphetamine, and phencyclidine) and DA-transport inhibitors (GBR 12909, mazindol, nomifensine, and amnfoleic acid) with potency values that correlate well with the those observed for competition of radiolabeled cocaine-congener binding. The pharmacological profile observed for inhibition of binding of the cocaine congeners RTI-55 and RTI-121 in human striatum agrees with the values observed for inhibition of DA uptake and [³H]WIN 35,428 binding to the cloned human DAT (Table 1).

The molecular structure of the rat DAT has been modified using site-directed mutagenesis to further characterize the regions of the transporter that mediate cocaine's interaction *(10,11)* (*see* also Chapter 3). In transmembrane (TM) 1, mutation of Asp 79 dramatically reduced the uptake of [³H]DA and decreased the binding affinity of the cocaine congener [³H]WIN 35,428. Replacement of serine residues 356 and 359 in TM 7 decreased [³H]DA uptake and had a minimal effect on [³H]WIN 35,428 binding *(10)*. These studies suggest that cocaine congeners and DA may interact with distinct yet overlapping domains within the DAT complex. Discrete domains for cocaine, "classical" tricyclic antidepressant and substrate binding, ionic

Table 1

Pharmacological Characterization of Cloned and Native Human Brain DAT

Inhibitor	Cloned human DAT		Human brain DAT		
	[^{3}H]DA uptake[*],[†] K_i, nM	[^{3}H]WIN 35,428[#],[†] K_i, nM	[^{3}H]WIN 35,428[†] K_i, nM	[^{125}I]RTI-55[§] IC$_{50}$, nM	[^{125}I]RTI-121[‡] IC$_{50}$, nM
Mazindol	11, 60	3, 24	15	8	7
GBR 12909	17, 14	n.d., 4	30	15	1
(−) Cocaine	58, 743	39, 240	68	151	42
Pimozide	344, n.d.	n.d.	n.d.	141	n.d.
Buproprion	330, 784	121, 950	560	n.d.	n.d.
Nomifensine	17, 53	n.d., 42	40	n.d.	18
Desipramine	13000, n.d.	3320, n.d.	n.d.	2014	1560

[*], *see refs. 6 and 12;* [†], *see ref. 7;* [#], *see ref. 5;* [§], *see ref. 15, 1994;* [‡], *see ref. 16.* n.d., not determined.

dependence, and substrate uptake have been delineated using chimeric dopamine-norepinephrine transporters *(9,25)*. Regions from the amino terminus through TM 5 are important for ionic dependence and DA uptake *(9,25)*, and regions within TM 6–8 confer tricyclic antidepressant and cocaine binding and cocaine interactions with DAT *(9)*. Transmembrane regions 4–8 are important for substrate translocation, and TM 9 through the C-terminus may be responsible for stereoselectivity and high-affinity substrate interactions *(9,25)*. Giros and colleagues *(9)* caution that the involvement of a particular region of the protein in a given function does not imply that no other regions of the protein are implicated in that function, but does suggest the possibility that certain properties of these functions are specified by modular structural entitities *(9)*. The discovery that cocaine binding can be virtually eliminated in some chimeras without interfering with DA reuptake, suggests that a specific determinant for cocaine binding must exist independent of the binding of DA. Since the DAT is the endogenous target of cocaine action *(3,4)* these findings raise the important possibility that selective antagonists of cocaine's interaction with the DAT could be developed for clinical use in the treatment of cocaine abuse.

3. IN VITRO AND IN VIVO MAPPING OF THE DAT IN THE HUMAN BRAIN

Brain dopaminergic systems consist of three distinct pathways, including the nigrostriatal, mesocortical, and mesolimbic projections. The nigrostriatal pathway originates in the substantia nigra (SN) pars compacta and terminates in the striatum. The mesolimbic pathway originates in the ventral tegmental area (VTA) and projects to the limbic sectors of the striatum, amygdala, and olfactory tubercle. The mesocortical pathway originates in the VTA and terminates within particular sectors of the cerebral cortical mantle, including the prefrontal, cingulate, and entorhinal cortices *(26)*. The human striatum is organized into distinct neurochemical compartments termed patch (striosome) and matrix *(27)*. The DAergic terminals within the patch arise from cell bodies localized to the SN pars compacta, whereas the DAergic projections to the matrix compartment originate from the VTA (rostrorubral area) and the SN pars compacta *(27,28)*. Autoradiographic visualization of the distribution of DATs indicates that the topographic distribution of the DAT correlates well with DA innervation, with high densities localized to nigrostriatal terminals, moderate densities within the mesolimbic terminals, and low densities within the mesocortical terminals *(29,30)*.

The regional distribution of the DAT in the human brain varies depending on the probe and the target (messenger RNA, receptor-binding sites, or immunoreactive protein). *In situ* hybridization histochemistry reveals high densities of DAT mRNAs localized to the ventral SN, VTA, and the retrorubral cell groups *(28).* Abundant DAT mRNA is found within the SN pars compacta, which contains cell bodies that project primarily to the motor sectors of the striatum corresponding to the increased gradient in DATs *(15).* Matrix-directed neurons have the lowest level of DAT mRNA, but the DAT terminals in the matrix of the striatum maintain the highest density of DA binding *(28).* DAT immunoreactivity is enriched in dendrites and cell bodies of SN pars compacta and VTA neurons, as well as in nerve terminals in the striatum and nucleus accumbens *(31).* Low DAT immunoreactivity is observed in scattered varicose and a few nonvaricose fibers in the motor, premotor, anterior cingulate, prefrontal, entorhinal/perirhinal, insular, and visual cortices. DAergic projections are visualized in the basolateral and central subnuclei of the amygdala, with sparser fibers in the lateral and basomedial subnuclei. Fine DAT-immunoreactive axons are scattered throughout the hypothalamus, and are particularly concentrated along the medial border, with more coarse axons present along the lateral border *(31).* Immunocytochemical mapping studies demonstrate the high levels of DATs are localized to most mesotelencephalic DA neurons throughout their entire somatodendritic and axonal domains, whereas mesencephalic dopaminergic neurons express very low levels of DATs *(31).*

The regional distribution of the DAT in the human brain has been mapped in vitro using ligand binding and autoradiography on postmortem brain sections and visualized in vivo using positron emission tomography (PET) and single photon emission computed tomography (SPECT). The first detailed regional maps of the distribution of the DAT in the postmortem human brain were generated using [³H]mazindol *(32).* When binding to the norepinephrine transporter was occluded with desipramine, [³H]mazindol labeling was most evident in the striatum, with local gradients corresponding to the pattern of DAergic projections. Moderate to low [³H]mazindol labeling was visualized over DA cell-body fields, including the SN compacta, VTA, and the cell group in the retrorubral field *(32).* Autoradiographic mapping of the regional distribution of the DAT in the human brain using [³H]cocaine *(33)* and the cocaine congeners [³H]WIN 35,428 *(23,34),* [¹²⁵I]RTI-55 *(15)* [¹²⁵I]RTI-121 *(16),* and [¹²⁵I]altropane *(35)* has demonstrated high densities of labeling over the striatum with moderate labeling over DAergic cell-body fields of the SN (Fig. 2). In addition to high densities of binding in DAergic projections to the striatum, the cocaine congener radioligands demonstrate low but widespread binding throughout the cerebral cortical mantle. The identity of these extrastriatal binding sites is not completely clear, since cocaine binds to recognition sites on both the serotonin (5-HT) and norepi-

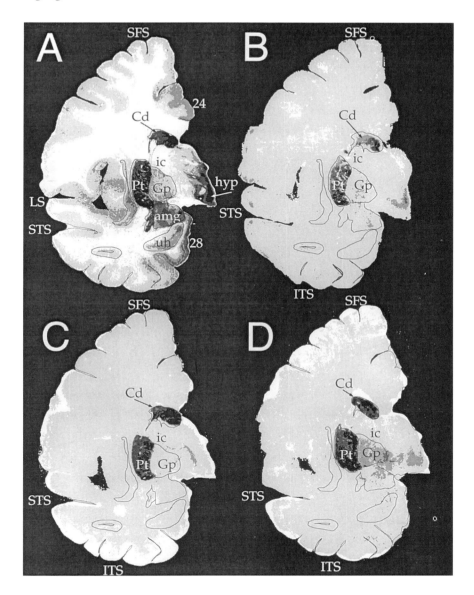

Fig. 2. In vitro autoradiography of radiolabeled cocaine congeners in human brain. **(A–D)** Gray-scale images of [125I]RTI-55 (50 p*M*); [125I]RTI-55 (50 p*M*) in the presence of 500 n*M* citalopram; [125I]RTI-121 (20 pM) and [3H]WIN 35,428 (2 n*M*), respectively. Abbreviations: amg, amygdala; Cd, caudate; Gp, globus pallidus; hyp, hypothalamus; ic, internal capsule; ITS, inferior temporal sulcus; LS, lateral sulcus; Pt, putamen; sn, substantia nigra; SFS, superior frontal sulcus; STS, superior temporal sulcus; th, thalamus; uh, uncus.

nephrine (NE) transporters. The cocaine congener radioligands demonstrate variable regional distributions that may be accounted for in part by the selectivity profiles across individual monoamine transporters.

For example, the distribution of [^{125}I]RTI-55 (β-CIT) binding is extensive, with binding sites prevalent throughout the cerebral cortex, striatum, thalamus, hypothalamus, and amygdala (Fig. 2A). This pattern of labeling correlates with the known distribution of monoaminergic nerve terminals and is consistent with binding of [^{125}I]RTI-55 (β-CIT) to both 5-HT and DATs *(15,36)*. Selective visualization of the DA transporter using [^{125}I]RTI-55 (β-CIT) has been achieved by occluding binding with the 5-HT reuptake inhibitors paroxetine or citalopram (Fig. 2B; *15,36*). Alternatively, the DAT may selectively labeled in vitro using the [^{125}I]RTI-121 *(16)* or [^{125}I]altropane *(35)*. The anatomical distribution of [^{125}I]RTI-121 (Fig. 2C) is similar to [^{125}I]altropane *(35)*, with a more restricted pattern that is well correlated with the known distribution of DA projections, and unlike the pattern of either [^{125}I]RTI-55 or [^{3}H]WIN 35,428 (Fig. 2D). Interestingly, the anatomical distribution of [^{3}H]WIN 35,428 labeling is more closely correlated with the pattern of the [^{3}H]cocaine labeling than [^{125}I]RTI-55 *(37)*.

In the living human brain, the regional distribution of the DAT has been visualized using[11]C-nomifensine *(38,39)* [11C]dl-threo-methylphenidate *(40)*, [11C]cocaine *(41–44)*, [11C]WIN 35,428 *(45)*, [11C]β-CIT, (also called RTI-55; *36*), [123I]-β-CIT *(46–48)*, [123I]β-CIT-FE *(49)*, [18F]FPCIT/([123I])β-CIT-FP *(50,51)*, [123I]altropane *(52)*, and [99mTc]TRODAT-1 *(53)*. All of the cocaine-like radiotracers demonstrate high uptake in the caudate, putamen, and nucleus accumbens, with moderate uptake in the thalamus and minimal uptake in other brain areas. Administration of DAT inhibitors significantly decreased striatal [11C]cocaine and [11C]dl-threo-methylphenidate uptake, confirming that these radioligands primarily label the DAT in the striatum. Interestingly, in vitro studies have clearly demonstrated significant radiolabeled cocaine binding in the thalamus, hippocampus, and amygdala *(37)*, these regions are barely detected in vivo. The lack of signal is caused by the low target to background ratios that occur because of the markedly lower expression of the protein in the region and the need to use "tracer" in vivo doses as opposed to saturating in vitro concentrations. Despite this limitation, extrastriatal cocaine binding was measured in a [11C]cocaine brain map created by averaging images from 17 healthy male subjects *(43)*. In this study, [11C]cocaine uptake was high in the caudate, putamen, and nucleus accumbens, intermediate in the hippocampus, amygdala, insular cortex, SN, and anterior cingulate, and low although detectable in the orbital cortex, posterior cingulate gyrus, dorsomedial thalamus, precuneus, and cerebellum *(43)*. The identity of the extrastriatal cocaine-binding sites has not yet been characterized. However, given the high affinity of [^{11}C]cocaine for the 5-HT and NETs, it is highly possible that some of the labeling in these brain areas is partially caused by binding to these monoamine transporters. The cocaine congeners [^{11}C]WIN 35,428 *(45)*, [^{11}C]β-CIT *(36)*, and [^{123}I]β-CIT *(46–48)* also demonstrated marked radiotracer accumulation in the striatum and low but measurable levels of binding to the thalamus, hypothalamus, midbrain, and pons. Radiotracer uptake in extrastriatal brain regions was occluded by pretreatment with mazindol and citalopram, indicating that the labeling was primarily to the 5-HT and NETs and not the DAT *(36,47,48)*. The [^{123}I]β-CIT uptake in the thalamus, hypothalamus, midbrain, and pons was significantly reduced in depressed patients treated with the selective 5-HT reuptake inhibitor citalopram *(47)*, further suggesting that the labeling of extrastriatal regions in vivo may reflect occupancies of the 5-HT transporter.

The second-generation cocaine-like radiotracers have demonstrated greater pharmacological specificity for the DAT and a more discrete regional distribution comparable to that observed for [^{123}I]β-CIT uptake in the presence of citalopram. Specific striatal uptake of [^{123}I]β-CIT-FP and [^{123}I]β-CIT-FE was approx 10% lower compared to [^{123}I]β-CIT and uptake was virtually undetectable in the thalamus, hypothalamus, or cerebral cortical areas. Considering the ratio of 5-HT to DA terminals in the human striatum is approx 1:10; the 10% decrease most likely corresponds to decreased labeling of striatal 5-HT transporters. The regional distribution for these radiotracers correlates well with the distribution of nigrostriatal and mesolimbic DAergic terminals and indicates that [^{123}I]β-CIT-FP and [^{123}I]β-CIT-FE *(49,50)* are highly promising imaging agents for measuring DAT regulation in vivo. In vitro and in vivo imaging of DAT-selective radiotracers do not detect appreciable numbers of DATs in allocortical and neocortical regions known to receive DA projections, including the amygdala, hippocampus, and the frontal, cingulate, and entorhinal cortices. The inability to detect DATs in corticolimbic regions may be caused by markedly lower densities of DATs in these brain areas, low sensitivity of the imaging agents, or the existence of a distinct pharmacological subtype of DAT associated with mesocortical and mesolimbic projections. Recent findings from functional brain-imaging studies have demonstrated that the neuronal activities in the amygdala and anterior cingulate correlate with the subjective feeling of "craving" *(54,55)*, a primary indicator of relapse to cocaine addiction. Thus, imaging extrastriatal DATs may prove important for identifying the neurochemical targets of cocaine and for clarifying the regulatory phenotype of the DAT in drug dependence.

The total numbers of cocaine-binding sites associated with the DAT in vivo have been measured using [^{11}C]cocaine and PET imaging. In nonhuman primates, the concentration of striatal [^{11}C]cocaine uptake ranged from 36 to 450 nM (41,56) to 2300 nM (57). The reasons for the high variability are unclear but might be related to choice of anesthesia used to sedate the animals, since in vivo labeling of the DA transporter is known to be differentially sensitive to anesthetics (58). Because of the addictive potential of cocaine, in vivo assessments of the total number of cocaine-binding sites in humans have been limited to studies of actively dependent cocaine abusers. In a living cocaine abuser, the average concentration of cocaine labeling of the striatal DAT ranged from of 627 to 765 nM which would correspond to in vitro densities of approx 627 to 765 pmol/g tissue, respectively (20). Overall, the total numbers of cocaine-binding sites are higher in living cocaine abusers compared to postmortem assessments measured in vitro in deceased cocaine abusers. The lower values measured in vitro may be caused in part to a decline with postmortem interval together with the effect of freezing of neuropathological specimens. Although the in vitro measures of cocaine-binding sites on the DAT are affected by postmortem interval and cryopreservation, in vivo measures also are unresolved because of the inherent limitations of the in vivo neurochemical imaging modalities. PET and SPECT imaging use tracer concentrations of radioligand which occupy less than 1% of the total cocaine-binding sites—ideal for studying the neurochemical state of the brain because there are no pharmacodynamic effects provoked by the administration of the radiotracer. However, because of the low dose of radiotracer, brain uptake primarily reflects cocaine binding to the "high"-affinity cocaine-recognition site (57). To estimate the total density of binding sites, high doses of unlabeled cocaine must be administered to reach a minimum of 50% occupancy of transport sites. Although the high concentrations of cocaine occupy a larger proportion of the low-affinity cocaine-recognition site (57), the accuracy of this "in vivo cold saturation" approach is constrained by the low signal-to-noise ratio of [^{11}C]cocaine, which is decreased even further in the presence of high concentrations of unlabeled cocaine, and which may overextrapolate the actual binding density of the radiotracer. Furthermore, high doses of cocaine provoke physiological effects in the living brain that may also affect quantitative estimates of the number of binding sites. A cocaine-induced allosteric change in the DAT may alter the number of high-affinity cocaine recognition sites and change the overall ratios of high-to-low affinity sites on the DAT, further limiting the ability to extrapolate in vivo measurements to absolute density of transporter sites (59).

4. TIME-COURSE OF BRAIN UPTAKE AND IN VIVO OCCUPANCY OF DAT BY COCAINE

The rate of cocaine's entry into and elimination from the brain and the level of occupancy of the DAT required to produce behavioral effects have been assessed in vivo using PET and SPECT imaging (see also Chapter 12). After intravenous administration, [¹¹C]cocaine rapidly enters the brain and reaches peak levels in the striatum between 6–8 min *(60)*. The half-time for cocaine occupancy of the DAT has been estimated to be between 20 and 35 min *(20,61)*. The rapid entry of cocaine into the brain and its ability to block DA reuptake and markedly elevate synaptic DA levels may explain the reinforcing potential of cocaine, distinguishing it from other DA reuptake blockers *(62)*.

The occupancy of the DAT by cocaine required to induce euphoria has been modeled based on PET measurements of brain concentrations of [¹¹C]cocaine. The model predicted that 40% (5 mg or 0.07 mg/kg cocaine assuming a 70-kg subject) occupancy of the DAT by cocaine was necessary for a human subject to "perceive" cocaine's effect with a "high" elicited at 80% (40 mg or 0.57 mg/kg cocaine) occupancy of the DAT *(63)*. Interestingly, the model further predicted that if DA reuptake was reduced in proportion to the fraction of transporter occupied by cocaine, then synaptic DA would rise supra-linearly, so that 40 and 80% occupancy would result in two- and 10-fold increases in synaptic DA levels, respectively *(63)*. This model further suggests that a given dose of cocaine would elevate DA to a similar degree regardless of the prior level of occupancy of the transporter by cocaine , which would explain why the intensity of the high is similar with either low or high DAT occupancies. These observations may explain in part why the therapeutic use of DA reuptake inhibitors has led to the adverse effect of increased cocaine use in some studies *(64–66)*.

In vivo assessment of the occupancy of the striatal DAT using [¹¹C]cocaine and PET imaging confirmed predictions that intravenous (iv) administration of cocaine at doses commonly abused by humans (0.05, 0.1, 0.3, and 0.6 mg/kg) occupy 41, 47, 60, and 77% of striatal DATs, respectively *(60)*. At least 50% of striatal DATs had to be blocked for cocaine's euphoriant effects to be reported reliably across subjects *(60,67)*. The degree of striatal DAT occupancy was dose related and correlated with the plasma concentration of cocaine and the self-reported high. These observations support the hypothesis that cocaine binding to the DAT mediates the reinforcing effects of cocaine. In contrast to these findings, estimates of cocaine occupancies of the DAT using the high-affinity cocaine analog, [¹²³I]β-CIT demonstrated significantly lower occupancies of the DAT by cocaine with higher cocaine doses (0.28 mg/kg and 0.56 mg/kg), decreasing binding of β-CIT by only 6–17% *(68)*. A cocaine dose of 7 mg/kg was required to decrease striatal [^{11}C]β-CIT uptake by 50% *(36)*. Differences in cocaine's occupancy of the DAT have been attributed to radiotracer kinetics and the time-course for peak striatal uptake. For example, β-CIT peaks between 1 and 2 h, and cocaine uptake peaks in 10 min *(67)*.The slower rate of β-CIT dissociation as compared to cocaine is likely also to be a factor *(20,69)*. These observations suggest that radiotracers with similar pharmacokinetic properties are necessary for accurate measurements of DAT occupancy. In keeping with this observation, striatal DATs labeled with [^{11}C]d-*threo*-methylphenidate and probed with GBR 12909, drugs which exhibit similar peak uptake and binding to the DAT in the brain, demonstrated occupancies that were more comparable to those observed for [^{11}C]cocaine *(67,71)*.

The accuracy of in vivo occupancy measurements of the DAT may be affected by limitations inherent in the methodology. Current PET and SPECT imaging methods suffer from poor spatial resolution and partial volume errors (i.e., an underestimation of regional activity) for brain regions smaller than the spatial resolution of the camera. This is particularly problematic for measures of the nucleus accumbens, a region which has different DA release and uptake dynamics as compared to caudate and putamen *(72)*. Also, PET and SPECT imaging do not measure occupancies of the high- and low-affinity cocaine-recognition sites, which may be differentially regulated in chronic cocaine abusers *(59)*. A multisite model of cocaine interactions with the DAT would require dose escalations and high concentrations of cocaine. In vivo imaging in living humans is limited usually to 2–3 concentrations of cocaine because of the high expense of PET imaging. Also, in vivo studies are conducted using "tracer" doses of the radioligand which would predominantly occupy the high-affinity, low-capacity recognition sites. The higher cocaine concentrations required for the Scatchard plot determination in vivo are assessed by the administration of nonradioactive cocaine, which when combined with the overall low signal-to-noise ratio for [^{11}C]cocaine, would limit accurate measurement of the low-affinity cocaine site. Furthermore, ethical considerations restrict studies to those who are already dependent on cocaine. Chronic cocaine abuse is known to cause neuroadaptations in DAT numbers with active dependence and during withdrawal which are different from that baseline state for nondependent drug-free subjects.

5. REGIONAL VARIABILITY AND THE EFFECT OF AGE AND SEX ON THE HUMAN DAT

Anatomical studies with different classes of radiolabeled DAT inhibitors and cocaine congeners have demonstrated distinct subregional rostrocaudal, mediolateral, and dorsoventral gradients within the human striatum. For example, autoradiographic localization of [³H]mazindol demonstrated the highest binding in ventral sectors of the striatum, with a decreasing rostral-to-caudal gradient in both the caudate and putamen *(32)*. The gradient for [³H]GBR 12935 binding has not been described in the human brain. However, in the rat brain, a decreasing rostral-to-caudal gradient was evident for striatal [³H]GBR 12935 binding *(73)*. Localization of [³H]WIN 35,428 binding demonstrated an increasing medial-to-lateral and dorsal-to-ventral gradient, yet binding was uniform throughout the anterior-to-posterior extent of the rat striatum *(73)*. In the human brain, total [¹²⁵I]RTI-55 binding demonstrated a decreasing lateral-to-medial gradient, and was uniform across the rostral-to-caudal boundaries of the human striatum. In contrast, when binding to the 5-HT transporter was occluded with citalopram, binding of [¹²⁵I]RTI-55 was high in the anterior sectors and lower in the posterior sectors of the striatum *(15)*. A similar topographic pattern for the distribution of the DAT was visualized using [¹²⁵I]RTI-121 *(16)*. The heterogeneity in topographic profiles obtained with different radioligands in the human striatum in vitro provides additional evidence to support the non-identity of pharmacological sites associated with the DAT. One possibility is that DATs may be differentially regulated, depending upon their signaling rates within discrete anatomical target regions. Additional functional studies are needed to relate different patterns of radioligand binding to the DAT with parameters of DA neurotransmission and innervation from SN and ventral tegmental DAergic projection neurons.

Pharmacological heterogeneity for binding of the radiolabeled DAT inhibitors may be caused by the labeling of binding sites that are not related to the DAT. A question has been raised regarding the DAergic nature of the diphenyl-substituted piperazine derivative ([³H]GBR 12935) binding in the human frontal cortex *(74–77)*. Although DA is a potent inhibitor of [³H]GBR 12935 binding in the human striatum, it is inactive in inhibition assays conducted in the human frontal cortex *(74–76)*. One possible explanation for these results is that DA- and [³H]GBR 12935-binding reflects distinct, non-overlapping binding domains on the DAT *(77–79)*. However, this hypothesis fails to explain the ability of DA to compete for binding to DATs in the striatum, but not the frontal cortex. Pharmacological evidence suggests a profile for [³H]GBR 12935 binding to the frontal cortex consistent with that of the piperazine-acceptor site on the cytochrome p450 complex *(75,76)*. DAT inhibitors exhibit significantly lower potencies for competition of [³H]GBR 12935 binding in the human frontal cortex and platelet membranes compared to striatal membranes, whereas piperazine analogs, including *cis*-flupentixol, proadifen, lobeline, budipine, and quinidine inhibit [³H]GBR

12935 binding with potency values characteristic of the piperazine-acceptor site in the human frontal cortex. Ligand-binding studies have shown that piperazine derivatives (GBR 12909 and GBR 12935) and antidepressants (buproprion and maprotiline) are substrates for cytochrome P450IID6 (debrisoquine 4-hydroxylase) *(80)*. In the frontal cortex, the binding of [^{3}H]GBR 12935 is biphasic, suggesting the involvement of multiple binding sites *(77–79)*. The overall higher density of the piperazine-acceptor sites in this brain region and the high affinity of the GBR analogs for these sites may render this radioligand unsuitable for assaying DAT regulation in the brain areas with low densities of DAT expression. The regional binding densities (B_{max}) reported for [^{3}H]GBR 12935 binding do not correlate well with the distribution and overall densities of DA neuronal projections, providing additional evidence for the labeling of sites unrelated to the DAT. The distribution of cocaine congeners corresponds to the regional density of DA presynaptic markers, with the highest densities localized in the caudate and putamen, and significantly lower densities in the cerebral cortex. In contrast, the regional densities for [^{3}H]GBR 12935 binding to the frontal cortex are equivalent to or higher than the density of DATs measured within the human striatum *(78)*. Nonspecific binding of [^{3}H]GBR 12935 in the frontal cortex defined with different DAT inhibitors yields variable percentages of specific binding, which are dependent upon the choice of inhibitor *(75,76)*. Although these results suggest that studies of [^{3}H]GBR 12935 binding should be interpreted with caution, they may disclose some novel regional variability in specific properties of the transport carrier.

Sex differences in DAT expression may partially explain the differences between males and females in the severity of cocaine-dependency syndrome and the pattern of abuse. The sex steroids estradiol and progesterone upregulate striatal DATs *(81,82)*. In keeping with this finding, recent in vivo imaging studies have observed higher DAT numbers in healthy females compared to age-matched male subjects *(83–85)*. However, a previous study failed to measure any sigificant effect of sex on transporter numbers *(86)*. Since these studies failed to control for the stage of the menstrual cycle, exposure to birth control pills, or estrogen-replacement therapy, additional studies are needed to determine the precise magnitude of the hormones effects on regulation of the DAT over the course of the menstrual cycle. This information is needed to design sex-effective treatment protocols targeted to the DAT for cocaine dependence.

Little information about the effects of chronic cocaine exposure on the human-brain DAergic system over the lifespan is currently available. Reductions in DA and DA-related synaptic markers in the striatum are known to contribute to the cognitive and motor deficits associated with normal aging. Cocaine, unlike methamphetamine *(70,87)* does not appear to be neurotoxic in experimental animals *(88)* or in living humans *(89,90)*. In the postmortem human striatum, a progressive decrease in DAT density with age has been demonstrated using [^{3}H]GBR 12935 *(91–94)*. Decreases in DAT density of 75 and 65% were reported for subjects ranging from 19–100 yr *(91)* and 18–88 yr *(92)*, respectively. In vivo imaging of the DAT with cocaine and other tropane analogs ([^{11}C]cocaine *[95]*, [^{123}I]β-CIT *[86,96]*, [^{123}I]FP-CIT *[84]*, [^{123}I]PE2I *[97]*, and [^{123}I]-IPT *[98]*) and "classical" DAT inhibitors ([^{11}C]nomifensine *[98]*) also demonstrated a decline of striatal DATs with increasing age. Using [^{11}C]cocaine, a gradual decline in the density of cocaine-recognition sites was detected over an age range of 21–63 yr *(95,96)*. Using [^{123}I]β-CIT, a 51% linear decline in DAT density was observed over an age range of 18–83 yr *(86)*. In contrast, using [^{123}I]-IPT the decline in DATs over the age range of 19–67 yr was best described by a nonlinear model *(98)*. A decrease in the [^{11}C]nomifensine striatum/cerebellum ratios also was observed over an age range of 24–81 years *(99)*. Taken together, in vivo imaging with a variety of radiotracers demonstrates age-related declines in DAT density that occur at a rate of 6–10% per decade. Surprisingly, this age-related loss in striatal [^{11}C]cocaine binding was not apparent in current and detoxified cocaine abusers *(100)*, suggesting that chronic cocaine abuse may lead to increased expression of the DAT, which is above the age-related loss in DA neuronal integrity.

In keeping with the marked decline in DATs with normal aging, studies of the mRNA encoding the DAT demonstrated a profound loss of DAT gene expression in DA-containing SN neurons with increasing age *(101)*. Although a precipitous age-related decline (> 95% in subjects > 57 yr old) was reported for DAT mRNA, the mRNA for tyrosine hydroxylase (another phenotypic marker of DA neurons) decreased linearly with age *(102)*. The abrupt decline in the DAT at the end of the fifth decade was surprising, considering that the decrease in DAT density was linear over the lifespan. This difference may reflect differential regulation of DAT mRNA and protein with normal aging. However, additional studies with more subjects may disclose a decline in DAT message that more closely correlates with the decrease in DAT densities.

It is presently unknown whether this decline in DAT density corresponds to a loss of DA nerve terminals or to a decrease in the number of DATs expressed by aging DAergic neurons. Comparable changes in DAergic pre- and postsynaptic markers and DAT densities suggest that the decline DAT labeling may be caused by reduced integrity of DAergic projections *(92)*. Age-related decreases have been shown for tyrosine hydroxylase *(102)*, striatal DA content *(103,104)*, D1 and D2 receptors *(92)*. This decrease in DA synaptic markers suggests that the observed decrease in DAT density with normal aging may be caused by an actual decline of DA neurons. Alternatively, if the DAT is regulated by synaptic DA content, the decrease in density may be a compensatory response to the age-related decline in neuronal DA content, with correspondingly abnormal rates of DA turnover. The marked decline in striatal DAT density with normal aging clearly demonstrates the importance of choosing successfully age-matched control subjects for assessing the effects of chronic cocaine abuse on the neuroadaptive regulation of synaptic DATs in cohorts of cocaine addicts.

6. REGULATORY CHANGES IN THE HUMAN DAT CAUSED BY COCAINE ABUSE

New tools from the neurosciences have led to the proliferation of research approaches aimed at understanding the neurobiological consequences of chronic cocaine use on the human brain. The recent development of radioligands with high specific activity and selectivity for neurotransmitter carriers have provided the tools neeeded to map and quantify neuroadaptations to cocaine exposure in living cocaine abusers and in the postmortem brain from cocaine fatalities. Since the DAT is a key regulator of DA neurotransmission, alterations in the number, affinity or allosteric regulation of the DAT by cocaine may lead to marked alterations in DAergic signaling. Understanding the sites of cocaine actions in the brain and the acute and long-term neurobiological effects may disclose the clinical relevance of regulatory changes in DAT function to the behavioral effects of chronic cocaine use.

The cocaine congeners [^{3}H]WIN 35,428 and [^{125}I]RTI-55 have been used to visualize the regional density of DATs postmortem in the human brain from victims of fatal cocaine overdose and cocaine-related deaths caused by homicide and motor-vehicle accidents. Postmortem radioligand binding and autoradiographic studies demonstrated significant increases in DA transporter densities using the cocaine congeners [^{125}I]RTI-55 and [^{3}H]WIN 35,428 throughout the caudate, putamen, and nucleus accumbens from cocaine-related deaths (24) and fatal cocaine-overdose victims (23,105; Figs. 3 & 4). Saturation-binding analysis confirmed the increase in [^{3}H]WIN 35,428 binding to putamen membranes of cocaine overdose victims as compared to drug-free and age-matched control subjects (23,105). Rosenthal plots of the saturation binding data demonstrated that the increase of [^{3}H]WIN 35,428 binding observed in the cocaine-overdose victims was caused by an elevation in the apparent density of the high-affinity cocaine-recognition site on the DAT (23,105). These studies suggest that the high-

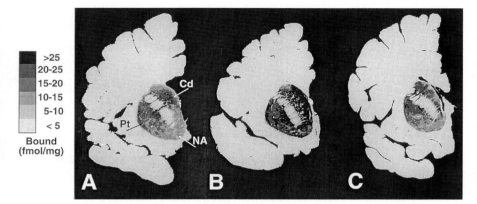

Fig. 3. Elevated densities of [³H]WIN 35,428 binding to the DAT in human cocaine-overdose victims. **(A–C)** Gray scale images of [³H]WIN 35,428 labeling in coronal sections of a drug-free control subject; cocaine-overdose victim; and excited delirium victim. A marked increase in the density of [³H]WIN 35,428 binding was observed throughout the ventral sectors of the anterior striatum in the cocaine-overdose victim as compared to a drug-free, age-matched control subject. The gray-scale bar at the left depicts the relationship between the [³H]WIN 35,428 binding-site density (femtomoles per mg of tissue equivalent) and the gray-scale intensity.

affinity cocaine-binding site may upregulate in the human striatum with chronic cocaine abuse as a compensatory response to elevated synaptic levels of DA. This compensatory effect on the DAT may result in an acute decrease in the intrasynaptic concentration of DA following cocaine challenge. If this regulatory change in high-affinity [³H] WIN 35,428 binding sites on the human DAT actually reflects an increased ability of the protein to transport DA, it may help to explain the addictive liability of cocaine. As the transporter upregulates its apparent density in the nerve terminal to more efficiently transport DA, more cocaine will be needed to experience cocaine's reinforcing effects and euphoria. Once cocaine is no longer present to block reuptake, increased DAT will result in a net DA deficit in the synapse that may explain the reports of anhedonia associated with the post-cocaine crash.

In contrast to the elevated densities of DATs observed with radiolabeled cocaine congeners, [³H]mazindol binding was lower in the caudate and putamen from cocaine-related death victims as compared to age-matched and drug-free controls *(106)*. The different regulatory profile observed with [³H]mazindol binding provides additional evidence for the non-identity of various classes of DAT radioligands. Consistent with this view, previous

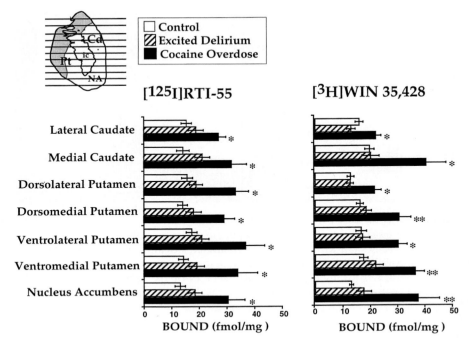

Fig. 4. Region-of-interest densitometric measurements of cocaine-recognition sites in striatum of human cocaine-overdose victims with and without preterminal excited delirium and drug-free control subjects. The density of DAT in the striatum was assessed by using [^{125}I]RTI-55 (50 pM); [^{3}H]WIN 35,428 (2 nM). Values (femtomoles per mg of tissue equivalent) representing the mean (± S.E.) for drug-free control subjects (white bars); cocaine overdose (black bars), and excited delirium victims (stripped bars) are shown. The drawing in the upper left-hand corner illustrates in diagrammatic form the regional sites sampled from the autoradiograms at the anterior level of the striatum. The quantitative densitometric measurements demonstrate elevated [^{3}H]WIN 35,428 and [^{125}I] RTI-55 binding throughout the anterior sectors of the striatum in the cocaine-overdose deaths (*$p < 0.01$; **$p < 0.001$). Abbreviations: Cd, caudate; ic, internal capsule; na, nucleus accumbens; Pt, putamen. Portions of the data reproduced with permission from ref. *104.*

studies in rats have also failed to demonstrate significant changes with chronic administration of cocaine in the binding of non-cocaine-like dopamine transport inhibitors, including [^{3}H]nomifensine *(107)*, [^{3}H]GBR 12935 *(108–111)*, and [^{3}H]mazindol (Mash, unpublished observations). Many experimental cocaine-treatment paradigms have been developed as animal models of human cocaine abuse, and it has been reported that the dose, route, and frequency of administration influence the effects of chronic cocaine on DAT densities. Chronic treatment of rats with intermittent doses of cocaine demonstrated a two- to fivefold increase in the apparent density of [^{3}H]cocaine and [^{3}H]BTCP binding in the striatum *(112)*. Rats allowed to self-administer cocaine in a chronic unlimited-access paradigm had significant increases in [^{3}H]WIN 35,428 binding, observed when the animals were sacrificed on the last day of cocaine access *(73)*. Rabbits and mice treated chronically with cocaine show an elevation in the density of [^{3}H]WIN 35,428 binding sites in the caudate *(113,114)*. Taken together, these results demonstrate that cocaine exposure leads to an apparent increase in the density of high-affinity cocaine-binding sites on the DAT, but not the recognition sites for non-cocaine-like DAT inhibitors.

Regulatory changes in DAT densities have been visualized in parallel with different classes of DAT inhibitors. In the brains of animals allowed to self-administer cocaine for 7 wk and sacrificed on the last day of access to cocaine, binding of [^{3}H]WIN 35,428 and [^{3}H]GBR 12935 to the DA transporter were elevated throughout the rostral sectors of the caudate-putamen and nucleus accumbens *(73)*. In the cell-body fields (SN and VTA), binding of both radioligands was elevated, yet only the density of [^{3}H]GBR 12935 reached significance. After a 3-wk withdrawal period, [^{3}H]WIN 35,428 binding to the DA transporter returned to baseline levels throughout the caudate, putamen, SN, and VTA, and was decreased significantly in the nucleus accumbens. In contrast, [^{3}H]GBR 12935 binding was significantly decreased throughout all sectors of the striatum. Similar decreases were reported in previous studies of cocaine-withdrawal models *(115,116)*. The results demonstrate that different patterns of DAT regulation are observed with cocaine-like and non-cocaine-like DAT inhibitors with chronic cocaine administration. The time since the last cocaine administration has a marked effect on measurements of DAT density.

In keeping with these findings, in vivo SPECT measurements of DAT densities in living cocaine abusers also vary depending on the time since the last cocaine administration (Figs. 5 and 6). When the DAT was imaged in vivo within 96 h of drug abstinence, [^{123}I]β-CIT uptake in the striatum was approx 20% higher in the cocaine abusers compared to age-matched drug-free control subjects *(89,117)*. After a prolonged period of drug abstinence (3–18 mo) [^{123}I]β-CIT measures were still elevated, but decreased in their level of statistical significance and demonstrated a trend toward a return to baseline values measured in drug-free control subjects *(69)*. Furthermore, at 2 wk abstinence and beyond, there was an increasing positive correlation between plasma HVA and striatal [^{123}I]β-CIT uptake *(118)*. In another study, chronic cocaine abusers had significantly lower [^{11}C]cocaine uptake in the basal ganglia and thalamus when screened 10–90 d after the last use of cocaine *(119)*. The uptake of [^{11}C]cocaine was negatively correlated with

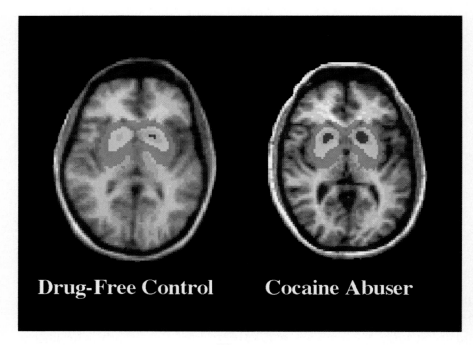

Fig. 5. In vivo SPECT Images of [^{123}I]β-CIT brain uptake in a drug-free control subject and a cocaine abuser. The drug-free control subject is a 35-yr-old caucasion male who was injected with 6.0 μCi and imaged 21.5 h later. The cocaine abuser is a 34-yr-old caucasion male who used approx 14 g/wk of cocaine prior to scanning. The cocaine abuser was injected with 6.1 μCi of [^{123}I]β-CIT and scanned 22.5 h later. Note that compared to the healthy drug-free control subject, the cocaine abuser demonstrates higher [^{123}I]β-CIT uptake throughout both the dorsolateral and ventromedial sectors of the striatum (Courtesy of Leslie Jacobsen, MD; Yale University, 2000).

"cocaine craving" and with depressive symptoms, suggesting an association between withdrawal symptoms and DAT densities. The results of human and rodent studies suggest that the DAT upregulates in response to acute "binges" of cocaine administration, but may gradually normalize or decrease with long periods of drug abstinence. The decrease in DAT densities in the striatal reward centers suggests that decreased DAT densities may reflect lower DAergic tone. A hypodopaminergic state may be one of the triggers of cocaine craving during acute abstinence, which causes the addict to relapse to previous patterns of drug use.

A case series of cocaine abusers who died following a syndrome of excited delirium was first described in 1985 *(120)*. The characteristics of victims of excited delirium distinguish them from other cocaine abusers who

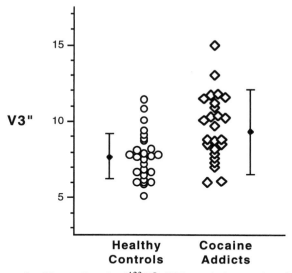

Fig. 6. Scatterplot illustrating the [^{123}I]β-CIT uptakein cocaine abusers and drug-free control subjects. Subjects received a bolus injection of 9.5 ± 1.4 μCi [^{123}I]β-CIT followed by SPECT imaging on d 2. The ratio of specific to nonspecific activity ([striatal–occipital]/occipital), a measure known as the specific to nonspecific equilibrium partition coefficient (V_3") and proportional to the binding potential (B_{max}/K_d), was used to estimate differences in DAT levels between groups. Results showed significantly ($p = 0.02$; unpaired two-tailed t-test) increased V_3" values (mean ± S.D.) in cocaine addicts (9.6 ± 2.1) vs healthy controls (7.8 ± 1.6). (Reproduced with permission from ref. *68.*).

were examined at autopsy. We have compared the regulatory patterns of DAT in cocaine abusers to the excited delirium subgroup, using ligand binding and in vitro autoradiography. Autoradiographic mapping with a single concentration of [^{3}H]WIN35,428 and [^{125}I]RTI-55 failed to demonstrate an elevation in the apparent density of the DAT in the striatum (*59,105*; Figs. 3C and 4) of excited delirium subjects as compared to drug-free and age-matched control subjects. Rosenthal analysis of saturation binding curves for [^{3}H]WIN35,428 and [^{125}I]RTI-55 demonstrated a significant decrease in total binding sites for the excited delirium subgroup of cocaine-overdose victims. Analysis of curvilinear Rosenthal plots demonstrated that there was no change in the apparent density of the high-affinity cocaine-recognition site in the cocaine abusers with premorbid excited delirium as compared to age-matched and drug-free control subjects. However, the density of the low-affinity cocaine-recognition site was significantly decreased in the excited delirium victims (*59,105*). The differential regulation of cocaine-recognition sites on the DAT in the excited delirium subgroup may indicate a diminished capacity for DA reuptake during a cocaine challenge or "crack binge." Since the concentration of synaptic dopamine is controlled by the reuptake mechanism(s), the lack of compensatory increase in cocaine-recognition sites could be the molecular defect that explains the paranoia and agitation associated with this syndrome. Paranoia in the context of cocaine abuse is common and several lines of evidence suggest that this phenomenon may be related to the function of the DAT protein *(12)*. Further studies are needed to determine if the failure of the DAT to upregulate with response to cocaine exposure plays a role in causing the syndrome. Genetic differences in the makeup of individuals who abuse cocaine may underlie some of these differences in susceptibility to adverse neuropsychiatric effects of chronic cocaine abuse.

7. CLINICAL RELEVANCE OF DAT REGULATION FOR UNDERSTANDING COCAINE ADDICTION

A unique advantage of in vivo imaging is that drug interactions within the brain can be assessed while simultaneously monitoring the behavioral effects and the plasma bioavailability of the drug *(121)*. In vivo imaging with [^{11}C]cocaine has demonstrated that the time-course of cocaine binding to the striatum correlates with the onset of the euphoric experience *(61,72)*. These studies support the hypothesis that the "high" or euphoria induced by psychostimulants is related to the rapid occupancy of DAT and the corresponding rise in the synaptic concentration of DA *(121)*. Similar studies with [^{11}C]dl-threo-methylphenidate also demonstrated rapid uptake, which correlates with the onset of the behavioral effects. When compared to [^{11}C]cocaine, [^{11}C]dl-threo-methylphenidate exhibited slower clearance from the striatum. Since methylphenidate has lower abuse liability as compared to cocaine, it is possible that kinetic differences in the rates of occupancy of the DAT by [^{11}C]dl-threo-methylphenidate may explain the lack of the euphoriant "rush" and lower arousal *(40)*. Rothman and colleagues *(120)* have suggested that DA transport inhibitors with slow rates of entry into the brain and slow onset of action may be candidate drugs for treating cocaine dependence.

Although the development of a cocaine antagonist that blocks the effects of cocaine—but not DA—at the transporter is supported on the basis of molecular biology and structural drug-protein interaction studies (*see* also Chapters 3 and 11), the pharmacotherapeutic efficacy of this approach may be limited for a number of reasons. First, transgenic mice deficient in the DAT self-administer cocaine at high rates of responding *(123)*, suggesting that the DAT may not mediate the addictive properties of cocaine as originally proposed in the DA hypothesis of cocaine addiction *(124)*. Most cocaine addicts are dependent on nicotine, which may diminish the therapeutic effects of a cocaine antagonist. When administered with cocaine, nicotine synergistically elevates synaptic DA levels *(125)*, through the indirect inhibition of DA reuptake *(126,127)*. Furthermore, a recent [^{11}C]cocaine and PET imaging study demonstrated that cocaine can elevate striatal DA and produce an "intense high" regardless of the prior level of occupancy (0 or 75–85%) of the DAT *(63)*. Insufficient DAT blockade may enhance cocaine effects, making patient compliance a prerequisite for optimal therapeutic response of even the best cocaine antagonist. Thus, the development of depot formulation of cocaine blockers may be the best approach for treating intractable cocaine abusers. However, the involvement of other neurotransmitters in cocaine dependence may limit the efficacy of a cocaine antagonist targeted solely to the DAT.

The investigation of compounds to reduce cocaine use by diminishing craving and reducing withdrawal symptoms has received the most attention in recent years *(128,129)*. Although there is still controversy regarding the long-term effects of cocaine abuse, pharmacological treatment strategies have been based on the assumption that chronic use leads to neuroadaptive changes in the sensitivity of DA pre- and postsynaptic markers with increased DA turnover and a resultant DA depletion *(130,131)*. The compensatory increase in high-affinity cocaine-recognition sites on the human DAT may indicate an enhanced ability of cocaine to inhibit DA transport in chronic cocaine users. An elevation in [^{3}H]methylphenidate binding to the DAT has been shown to occur as early as 1 h after the acute administration of cocaine to drug-naive rats *(132)*, indicating that DATs may undergo rapid regulatory changes in the membrane regulate synaptic levels of neurotransmitter. Persisting increases in the apparent density of the DAT after cocaine levels have fallen in blood and brain may result in an acute decrease in the intrasynaptic concentration of DA and lower DAergic tone. If this regulatory change in high-affinity [^{3}H]WIN 35,428 binding sites on the human DAT reflects increased DAT function, it may help to explain the addictive liability of cocaine. As the transport carrier upregulates its apparent density in the nerve terminal to more efficiently transport DA, more cocaine will be needed to experience the reinforcing effects and euphoria. During acute abstinence from cocaine, enhanced function of the DAT could lead to net depletion in synaptic DA. This depletion of DA may serve as a biological substrate for anhedonia, a cardinal feature of cocaine withdrawal. Further studies are needed to define the relevance of cocaine-induced alterations in radioligand binding sites to the molecular mechanisms regulating DAT function. The newly developed tools to map and characterize the regulation and function of the DAT in vivo and in vitro will help clarify the precise role of the DA transporter in cocaine dependence.

REFERENCES

1. Baker S. P. (1992) *The Injury Fact Book.* 2nd ed. New York Oxford University Press, New York, NY.
2. Galloway, M. P. (1988) Neurochemical interactions of cocaine with the dopaminergic system. *Trends Pharmacol. Sci.* **9,** 451–454.
3. Kuhar, M. J., Ritz, M. C., and Boja, J. W. (1991) The dopamine hypothesis of the reinforcing properties of cocaine. *TINS* **14,** 299–302.
4. Ritz, M. C., Lamb, R. J., Goldberg, S. R., and Kuhar, M. J. (1987) Cocaine receptors on dopamine transporters are related to self-administration of cocaine. *Science* **237,** 1219–1223.
5. Giros, B., Mestikawy, S. E., Godinot, N., Zheng, K., Han, H., and Yan-Feng, T. (1992) Pharmacological characterization and chromosome assignment of the human dopamine transporter. *Mol. Pharm.* **42,** 383–390.
6. Pristupa, Z. B., Wilson, J. M., Hoffman, B. J., Kish, S. J., and Niznik, H. B. (1994) Pharmacological heterogeneity of the cloned and native human dopamine transporter: dissociation of [^{3}H]WIN 35,428 and [^{3}H]GBR 12935 binding. *Mol. Pharm.* **45,** 125–135.
7. Vandenbergh, S. J., Persico, A. M., and Uhl, G. R. (1992) A human dopamine transporter cDNA predicts reduced glycosylation, displays a novel repetitive element and provides racially-dimorphic Taq I RFLPS. *Mol. Brain Res.* **15,** 161–166.
8. Rothman, R. B., (1990) High affinity dopamine reuptake inhibitors as potential cocaine antagonists: A strategy for drug development. *Life Sci.* **46,** PL17–PL21.
9. Giros, B., Wang, Y. M., Suter, S., McLeskey, S. B., Pifl, C., and Caron, M. G. (1994) Delineation of discrete domains for substrate, cocaine, and tricyclic antidepressant interactions using chimeric dopamine-norepinephrine transporters. *J. Biol. Chem.* **269,** 15,985–15,988.
10. Kitayama, S., Shimada, S., Xu, H., Markham, L., Donovan, D. M., and Uhl, G. R. (1992) Dopamine transporter site-directed mutations differentially alter substrate transport and cocaine binding. *Proc. Natl. Acad. Sci. USA* **89,** 7782–7785.
11. Kitayama, S., Wang, J.-B., and Uhl, G. R. (1993) Dopamine transporter mutants selectively enhance MPP+ transport. *Synapse* **15,** 58–62.
12. Giros, B. and Caron, M. G. (1993) Molecular characterization of the dopamine transporter. *Trends. Pharmacol. Sci.* **14,** 43–49.
13. Boja , J. W., Vaughen, R., Patel A., Shaya, E. K., and Kuhar, M. J. (1994) The dopamine transporter, in *Dopamine Receptors and Transporters,* (Niznik, H., ed.), Marcel Dekker, New York, NY, pp. 611–644.
14. Madras, B. K., Spealman, R. D., Fahey, M. A., Neumeyer, J. L., Saha, J. K. and Milius, R. A. (1989) Cocaine receptors labeled by [^{3}H]2β-carbomethoxy-3β-(4-fluorophenyl) tropane. *Mol. Pharmacol.* **36,** 518–524.

15. Staley, J. K., Basile, M., Flynn, D. D., and Mash, D. C. (1994) Visualizing dopamine and serotonin transporters in the human brain with the potent cocaine analogue [^{125}I]RTI-55: in vitro binding and autoradiographic characterization. *J. Neurochem.* **62,** 549–556.

16. Staley, J. K., Boja, J. W., Carroll, F. I., Seltzman, H. H., Wyrick, C. D., Lewin, A. H., et al. (1995) Mapping dopamine transporters in the human brain with novel selective cocaine analog [^{125}I]RTI-121. *Synapse* **21,** 364–372.

17. Boja, J. W., Markham, L., Patel, A., Uhl, G., and Kuhar, M. J. (1992) Expression of a single dopamine transporter cDNA can confer two cocaine binding sites. *Neuro. Rep.* **3,** 247–248.

18. Reith, M. E. A., De Cosata, B., Rice, K. C., and Jacobsen, A. E., (1992) Evidence for mutually exclusive binding of cocaine, BTCP, GBR 12935, and dopamine to the dopamine transporter. *Eur. J. Pharmacol.* **227,** 417–425.

19. Reith, M. E. A. and Selmeci, G. (1992) Radiolabeling of dopamine uptake sites in mouse striatum-comparision of bindings -sites for cocaine, mazindol, and GBR 12935. *Nauym-Schmeidebergs's Arch. Pharmacol.* **345,** 309–318.

20. Logan, J., Volkow, N. D., Folwer, J. S., Wang, G.-J., Fischman, M. W., Foltin, R. W., et al. (1997) Concentration and occupancy of dopamine transporters in cocaine abusers with [11C]cocaine and PET. *Synapse* **27,** 347–356.

21. Little, K. Y., Kirkman, J. A., Carroll, F. I., Breese, G. R., and Duncan, G. E. (1993) [^{125}I]RTI-55 binding to cocaine-sensitive dopaminergic and serotonergic uptake sites in the human brain. *J. Neurochem.* **61,** 1996–2006.

22. Marcusson, J. and Ericksson, K. (1988) [^{3}H]GBR 12935 binding to dopamine uptake sites in the human brain. *Brain Res.* **457,** 122–129.

23. Staley, J. K., Hearn, W. L., Ruttenber, A. J., Wetli, C. V., and Mash, D. C. (1995) High affinity cocaine recognition sites on the dopamine transporter are elevated in fatal cocaine overdose victims. *J. Pharm. Exp. Therap.* **271,** 1678–1685.

24. Little, K. Y., Kirkman, J. A., Carroll, F. I., Clark, T. B., and Duncan, G. E. (1993) Cocaine use increases [^{3}H]WIN 35,428 binding sites in human striatum. *Brain Res.* **628,** 17–25.

25. Buck, K. J. and Amara, S. G. (1994) Chimeric dopamine-norepinephrine transporters delineate structural domains influencing selectivity for catecholamines and 1-methyl-4-phenylpyridinium. *Proc. Natl. Acad. Sci. USA* **91,** 12,584–12,588.

26. Bjorklund A. and Lindvall, O. (1994) Dopamine-containing systems in the CNS, in H*andbook of Chemical Neuroanatomy, Vol 2. Classical Transmitters in the CNS, Part I.* (Bjorklund, A. and Hokfelt, T., eds.), pp. 55–122.

27. Graybiel, A. M. and Ragsdale, C. W. (1978) Histochemically distinct compartments in the striatum of human, monkey and cat demonstrated by acetylcholinesterase staining. *Proc. Natl. Acad. Sci. USA* **75,** 5723–5726.

28. Hurd, Y. L., Pristupa, Z. B., Herman, M. M., Niznik, H. B., and Kleinman, J. E. (1994) The dopamine transporter and dopamine D2 receptor messenger RNAs are differentially expressed in limbic- and motor-related subpopulations of human mesencephalic neurons. *Neuroscience* **63,** 357–362.

29. Graybiel, A. M. and Moratalla, R. (1989) Dopamine uptake sites in the striatum are distributed differentially in striosome and matrix compartments. *Proc. Natl. Acad. Sci. USA* **86,** 9020–9024.

30. Lowenstein, P. R., Joyce, J. N., Coyle,J. T., and Marshall, J. F. (1990) Striosomal organization of cholinergic and dopaminergic uptake sites and cholinergic M1 receptors in the adult human striatum: a quantitative receptor autoradiographic study. *Brain Res.* **510,** 122–126.

31. Ciliax, B. J., Drash, G. W., Staley, J. K., Haber, S., Mobley, C. J., Miller, G. W., et al. (1999) Immunocytochemical localization of the dopamine transporter in human brain. *J. Comp. Neurol.* **409,** 38–56.

32. Donnan, G. A., Kaczmarczyk, S. J., Paxinos, G., Chilco, P. J., Kalnins, R. M., Woodhouse, D. G., et al. (1991) Distribution of catecholamine uptake sites in human brain as determined by quantitative [^{3}H]mazindol autoradiography. *J. Comp. Neurol.* **304,** 419–434.

33. Biegon, A., Dillon, K., Volkow, N. D., Hitzemann, R. J., Fowler, J. S., and Wolf, A. P. (1992) Quantitative autoradiography of cocaine binding sites in human brain postmortem. *Synapse* **10,** 126–130.

34. Kaufman, M. J. and Madras, B. K. (1991) Severe depletion of cocaine recognition sites associated with the dopamine transporter in Parkinson's diseased striatum. *Synapse* **9,** 43–49.

35. Madras, B. K., Gracz, L. M., Fahey, M. A., Elmaleh, D., Meltzer, P. C., Liang, A. Y., et al. (1998) Altropane, a SPECT or PET imaging probe for dopamine neurons: III. Human dopamine transporter in postmortem normal and Parkinson's diseased brain. *Synapse* **29,** 116–1127.

36. Farde, L., Halldin, C., Muller, L., Suhara, T., Karlsson, P., and Hall, H. (1994) PET Study of [^{11}C]β-CIT binding to monoamine transporters in the monkey and human brain. *Synapse* **16,** 93–103.

37. Madras, B. K. and Kaufman, M. J. (1994) Cocaine accumulates in dopamine-rich region of primate brain after i.v. administration: comparison with mazindol distribution. *Synapse* **18,** 261–275.

38. Aquilonius, S. M., Bergstrom, K., Eckernas, S. A., Hartvig, P., Leenders, K. L., Lundquist, H., et al. (1987) In vivo evaluation of striatal dopamine reuptake sites using ^{11}C-nomifensine and positron emission tomography. *Acta. Neurol. Scand.* **76,** 283–287.

39. Salmon, E., Brooks, D. J., Leenders, K. L., Turton, D. R., Hume, S. P., Crmer, J. E., et al. (1990) A two-compartment description and kinetic procedure for measuring regional cerebral [^{11}C]nomifensine uptake using positron emission tomography. *J. Cereb. Blood Flow Metab.* **10,** 307–316.

40. Ding, Y. -S., Fowler, J. S., Volkow, N. D., Gatley, S. J., Logan, J., Dewey, S. L., et al. (1994) Pharmacokinetics and in vivo specificity of [^{11}C]dl-threo-methylphenidate for the presynaptic dopaminergic neuron. *Synapse* **18,** 152–160.

41. Fowler, J. S., Volkow, N. D., Wolf, A. P., Dewey, S. L., Schyler, D. J., Macgregor, R. R., et al. (1989) Mapping cocaine binding sites in human and baboon brain in vivo. *Synapse* **4,** 371–377.

42. Gatley, S. J., Volkow, N. D., Fowler, J. S., Dewey, S. L., and Logan, J. (1995) Sensitivity of striatal [^{11}C]cocaine binding to decreases in synaptic dopamine. *Synapse* **20,** 137–144.

43. Telang, F. W., Volkow, N. D., Levy, A., Logan, J., Fowler. J. S., Felder, C., et al. (1999) Distribution of tracer levels of cocaine in the human brain as assessed with averaged [^{11}C]cocaine images. *Synapse* **31,** 290–296.

44. Wang, G.-J., Volkow, N. D., Fowler, J. S., Ding, Y.-S., Logan, J., Gatley, J. S., et al. (1995) Comparision of two PET radioligands for imaging extrastriatal dopamine transporters in human brain. *Life Sci.* **57,** 187–191.

45. Wong, D. F., Yung, B., Dannals, R. F., Shaya, E. K., Ravert, H. T., Chen, C. A. et al. (1993) In vivo imaging of baboon and human dopamine transporters by positron emission tomography using [^{11}C]WIN 35,428. *Synapse* **15,** 130–142.

46. Neumeyer, J. L., Wang, S., Gao, Y., Milius, R. A., Kula, N. S., Campbell, A., et al. (1994) N-w-Fluoroalkyl analogs of (1R)-2β-carbomethoxy-3β-(4-iodophenyl)-tropane (β-CIT): Radiotracers for positron emission tomography and single photon emission tomography and single photon emission computed tomography imaging of dopamine transporters. *J. Med. Chem.* **37,** 1558–1561.

47. Pirker, W., Asenbaum, S., Kasper, S., Walter, H., Angellberger, P., Koc, G., et al. (1995) β-CIT SPECT demonstrates blockade of 5HT-uptake sites by citalopram in the human brain in vivo. *J. Neural. Transm.* **100,** 247–256.

48. Seibyl, J. P., Wallace, E., Smith, E. O., Stabin, M., Baldwin, R. M., Zoghbi,, S., et al. (1994) Whole-body biodistribution, radiation absorbed dose and brain SPECT imaging with iodine-123-β-CIT in healthy human subjects. *J. Nucl. Med.* **35,** 764–770.

49. Kuikka, J. T., Akerman, K., Bergstrom, K. A., Karhu, J., Hiltunen, J., Haukka, J., et al. (1995) Iodine-123 labelled N-(2-fluoroethyl)-2β-carbomethoxy-3β-(4-iodophenyl) nortropane for dopamine transporter imaging in the living human brain. *Eur. J. Nucl. Med.* **22,** 682–686.

50. Kazumata, K, Dhawan, V., Chaly, T., Antonin, A., Margouleff, C., Belakhlef, A., et al. (1998) Dopamine transporter imaging with fluorine-18-FPCIT and PET. *J. Nucl. Med.* **39,** 1521–1530.

51. Kuikka, J. T., Bergstrom, K. A., Ahonen, A., Hiltunen, J., Haukka, J., Lansimies, E., et al. (1995) Comparison of iodine-123 labelled PE2I for dopamine transporter imaging: influence of age in healthy subjects. *Eur. J. Nucl. Med.* **26,** 1486–1488.

52. Fischman, A. J., Bonab, A. A., Babich, J. W., Palmer, E. P., Alpert, N. M., Elmaleh, D. R., et al. (1998) Rapid detection of Parkinson's disease by SPECT with Altropane: a selective ligand for dopamine transporters. *Synapse* **29,** 128–141.

53. Kung, H. F., Kim, H. J., Kung, M. P., Meegalla, S. K., Plossl, K., and Lee, H. K. (1996) Imaging of dopamine transporters in humans with technetium-99m TRODAT-1. *Eur. J. Nucl. Med.* **23,** 1527–1530.

54. Childress, A. R., Mozley, P. D., McElgin, W., Fitzgerald J., Reivich, M., and O'Brien, C. P. (1999) Limbic activation during cue-induced cocaine craving. *Am. J. Psychiatry* **156,** 11–18.

55. Grant, S., London, E. D., Newlin, D. B., Billemagne, V. L., Liu, X., Contoreggis, C., et al. (1996) Activation of memory circuits during cue-elicited cocaine craving. *Proc. Natl. Acad. Sci. USA* **93,** 12,040–12,045.

56. Morris, E. D., Babich, J. W., Alpert, N. M., Bonab, A. A., Livini, Weise, S., et al. (1996) Quantification of dopamine transporter density in monkeys by dynamic PET imaging of multiple injections of [^{11}C]-CFT. *Synapse* **24,** 262–272.
57. Volkow, N. D., Fowler, J. S., Gatley, J. S., Dewey, S. L., MacGregor, R. R., Schlyer, D. J., et al. (1995b) Carbon-11-cocaine binding compared at subpharmacological and pharmacological doses: A PET study. *J. Nucl. Med.* **36,** 1289–1297
58. Tsukada, H., Nishiyama, T., Kakiuchi, H., Ohba, K., Sato, N., Harada, N., et al. (1999) Isoflurane anesthesia enhances the inhibitory effects of cocaine and GBR 12909 on dopamine transporter: PET studies in combination with microdialysis in the monkey brain. *Brain Res.* **849,** 85–96.
59. Staley, J. K., Wetli, C. V., Ruttenber, A. J., Hearn, W. L., and Mash, D. C. (1995) Altered dopaminergic synaptic markers in cocaine psychosis and sudden death. *NIDA Res. Mono. Ser.* **153,** 491.
60. Volkow, N. D., Wang, G. J., Fischman, M. W., Foltin, R. W., Fowler, J. S., Abumrad, N. N., et al. (1997) Relationship between subjective effects of cocaine and dopamine transporter occupancy. *Nature* **386,** 827–830.
61. Cook, C. E., Jeffcoat, A. R., and Perez-Reyes, M. (1985) Pharmacokinetic studies of cocaine and phencyclidine in man, in *Pharmacokinetics and Pharmacodynamics of Psychoactive Drugs.* (Barnett, G. and Chiang, C. N. eds.), Biomedical Publications, Foster City, CA, pp. 48–74.
62. Stathis, M., Scheffel, U., Lever, S. Z., Boja, J. W., Carroll, F. I., and Kuhar, M. J. (1995) Rate of binding of various inhibitors at the dopamine transporter in vivo. *Psychopharmacology* **119,** 376–384.
63. Gatley, S. J., Volkow, N. D., Gifford, A. N., Ding, Y. S., Logan, J., and Wang, G. J. (1996) Model for estimating dopamine transporter occupancy and subsequent increases in synaptic dopamine using positron emission tomography and carbon-11-labeled cocaine. *Biochem. Pharmacol.* **53,** 43–52.
64. Alim T. N., Rosse R. B., Vocci F. J., Lindquist T., and Deutsch S. I. (1995) Diethylpropion pharmacotherapeutic adjuvant thereapy for inpatient treatment of cocaine dependence. A test of the cocaine-agonist hypothesis. *Clin. Neuropharmacol.* **18,** 183–195.
65. Gawin, F. H., Riordan, C., and Kleber, H. D. (1985) Methylphenidate use in non-ADD cocaine abusers: a negative study. *Am. J. Drug Alcohol Abuser* **11,** 193–197.
66. Freston, K. L., Sullivan, J. T., Berger, P., and Bigelow, G. E. (1993) Effects of cocaine alone and in combination with mazindol in human cocaine abusers. *J. Pharmacol. Exp. Therap.* **258,** 296–307.
67. Fowler, J. S., Volkow, N. D., Logan, J., Gatley, S. J., Pappas, N., King, P., et al. (1998) Measuring dopamine transporter occupancy by cocaine *in vivo*: radiotracer considerations. *Synapse* **28,** 111–116.
68. Malison, R. T., Best, S. E., Wallace, E. A., McCance, E., Laruelle, M., Zogbhi, S. S., et al. (1995) Euphorigenic doses of cocaine reduce [^{123}I]β-CIT SPECT measures of dopamine transporter availability in human cocaine addicts. *Psychopharmacology* **122,** 358–362.
69. Malison, R. (1995) SPECT imaging of DA transporters in cocaine dependence with [^{123}I]β-CIT. *Natl. Inst. Drug Abuse Res. Monogr.* **152,** 60.

70. Villemagne, V. Yuan, J., Wong, D. F., Dannals, R. F., Hatzidimitriou, G., Mathews, W. B., et a.l (1998) Brain dopamine neurotoxcity in baboons treated with doses of methamphetamine comparable to those recreationally abused by humans: evidence from [^{11}C]WIN35428 positron emission tomgoraphy studies and direct in vitro determination. *J. Neurosci.* **18,** 419–427.

71. Volkow, N. D., Wang, G. J., Fowler, J. S., Gatley, J. S., Logan, J., Ding, Y. S., et al. (1999a) Blockade of striatal dopamine transporters by intravenous methylphenidate is not sufficient to induce self-reports of "high". *J. Pharmacol. Exp. Therap.* **288,** 14–20.

72. Cragg, S. J., Hille, C. J., and Greenfield, S. A. (2000) Dopamine release and uptake dynamics within nonhuman primate striatum *in vitro. J. Neurosci.* **29,** 8209–8217.

73. Wilson, J. M., Nobrega, J. N., Carroll, M. E., Niznik, H. B., Shannak, K., Lac, S. T., et al. (1994) Heterogenous subregional binding patterns of ^{3}H-WIN 35,428 and ^{3}H-GBR 12,935 are differentially regulated by chronic cocaine self-administration. *J. Neurosci.* **14,** 2966–2974.

74. Allard, P. (1994) Questions about the dopaminergic nature of [^{3}H]GBR 12935 binding in the human frontal cortex. *J. Neurochem.* **63,** 1182,1183.

75. Allard, P., Marcusson, J. O., and Ross, S. B. (1994) [^{3}H]GBR 12935 binding to cytochrome P450 in the human brain. *J. Neurochem.* **62,** 342–348.

76. Allard, P., Danielsson, M., Papworth, K., and Marcusson, J. O. (1994) [^{3}H]GBR 12935 binding to human cerebral cortex is not to dopamine uptake sites. *J. Neurochem.* **62,** 338–341.

77. Hitri, A. and Wyatt, R. J. (1994) Questions about the dopaminergic nature of [^{3}H]GBR 12935 binding in the human frontal cortex. *J. Neurochem.* **63,** 1181,1182.

78. Hitri, A., Venable, D., Nguyen, H.Q., Casanova, M. F., Kleinman, J. E., and Wyatt, R. J., (1991) Characteristics of [^{3}H]GBR 12935 binding in the human and rat frontal cortex. *J. Neurochem.* **56,** 1663–1672.

79. Hitri, A., Hurd, Y. L., Wyatt, R. J., and Deutsch, S. I. (1994) Molecular, functional, and biochemical characteristics of the dopamine transporter: regional differences and clinical relevance. *Clin. Neuropharm.* **17,** 1–22.

80. Niznik, H. B., Tyndale, R. F., Sallee, F. R., Gonzalez, F. J., Hardwick, J. P., Inaba, T., et al. (1990) The dopamine transporter and cytochrome p450IIDI (debrisoquine 4-hydroxylase) in brain: resolution and identification of two distinct [^{3}H]GBR 12935 binding proteins. *Arch. Biochem. Biophys.* **276,** 424–432.

81. Morissette, M., Di Paolo (1993a) Effect of chronic estradiol and progesterone treatments on ovarectimized rats on brain dopamine uptake sites. *J. Neurochem.* **60,** 1876–1883.

82. Morissette, M., Di Paolo, T. (1993) Sex and estrous cycle variations of rat striatal dopamine uptake sites. *Neuroendocrinology* **58,** 16–22.

83. Kuikka, J. T., Tiihonen, J., Karhu, J., Bergstroom, K. A., and Rasanen, P. (1997) Fractal analysis of striatal dopamine re-uptake sites. *Eur. J. Nuc. Med.* **24,** 1085–1090.

84. Lavalaye, J., Booij, J., Reneman, L., Habraken, J. B. A., and van Royen, E. A. (2000) Effect of age and gender on dopamine transporter imaging with [^{123}I]FP-CIT SPET in healthy volunteers. *Eur. J. Nucl. Med.* **27,** 867–869.

85. Staley, J. K., Krishnan-Sarin, S., Zoghbi, S., Tamagnan, G., Fujita, M., Seibyl, J. P., et al. (2001) Sex differences in [^{123}I]β-CIT SPECT measures of dopamine and serotonin transporter availability in healthy smokers and nonsmokers. *Synapse* **41**, 275–284.

86. van Dyck, C. H., Seibyl, J. P., Malison, R. T., Laruelle, M., Wallace, E., Zoghbi, S. S., et al. (1995) Age-related decline in striatal dopamine transporter binding with iodine-123-β-CIT SPECT. *J. Nucl. Med.* **36**, 1175–1181.

87. McCann, U. D., Wong, D. F., Yokoi, F., Villemagne, V., Dannals, R. F., and Ricaurte, G. A. (1998) Reduced striatal dopamine transporter density in abstinent methamphetamine and methcathinione users: evidence from positron emission tomography studies with [^{11}C]WIN-35,428. *J. Neurosci.* **18**, 8417–8422.

88. Kleven, M. S., Woolverton, W. L., and Seiden, L. S. (1988) Lack of long-term monoamine depletions following repeated or continuous exposure to cocaine. *Brain. Res. Bull.* **21**, 233–237.

89. Malison, R. T., Best, S. E.; van Dyck, C. H., McCAnce, E. F., Wallace, E. A., Laruelle, M., et al. (1998) Elevated striatal dopamine transporters during acute cocaine abstinence as measured by [^{123}I]β-CIT SPECT. *Am. J. Psychiatry* **155**, 832–834.

90. Staley, J. K., Talbot, J. Z., Ciliax, B. J., Miller, G. W., Levey, A. I., Kung, M. P., et al. (1997) Radioligand binding and immunoautoradiographic evidence for a lack of toxicity to dopaminergic nerve terminals in human cocaine overdose victims. *Brain Res.* **747**, 219–229.

91. Allard, P. and Marcusson, J. O. (1989) Age-correlated loss of dopamine uptake sites with [^{3}H]GBR 12935 in human putamen. *Neurobiol. Aging.* **10**, 661–664.

92. DeKeyser, J., Ebinger, G., and Vauquelin, G. (1990) Age-related changes in the human nigrostriatal dopaminergic system. *Ann. Neurol.* **27**, 157–161.

93. Hitri, A., Casanove, M. F., Kleinman, J. E., Weinberger, D. R., and Wyatt, R. J. (1995) Age-related changes in [^{3}H]GBR 12935 binding site density in the prefrontal cortex of controls and schizophrenics. *Biol. Psych.* **37**, 175–182.

94. Zelnik, N., Angel, I. Paul, S. M., and Kleinman, J. E. (1986) Decreased density of human striatal dopamine uptake sites with age. *Eur. J. Pharmacol.* **126**, 175,176.

95. Volkow, N. D., Fowler, J. S., Wang, G.-J., Logan, ?., Schlyer, D., MacGregor, R., et al. (1994) Decreased dopamine transporters with age in healthy human subjects. *Ann. Neuro.* **36**, 237,238.

96. Pirker, W., Asenbaum, S., Hauk, M., Kandlhofer, S., Tauscher, J., Willeit, M., et al. (2000) Imaging serotonin and dopamine transporters with 123I-beta-CIT SPECT: binding kinetics and effects of normal aging. *J. Nucl. Med.* **41**, 36–44.

97. Kuikka, J. T., Tupala, E., Bergstrom, K. A., Hiltunen, J., and Tiihonen, J. (1999) Iodine-123 labelled PE2I for dopamine transporter imaging: influence of age in healthy subjects. *Eur. J. Nucl. Med.* **26**, 1486–1488.

98. Mozley, P. D., Kim, H. J., Gur, R. C., Tatsch, K., Muenz, L. R., McElgin, W. T., et al. (1996) Iodine-123-IPT SPECT imaging of CNS dopamine trans-

porters: Nonlinear effects of normal aging on striatal uptake values. *J. Nucl. Med.* **37,** 1965–1970.

99. Tedroff, J., Aquilonius, S. M., Hartvig, P., Lundquist, H., Gee, A. G., Uhlin, J., et al. (1988) Monoamine reuptake sites in the human brain evaluated in vivo by means of [^{11}C]nomifensine and positron emission tomography:the effects of age and Parkinson's disease. *Acta. Neurol. Scand.* **77,** 192–201.

100. Wang, G. J., Volkow, N. D., Fowler, J. S., Fischman, M., Foltin, R., Abumrad, N. N., et al (1997) Cocaine abusers do not show loss of dopamine transporters with age. *Life Sci.* **61,** 1059–1065.

101. Bannon, M. J., Poosch, M. S., Xia, Y., Goebel, D. J., Cassin, B., and Kapatos, G. (1992) Dopamine transporter mRNA content in human substantia nigra decreases precipitously with age. *Proc. Natl. Acad. Sci. USA* **89,** 7095–7099.

102. McGeer, P. L., McGeer, E. G., and Suzuki, J. S. (1977) Aging and extrapyramidal function. *Arch Neurol.* **34,** 33–35.

103. Adolfsson, R., Gottfries, C.-G., Roos, B.-E., and Winblad, B. (1979) Postmortem distribution of dopamine and homovanillic acid in human brain, variations related to age, and a review of the literature. *J. Neural. Transm.* **45,** 81–105.

104. Hornykiewicz, O. (1983) Dopamine changes in the aging human brain, in *Aging Brain and Ergot Alkaloids*, Vol. 23, (Agnoli A., Grepaldi, G., Spano P. F., and Trabucchi, M. eds.), Raven Press, New York, NY, pp. 9–14.

105. Staley, J. K., Basile, M., Wetli, C. V., Hearn, W. L., Flynn, D. D., Ruttenber, A. J., et al. (1994) Differential regulation of the dopamine transporter in cocaine overdose deaths. *NIDA Res. Mono. Ser.* **141,** 32.

106. Hurd, Y. L. and Herkanham, M. (1993) Molecular alterations in the neostriatum of human cocaine addicts *Synapse* **13,** 357–369.

107. Peris, J., Boyson, S. J., Cass, W. A., Curella, R., Dwoskin, L. P., Larson, G., et al. (1990) Persistence of neurochemical changes in dopamine systems after repeated cocaine administration. *J. Pharmacol. Exp. Therap.* **253,** 38–44.

108. Izenwasser, S. and Cox, B. M. (1990) Daily cocaine treatment produces a persistent reduction of [^{3}H]dopamine uptake in vitro in rat nucleus accumbens but not in the striatum. *Brain Res.* **531,** 338–341.

109. Katz, J. L., Griffiths, J. W., Sharpe, L. G., De Souza, E. B., and Witkin, J. M. (1993) Cocaine tolerance and cross-tolerance. *J. Pharm. Exp. Therap.* **264,** 183–192.

110. Kula, N. S. and Baldessarini, R. J. (1991) Lack of increase in dopamine transporter binding or function in rat brain tissue after treatment with blockers of neuronal uptake of dopamine. *Neuropharmacology* **30,** 89–92.

111. Yi, S.-J. and Johnson, K. M. (1990) Effects of acute and chronic administration of cocaine on striatal uptake, compartmentalization and release of [^{3}H]dopamine. *Neuropharmacology* **29,** 475–486.

112. Alburges, M. E., Narang, N., and Wamsley, J. K. (1993) Alterations in the dopaminergic receptor system after chronic administration of cocaine. *Synapse* **14,** 314–323.

113. Aloyo, V. J., Harvey, J. A., and Kirifides, A. L. (1993) Chronic cocaine increases WIN 35,428 binding in rabbit caudate. *Soc. Neurosci. Abstr.* **19,** 1843.

114. Koff, J. M., Shuster, L., and Miller, L. G. (1994) Chronic cocaine administration is associated with behavioral sensitization and time-dependent changes in striatal dopamine transporter binding. *J. Pharm. Exp. Therap.* **268,** 277–282.
115. Farfel, G. M., Kleven, M. S., Woolverton, W. L., Seiden, L. S., and Perry, B. D. (1992) Effects of repeated injections of cocaine on catecholamine receptor binding sites, dopamine transporter binding sites and behavior in rhesus monkey. *Brain Res.* **578,** 235–243.
116. Sharpe, L. G., Pilotte, N. S., Mitchell, W. M., and De Souza, E. B. (1991) Withdrawal of repeated cocaine decreases autoradiographic [^{3}H]mazindol-labelling of dopamine transporter in rat nucleus accumbens. *Eur. J. Pharm.* **203,** 141–144.
117. Jacobsen, L. K., Staley, J. K., Malison, R. T., Baldwin, R. M., Seibyl, J. P., Kosten, T. R., et al. (2000) Elevated central serotonin transporter binding availability in acutely abstinent cocaine dependent patients. *Am. J. Psychiatry* **157,** 1134–1140.
118. Bowers, M. B., Malison, R. T., Seibyl, J. P., and Kosten, T. R. (1998) Plasma homovanillic acid and dopamine transporter during cocaine withdrawal. *Biol. Psychiatry* 278–281.
119. Volkow, N. D., Fowler, J. S., Logan, J., Wang, G.-J. Hitzemann, R., MacGregor, R., et al. (1992) Decreased binding of 11-C-cocaine in the brain of cocaine addicts. *J. Nucl. Med.* **33,** 888.
120. Wetli, C. V. and Fishbain D. A. (1985) Cocaine-Induced psychosis and sudden death in recreational cocaine users. *J. Foresci. Sci.* **30,** 873–880.
121. Volkow, N. D., Ding, Y.-S., Fowler, J. S., Wang, G.-J., Logan, J., Gatley, J. S., et al. (1995a) Is methylphenidate like cocaine. *Arch. Gen. Psychiatry* **52,** 456–463.
122. Rothman, R. B., and Glowa, J. R. (1995) A review of the eddects of dopaminergic agents on humans, animals, and drug-seeking behavior, and its implications for medication development. *Mol. Neurobiol.* **11,** 1–19.
123. Rocha, B. A., Fumagalli, F., Gainetdinov, R. R., Jones, S. R., Ator, R., Giros, B., et al. (1998) Cocaine self-administration in dopamine-transporter knockout mice. *Nature Neurosci.* 132–137.
124. Kuhar, M. J., Ritz, M. C., and Boja, J. W. (1991) The dopamine hypothesis of the reinforcing properties of cocaine. *Trends Neurosci.* **14,** 299–302.
125. Gerasimov, M. R., Franceschi, M., Volkow, N. D., Rice, O., Schiffer, W. K., and Dewey, S. L. (2000) Synergistic interactions between nicotine and cocaine or methylphenidate depend on the dose of dopamine transport inhibitor. *Synapse* **38,** 432–437.
126. Izenwasser, S., Jacocks, H. M., Rosenberger, J. G., and Cox, B. M., (1991) Nicotine indirectly inhibits [3H]dopamine uptake at concentrations that do not directly promote [3H]dopamine release in rat striatum. *J. Neurochem.* **56,** 603–610.
127. Izenwasser, S. and Cox, B. M. (1992) Inhibition of dopamine uptake by cocaine and nicotine: tolerance to chronic treatments. *Brain Res.* **573,** 119–125.

128. Kleber, H. D. (1995) Pharmacotherapy, current and potential, for the treatment of cocaine dependence. *Clin. Neuropharm.* **18,** S96–S109.

129. Weiss, R. D. and Mirin, S. M. (1990) Psychological and pharmacological treatment strategies in cocaine dependence. *Ann. Clin. Psych.* **2,** 239–243.

130. Dackis, C. A. and Gold, M. S. (1985) Bromocriptine as a treatment of cocaine abuse. *Lancet* **1,** 1151,1152.

131. Gawin, F. H. and Ellinwood, E. H. (1988) Cocaine and other stimulants. *N. Engl. J. Med.* **318,** 1173–1182.

132. Schweri, M. M. (1993) Rapid increase of stimulant binding to the dopamine transporter after acute cocaine administration: physiological basis of drug craving. *Soc. Neurosci. Abstr.* **19,** 936.

Index

A

Acetylcholine transporter,
316–317, *see also* VAchT
Adrenergic receptors
NET knockouts, 182
Alzheimer's
involvement SERT, 149
Amino acid transporter B^{0+}, *see* B^{0+}
Amino acid transporter KAAT1,
see KAAT1
Amino acid transporter
CAATCH1, *see* CAATCH1
Amphetamine derivatives, *see
also* p-Chloroamphetamine,
Methamphetamine, 3,4-Methyl-
enedioxymethamphetamine
permeability paradox, 38
inhibitor of vesicular
monoamine transporters, 38
weak base effect at vesicles, 36
Amphetamine, *see also* Amphetamine derivatives
DAT knockout mice, 173
inhibitor of DAT, 36, 60, 471
inhibitor of NET, 36, 60
inhibitor of SERT, 36, 60
inhibitor of vesicular
monoamine transporters,
36, 38
substrate for monoamine
transporters, 36, 60
uncoupling effect at vesicles, 38
synaptic vesicles, 39

Amyotrophic lateral sclerosis,
292–294
GLT-1 mutation, 294
oxidative stress, 293
Anesthetics, 207
Antidepressants
effect on norepinephrine, 115
inhibitors of monoamine
transporters, 60
inhibitors of SERT, 60, 80, 81
NET knockout mice, 115
SSRIs and SERT, 80, 81, 135
Antiepileptic drugs, 206, 236
Anxiety
involvement SERT, 141
Arachidonic acid
glutamate neurotoxicity, 272
regulation of glutamat
transporters, 294–296
regulation of glutamate
uptake, 294–296
ASCT1 or ASCT-1, 242, 259
exchange, 268
ion coupling, 260, 264
localization, 280, 281
K^+ countertransport, 269
molecular mass, 242
topology, 260, 261
ASCT2 or ASCT-2, 242, 260
exchange, 268
ion coupling, 260, 264
K^+ countertransport, 269
molecular mass, 242
topology, 260, 261